# Production of Biofuels and Numerical Modeling of Chemical Combustion Systems

# Production of Biofuels and Numerical Modeling of Chemical Combustion Systems

Editors

**Miguel Torres García**
**Juan Francisco García Martín**

MDPI • Basel • Beijing • Wuhan • Barcelona • Belgrade • Manchester • Tokyo • Cluj • Tianjin

*Editors*

Miguel Torres García
Department of Energy
Engineering
University of Seville
Seville
Spain

Juan Francisco García Martín
Department of Chemical
Engineering
University of Seville
Seville
Spain

*Editorial Office*
MDPI
St. Alban-Anlage 66
4052 Basel, Switzerland

This is a reprint of articles from the Special Issue published online in the open access journal *Processes* (ISSN 2227-9717) (available at: www.mdpi.com/journal/processes/special_issues/Biofuel_combustion).

For citation purposes, cite each article independently as indicated on the article page online and as indicated below:

LastName, A.A.; LastName, B.B.; LastName, C.C. Article Title. *Journal Name* **Year**, *Volume Number*, Page Range.

**ISBN 978-3-0365-1332-4 (Hbk)**
**ISBN 978-3-0365-1331-7 (PDF)**

# Contents

# About the Editors

**Miguel Torres García**

Miguel Torres García is an associate professor in mechanical engineering. He obtained his Tech Degree in Industrial Engineering from the University of Seville in 2000. Between 2001 and 2005, he worked on various projects in Spain and Italy. He joined the Thermal Power group of the Department of Energy Engineering of the University of Seville in 2005, where he obtained his Ph.D. degree in Industrial Engineering from the University of Seville in 2007. He has eight years of experience in teaching laboratories. His main research interest is a new mode of combustion called homogeneous charge compression ignition applied in alternative internal combustion engines.

**Juan Francisco García Martín**

Dr. Juan Francisco García Martín obtained a Ph.D. summa cum laude (European Doctorate Mention) degree in 2007 from the University of Jaén (Spain). Subsequently, he became a postdoctoral researcher at University of Granada (Spain), University College Dublin (Ireland) and Instituto de la Grasa (Spanish National Research Council), and meanwhile did several research stays at Institute of Catalysis and Petrochemistry (Spanish National Research Council) and the Universities Paul Sabatier Toulouse III (France), Claude Bernard Lyon I (France) and Geisenheim (Germany). Dr. García Martín was also lecturer at the Universities of Jaén (Spain), Granada (Spain) and Málaga (Spain). Since 2016, he has been an associate professor at the Department of Chemical Engineering of the University of Seville (Spain).

*Editorial*

# Special Issue "Production of Biofuels and Numerical Modelling of Chemical Combustion Systems"

Miguel Torres-García [1], Paloma Álvarez-Mateos [2] and Juan Francisco García-Martín [2,*]

[1] Escuela Técnica Superior de Ingenieros Industriales, University of Seville-Thermal Power Group, Camino de los descubrimientos s/n, 41092 Sevilla, Spain; migueltorres@us.es
[2] Departamento de Ingeniería Química, Facultad de Química, Universidad de Sevilla, 41012 Sevilla, Spain; palvarez@us.es
* Correspondence: jfgarmar@us.es



**Citation:** Torres-García, M.; Álvarez-Mateos, P.; García-Martín, J.F. Special Issue "Production of Biofuels and Numerical Modelling of Chemical Combustion Systems". *Processes* **2021**, *9*, 829. https://doi.org/10.3390/pr9050829

Received: 5 May 2021
Accepted: 5 May 2021
Published: 9 May 2021

**Publisher's Note:** MDPI stays neutral with regard to jurisdictional claims in published maps and institutional affiliations.



Biofuels have recently attracted a lot of attention, mainly as alternative fuels for applications in energy generation and transportation. The utilization of biofuels in such controlled combustion processes has the great advantage of not depleting the limited resources of fossil fuels, but leads to emissions of greenhouse gases and smoke particles similar to traditional combustion processes, i.e., those of fossil fuels. On the other hand, a vast amount of biofuels is subjected to combustion in small-scale processes, such as for heating and cooking in residential dwellings, as well as in agricultural operations, such as crop residue removal and land clearing. In addition, large amounts of biomass are consumed annually during forest and savanna fires in many parts of the world. These types of burning processes are typically uncontrolled and unregulated. Consequently, the emissions from these processes may be larger compared to industrial-type operations. Aside from direct effects on human health, especially due to a sizeable fraction of the smoke emissions remaining inside residential homes, the smoke particles and gases released from uncontrolled biofuel combustion imposes significant effects on the regional and global climate. Estimates have shown the majority of carbonaceous airborne particulate matter to be derived from the combustion of biofuels and biomass.

This Special Issue on "Production of Biofuels and Numerical Modelling of Chemical Combustion Systems" contains sixteen high-quality studies (fifteen research papers and one review paper) addressing techniques and methods for bioenergy and biofuel production as well as challenges in the broad area of process modeling and control in combustion processes.

First at all, García Martín et al. comprehensively review the latest advances focused on energy production from different olive biomasses, including the production of biofuels such as bioethanol and biodiesel and processes such as combustion, gasification and pyrolysis [1]; all of them can be applied to any type of biomass. For lignocellulose biomasses, mainly composed of cellulose, hemicellulose and lignin, the determination of their content in these fibers and moisture is of major importance to select the most suitable process for energy production. In this sense, Díez et al. provide a new efficient, low-cost and fast method for the determination of these contents in different types of biomasses from agricultural by-products to softwoods and hardwoods [2]. The proposed method is based on applying deconvolution techniques on the derivative thermogravimetric pyrolysis curves obtained by thermogravimetric analysis through a kinetic approach based on a pseudocomponent kinetic model.

Biofuels cannot replace our current dependence on coal, oil, and natural gas, but they can complement other renewable energies such as solar and wind energies. Thus, due to the merits of biofuel energy for environmental sustainability, biofuel and bioenergy technologies play a crucial role in the renewable energy development and replacement of chemicals from highly functional biomass. The following six research papers investigate the production of biofuels from biomass.



*Processes* **2021**, *9*, 829. https://doi.org/10.3390/pr9050829                    https://www.mdpi.com/journal/processes

Starting with liquid biofuels, Cuevas et al. obtain bioethanol from the acid hydrolysis at high temperature of olive stones followed by enzymatic hydrolysis and fermentation of the released sugars with the yeast *Pachysolen tannophilus* [3]. The pretreatment was optimized by means of response surface methodology with two independent variables (temperature and reaction time). With regard to the production of biodiesel by transesterification, García Martín et al. deal with the issue caused by waste cooking oils with high acidity, which cannot be directly transformed into fatty acid methyl esters by transesterification due to the soap formation [4]. The solution proposed, within the biodiesel production framework and the circular economy concept, is the esterification of these high acidity oils with the residual glycerol from transesterification. On the other hand, the existence of excess water content in the starting oil can retard the transesterification rate and make the resulting biodiesel not comply to legal specifications for its use as biofuel. Hence, Lin and Ma assess the optimum water content in the raw oil and its effects on burning characteristics of the resulting biodiesel [5].

Biomethane and syngas stand out among the gaseous biofuels. Alonso-Fariñas et al. obtain biomethane from the anaerobic digestion of olive pomaces within an olive oil mill framework [6]. These authors study the life cycle assessment of a scheme consisting of the production of biogas from the anaerobic digestion of the olive pomace, heat and electricity cogeneration by the combustion of the generated biogas, and composting of the anaerobic digestate. Syngas can be obtained from biomass as a renewable energy source though gasification. However, syngas from biomass contains some impurities, such as organic tars, which must be removed before its application. To perform this, Díez et al. synthesize a catalyst based on layered double hydroxides with a molar cation concentration Ni/Cu/Fe/Mg/Al of 30/5/5/40/20 for the steam reforming of toluene as a model compound of biomass tar [7], yielding high concentrations of $H_2$.

Finally, the production of solid biofuels, such as pellets, is also highlighted. Notwithstanding, pellets from biomass cannot compete with those from fossil fuel sources because the biomass densifying process and the raw materials price make pellet production economically unfeasible. To overcome these issues, Clavijo et al. propose the use of eucalyptus kraft lignin as an additive for tree-leaf pellet production [8]. The resulting pellets fulfilled all requirements of European standards for certification except for ash content.

The following four research papers are focused on the production of biomass and biocrude from microalgae. Advanced biofuels obtained from microalgae have attracted great interest because they do not compete with food production since they can grow on non-arable land. The production of biomass is a key prerequisite in the production process of biocrude and other valuable compounds from microalgae. Thus, Fuentes et al. produce biomass at a large scale from the microalgae *Coccomyxa onubensis* in an 800-dm$^3$ tubular photobioreactor [9]. These authors obtained a biomass productivity of 0.14 g dm$^{-3}$ day$^{-1}$ along with high contents of lutein, an interesting food colorant, when growing the microalga under outdoor conditions. Hydrothermal liquefaction has proven to be an attractive process for the production of biofuels from microalgae, rendering biocrude (the liquid organic phase), aqueous phase compounds, solid residue, and gas phases, as illustrated in three works published in this Special Issue by Dr. Vicente's research group. In the first, Megía-Hervás et al. assess the temperature, reactor loading and time of the hydrothermal liquefaction of the microalga *Nannochloropsis gaditana* at mild temperatures in order to reduce the N and O content in the biocrude and to maximize the yield of the pretreated biomass [10]. Subsequently, Sánchez-Bayo et al. improve the hydrothermal liquefaction of the same microalgae using heterogeneous catalysts [11]. To be specific, the catalysts were based on metal oxides (CaO, CeO$_2$, La$_2$O$_3$, MnO$_2$, and Al$_2$O$_3$), yielding remarkable amounts of biocrude, high values of C, H and heating value and low contents of N, O and S. In the third paper, Megía Hervás et al. perform the hydrothermal liquefaction of another microalgae (*Phaeodactylum tricornutum*) in a scale-up photoreactor [12]. The biocrude yields obtained in this scale-up photoreactor were lower than the ones obtained from the biomass cultivated at laboratory scale because of the lower lipid and high ash contents in this biomass. However, the culture scaling-up did not have influence on the heteroatom concentrations in the biocrudes.

The last four research papers study the modelling and design of combustion processes of biofuels. Since biodiesel has higher viscosity than petroleum diesel, which leads to lower performances of compression ignition engine and higher emissions, Hamid et al. perform a numerical investigation of fluid flow and in-cylinder air flow characteristics [13]. As a result, the authors propose to install a guide vane design in the intake manifold, with shallow depth re-entrance combustion chamber pistons, in order to promote better diffusion, evaporation and combustion processes. Next, Orihuela et al. focus on the performance loss of the post-combustion system due to the variability that biofuels introduce in the exhaust particle distribution [14]. These authors use a well-validated particulate filter model available as commercial software to predict the filtration performance of a wall-flow particulate filter made of biomorphic silicon carbide (a bioceramic material made from vegetal waste) with a systematic procedure that allows one to eventually fit different fuel inputs. On the other hand, dual fuel engines (those that use diesel and fuels that are gaseous at normal conditions) are receiving increasing attention because they achieve the same (or better) power density and efficiency, steady state and transient performances than petroleum diesel. In his research paper, Boretti develops a numerical analysis of a novel, high-pressure ($1.6 \times 10^8$ Pa) liquid phase injector for liquefied natural gas in a high compression ratio, high boost engine featuring two direct injectors per cylinder: one for diesel and one for liquefied natural gas [15]. Finally, Guo et al. assess the effects of the diameter and the angle of the pre-combustion chamber nozzle on the performance of a marine two-stroke dual fuel engine [16], finding that both parameters have influence on the flame propagation in the combustion chamber.

To sum up, this Special Issue on "Production of Biofuels and Numerical Modelling of Chemical Combustion Systems" comprehensively overviews and includes in-depth technical research papers addressing recent progress in biofuels production and combustion processes. All these manuscripts contributed—with their topics and their high quality—to the success of the present Special Issue.

**Funding:** This review paper received no external funding.

**Acknowledgments:** The Guest Editors are grateful to all the authors that have contributed to this Special Issue, all the reviewers for their excellent work in evaluating the submitted manuscripts and the editorial staff of Processes, especially Tami Hu and Milica Ma, for their assistance in the success of this Special Issue.

**Conflicts of Interest:** The authors declare no conflict of interest.

## References

1. García Martín, J.F.; Cuevas, M.; Feng, C.H.; Álvarez Mateos, P.; Torres García, M.; Sánchez, S. Energetic valorisation of olive biomass: Olive-tree pruning, olive stones and pomaces. *Processes* **2020**, *8*, 511. [CrossRef]
2. Díez, D.; Urueña, A.; Piñero, R.; Barrio, A.; Tamminen, T. Determination of hemicellulose, cellulose, and lignin content in different types of biomasses by thermogravimetric analysis and pseudocomponent kinetic model (TGA-PKM method). *Processes* **2020**, *8*, 1048. [CrossRef]
3. Cuevas, M.; Saleh, M.; García-Martín, J.F.; Sánchez, S. Acid and enzymatic fractionation of olive stones for ethanol production using *Pachysolen tannophilus*. *Processes* **2020**, *8*, 195. [CrossRef]
4. García Martín, J.F.; Carrión Ruiz, J.; Torres García, M.; Feng, C.-H.; Álvarez Mateos, P. Esterification of free fatty acids with glycerol within the biodiesel production framework. *Processes* **2019**, *7*, 832. [CrossRef]
5. Lin, C.Y.; Ma, L. Influences of water content in feedstock oil on burning characteristics of fatty acid methyl esters. *Processes* **2020**, *8*, 1130. [CrossRef]
6. Alonso-Fariñas, B.; Oliva, A.; Rodríguez-Galán, M.; Esposito, G.; García Martín, J.F.; Rodríguez-Gutiérrez, G.; Serrano, A.; Fermoso, F. Environmental assessment of olive mill solid waste valorization via anaerobic digestion versus olive pomace oil extraction. *Processes* **2020**, *8*, 626. [CrossRef]
7. Díez, D.; Urueña, A.; Antolín, G. Investigation of Ni–Fe–Cu-layered double hydroxide catalysts in steam reforming of toluene as a model compound of biomass tar. *Processes* **2021**, *9*, 76. [CrossRef]

8. Clavijo, L.; Zlatanovic, S.; Braun, G.; Bongards, M.; Dieste, A.; Barbe, S. Eucalyptus kraft lignin as an additive strongly enhances the mechanical resistance of tree-leaf pellets. *Processes* **2020**, *8*, 376. [CrossRef]

9. Fuentes, J.; Montero, Z.; Cuaresma, M.; Ruiz-Domínguez, M.; Mogedas, B.; Nores, I.G.; González del Valle, M.; Vílchez, C. Outdoor large-scale cultivation of the acidophilic microalga *Coccomyxa onubensis* in a vertical close photobioreactor for lutein production. *Processes* **2021**, *8*, 324. [CrossRef]

10. Montero-Hidalgo, M.; Espada, J.J.; Rodríguez, R.; Morales, V.; Bautista, L.F.; Vicente, G. Mild hydrothermal pretreatment of microalgae for the production of biocrude with a low N and O content. *Processes* **2019**, *7*, 630. [CrossRef]

11. Sánchez-Bayo, A.; Rodríguez, R.; Morales, V.; Nasirian, N.; Bautista, L.F.; Vicente, G. Hydrothermal liquefaction of microalga using metal oxide catalyst. *Processes* **2020**, *8*, 15. [CrossRef]

12. Megía-Hervás, I.; Sánchez-Bayo, A.; Bautista, L.F.; Morales, V.; Witt-Sousa, F.G.; Segura-Fornieles, M.; Vicente, G. Scale-up cultivation of *Phaeodactylum tricornutum* to produce biocrude by hydrothermal liquefaction. *Processes* **2020**, *8*, 1072. [CrossRef]

13. Hamid, M.F.; Idroas, M.Y.; Sa'ad, S.; Heng, T.Y.; Mat, S.C.; Alauddin, Z.A.Z.; Shamsuddin, K.A.; Shuib, R.K.; Abdullah, M.K. Numerical investigation of fluid flow and in-cylinder air flow characteristics for higher viscosity fuel applications. *Processes* **2020**, *8*, 439. [CrossRef]

14. Orihuela, M.P.; Haralampous, O.; Chacartegui, R.; Torres García, M.; Martínez-Fernández, J. Numerical simulation of a wall-flow particulate filter made of biomorphic silicon carbide able to fit different fuel/biofuel inputs. *Processes* **2019**, *7*, 945. [CrossRef]

15. Boretti, A. Numerical analysis of high-pressure direct injection dual-fuel diesel-liquefied natural gas (LNG) engines. *Processes* **2020**, *8*, 261. [CrossRef]

16. Guo, H.; Zhou, S.; Shreka, M.; Feng, Y. Effect of pre-combustion chamber nozzle parameters on the performance of a marine 2-stroke dual fuel engine. *Processes* **2019**, *7*, 876. [CrossRef]

*Review*

# Energetic Valorisation of Olive Biomass: Olive-Tree Pruning, Olive Stones and Pomaces

**Juan Francisco García Martín** [1,2,*], **Manuel Cuevas** [2,3], **Chao-Hui Feng** [4],
**Paloma Álvarez Mateos** [1], **Miguel Torres García** [5] and **Sebastián Sánchez** [2,3,*]

1   Departamento de Ingeniería Química, Facultad de Química, Universidad de Sevilla, 41012 Sevilla, Spain;
    palvarez@us.es
2   Center for Advanced Studies in Olive Grove and Olive Oils, Science and Technology Park GEOLIT,
    23620 Mengibar, Spain; mcuevas@ujaen.es
3   Department of Chemical, Environmental and Materials Engineering, University of Jaén, 23071 Jaén, Spain
4   RIKEN Centre for Advanced Photonics, RIKEN, 519-1399 Aramaki-Aoba, Aoba-ku, Sendai 980-0845, Japan;
    chaohui.feng@riken.jp
5   Escuela Técnica Superior de Ingenieros Industriales, University of Seville-Thermal Power Group,
    Camino de los descubrimientos s/n, 41092 Sevilla, Spain; migueltorres@us.es
*   Correspondence: jfgarmar@us.es (J.F.G.M.); ssanchez@ujaen.es (S.S.)

Received: 20 March 2020; Accepted: 21 April 2020; Published: 26 April 2020

**Abstract:** Olive oil industry is one of the most important industries in the world. Currently, the land devoted to olive-tree cultivation around the world is ca. $11 \times 10^6$ ha, which produces more than $20 \times 10^6$ t olives per year. Most of these olives are destined to the production of olive oils. The main by-products of the olive oil industry are olive-pruning debris, olive stones and different pomaces. In cultures with traditional and intensive typologies, one single ha of olive grove annually generates more than 5 t of these by-products. The disposal of these by-products in the field can led to environmental problems. Notwithstanding, these by-products (biomasses) have a huge potential as source of energy. The objective of this paper is to comprehensively review the latest advances focused on energy production from olive-pruning debris, olive stones and pomaces, including processes such as combustion, gasification and pyrolysis, and the production of biofuels such as bioethanol and biodiesel. Future research efforts required for biofuel production are also discussed. The future of the olive oil industry must move towards a greater interrelation between olive oil production, conservation of the environment and energy generation.

**Keywords:** bioethanol; combustion; gasification; olive; olive oils; olive-pruning debris; olive stones; olive pomaces; pyrolysis

---

## 1. Introduction

With the dramatic development of society, biofuels produced from biomass are regarded as a potential alternative source for energy and therefore solve the increasing energy crisis. Besides efficient food processing [1] and food safety inspection [2], an affordable, reliable, sustainable and modern energy is also an important sustainable development goal requested to be achieved.

The olive tree (*Olea europaea* L., subsp. *europaea*) belongs to the Oleaceae family. Over 11 million ha are devoted to olive tree cultivation around the world [3], resulting in more than 20 million t of olives per year (Figure 1). The olive tree predominates in the agricultural production in the Mediterranean countries such as Spain, Italy, Greece, Turkey, Tunisia, Morocco, Syrian Arab Republic, Portugal, Egypt and Algeria (Figure 2), and it has been expanded in recent decades to other regions such as West Coast of the USA, Argentina, Australia, Chile, Peru, Uruguay and China [4]. Therefore, these countries are responsible for most of the worldwide production of olives (Figure 3). Olive tree not only supplies important olive oils providing us the healthiest vegetable fats, but also produces table olives, which are famous pickles for dishes starter.

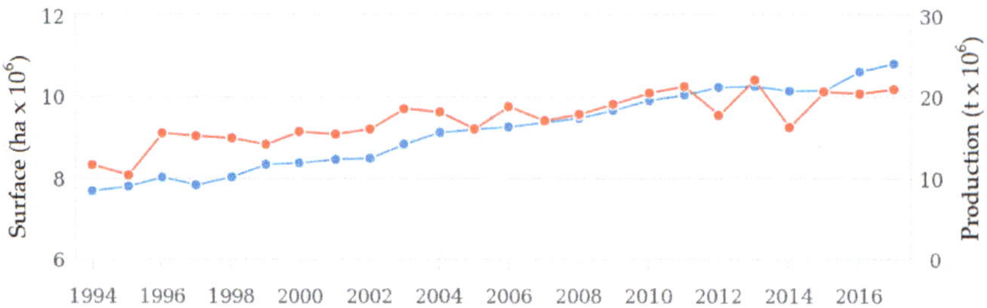

**Figure 1.** Total olive-tree cultivation surface (blue points) and worldwide olive production (red points) from 2001 to 2017. Reproduced with permission from FAOSTAT [5].

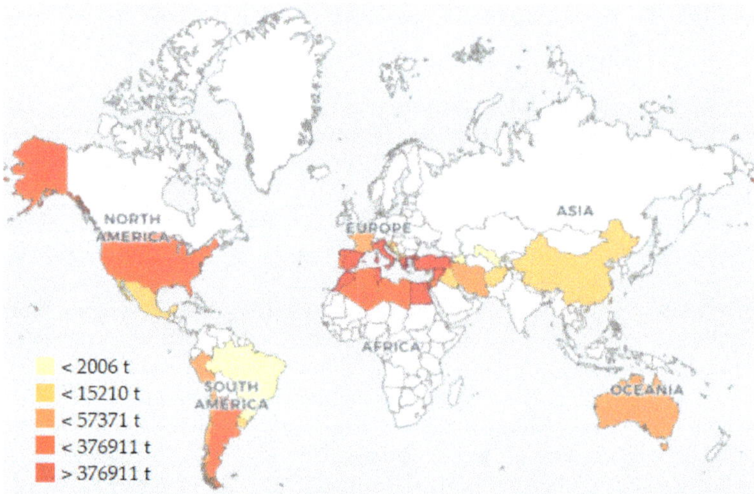


**Figure 2.** Average production of olives by country between 1994 and 2017. Reproduced with permission from FAOSTAT and OpenStreetMap Foundation [5].

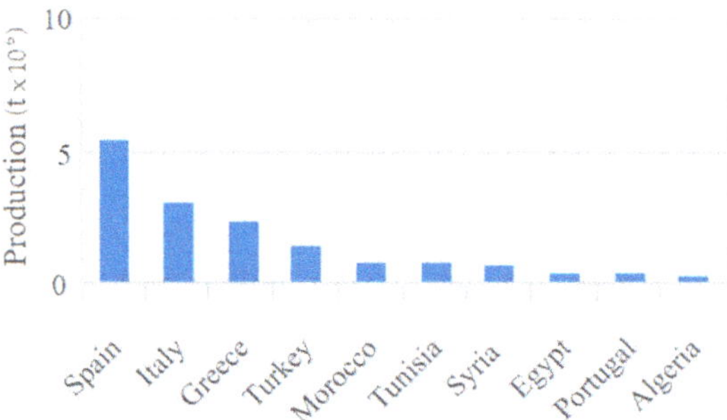


**Figure 3.** Average olive production (1994–2017) of the top 10 olive producers. Reproduced with permission from FAOSTAT [5].

The generation of by-products from the olive oil industry per ha of culture is illustrated in Figure 4. As it can be observed, the main by-products are olive-pruning debris (OP), olive stones (OS) and extracted pomace dry by-products.

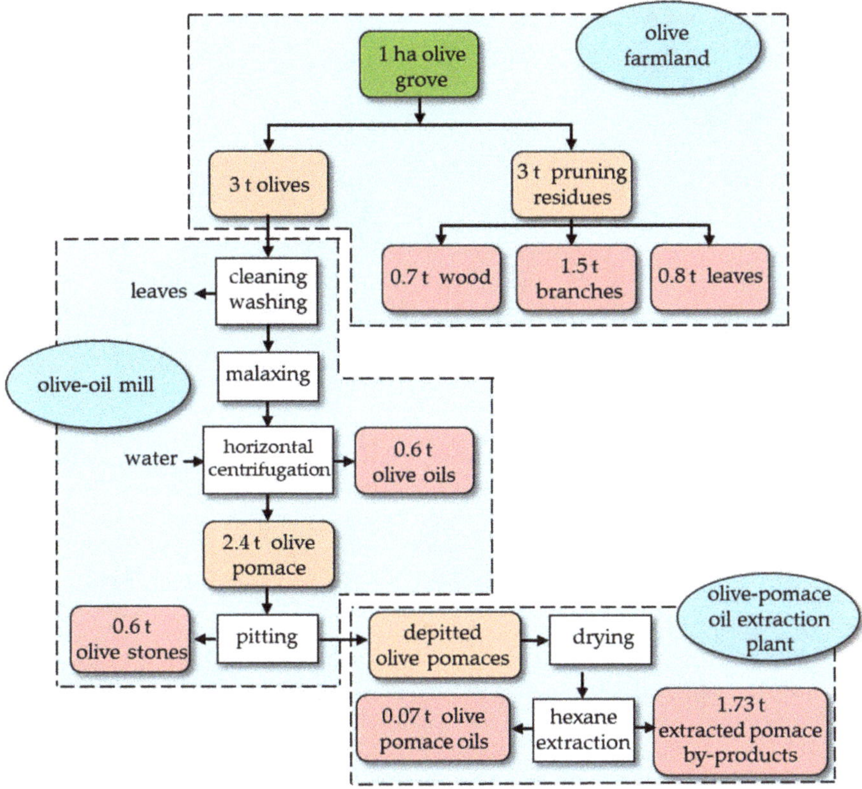


**Figure 4.** By-products obtained from olive orchards and olive mills.

Pruning is an important biennial operation carried out by farmers for removing old branches thus regenerating the tree. The olive-pruning debris consists of thin branches (usually <5 cm diameter) and leaves. Leaves can be removed from the pruning debris by means of a densimeter machine for industrial applications. Farmers usually use the bigger branches and trunks in small home boilers. An average of 3 tons of olive-pruning debris is generated from a hectare of olive orchard [6], leading to producing more than $3.3 \times 10^7$ t biomass. Generally, the olive-pruning debris is either grinded and ploughed into soil or left on the land to be incinerated (Figure 5), which not only causes air pollution ($CO_2$ emissions) but also mineralizes the soil and increases the risks of pest propagation and fire accidents. The higher heating value (HHV) of the olive-pruning debris ranges between 16.7 MJ/kg and 19.8 MJ/kg [3,7–9], and its bulk density between 272 kg/m$^3$ [9] and 347.9 kg/m$^3$ [3]. The lower heating value (LHV) has been reported to be 16.55 MJ/kg [7]. If the olive pruning-debris were utilised to produce energy, it will not only maximise the reuse and exploitation of the value-added by-products, but also solve the environmental contamination and supply for the clean renewable energy. Its current price in Andalusia is 30–40 €/t [3]. This low price makes OP an attractive material to produce pellets and other solid or liquid biofuels.

**Figure 5.** Olive-tree pruning piled up in the field.

Olive-tree pruning is a lignocellulose material and, therefore, is mainly composed of cellulose, hemicellulose and lignin (Table 1). Cellulose consists of thousands of chains, each chains made up of hundreds of D-glucose units linked up together by β(1,4) glycosidic bonds with reducing and non-reducing ends. The presence of hydrogen bonds results in a crystalline structure [10]. The cellulose is wrapped in a sheath of hemicellulose and lignin, which protects cellulose from being broken down. Hemicellulose, which can account for between 15% and 35% of a lignocellulose material on a dry basis, is a heterogeneous polymer containing pentoses (D-xylose, L-arabinose), hexoses (D-mannose, D-glucose, D-galactose, and D-fructose), acetyl groups and uronic acids. This heteropolymer is easier to break down than cellulose. Lignin is mainly composed of guaiacyl, *p*-hydroxyphenyl and syringyl units polymerized by ether bonds or carbon-carbon linkages. Lignin provides chemical, mechanical and biological resistance to the lignocellulose material, and it is a potential source of aromatic compounds.

**Table 1.** Olive-pruning debris fiber composition.

| Composition (wt.%) | | | Reference |
|---|---|---|---|
| Cellulose | Hemicellulose | Lignin | |
| 36.6 | 19.7 | 20.8 | [6] |
| 30.3 | 17.9 | 24.1 | [4] |
| 39.1 | 25.7 | 14.3 | [10] |
| 36.4 | 21.5 | 17.1 | [11] |
| 25.4 | 19.0 | 18.5 | [12] |
| 25.0 | 18.3 | 18.8 | [13] |
| 36.5 | 20.2 | 22.5 | [14] |
| 36.5 | 20.8 | 21.3 | [15] |

Regarding the elemental composition, scarce information is available in literature. However, the data provided by some authors confirms that sulphur is not detected in the olive-pruning debris (Table 2), which is characteristic of the olive-tree by-products.

**Table 2.** Elemental analysis of the olive-pruning debris.

| Element (%) | | | | | Reference |
|---|---|---|---|---|---|
| C | H | O | N | S | |
| 44.6 | 6.7 | 47.9 | 0.8 | 0.0 | [6] |
| 46.1 | 6.4 | 47.2 | 0.4 | 0.0 | [16] |

Olive stones (Figure 6) are another biomass that is generated in olive oil mills. Olive table industry annually generates around 30,000 t olive stones [17]. Besides, olive stones are also generated in olive oil and olive pomace oil extractor industries after separating the olive stones from the olive pulp to obtain olive oil or olive pomace oil, respectively. In both cases, the size of the crashed endocarps never exceeds 7 mm in length. Most of the olive stones recovered in olive mills have sizes larger than 1 mm, the percentage of thinner solids increasing when increasing the weight of the pulp and inorganic matter in the olive. Saleh et al. (2014) reported that 97.3% (wt.%) of the olive stones collected in an olive mill had sizes greater than 1.2 mm [18]. Mata-Sánchez et al. (2015) found that 96.3% of their olive stones samples had diameters higher than 1.4 mm [19]. Furthermore, Fernández-Bolaños et al. (1999) and Barreca and Fichera (2013) reported maximum olive stones sizes of 6.2 mm and 1.6 mm, respectively [20,21]. This small particle size of olive stones is an advantage over other biomasses with larger sizes (e.g., olive prunings), avoiding the need to resort costly milling stages for subsequent use. Notwithstanding, its current price (80–100 €/t in Andalusia, Spain) is much higher than that of olive-tree pruning [3].

**Figure 6.** Endocarps from olive fruit.

Olive stones account for roughly 20% of the olive weight. Thus, it was reported that 100 kg fresh olives contain 22 kg olive stones (4 kg seeds plus 18 kg endocarps) [22]. As a lignocellulose material, its main components are cellulose, hemicellulose and lignin (Table 3). As it can be observed, the percentage of hemicellulose in OS is higher than in OP. The different olive varieties, geographical location, the presence of traces of pulp and pericarp, and the analytical techniques used by the different authors could account for the different results shown of results in Table 3.

**Table 3.** Fiber composition of olive stones.

| Composition (wt.%) | | | Reference |
|---|---|---|---|
| Cellulose | Hemicellulose | Lignin | |
| 29.9 | 28.1 | 27.7 | [23] |
| 33.5 | 24.5 | 23.1 | [22] |
| 27.1 | 32.2 | 40.4 | [24] |
| 36.4 | 26.8 | 26.0 | [25] |

Due to its high lignin content, it was reported to be suitable for thermal utilization. Olive stones were reported to have a great potential as solid biofuel for combustion in comparison with other lignocellulose materials. An average lower heating value of 19,167 kcal/kg was obtained from 30 different olive stones obtained from different regions of Andalusia (Spain) [17]. Concerning the higher heating value, different authors have pointed out that the HHV of olive stones ranged between 18.8 MJ/kg and 20.9 MJ/kg [3,17], these HHV measured by different methods. As a result, 99% of olive stones produced are currently used as solid biofuel to thermal power generation.

Olive stones possess a relatively low ignition point (approximately 215 °C) while the maximum combustion rate is 0.341 $dm^3$/min and is achieved at 284 °C. As little mineral matter is contained in olive stones and their ash melting temperature was over 1400 °C, it thus reduces costs of burner cleaning. The bulk density of olive stones is 721.6 $kg/m^3$ [3], double than that of olive-pruning debris, which accounts for the aforementioned potential of olive stones for combustion. On the other hand, as a type of lignocellulose material, it has been proposed as a source of fermentable sugars, antioxidants and other applications [26]. If added-value products can be obtained from olive stones from thermochemical and biochemical points of view, it will greatly solve the problem of environmental contamination.

With regard to elemental composition, sulphur is barely detected in olive stones, as it is characteristic of olive tree by-products (Table 4). Among the different trace elements present that can be found, chlorine and copper can be highlighted, with concentrations ranging from 90 mg/kg to 435 mg/kg and from 0.6 mg/kg to 2.3 mg/kg, respectively [19,27]. Ash percentages are usually lower than 2% (wt.) (Table 4). The main inorganic compounds found in olive stones ash are $Al_2O_3$, CaO, $Fe_2O_3$, $K_2O$, MgO, and $SiO_2$ [28–30].

**Table 4.** Elemental analysis and ash content of olive stones.

| Composition (wt.%). | | | | | | Reference |
|---|---|---|---|---|---|---|
| C | H | O | N | S | Ash | |
| 51.2 | 6.0 | 41.9 | 0.15 | 0.02 | 0.78 | [19] |
| 50.1 | 5.9 | 42.0 | 0.6 | 0.02 | 1.33 | [31] |
| 46.6 | 6.3 | 45.2 | 1.8 | 0.10 | 1.40 | [32] |
| 48.6 | 5.7 | 44.1 | 1.6 | 0.05 | 1.90 | [33] |

The extraction process of olive oils has evolved over the years from discontinuous to continuous methods. In the first method, olive oil was obtained by applying hydraulic pressure (press method). Nevertheless, the olive oil industry has been modernized with the introduction of continuous methods by centrifugation. In the first instance, process with decanter of three outlets (olive oil, pomace and wastewater) were used. From the 1990s and with the aim of reducing the environmental impact of the wastewater generated by this process, the number of outlets in the horizontal centrifuge was reduced from three to two outlets, one for olive oil and the other for olive pomace and vegetable water (plus the added water during the process). Therefore, the type of pomaces and their physicochemical characteristics depend on the kind of process (Table 5).

**Table 5.** Physicochemical characteristics of the pomaces based on the process used.

| Process | Moisture and Volatile Matter * % | Fat Matter * % (In Wet Basis) | Fat Matter ** % (In Dry Basis) | Production * kg/t Olives |
|---|---|---|---|---|
| Pressure | 22–35 | 5.0–8.0 | 7.0–11.0 | 250–350 |
| Three outlets | 45–55 | 3.0–4.5 | 6.0–8.0 | 450–520 |
| Two outlets | 65–75 | 2.0–3.5 | 6.0–7.5 | 800–850 |

* Data from [34]; ** Own data.

In Spain, the country with the highest olive production, the most widely used olive oil extraction process is the continuous centrifugation system using two-outlet decanters (Figure 7). In Italy, the second olive producing country, pressure systems, centrifugation with a three-outlet decanter and, to a lesser extent, the system with a two-outlet decanter are used. In Greece and the other countries of the Mediterranean basin the three systems are used, although the most widespread is the centrifugation process with two-outlet decanter. The same happens in the rest of the olive oil producing countries (Portugal, Argentina, Uruguay, Chile, Peru, Australia, USA and China).

**Figure 7.** Olive pomace from oil mills that use the centrifugation process with a two-outlet decanter. These pomace rafts are located in the extraction plants of pomace oils.

In general, the pomaces produced are transported to the pomace extraction plants to separate the residual oil that they still contain. In these facilities, the extraction of residual oil is carried out using a solid-liquid extraction process, being technical hexane (mixture of alkanes) the most widely used solvent. From this process, crude pomace oil and a by-product that is extracted pomace are obtained. Crude pomace oils are sent to oil refining plants to obtain olive pomace oil. On the other hand, the extracted pomace (by-product from extraction process) constitutes a very interesting solid biofuel and with HHV in the range of 13.8 to 15.8 MJ/kg.

All these studies highlighted an unyielding interest in exploiting biomass obtained from olive tree. Therefore, the objective of this paper is to comprehensively review the latest advances focused on energy production from olive-pruning debris, olive stones and olive pomaces. The different routes to produce energy from these by-products are summarized in Figure 8 and discussed in the following sections. Future research efforts required to biofuel production are also discussed.

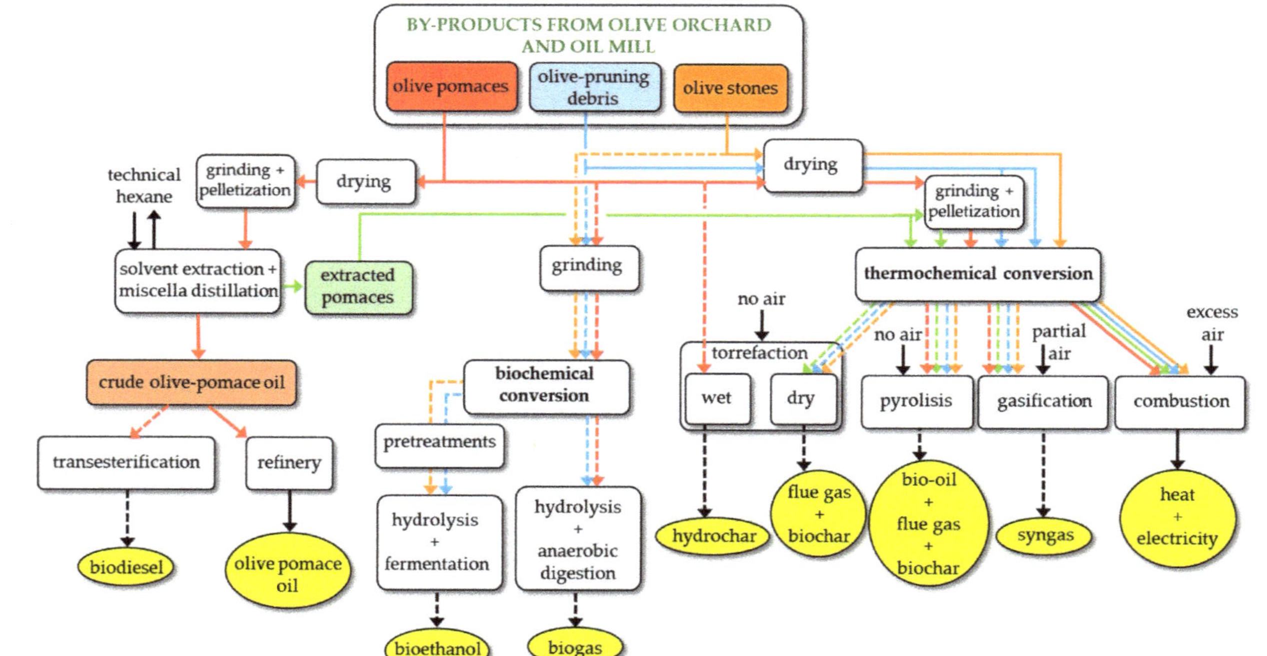


**Figure 8.** Valorisation routes of olive mills by-products towards energy. Solid lines stand for actual industry processes while dashed lines represent potential applications at research stage.

## 2. Olive-Pruning Debris

As the rest of lignocellulose wastes, the olive-pruning debris is a major potential source for renewable energy and high-value chemical products. Currently, the most direct application of the pruning debris is for combustion. Thicker branches are sometimes separated and used as firewood for home or small industries. However, this direct combustion is an inefficient process for energy conversion that is responsible of large $CO_2$ emissions. Assuming complete combustion under ideal conditions, extremely unlikely to occur, and an average carbon content in OP of 45.35% (Table 2), 1.7 kg of $CO_2$ would be emitted to the atmosphere per kg of burnt pruning. The combustion reactions and kinetics of OP can be found elsewhere [8]. As no sulphur is detected in the pruning and its nitrogen content is small, which are major advantages from the environmental point, other energy production processes have been researched in the last decades, as below indicated.

### 2.1. Bioethanol Production

As a lignocellulose waste, and thus a sugar-rich substrate, OP pruning is suitable for bioethanol production. Furthermore, its high cellulose content and low lignin content (Table 1), compared to other lignocellulose waste, makes OP an excellent substrate for bioethanol production. Currently, the bioethanol production includes 4 stages: (1) pre-treatment, (2) hydrolysis of polysaccharides and oligosaccharides into monomer sugars, (3) fermentation of sugars to ethanol, and (4) ethanol concentration to absolute alcohol. These stages and their applications to OP are comprehensively detailed in the following subsections.

#### 2.1.1. Pre-Treatment

Since lignocellulose biomass is a mixture of carbohydrate polymers from plant cell walls, the pre-treatment process is thus required to reduce the feedstock size, break down the hemicellulose to sugars and open the structure of the cellulose. Two parallel phases, i.e., a hemicellulose-rich liquid and a cellulose-rich solid, will be formed. Following this, the cellulose is hydrolyzed by enzymes into D-glucose that could be fermented into ethanol. Likewise, the D-xylose, mainly from the hemicellulose, could be fermented to ethanol or xylitol. Lignin can be extracted from the cellulose-rich solid by using an alkaline solution. As a result, an alkaline residue containing lignin is formed, which can be extracted by oxidizing agents, such a $H_2O_2$ [35], thus reducing the volume of the hydrolysis reactor, increasing sugar concentration, and reducing the energy consumption the subsequent cellulose hydrolysis. In spite of these advantages, this step is not mandatory to improve the enzymatic digestibility of cellulose. Among the pre-treatments applied to OP, the most widely used are:

- Ultrasound pre-treatment

Ultrasound technology, which produces cavitation and acoustic streaming, has also been regarded as a promising pre-treatment for the degradation of lignin. The particles nearby can be crumbled because the powerful hydro-mechanical shear forces in the bulk liquid is generated. Ultrasound irradiation disrupted lignocellulose structure and breakdown the crystalline nature of cellulose structure, apart from provoking cellulose depolymerisation and solubilisation, lysis of cell walls and membranes as well as improvement of the solubilisation of organic matter [36]. It was demonstrated that a hydrodynamic shear force in aqueous phase was generated from ultrasound, which increased the disintegration of coarse particles and so promoted the surface area for enzyme activity [37]. As the cavitation can generate high temperatures and high pressures (often referred as "hot spots"), the reaction times in heterogeneous solid-liquid systems can be reduced and the chemical degradation could be accelerated via the production of oxidative species. Combined with other biomass pre-treatments such as ozone or alkaline pre-treatment, a decrease of hemicellulose and lignin content was observed when sugarcane bagasse was pre-treated by ultrasound [38].

Notwithstanding, there is a lack of papers concerning the application of ultrasound to OP as a pre-treatment in the bioethanol production scheme. Nevertheless, some works can be found in literature. Thus, The application of ultrasound to OP to extract antioxidant compounds has been assessed [39]. Different ethanol/water ratios (20%, 50% and 80% of ethanol concentration), amplitude percentages (30%, 50% and 70%) and process times (5 min, 10 min and 15 min) were assessed. From the results on total phenolic compounds, total flavonoid content and antioxidant activities, these authors concluded that ultrasound pre-treatment could be a first step of the process within a biorefinery context and OP a potential source of natural antioxidants [39]. In the available literature it can be also found the effects of ultrasonic treatments on organosolv black liquor from olive tree pruning residues [40]. Up to 20% monosaccharides were degraded when 15 min of ultrasound treatment was applied. The monomeric sugars were increased from 3% to 16% as result of the lignin-carbohydrate complex rupture that was caused by ultrasonic irradiation [40].

- Ozonation

Ozone, the strongest chemical oxidant after fluorine, can be utilised as pre-treatment for removing the lignin from the lignocellulose biomass and finally improving the enzymatic degradation. Compared with other conventional pre-treatment methods, no harm residue such as acids or mineral bases are produced after ozonation [41]. Unlike ultrasound pre-treatment that is carried out at extremely high temperatures and pressures, ozone reactions can occur even at room temperature and pressure. Besides, ozone could be generated in situ in the bioethanol plant, thus avoiding transportation and storage issues.

Although ozonation has been demonstrated to be an effective pre-treatment, it is regarded as uneconomical due to the assumed need for lignin mineralization [42]. In order to overcome this issue, ozonation has been applied to tannic acid to simulate the effects of ozonation process on biomass pre-treatment for bioethanol production [42]. A high concentration of tannic acid solution (60 g/dm$^3$), as a lignin model, was treated by ozonation. Most of tannic acid was disappeared after 3.5 h treatment. As a negative result, tannic acid negatively affected cellulases activity. Despite the aforementioned doubt of widespread application to biomass as a pre-treatment, it was stated that a short-time ozonation, as an alternative, could reduce the pre-treatment energy costs from about ~$104.28 to ~$2.22 per ton of biomass and thus cut down the labor costs and operation times [42]. In spite of these advantages, to the best of our knowledge there are not available works on the use of ozonation as OP pre-treatment for bioethanol production.

- Steam explosion

Steam explosion, where the biomass is subjected to pressurised steam injected to a high temperature (180–240 °C) for a short time (from 10 s to several minutes) and ends with a sudden decompress of the system, is a widely used pre-treatment of biomass. The cellulose and lignin degraded by the high temperature whilst the tissue structure damaged during the rapid pressure release. In this way, the biomass is easier to be hydrolysed and fermented. A high D-xylose yield can be achieved especially when an acid catalyst is applied. The role of sulphuric acid (as catalyst) is to hydrolyse the hemicellulose into monomers without degrading them furfural and 5-hydroxy-methyl furfural (5-HMF). As a result, the addition of an acid catalyst can reduce the processing temperature to 150–200 °C, thus improving the subsequent enzymatic hydrolysis. Steam explosion has been applied to OP and these facts have been verified. In this way, it was reported the solubilisation of hemicellulose from olive tree pruning was improved by using steam explosion at temperatures between 190 °C and 240 °C [13]. Other acids have been used to improve the steam explosion of OP, such as phosphoric acid [43]. Furthermore, the steam explosion of olive tree pruning has been investigated at pilot scale to maximize the glucose yield in the subsequent enzymatic hydrolysis [44].

- Autohydrolysis or liquid hot water (LHW) pre-treatment

This technique is quite similar to the steam explosion and leads to similar results. LHW refers to the process of solubilisation of hemicellulose in pressurized water at temperatures ranging between

165 °C and 225 °C [45]. The hemicellulosic acetyl groups are rapidly released, thus hydrolyzing the rest of hemicellulose (autohydrolysis). To determine the effect of this pre-treatment, the severity parameter, $R_0$, is calculated as follows:

$$R_0 = \int_0^t exp^{\frac{T(t)-T_R}{w}} dt$$

where $T(t)$ is the temperature (°C)–time (min) function, calculated graphically, and $T_R$ a reference temperature (100 °C) below which the autohydrolysis can be considered of scarce significance [46]. The value of 14.75 is the generally used for the parameter $w$ in the autohydrolysis of olive pruning debris [47].

LHW pre-treatment has been applied to OP in the temperature range 150–210 °C for short reaction times (0–5 min) at the selected temperature (average $R_0 = 3.54 \pm 0.06$), resulting in a complete solubilisation of hemicellulose [4,47]. Similarly to steam explosion, LHW leads to a hemicellulosic oligomers solution which requires a further acid hydrolysis to release the monomeric sugars, if they are intended to be fermented [48]. To avoid the application of two pre-treatments, it has been assayed the addition of dilute acid into the pressurized water reactor, thus performing simultaneously both autohydrolysis and dilute-acid hydrolysis. Although it is performed in pressurized water reactors, this combined technique is generally referred as dilute-acid hydrolysis in spite of using high pressures, achieving high sugars yields and complete hemicellulose solubilisation when applied to OP [4].

- Extrusion

Continuous extrusion process, which involves a single or twin-screw extruder combining thermal, mechanical and chemical action, is regarded as a cost-effective pre-treatment method for enzymatic saccharification. What is more, the addition of enzymes during the extrusion process (bioextrusion) as new biomass pre-treatment technique for second generation bioethanol production has been proposed, resulting in a better sugar production [49]. With regards to its application to OP, extrusion in a twin-screw extruder was applied to OP to remove the extracts and the amorphous cellulose at 70 °C and sulphuric acid concentrations lower than 0.5 mol/dm$^3$ [10]. The yield of sugars was low. However, combined with a subsequent dilute-acid hydrolysis with 1 mol/dm$^3$ H$_2$SO$_4$ of the resulting solid in a stirred-tank reactor, led to a sugar-rich hydrolysate [14]. Extrusion operation still needs to be optimised for the extrusion to be regarded as a single hydrolysis stage for bioethanol production.

- Dilute-acid hydrolysis

Dilute-acid pre-treatment, as one of the most important pre-treatments, has been widely applied to OP within the biorefinery concept. The aim of dilute acid is to solubilise the hemicellulose fraction without degrading cellulose as far as possible. With regards to the application of dilute-acid pre-treatment to OP, several authors have reported that hemicellulose depolymerised into a mixture of sugar oligomers and monomers under the acidic, thermal process while scarce alteration took place in lignin and cellulose structures [4,11,50,51]. In addition, cellulose porosity increased with the removal of hemicellulose and so enhanced enzymatic digestibility of the cellulose. The effects of temperature and acid concentration of dilute-acid pre-treatment on the subsequent simultaneous saccharification and fermentation of olive-pruning debris using response-surface methodology (RSM) have been studied. The pre-treatment led to a complete solubilisation of the hemicellulose. As a result, the cellulose percentage in the resulting solid was roughly 1.5 times as much as that for raw material [4]. According to the RSM, the highest overall ethanol yields would be obtained when the pre-treatment for olive-pruning debris were performed with 0.059 kmol/m$^3$ and 0.030 kmol/m$^3$ H$_2$SO$_4$ at 185 °C. Under these conditions, 15.3 kg and 14.5 kg ethanol would be generated from 100 kg olive-pruning debris, respectively [4]. Likewise, the effects of the reaction time (0–300 min), temperature (70–90 °C) and sulphuric acid concentration (0–0.05 kmol/m$^3$) on the formation of D-glucose and D-xylose were evaluated by RSM [50]. Results showed that there were interactive effects between the three parameters on sugars production. The highest concentrations of D-glucose and D-xylose were achieved when the highest temperature, acid concentration and residence time applied. The optimal conditions for generating D-xylose were 90 °C, 0.05 kmol/m$^3$ H$_2$SO$_4$ and 300 min reaction time. Under these conditions, it was predicted that approximately 40% of the maximum attainable D-glucose and 60% of the potential D-xylose would be obtained [50]. The olive-pruning debris has been also hydrolyzed by 0.050–0.100 kmol/m$^3$ oxalic acid at 130–170 °C for 30 min using 1:10 dry raw material to organic acid ratio. However, the hydrolysate was not able to be fermented by the yeast *Pichia stipitis* CBS 6054 [14]. Similarly, the fermentation with *Pachysolen tannophilus* of the hydrolysates resulting from the hydrolysis with concentrated phosphoric acid of OP at 90 °C for 240 min led to very low bioethanol yields [52].

Dilute-acid pre-treatment has been assayed in combination with other pre-treatments, such as fungal pre-treatment [51] and autohydrolysis [15,48], to enhance the enzymatic hydrolysis of OP. For instance, the combination of fungal pre-treatment with a dilute-acid pre-treatment was studied [51]. It was found that the order of the pre-treatment combination has a relevant effect on the D-glucose yield of the subsequent enzymatic hydrolysis. The application of the best pre-treatments combination plus enzymatic hydrolysis to OP achieved 51% of the theoretical sugar yield. The afore mentioned best sequential pre-treatments were fungal pre-treatment with *Irpex lacteus* for 28 days followed by diluted-acid pre-treatment with 2% (*w/v*) H$_2$SO$_4$ at 130 °C for 90 min, which enhanced 34% the enzymatic hydrolysis yield compared with that of the application of solely the dilute-acid pre-treatment [51]. On the other hand, the application of an dilute-acid hydrolysis (90 °C, 0.05 kmol/m$^3$ H$_2$SO$_4$) to the solid obtained after the autohydrolysis (200 °C, 0 min) of olive-pruning debris, neither increased the cellulose conversion nor led to lignin degradation [48]. By contrast, the application of the same dilute-acid pre-treatment to the resulting liquid from autohydrolysis rocketed the D-glucose and D-xylose extraction from OP. As a result, this sequential pre-treatments led to a phenolic compound-free pre-hydrolysate containing 114 g D-glucose and 78 g D-xylose per pre-treated kg of olive-tree pruning [48]. This fact has been verified by other authors, who stated that in all their experiments the percentage of lignin recovered was close to 100% [15]. These authors concluded that the dilute-acid hydrolysis of OP at 90 °C for 180 min leads to an almost complete hydrolysis of the hemicellulose when using a concentration of hydrochloric acid (HCl) exceeding 3.77%.

- Alkaline peroxide pre-treatment

Alkaline peroxide, during which effective radicals perform the delignification without degrading sugars, is also widely used as pre-treatment for biomass saccharification [53]. Since it is difficult for chemicals and enzymes to access to cellulose due to its complex glycosidic and hydrogen bonds network, an alternative strategy is to remove lignin to easier access to cellulose carbohydrates in the subsequent enzymatic hydrolysis.

Under alkaline conditions, hydrogen peroxide can attack the phenolic compounds, which is quite different from that under the normal conditions where hydrogen peroxide can only react with the aliphatic part of lignin without degrading the phenolic compounds [54]. Lignin was reported to be effectively removed from the olive-tree pruning biomass when treated with formic acid and alkaline hydrogen peroxide, allowing the production of ethanol up to 46 g/dm$^3$ in the subsequent enzymatic hydrolysis and fermentation [12].

### 2.1.2. Hydrolysis

Acid hydrolysis was the main technique for lignocellulose hydrolysis in the 20th century for ethanol production because mineral acids can penetrate lignin without pre-treatment and the rate of acid hydrolysis is faster than enzymatic hydrolysis. Notwithstanding, sugars also degrades rapidly under acidic conditions to compounds such as furfural, a product of dehydration of pentoses, and 5-hydroxymethylfurfural, a product of the dehydration of hexoses (Figure 9). These compounds, along with the acetic acid released during initial decomposition of the hemicellulose, inhibit the later fermentation, leading to reduced ethanol yields [55]. Furthermore, the use of high temperatures and acid concentrations is dangerous and leads to severe corrosion of equipment and the need to recover the acid used, after hydrolysis, to make it economically feasible [56]. As a result, this technique has been replaced by the enzymatic hydrolysis when the goal is to obtain D-glucose from cellulose, and it is usually used as a pre-treatment, under mild conditions, of the enzymatic process as aforementioned.

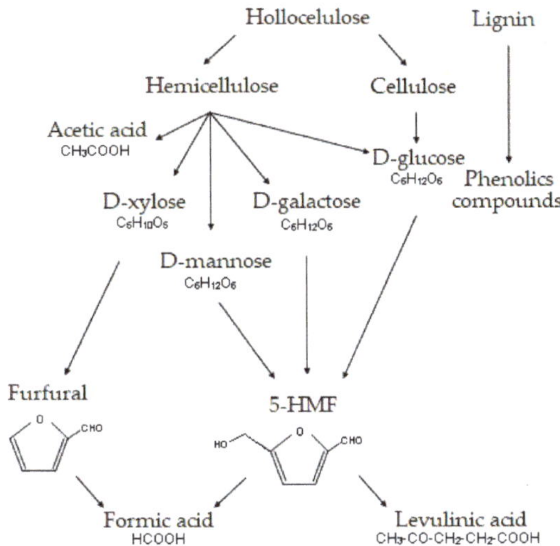


**Figure 9.** Products and by-products obtaining during the acid hydrolysis of lignocellulose biomass.

Enzymatic hydrolysis began to be researched from the seventies of last century with the first commercial preparations of biocatalysts. Its main advantages are that the enzymatic reactions are carried out under mild conditions (40–50 °C and pH around 5.0) and its selectivity. The cellulases used for breaking down cellulose are originated from fungi, especially from *Trichoderma reesei* or *Trichoderrma viride*. A single enzyme is unable to hydrolyse the whole lignocellulose biomass and most cellulases are complexes of three type of enzymes that hydrolyse cellulose together. First, an endoglucanase breaks one of the chains within the cellulose crystalline structure. Then, an exoglucanase attaches to one of the loose ends, pulls the cellulose chain out of the crystal structure, and works its way down the chain, breaking off units of cellobiose (two linked D-glucose units). Finally, a β-glucosidase splits the cellobiose into two D-glucose molecules [56]. Cellulases and β-glucosidases are inhibited by cellobiose and D-glucose, respectively.

Aiming to a full valorisation of the sugars contained in the olive-pruning debris, it is mandatory to hydrolyse hemicellulose. To this end, xylanases are required to break down the linear β-1,4-xylan polysaccharide to D-xylose. Generally, both xylanases and cellulases are added to the resulting mixture from the pre-treatment, i.e., the pre-hydrolysate (rich in oligomers from hemicellulose) and the cellulose-rich solid are simultaneously hydrolysed to cut costs. Industrial cellulases and xylanases are well-known commercial products, tailor-made according to the desired commercial application [56].

The so-called simultaneous saccharification and fermentation (SSF) greatly improves D-glucose yield, reducing the concentration of required enzyme. During SSF, enzymes hydrolyse cellulose chains and yeasts ferment D-glucose to ethanol at the same time, so that the inhibition by degradation products will be lessened when sugars fermented immediately after generation. The ethanol production from OP by SSF of the pre-treated solid has been already investigated using a commercial cellulolytic complex (Celluclast 1.5 L) and the yeast *Saccharomyces cerevisiae* [4,13]. Using as pre-treatment steam explosion [13] or autohydrolysis [4] acid-catalyzed, the ethanol yields were 7.2 and 9.9 kg/kg OP, respectively.

### 2.1.3. Fermentation

The conventional yeast *Saccharomyces cerevisiae* is the most common yeast for alcoholic fermentation because it can easily convert D-glucose into ethanol, reaching fermentation yields (expressed as kg ethanol per kg total sugars) close to the maximum theoretical ethanol yield (0.51 kg/kg) with volumetric productivities of up to 5 kg/m$^3$ h and final ethanol concentrations of roughly 10% (*v/v*) [57]. Moreover, it can be used for continuous industrial processes. Notwithstanding, this yeast is unable to ferment D-xylose and the rest of pentoses from hemicellulose. Therefore, other yeasts able to ferment both hexoses and pentoses into ethanol have been assayed, such as *Pichia stipitis*, *Candida shehatae*, *Candida tropicalis* and *Pachysolen tannophilus*. Some of them ferment D-xylose to xylitol instead of to ethanol (*C. tropicalis*, *P. tannophilus*), depending on the oxygen conditions [47,58,59].

Olive-pruning-debris hydrolysates have been fermented with several of these yeasts. In order to reduce cost, a scheme consisting in dilute-acid hydrolysis plus fermentation, thus eliminating the costly enzymatic hydrolysis stage, has been assayed by several authors [15,16,52,60]. Hydrolysates obtained from the dilute-acid hydrolysis of OP at 90 °C with hydrochloric, sulphuric and trifluoroacetic acids in concentrations between 0.1 N and 1.0 N were fermented with *P. tannophilus*. The highest ethanol yield (0.37 kg ethanol/kg total sugars) was achieved in the fermentation of hydrolysates obtained with 0.1 N C$_2$HF$_3$O$_2$ [60] while the highest ethanol concentration (6.34 g/dm$^3$) was found in the hydrolysate from hydrolysis with 1.0 N HCl [60]. Lower yields were found by these authors using phosphoric acid hydrolysates [60]. Anyway, with regard to the overall process, the production of ethanol from initial dry olive-pruning debris was almost negligible (Table 6). This was probably because dilute-acid hydrolysis solely solubilized sugars from hemicellulose. Therefore, cellulose remained in the solid and its D-glucose was not fermented. Similarly, the fermentation with *P. tannophilus* of the hydrolysates resulting from the hydrolysis with concentrated phosphoric acid of OP at 90 °C for 240 min led to very low bioethanol yields instead of the high fermentation yield (0.38 kg ethanol/kg total sugars) [52]. Furthermore, the hydrolysates obtained from dilute-acid hydrolysis using 0.050–0.100 kmol/m$^3$ oxalic acid at 130–170 °C for 30 min could not be fermented by the yeast *P. stipitis* CBS 6054 [14], probably to production of yeast inhibitors from sugar degradation. The best results of the application of this scheme to OP has been achieved using 1 N H$_2$SO4 at 90 °C for 240 min [16]. These authors obtained 70 g ethanol plus 34 g xylitol (a high-added value by-product) per kg OP, reaching fermentation yields higher that the aforementioned (Table 6).

Another interesting scheme applied to OP is autohydrolysis plus fermentation without enzymatic hydrolysis [47]. As it can be seen in Table 6, this scheme resulted in the production of 7.2 g of ethanol from 100 g of olive pruning debris along with an instantaneous ethanol yield of 0.44 kg/kg total sugars. In spite of the fact that these two schemes, which do not involve enzymatic hydrolysis, do not take advantage of the most of D-glucose of cellulose, they are less costly and provide a solid residue composed of cellulose and lignin that is an excellent raw material for pellet production, thus leading to a circular economy of the olive-tree pruning.

With regard to the last scheme, consisting of pre-treatment, enzymatic hydrolysis and subsequent fermentation, it is worth noting that the ethanol yield achieved using steam explosion plus SSF with *S. cerevisiae* of the pre-treated cellulose [13] was similar to that obtained by autohydrolysis and fermentation of the prehydrolysate [47], which had the advantage of the non-use of enzymes (Table 6). The use of an acid catalyst in the pre-treatment increased the ethanol yield in the SSF of the pre-treated cellulose up to 9.9 kg/kg [4]. To improve these ethanol yields, it is mandatory to ferment the pre-hydrolysate, rich in hemicellulosic sugars. The yeasts used by several authors to ferment pre-hydrolysates from OP were *Scheffersomyces stipites* [43] and *Escherichia coli* [12]. The ethanol production from both fermentation of hemicellulosic sugars solubilized in the pre-treatment and SSF of pre-treated cellulose was roughly 16 kg/kg OP. [4,12,43]. Notwithstanding, the use of enzymes and 2 fermentation stages make this scheme infeasible from an economic point of view.

**Table 6.** Ethanol yields obtained from olive-pruning debris (OP) using different hydrolysis plus fermentation schemes.

| Pre-Treatment | Commercial Enzyme Preparation | SSF Yeast | Fermentation Yeast | Fermentation Yield (kg/kg) | Overall Yield (kg/kg OP) | Reference |
|---|---|---|---|---|---|---|
| 0.5 N $H_3PO_4$, 90 °C, 240 min | - | - | *P. tannophilus* | 0.38 | NP | [52] |
| 1 N $C_2HF_3O_2$, 90 °C, 240 min | | | *P. tannophilus* | 0.37 | NP | [60] |
| 75 mM $C_2H_2O_4$, 150 °C, 240 min | - | - | *P. stipitis* | 0.00 | 0.00 | [14] |
| 1 N $H_2SO_4$, 90 °C, 300 min | - | - | *C. tropicalis* | 0.42 | 0.070 | [16] |
| * log $R_0$ = 3.54 | - | | *C. tropicalis* | 0.44 | 0.072 | [47] |
| ** 230 °C, 1% (wt.) $H_2SO_4$ impregnation | Novozymes Celluclast 1.5 L and Novozymes 188 | *S. cerevisiae* | - | - | 0.072 | [13] |
| * 0.08 N H2SO4, 107 °C, 5 min | Novozymes Celluclast 1.5 L and Novozymes 188 | *S. cerevisiae* | - | - | 0.099 | [4] |
| 2.4% (wt.) $H_2SO_4$ 130 °C, 84 min + 7% $H_2O_2$ (*w/v*), 80 °C, 90 min | Novozymes Cellic CTec3 and Novozymes 50010 | *S. cerevisiae* | *E. coli* | 0.43 | 0.15 | [12] |
| ** 1% (wt.) $H_3PO_4$, 195 °C, 10 min | Novozymes 50013 and Novozymes 50010 | *S. cerevisiae* | *S. stipitis* | 0.35 | 0.16 | [43] |

* Liquid hot water; ** Steam explosion; NP = not provided.

*2.2. Methane Generation*

Scarce information about methane production from OP can be found in literature. The finest OP particles have been reported as the most suitable for methane production due to their high concentration of nitrogenous matter and low C/N ratios, which make these OP suitable for anaerobic digestion [61]. After the application of a fractionation process to obtain the finest OP particles and the subsequent batch anaerobic digestion at 38 °C, the highest methane yield achieved by these authors was 176.5 Nm$^3$ per t volatile solids, which accounted for 93.5% gains over the untreated OP. The authors described their process as highly energy efficient, since energy consumption was low compared with the energy required by the residues of fuel upgrading, intended for anaerobic digestion [61]. Nevertheless, there is other olive mill wastes with higher potential for anaerobic digestion and therefore methane production, such as olive pomace and, mainly, olive mil wastewater. Thus, about 10–12 million cubic meters of olive mill wastewater are generated every year from the olive oil extraction in 2- and 3-outlet decanters in a relatively short period of time, so that researchers have put a lot of effort to cope with this issue, mainly through biogas production. This could account for the scarce available information on methane production from OP [62].

*2.3. Pellets Production*

The production of pellets from the olive-pruning debris is a low-cost alternative that would allow the conversion of this waste into an energy resource that could be used in the surrounding areas where the OP was collected or be packaged for external sales. In both cases, it would suppose a benefit for farmers. Notwithstanding, it has to be taken into account that some authors have pointed out that OP meets the specification for industrial pellets given in the European Standard EN ISO 17225-2, but not for residential pellets based on this standard because of its high ash and nitrogen content [7,61]. Another hindrance for pellets production from OP is its low bulk density (347.9 kg/m$^3$ [3]). The use of denser biofuels reduces the costs associated with handling, storage and transportation [7]. One suggested alternative is to mix OP with other raw materials in order to reduce the percentages of these chemical components in the final biomass and increase bulk density [7]. The effects of main process parameters (pressure and temperature) and OP properties (moisture content and particle size) on pellet density and durability have been previously investigated [63]. High process temperatures, low moisture contents (less than 15% [64]) and reduced particle sizes were reported as the conditions to obtain high-quality pellets from OP, while the compression force had scarce effect on pellet quality [63]. In this sense, 9% moisture content along with short compression lengths (20–24 mm) and temperatures higher than 40 °C were reported as the most suitable pelleting conditions for the residual biomass from olive trees (leaves, prunings and wood), although these parameters varied among the raw materials analyzed [7]. The pelleting process slightly increase the calorific power of this biomass. Thus, the HHV and LHV of pellets from OP have been reported to be 19.47 and 16.17 MJ/kg, respectively [64].

*2.4. Torrefaction*

Torrefaction is a pre-treatment applied to enhance biomass properties for its further use in combustion or pyrolysis processes. Concerning to OP, torrefaction was carried out under an inert atmosphere at low temperature (200–300 °C) and a low heating rate (residence time between 10 min and 60 min) to convert this biomass into a charcoal-like carbonaceous material with enhanced energetic properties [65]. For example, torrefaction at 300 °C for 60 min led to an increase on the ratio of fixed carbon to volatiles (from 0.23 to 0.39), improving the fuel quality of OP. Under these conditions, the OP surface structure was broken, hemicellulose was partially or totally decomposed, and under the severest torrefaction conditions, a modification on the thermal stability of cellulose was observed. These authors also proposed a pyrolysis kinetic model of OP through thermogravimetric measurements under N$_2$ atmosphere during the torrefaction of OP under different conditions [66]. Three stages in which the activation energy kept approximately constant were observed. These stages were related to the thermal degradation of the 3 main constituents of any lignocellulose biomass: cellulose, hemicellulose and lignin.

*2.5. Pyrolysis*

Pyrolysis is a thermal process carried out under partial or complete absence of oxygen and is based on capturing the off-gases from thermal decomposition of the organic material. Three main products are obtained from biomass pyrolysis: Syngas (non-condensable gases), bio-oil (hydrocarbon molecules) and a solid residue rich in carbon. Syngas and bio-oils are regarded as ordinary energy carriers. Besides, bio-oils from pyrolysis are of major interest, as they can replace diesel in internal combustion engines [67]. Among these different products obtained from biomass pyrolysis, the solid, carbon-rich material can be highlighted. This solid can be biochar or charcoal. Biochar is mainly used for soil applications for both agriculture and environment, while the charcoal is used as a fuel for heat, absorbent material, or reducing agent in metallurgical processes [64]. Slow pyrolysis is applied for retaining up to 50% of the carbon feedstock in the resulting biochar so that it can be used as soil fertilizer while high-temperature pyrolysis (>550 °C) led to biochar with high aromatic content from lignin decomposition and, therefore, recalcitrant to decomposition [64]. The biochar obtained by these latter authors from the pyrolysis or OP pellets showed HHV and LHV of 31.71 and 30.48 MJ/kg, respectively [64], far higher than those reported for OP [3,7–9] and pellets from OP [64]. This biochar can be certified as Biochar Premium according to the regulations of the European Biochar Certificate, based on the analysis of nutrients and trace elements, polycyclic aromatic hydrocarbons composition, elemental analysis, pH, electrical conductivity and density performed by the authors [64].

In an earlier work, leaves, branch barks, twigs (small branch of 1 cm) and olive wood (sawdust particle size 0.8–1 mm, cubes of 1, 2, 3, or 4 cm edge) were subjected to pyrolysis at 400, 500 and 600 °C (10 °C/min heating rate, 20 min residence time at the final temperature, 200 cm$^3$/min $N_2$ stream) [68]. Based on its high content of volatile matter and ash and the low process yield, the charcoal obtained from the pyrolysis of olive leaves was regarded as low-quality charcoal. These authors pointed out that since olive leaves are used to feed cattle, this is a more appropriate alternative to pyrolysis [68]. By contrast, in spite of the fact the charcoals obtained from the pyrolysis of branch bark, twig and wood (sawdust or cubes) also had a high ash content, their volatile matter and fixed carbon contents made them suitable for the production of B category briquettes in accordance with the French and Belgian regulations in force at the date of publication of that paper [68], both regulations were dating from 1984. With regards to the effect of temperature pyrolysis on the resulting charcoals, the fixed carbon content increased and that of the volatile matter decreased as the temperature in which the process was performed rose from 400 °C to 600 °C. Finally, the Brunauer–Emmett–Teller (BET) surface area of the charcoals obtained from leaves and branch barks were very low (5 and 16 m$^2$/g, respectively), while those of twigs and different olive woods ranged between 19 m$^2$/g and 198 m$^2$/g [68]. These surface area values are lower than those obtained for charcoals of other biomasses. For instance, BET surface area of coconut shell biochar was reported to be 244.2 m$^2$/g [69] and that of biochars obtained from the pyrolysis of roots from *Jatropha curcas L.* plants used for the phytoremediation of contaminated soils with different heavy metal concentrations soils was 447 m$^2$/g [70]. Notwithstanding, scanning electron microscopy (SEM) analysis indicated that the surface of biochar from OP was higher than that of OP surface, and that it had a very porous structure, indicating that it could be regarded as an ideal support for metal impregnation [9].

The pyrolysis of OP has been modelled from thermogravimetric experimental data of each obtained fraction of the lignocellulose material (cellulose, hemicellulose and lignin) [8]. It was reported that the pyrolysis of OP could be divided into three stages: those of hemicellulose, cellulose and lignin, respectively. Lignin was the component most difficult to decompose and it occurred slightly over a wide temperature range, with a very low mass loss [8]. Therefore, the pyrolysis of OP is strongly influenced by its lignin content.

The pyrolysis under inert atmosphere ($N_2$) and under oxidizing atmosphere (20% $O_2$) of OP, OP soaked with Ni solutions and OP soaked with Pb solutions has been investigated with a heating rate of 10 °C/min [9]. For the 3 biomasses, the greatest mass loss under was found in the ranges 150–400 °C and 150–480 °C under inert atmosphere and oxidizing atmosphere, respectively. In contrast with the OP impregnated with nickel, the impregnation with lead did not show very significant effects. The maximum volatilization rate was observed for the nickel-impregnated OP. As a result, there was a higher emission of volatile compounds and a higher mass loss in the gasification of OP soaked with nickel [9].

A pyrolysis-based circular system from the OP obtained from 10 ha olive grove, i.e., 25 t twigs (32 wt.% moisture) and 10 t 4-cm diameter wooden branches (40 wt.% moisture), has been proposed [67]. This scheme is based on the OP pyrolysis at 600 °C with an approximate heating rate of 200 °C/s under He atmosphere. The pyrolysis of this OP led to 8.5 t of bio-oil with (average LHV of 31 MJ/kg), 9.9 t of gas and 7.4 t of bio-char (average LHV of 29 MJ/kg). The gross energy obtained from the combustion of these 8.5 t bio-oil as a fuel (assuming 30% combustion efficiency) and the heat recovery from the dryer was 23.43 MWh [67]. These authors stated that the 7.4 t of biochar obtained from 10 ha olive grove could be used as soil amendment and carbon sequestration tool, which could make pyrolysis a suitable process for olive groves with poor soil and in conditions where soil management needs are preeminent [67].

Finally, low-temperature microwave-assisted pyrolysis of olive pruning residue using various absorbers has been assayed [71]. The bio-oils obtained contained interesting bio-chemical compounds (mainly acetic acid, aromatics and furans) while biochars showed calorific values close to that of commercial pellets (up to 25 MJ kg$^{-1}$). These authors concluded that the microwave-assisted pyrolysis of OP can be a sound method for obtaining useful chemicals and fuels, thus representing a potential possibility for reducing all of the environmental risks involved in its disposal [71].

*2.6. Gasification*

Gasification is a process based on the partial oxidation of the biomass with a controlled amount of oxygen or steam carried out at high temperatures (>700 °C), i.e., gasification does not involve combustion, resulting in the production of syngas ($H_2$ and CO, mainly) and condensable organic compounds as gasification byproducts. Gasification is a mature technology since it has been extensively applied for decades in the gas industry. Notwithstanding, some authors has pointed out that gasification is having its own rejuvenation due to its application to biomass, which is regarded as a new substrate for this process [67].

The gasification of OP has been studied in a laboratory fluidized bed [72]. The influence of temperature (800–900 °C), equivalence ratio (0.12–0.35), fuel particle size (from 0.5–4 mm to 2.5–4 mm), biomass throughput (485–725 kg/h m$^2$) and $O_2$ (21 and 40% $O_2$ in air) was investigated. These authors analyzed the results with the assistance of a previously developed fluidized bed gasification model. The application of this model to the laboratory-scale results allowed predicting the gas composition of industrial-scale fluidized bed gasifiers [72].

The gasification kinetics of olive tree pruning pellets in fluidized bed at temperatures between 760 °C and 900 °C has been described [73]. These authors prepared cylindrical pellets from OP using a pelletizing machine, which were reduced to 1–2.8 mm (average 1.9 mm) and 2.8–4 mm (average 3.4 mm) particle size by grinding. Experiments were carried out using gas mixtures containing $H_2O$, $CO_2$, $H_2$, CO and $N_2$ in various proportions at a superficial gas velocity of approximately 0.50 m/s. CO and $H_2$ inhibited the gasification process in a significant way. The authors obtained 2 kinetic equations, one for the gasification rate with $H_2O$, which took into account the inhibition effect of $H_2$, and another for gasification with $CO_2$, which included the inhibition effect of CO. The reaction rate of the former one (gasification with $H_2O$) was from 3 to 4 times faster than that of the latter one (gasification with $CO_2$) [73]. These kinetic models accounted for the effect of temperature, gas composition and the extent of carbon conversion.

The effects of nickel and lead as catalyst on the gasification of olive pruning under oxidizing and inert atmosphere have been assessed [9]. To do that, samples were soaked with nickel and lead solutions. Under nitrogen inert atmosphere, the effect of both metals catalyst was negligible and they were concentrated in the ash, thus increasing the concentration of metals in ash. The same occurs with lead under oxidizing atmosphere (20% $O_2$). By contrast, an increase in thermal decomposition was observed due to the presence of nickel under oxidizing atmosphere, probably because of the formation of NiO, which catalyzed decomposition reactions [9].

A gasification-based circular system has been proposed based on a bubbling fluidized bed gasifier working at about 800 °C and using a constant equivalence ratio (0.3) during the operation (50 t of air in the gasifier) [67]. As substrate, these authors used the annual OP obtained from 10 ha olive grove in Foggia province (Southern Italy, Apulia region): 25 t twigs (32 wt.% moisture) and 10 t larger wooden branches of at least 4 cm of diameter (40 wt.% moisture). From these, 62 t of syngas with an energy content of 88 MWh were obtained. Considering an electricity conversion efficiency for the syngas obtained from gasification of 37.5% for electricity production in a micro turbine (Brayton Cycle) and taking also into account the recovery of heat from the dryer, the total gross amount of energy reached 34.4 MWh [67]. These authors concluded this electricity generated is far more enough to supply the energy requirements of the olive-olive mill that produces the olive oil from the olives collected in that 10-ha olive grove.

The gasification of OP through a downdraft fixed bed reactor showed that the composition of syngas was strongly influenced by the air flow rate [74]. The $H_2$ and CO contents were roughly 16% (*v/v*) and over 13% (*v/v*), respectively, while the $CH_4$ content was lower than 3% (*v/v*) and $N_2$ content was higher than 50% (*v/v*) due to the use of air as oxidizing agent. The syngas had a LHV ranging between 3.6 MJ/Nm$^3$ and 4.6 MJ/Nm$^3$. The resulting biochar contained about 74 wt.% carbon.

The modeling of a small-scale plant based on a downdraft gasifier and a gas engine connected to the grid using OP is available in literature [75]. The power plant is able to produce 70 kW of electric power and 110 kW of thermal power when fuelled with 105 kg/h biomass and the gasifier operated in steady state conditions. The LHV of the gas produced in the gasifier was 3.7 MJ/kg. This relative low value was explained by the high air to OP ratio (2.72), which increased the $N_2$ formation, and the high ash content in OP (8.71%).

Finally, an experimental and feasibility study of a pilot plant installed in an olive mill located in Andalusia (Spain) for the conversion of olive tree pruning and into electrical and thermal power is described elsewhere [31]. The pilot plant was composed of a downdraft gasifier, gas cooling-cleaning stage and spark ignition engine with a modified carburetor. OP was regarded as a suitable substrate for the downdraft gasifier because of its low moisture and ash content. The cold gas efficiency was in the range 70.7–75.5% and the OP gasification led to a syngas with a LHV of 4.8 MJ kg$^{-1}$. The electric efficiency of the gas engine reached 21.3%. Besides, the plant achieved acceptable values for the electric (15%) and combined heat and power (almost 50%) efficiency [31]. As a result, the project investment would provide a whole benefit of around 300,000 € (net present value), with a payback period of 5–6 years.

## 3. Olive Stones

Currently, there are two industrial processes that are applied to the olive to obtain food products and that allow the recovery of olive stones in two different ways. Thus, in the olive mills, virgin olive oil is produced and, as a by-product, fragmented endocarps, since in the milling stage the olive is passed through a mill that breaks down the structure of the fruit. On the other hand, whole olive stones are recovered in the pitted table olive processing industries; that is, the endocarp plus the seed. Taking into account that the available volume of fragmented endocarps is much higher than that of whole olive stones, most of the studies carried out on the energy use of this type of biomass have been carried out using the former one. Notwithstanding, most of papers available in literature on olive stones do not clearly indicate whether fragmented endocarps or the whole olive stones have been used. Therefore, in this review paper the term olive stones (OS) is indistinctively used for both.

### 3.1. Preliminary Treatments: Cleaning and Drying

The most suitable pathway for the transformation of a lignocellulose material towards a specific product is strongly conditioned by the composition and moisture of the raw material. In the case of fragmented endocarps of olive, attention should be paid to the pulp, a material that is often found with them. According to some authors, the average percentage of pulp in OS is 3 wt.% [76]. Therefore, a first treatment to be applied to the biomass would aim to remove the pulp fraction, which can be easily achieved by sieving the OS, since an analysis of the particle size distribution showed that most of the fragmented endocarps (91.7 wt.%) had an average diameter of between 2 mm and 3.15 mm [77] and most of the pulp and mineral matter tends to accumulate in a fraction of smaller size (<1 mm) [76]. A direct consequence of the cleaning of the fragmented endocarps is the decrease in the ash content (from 0.78% to 0.55%) and the increase in the bulk density of the clean biomass (from 721.5 kg/m$^3$ to 764.2 kg/m$^3$). This can be explained by the fact that the pulp, which is the component with the lowest bulk density (129.8 kg/m$^3$, [76]) and the highest ash content (5.6%, [78]), remains in the fines fraction.

On the other hand, the fragmented endocarps of olives need to be dried in order to be used in certain applications. The moisture of 15 Spanish samples of OS ranged between 10.2% and 30.5%, with an average value of 22.3% [19]. Other authors have reported moistures of 19% [79] and 23% [80], and equilibrium moistures of 7% and 8%, respectively. In a previous work, we reported equilibrium moistures for OS between 9.1% and 11.5%, calculated at 30 °C and relative humidity between 43.2% and 73.1% [3]. The study of the drying kinetics of OS was addressed in the aforementioned last three references [3,79,80].

### 3.2. Biochemical Conversion

Numerous research works that addressing the biochemical use of the olive stones can be found in literature. OS contain hemicellulose very rich in xylans, a low cellulose fraction but with high crystallinity and a relatively high percentage of lignin [81]. The main products obtained in the biochemical conversion of olive stones are xylitol and bioethanol. Both can be achieved simultaneously in the same fermentation stage using yeasts able to ferment D-xylose to xylitol and D-glucose to ethanol. The common denominator of all the biochemical pathways is the application of several stages (pre-treatment, hydrolysis of polysaccharides, detoxification of hydrolysates, fermentation of sugars, separation of bioproducts...), some of which involve the use of microorganisms or enzymes. Among the pre-treatments described in Section 2.1.1., liquid hot water [82–84], steam explosion [25], organosolv [84] and dilute-acid hydrolysis [18,23,84–86] pre-treatments are the most widely ones applied to olive stones. Different schemes and the ethanol and xylitol yields (when available) achieved by several authors from 1 kg OS are illustrated in Table 7.

**Table 7.** Ethanol and xylitol yields obtained from olive stones (OS) using different process schemes.

| Pre-Treatment | Commercial Enzyme Preparation | Product | Fermentation Yeast | SSF Yeast | Overall Yield (kg/kg OS) | Reference |
|---|---|---|---|---|---|---|
| 0.01 M $H_2SO_4$, 201 °C, 5.2 min | Novozymes Celluclast 1.5 L | Ethanol<br>Xylitol | *P. tannophilus*<br>*P. tannophilus* | -<br>- | 0.122<br>0.067 | [23] |
| *log $R_0$ = 4.39 | Cellulase from *T. reesei* EC 3.2.1.4 | Ethanol<br>Xylitol | *P. tannophilus*<br>*P. tannophilus* | -<br>- | NP<br>NP | [82] |
| 0.025 M $H_2SO_4$, 195 °C, 5 min | Novozymes Celluclast 1.5 L | Ethanol<br>Xylitol | *P. tannophilus*<br>*P. tannophilus* | -<br>- | 0.103<br>0.092 | [18] |
| 1.5% (*w/v*) $H_2SO_4$, 121 °C, 60 min | Novozymes Celluclast 1.5 L and Novozymes 188 | Ethanol | - | *P. tannophilus* | 0.068 | [86] |
| Organosolv, 220 °C | - | Ethanol | - | *S. cerevisiae* | 0.131 | [84] |

* Liquid hot water; NP = not provided.

The overall process is fairly expensive and, therefore, the biochemical conversion of OS would only be economically viable if, in addition of products with low sale price (in relation to their production process), such as bioethanol or furfural [87], products with very high-added value were obtained, such as antioxidants [26], oligosaccharides [88] or other bioactive molecules. In this sense, the integrated production of xylitol, furfural, ethanol and poly-3-hydroxybutyrate from OS within the biorefinery concept, and a cogeneration system for producing bioenergy from the solid residues resulting from the obtaining of these products has been described [89]. On the other hand, an integrated biorefinery concept for OS management is available in literature, composed of supercritical fluid extraction to recover polyphenols, followed by pyrolysis of the solid to produce bio-renewable fuels and, finally, activation of the biochar to yield high surface area adsorbents for heavy-metals removal from water [90].

In a recent article, we described the fractionation of OS using dilute-sulphuric-acid pre-treatment at high temperature (201 °C) followed by, on one hand, the enzymatic hydrolysis of the pre-treated solids and, on the other hand, the detoxification of the liquid prehydrolysate by means of vacuum distillation. In this way, both the enzymatic hydrolysate, rich in D-glucose, and the detoxified acid prehydrolysate, rich in D-xylose, were easily fermented with *P. tannophilus* to bioethanol and xylitol. The mass balance for the complete ethanol production process is illustrated in Figure 10.

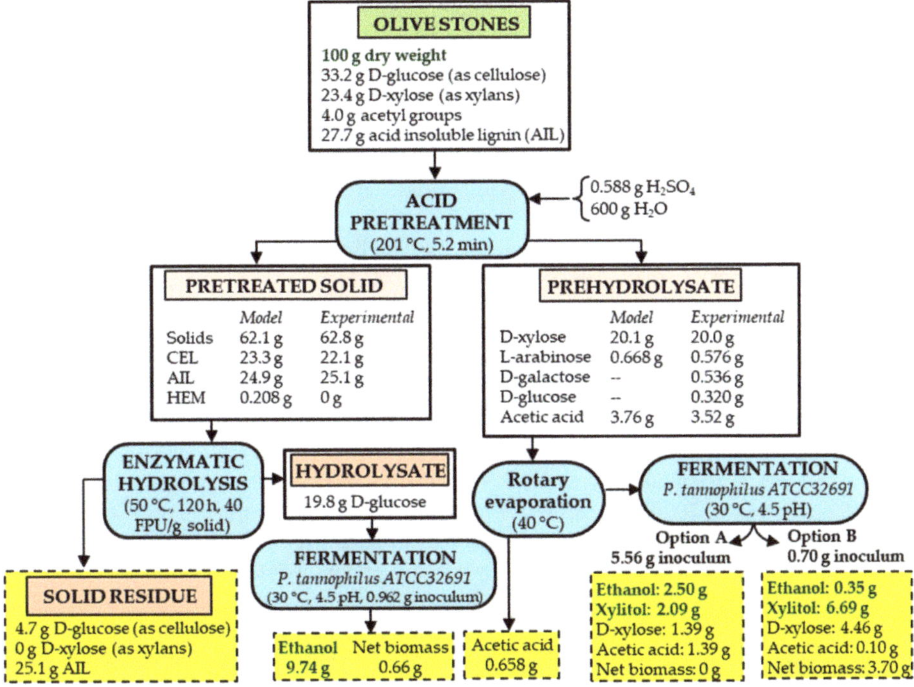


**Figure 10.** Mass macroscopic balance for the ethanol production flowsheet composed of dilute-acid pre-treatment of OS, enzymatic hydrolysis of pre-treated solids, detoxification with rotary evaporator and fermentation of hydrolysates using *P. tannophilus*. Reproduced with permission from Cuevas et al., Processes; published by MDPI, 2020 [23].

### 3.3. Thermochemical Conversion of Olive Stones

Among the thermochemical treatments that have been applied to OS, torrefaction, pyrolysis, combustion and gasification can be highlighted.

### 3.3.1. Torrefaction

Torrefaction is a process that takes place under a non-oxidizing atmosphere through heating biomass to maximum temperatures of the order of 300 °C. In this way, improvements in the energy characteristics of the resulting solids can be achieved such as reduction of the moisture content, increase in calorific value, and improvement of the subsequent pelletizing process [91,92].

Torrefaction can be conducted in an aqueous (wet torrefaction or hydrothermal carbonization) or dry (dry torrefaction) environment. A study on the rapid hydrothermal carbonization of OS at temperatures between 175 °C and 250 °C pointed out that the solid residue (hydrochar) obtained at 225 °C-10 min had the highest HHV (23.4 MJ/kg), energy efficiency (74.7%) and comprehensive combustibility index ($6.19 \times 10^{-7}$ $min^{-2}$ $°C^{-3}$) [93]. Furthermore, the short-time hydrothermal carbonization caused an increase in the hydrophobic capacity of the hydrochar, which contributes to improving the microbial stability of the material. These authors [93] also indicated that the hydrothermal carbonization has a greater capacity than pyrolysis to eliminate the OS components with lower calorific value (extractives and hemicelluloses) at low temperatures from the raw material.

On the other hand, the effect of the biomass to water ratio (1.1–12.3%), temperature (150–250 °C), and residence time (3.2–36.8 h) on the hydrothermal carbonization of OS has been investigated. The HHV ranged from 22.2 MJ/kg to 29.59 MJ/kg, being the maximum HHV achieved at 230 °C-30 h [94]. In relation to dry torrefaction, the optimal conditions found for OS were obtained at 278 °C-15 min and resulted in a solid product with 68% volatile matter, 29% fixed carbon, and 23.4 MJ/kg HHV [95].

### 3.3.2. Pyrolysis

Pyrolysis of olive stones is a promising technology for the production of renewable energy since it produces a gas with medium calorific value and byproducts (biochar and tar) which can be used to provide more energy to the process.

The thermogravimetric analyses applied to OS under inert atmosphere are of great interest to understand the pyrolysis of this biomass. The main weight losses in dry OS occur between 200 °C and 400 °C. Two large peaks are highlighted in the DTG curves. The first in the temperature range 264–323 °C and the second in the range 327–378 °C, which would correspond, respectively, to the decomposition of hemicellulose and cellulose [29,95,96]. Weight losses that take place above 400 °C would produce a long tail in the thermogravimetric curve which correspond to the slow process of lignin decomposition.

Regarding the products yield, the pyrolysis of OS at 600 °C-30 min generated 44.2 wt.% syngas, 24.9 wt.% tar, 14 wt.% water and 17.0 wt.% biochar [97]. The resulting gas was rich in $H_2$ (29.5% *v/v*), CO (31.5% *v/v*) and $CH_4$ (20.5% *v/v*), with LHV of 16.0 MJ/kg. Notwithstanding, other authors, when studying the same process with the same raw material and testing temperatures of up to 900 °C, were unable to achieve syngas yields greater than 25 wt.% related for any of the temperatures tested [95]. Thus, the gas, liquid (tar + water) and biochar fractions were 20.2 wt.%, 50.3 wt.% and 29.5 wt.%, respectively, at 500 °C-15 min [95]. The previous percentages are similar to those found in other studies when olive stones were pyrolyzed at 550 °C-15 min: 35 wt.% biochar and 55 wt.% liquid fraction [29]. In that study, the pyrolysis gas formed at 600 °C was composed of 17% *v/v* methane and over 10% *v/v* of both carbon monoxide and hydrogen. The significant deviations in the products yield data available in literature highlight the complexity of the pyrolysis process of OS, which is affected, among others, by factors such as particle size or mineral content [98]. Several studies have demonstrated that other biochar parameters such the actual and bulk densities are strongly affected by pyrolysis, which present minimum values for temperatures in the range 400–500 °C [95,99].

As the pyrolysis temperature increases, the biochar produced from OS contains less oxygen and more carbon, which results in an increase in its calorific value (for example, HHV = 31.1 MJ/kg at 900 °C) [95]. Besides, an increase of the temperature during biomass carbonization decreases the biochar reactivity, with higher ignition and burnout temperatures related to the decrease of volatile matter content [99]. The reactivity of the biochars obtained from five biomasses, including OS, can be found elsewhere [100].

Finally, apart from combustion, the biochar generated from olive stones could be used as hydroponic growing medium [101] or, after its activation [102], as adsorbent of different chemical compounds [103,104], in the manufacture of battery electrodes [105,106], etc.

### 3.3.3. Gasification

The air gasification of OS in a 5 kW bench scale, bubbling fluidized bed gasifier produced a medium heating value gas (LHV = 6.54 MJ/Nm$^3$). The maximum $H_2$ and CO concentrations (24% *v/v* and 14.3% *v/v*, respectively) were obtained at 750 °C and an equivalence ratio (ER, under stoichiometric amount of air inserted into the reactor to that necessary for complete combustion) of 0.2 [33]. The use of a 150 kW bubbling-fluidized-bed gasifier with ER = 0.22 led to a syngas with 12% *v/v* both $H_2$ and CO (LHV = 5.31 MJ/Nm$^3$) [107]. These authors also pointed out bed agglomeration problems due to the partial ash melting as a consequence of its high potassium content. Notwithstanding, these results improve those obtained in the gasification of OS in a countercurrent fixed-bed gasifier [108]. For instance, a gas containing 6.4% *v/v* $H_2$ and 26% *v/v* CO (HHV = 4.8 MJ/Nm$^3$) was achieved for ER = 0.22 in the countercurrent fixed-bed gasifier. The latter study concluded that olive husk gasification is a more complex process than wood gasification owing to non-uniform flow distribution across the high density bed and partial ash agglomeration. ER values of 0.31, much higher than those previously described, were used in the gasification of OS in a downdraft gasifier to obtain a syngas with 16% *v/v* $H_2$ and 22.8% *v/v* CO [31]. The gasification of OS with high temperature steam (750–1050 °C) has also been investigated [109]. This process allowed obtaining a syngas with notably high LHV (13.6 MJ/Nm$^3$) along with a biochar that contained 79 wt.% fixed carbon, which enables it for further reuses for energetic and agriculture purposes. This type of gasification also makes it possible to considerably reduce tar production [109].

Finally, a 100 kg/h downdraft gasifier system was used to produce clean synthesis gas production from OS in densified forms [110]. An average of 2.5 Nm$^3$ of product syngas per kg OS with a calorific value ranging from 4.5 MJ/Nm$^3$ to 5.0 MJ/Nm$^3$ was achieved, reaching over 97% conversion of carbon to product gas. A prototype of downdraft gasification system (70 kW$_e$-245 kW$_{th}$) combined with a gas engine was assayed for OS [111]. The prototype gasification system consisted of feeding system, hopper, reactor and gas cleaning and cooling system, producing a syngas with an average HHV of 16.1 MJ/Nm$^3$. The authors concluded that this prototype produced a good quality gas from OS with an electric efficiency of 16.1%. In a later work of these authors, the downdraft gasifier was coupled to a gas engine connected to the grid and fuelled with olive stones [75]. The system provided 70 kW$_e$ and 110 kW$_{th}$ with a consumption of 105 kg/h OS. The LHV of the gas produced in the gasifier was 5.1 MJ/kg. The electric efficiency and the overall efficiency of the small-scale plant were 14% and 36%, respectively.

### 3.3.4. Combustion

Combustion is currently the most important application of OS. Thus, in Spain there are a large number of facilities that burn OS to produce thermal energy in industries, public buildings, neighborhood communities and private homes [112].

The combustion of olive stones can be studied by means of thermogravimetry under oxidizing atmosphere. From the thermogravimetric data it can be verified that there are two different combustion stages. The first stage corresponds to the oxidation of the volatile matter removed from the solid at low relative temperatures and can be clearly observed in a differential thermogravimetric analysis (DTG analysis) in the peaks found in the range 200–400 °C, which are related to the combustion of the volatile matter and the breakdown of hemicellulose and cellulose [93,99]. To be specific, the decomposition of hemicellulose occurs from 220 °C to 315 °C while the breakdown of cellulose takes place between 320 °C and 400 °C. The second stage consists of the heterogeneous combustion of the char obtained from the pyrolysis of the polysaccharides and lignin. The two DTG peaks for those two reactions are in the range 400–600 °C.

The combustion of OS in fixed bed has been studied by several authors [113]. Besides, the combustion process of three olive-tree-derived biomasses (OS, pulp and residual olive pomace) was compared and the best results were achieved with OS, obtaining the lowest emission levels of unburned hydrocarbons and the highest efficiency (91.1%) [78]. The combustion of different biomasses in a mural boiler used for domestic heating (11.6 kW) was also studied. The authors showed that the use of OS pellets, characterized by lower percentages of sulphur and nitrogen contents, led to a significant decrease in NOx and SO$_2$ emissions [114]. Other authors carried out the combustion of OS pellets in a 40 kW counter-current fixed bed reactor and assessed some combustion parameters. The results achieved with OS pellets were quite similar to those obtained with the standard wood pellets that are used currently in European biomass markets [115].

In relation to the combustion of OS in fluidized bed, a numerical model was developed to calculate the burn-out time of different biomasses [116]. The results were greatly influenced by the temperature of the surrounding gas and, to a lesser extent, by the particle diameter, moisture content of the fuel, and bulk oxygen concentration. The burn-out time of OS reached values between 0.5 s and 50 s. The combustion of OS was reported to have bed agglomeration problems during combustion, which was related to the high percentage of potassium in the ash and the low melting point of the ash [117]. Thus, ash is generated during the combustion of OS with a high adhesion capacity on surfaces [118]. It was pointed out that OS combustion ash had an initial deformation temperature of 714 °C, so this biomass has a high tendency to form slag [77]. Ash generated in the combustion of olive stones have been used, with good results, in the production of biopolymers [119,120].

Finally, it can be highlighted that OS have also been used as fuel for combustion together with lignite [121] or animal sludge [122].

## 4. Olive Pomace from Olive Mills and Pomace Extracted from Extraction Plants

Currently, the most important use of olive pomaces involves a thermochemical conversion using a combustion reactor to generate electrical energy. For this use it is necessary to differentiate between olive pomace from olive mills and extracted pomace from pomace extraction plants.

Normally, the olive pomace from the oil mill that arrives to the extraction plant is subjected to a new process of separation of pulp and stone. This pomace with fewer stone fragments is dried using rotary driers until the equilibrium relative humidity (8–12%), then it is pelletized (after a previous grinding) and finally subjected to a solvent extraction, obtaining crude pomace oil and a solid residue that is the extracted pomace. The final objective set in any extraction plant is that this extracted pomace contains less than 0.5% in fat matter. This extracted pomace constitutes an excellent solid biofuel that is used in industrial boilers and in electric power generation industries (Figure 11).

However, if the olive pomace from the oil mill contains a low concentration of residual oil, in some cases, only the olive stone fragments are separated and the rest is dried in the rotary driers. In this case, this dry residue is directly used in the generation of electrical energy similarly to the extracted pomace.

In this last decade, tests are being carried out on the extracted pomace that involve a biochemical conversion aiming to obtaining bioethanol and other products of higher added value, including mannitol and phenolic compounds [123]. Nevertheless, other proposed applications of olive pomaces in which researchers have put more attention are pellets and biodiesel production.

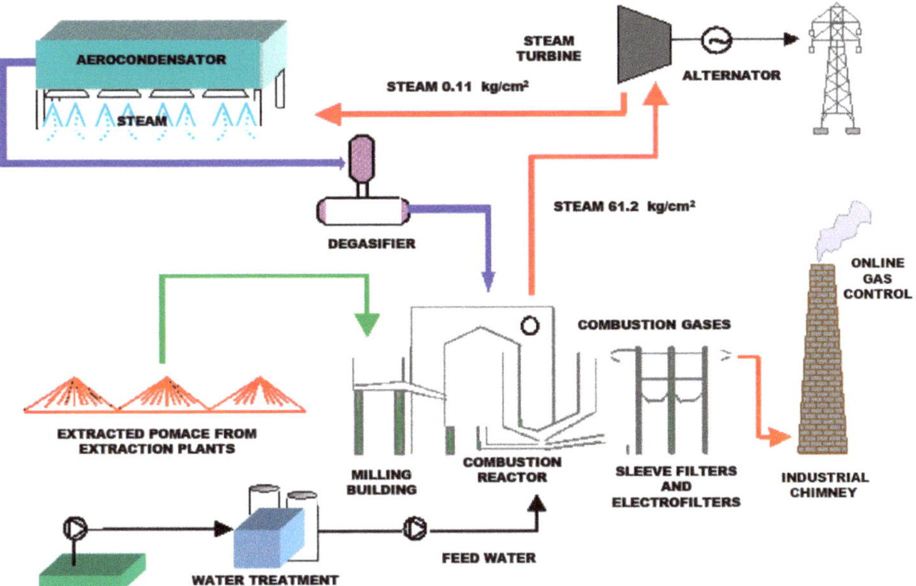


**Figure 11.** Industrial production of electrical energy using extracted pomace from extraction plants in Spain.

## 4.1. Biodiesel Production

Biodiesel is a promising, renewable, clean-burning fuel composed of long-chain mono alkyl esters from vegetable oils or animal fats through an esterification process typically carried out with methanol and NaOH as catalyst [124]. Biodiesel can also be obtained from agri-food residues (non-food biomass) such as olive pomace or waste cooking oils, constituting the second generation biodiesel, and from algae, bacteria, filamentous fungi and oil yeasts constituting the third generation biodiesel [125].

Similarly to olive oil, the main fatty acid present in olive pomace is oleic acid. The oil extracted from olive pomace can thus be used for the production of biodiesel through an esterification/transesterification process. Several authors have studied the optimal conditions for producing biodiesel from olive pomace, varying the type of oil extraction, alcohol to oil molar ratio, working temperature, type of catalyst, type of reactor and residence time for both esterification and transesterification. In some cases, the phenolic compounds present in olive pomace have been removed by steam-explosion [126] while in others they have been added to biodiesel to enhance both oxidation and microbial stability of the biofuel during storage [127]. Table 8 summarizes the main schemes available in literature for biodiesel production from olive pomace and the content in fatty acid methyl esters of the produced biodiesels. All these processes were performed in batch reactors. In some of the schemes there was a previous oil extraction while in others the olive pomace was directly subjected to esterification/transesterification. As illustrated in Table 8, the main problem of producing biodiesel from olive pomace is the low overall yields (kg biodiesel/100 kg olive pomace in dry basis) due to its low content in fat matter (Table 5).

**Table 8.** Procedures and fatty acid methyl esters (FAME) content achieved in the production of biodiesel from olive pomaces.

| Extraction/Solvent | Esterification | | | | Transesterification | | | | FAME Content (wt.%) | Overall Yield (wt.%) | Reference |
|---|---|---|---|---|---|---|---|---|---|---|---|
| | Catalyst | MR | t (min) | T (°C) | Catalyst | MR | t (min) | T (°C) | | | |
| Soxhlet/hexane | 1 wt.% $H_2SO_4$ | 6:1 | 80 | 60 | 1 wt.% NaOH | 6:1 | 80 | 60 | 98.9 | - | [126] |
| Stirring/hexane | - | - | - | - | 0.54 wt.% KOH | 1:2 | 60 | 60 | 94.7 | 4.0 ± 0.0 | [128] |
| - | - | - | - | - | 1.3 wt.% Ba(OH)$_2$ | 5:1 | 80 | 65 | 93.4 | - | [129] |
| - | - | - | - | - | 5.8 wt.% biocatalyst * | 5.3:1 | 1440 | 40 | 93.7 | - | [130] |
| Stirring/hexane | 20 wt.% $H_2SO_4$ | 35:1 | 60 | 40 | 0.6 wt.% KOH | 9:1 | 120 | 60 | 97.8 | 3.8 ± 1.8 | [131] |
| Soxhlet/hexane | - | - | - | - | 1 wt.% NaOH | 9:1 | 60 | 60 | 95.5 | 6.6 ± 1.6 | [132] |
| - | 3 wt.% Zn stearate | 30:1 | 30 | 140 | 3 wt.% Zn stearate | 30:1 | 30 | 140 | 84.1 | - | [133] |

MR: Methanol to oil molar ratio; * Microbial lipase from *Thermomyces lanuginosus* immobilized onto polyglutaraldehyde-activated olive pomace powder.

With regard to the third generation biodiesel, the soap stock from the chemical refining of crude olive pomace oil has been used as carbon source for the growth of several microorganisms used for biodiesel production. In general, filamentous fungi, yeasts and bacteria can contain more than 20 wt.% lipids. The problem for the cultivation of these microorganisms is the carbon source, which is usually quite expensive. The use of agro-industrial waste as carbon source, such as the soap stock obtained from the neutralization with NaOH during the refining of olive pomace oil, can greatly reduce its cost. For instance, 6 types of oleaginous yeasts showed lipid content between 60.16% and 64% when using ammonia as nitrogen source and soap stock from olive pomace oil chemical refining as carbon source [134]. The triglycerides produced had a high oleic content. The fatty acid composition of these yeasts was very similar to the composition of vegetable oils, mainly olive oil and soybean oil. In another study, lactose, oleic acid, maltose, D-xylose, D-galactose, glycerol, D-glucose, sucrose and residues such as soap stock from soybean oil chemical refinement, wheat bran, soap stock from olive pomace oil chemical refinement and waste cooking oil were used as carbon sources for the growth of the fungus *Fusarium verticillioides* [135]. This fungus produces oil with a major content in palmitic, oleic and linoleic acids. After studying the effects of various parameters such as nitrogen source, carbon source, pH and temperature, these authors stated that the fungus can not only grow on the low-cost carbon sources but also the produced lipids can be used as a potential feedstock for biodiesel production. Similarly, glucose, glycerol, soap stock from soybean oil chemical refinement, wheat bran, soap stock from olive pomace oil chemical refinement and waste cooking oil were used as carbon source for the fungus *Mucor circinelloides* [136]. The carbon source influenced the degree of unsaturation of produced lipids. Furthermore, it was demonstrated that the use of soap stock from soybean oil chemical refinement, waste cooking oils and soap stock from olive pomace oil chemical refinement as carbon source for *M. circinelloides* improved lipid yield and the lipid content [136].

*4.2. Anaerobic Digestion*

As stated in Section 2.2, the anaerobic digestion of olive pomaces has been investigated. Biological materials with high humidity and adequate biodegradability can be subjected to an anaerobic digestion process to generate a methane-rich biogas [137], which could be used in various combustion equipment. The wet pomaces produced in the olive mills fulfil these two requirements, although their high lignin content entails a lower digestibility in relation to other organic materials. For this reason, researches have focused on mixing wet pomaces from olive mills with other organic materials, such as olive mill wastewater [138], pig slurry [139], cow slurry and apple pulp [140], or food waste [141], to improve its biodegradability.

The application of different types of pre-treatment on wet pomace would be another possibility to improve methane production. Thus, methane production was enhanced by subjecting olive pomace to hydrothermal pre-treatment at 180 °C for 3 h [142]. The pre-treatment provoked a partial breakage of the polymers of the material, which improved its biodegradability. On the other hand, the effects of microwave, ultrasound and alkaline pre-treatments on olive pomace properties and its biomethane potential by anaerobic digestion was researched [143]. The alkaline pre-treatment at room temperature showed a positive effect on fiber degradation and lipid solubilisation as well as the best impact on methane production.

*4.3. Pellets Production*

Olive pomace is an olive mill by-product regarded as a potential raw material for densified biomass products. Notwithstanding, pellets from olive pomace presents out of limits values of nitrogen, which it is a hindrance. To overcome this problem, pellets of mixtures of olive pomace (from 2- and 3-outlet decanters), once dried, and olive -pruning debris were manufactured [144]. The best blends in terms of physical, chemical and mechanical characteristics, and of potential to be used as fuel for combustion and gasification were 75% OP and 25% olive pomace from 2-outlet decanter, and 50% OP and 50% olive pomace from 3-outlet decanter.

A study of the physico-chemical characteristics of olive pomaces demonstrated that both olive pomace from olive mills and pomace extracted from extraction plants were suitable to be combined with the residues of the sawmills of *Pinus radiata* and *Pinus Populus* spp. for pellet formation [145]. These authors performed a comparative cost analysis with traditional pellets from *Pinus radiata* to assess the profitability of the pellets produced, which showed that profitability increased by 11% mainly due to the decrease in acquisition, sawing and transportation costs. In addition, it was pointed out that the produced pellets could decrease $CO_2$ emissions by 5.0% (78,780 t $CO_2$ per year). The most profitable and environmental friendly pellet was that obtained from pomaces from extraction process with two-outlet decanter and *Pinus radiata* in 90/10 (wt.%) ratio [145].

Finally, pellets produced from olive mill wastewater and extracted pomace (obtained from olive pomace from 3-outlet decanter after the second extraction of the 3–5% of residual oil) have been investigated [146]. To be specific, these authors studied 3 pellets samples: extracted pomace, extracted pomace impregnated with olive mill wastewater, and pine sawdust impregnated with olive mill wastewater. The HHV of these pellets were similar to those of other biomasses, being the pellets produced from impregnated extracted pomace the ones with the highest LHV (19.8 MJ/kg) [146].

## 4.4. Wet Torrefaction

As shown in Table 5, the moisture and volatile matter of olive pomace issued from 2-oulet decanter is between 75% and 85%, which is a hindrance for direct combustion. Therefore, to enhance the energy production from olive pomace, wet torrefaction could be applied to overcome this limitation. It has been demonstrated that torrefaction can improve the carbon content, heating value, ash content and dewatering properties of olive pomace [147]. Thus, the LHV of raw olive pomace was 24.8 MJ/kg while that of the resulting hydrochar from torrefaction at 225 °C for 24 h was 31.4 MJ/kg. Besides, the hydrochars became more hydrophobic, allowing physical dewatering to occur easier and leading to a decrease of the moisture of the hydrochars. As a result, the moisture content decreased from over 70% to less than 30% and ash content was far lower than the initial ash content of the olive pomace [147].

## 4.5. Pyrolysis

Fast pyrolysis coupled with steam gasification of latter aforementioned pellets in Section 4.3, produced from olive mill wastewater and extracted olive pomace, have been assessed [146]. The percentage of char obtained in the three assayed pellets after pyrolysis was practically the same, being between 17% and 18%. Results showed that the highest pyrolysis time at 850 °C was obtained for the pellets obtained from the impregnation of the olive mill wastewater on the extracted olive pomace (154 s), while they were comparable for the two other pellets (104 s and 114 s, respectively). Besides, the pyrolysis rate of the impregnated extracted pomace pellets was significantly affected by the temperature in the range of 750–950 °C. Such behavior was attributed by the authors to the presence of organic compounds in the impregnated olive pomace, such as polyphenols [146].

## 4.6. Gasification

The steam gasification at 850 °C under 20% $H_2O$/80% $N_2$ with a flow rate of 13 dm$^3$/min of the resulting biochars from the pyrolysis of aforementioned pellets produced from olive mill wastewater and extracted olive pomace showed that impregnated olive pomace pellets achieved lower gasification time (755 s) than olive pomace pellets (830 s) [146]. The authors pointed out that metals present in the olive mill wastewater produced a catalytic effect in the gasification.

Finally, the gasification of pellets and the resulting charcoals from the pyrolysis in a fixed bed reactor of olive pomaces and pine sawdust under different water steam to nitrogen percentages (10%, 20% and 30% *v/v*) atmospheres at different temperatures (750 °C, 800 °C, 820 °C, 850 °C and 900 °C), using a macro-thermogravimetric equipment, resulted in a $H_2$-enriched syngas [148]. The pellets and biochars assayed were made of 100% pine sawdust, 50% pine sawdust and 50% olive pomace extracted from extraction plants, 50% pomace extracted from extraction plants and 50% olive mill wastewater, and 100% pomace extracted from extraction plants. The rate of conversion improved in pellets with olive mill wastewater, perhaps due to the content in K, Ca and Na.

## 5. Biofuels from Olive Biomass. Potential Application in Engines

Bioethanol offers numerous advantages over sophisticated fossil fuels, with the most significant being without doubt the environmental advantages. Bioethanol can allow a better oxidation of hydrocarbons in gasoline blends, which can greatly reduce the amounts of greenhouse gas emissions into the atmosphere. However, the lower heating value (6381.45 kcal/kg at ambient temperature) and comparably higher production cost (1.16 euro/dm$^3$ from lignocellulose biomass) hurdles its widespread application [149]. Thus, the energy generated from ethanol is only approximately two-thirds about that generated from gasoline, so engine consumption increases (in litres per 100 km).

Beyond the environmental benefits, the production of bioethanol from olive biomass would contribute to the development of the rural economy and the maintenance of jobs in the olive mill sector now that rural depopulation is so important. It would not only be a solution towards a new low-carbon.

On the other hand, power and torque do not decrease despite the fact that LHV of bioethanol is lower than that of fossil fuels. In some cases, power and torque even increase along with thermal efficiency for the same operating conditions. One explanation for this phenomenon is that the temperature for bioethanol self-ignition is higher than that of gasoline and diesel, as well as its heat of vaporization and octane number.

With respect to emissions produced by combustion, the use of bioethanol in engines shows a reduction in CO and unburned emissions, with an increase in the $CO_2$ emitted into the environment, mainly because ethanol is a fuel that already incorporates oxygen in the process and facilitates complete combustion. The $CO_2$ produced has no direct effect as a greenhouse gas considering the complete carbon cycle, since it comes from the combustion process of a fuel which raw material is biomass.

NOx emissions are directly related to engine operating conditions, especially temperature. However, it is difficult to predict the trend of NOx emissions when using bioethanol from olive biomass since there is no pre-established relationship between the fuel used and the mentioned parameters. On the other hand, in engines with constant pressure cycle or constant volume cycle, particulate emissions decrease as the proportion of ethanol in the mixture increases [150]. In addition, unburned bioethanol residues can occur in the intake and exhaust manifolds. It is also important to highlight unregulated emissions, such as aldehydes, which increase with respect to when gasoline is used.

With regard to biodiesel, it is usually found mixed with commercial diesel in varying percentages (between 5% and 30%), as well as 100% biodiesel in some cases. The presence of biodiesel in the fuel affects its physical and chemical properties. Biodiesel is a much more suitable fuel than vegetable oils for use in current diesel engines, due especially to its lower viscosity, lower cold filter plugging point, lower carbonaceous residue and higher cetane number. Therefore, biodiesel from olive pomace (i.e., biodiesel from the oil extracted from olive pomace, thus involving an extraction stage) would be more suitable for diesel engines than, for instance, olive oil. Nevertheless, biodiesel from olive pomace would have to compete with that from waste cooking oils. Biodiesel production from waste cooking oils has a huge development and its production using different reactors and the study of the exhaust gases are available in literature [124]. Furthermore, the overall biodiesel yields from raw material are far higher from waste cooking oils than from olive pomace, which are extremely low (Table 8).

In spite of these drawbacks, the combustion characteristics of biodiesel obtained from olive pomace oils have been studied at laboratory scale, without taking into account the economic feasibility [151–153]. In one of these works, biodiesel from olive pomace oil was mixed with diesel at volumetric ratios of 5/95, 7/93, 10/90 and 100/0 [153]. The engine performance (power, torque, specific fuel consumption) and combustion characteristic in cylinder (in-cylinder pressure values) of these mixtures were assessed. In another work, brake-specific fuel consumption and main exhaust emissions (NOx, $SO_2$ and CO) using three different load conditions (50%, 75% and 100%) were evaluated using different blends of biodiesel from olive pomace oil and diesel, aiming of selecting the optimal operating conditions of a diesel engine [152]. Finally, exhaust emissions, noise and sound quality of a direct injection diesel engine using blends of biodiesel from olive pomace oil and diesel were studied at several steady-state engine operating conditions [151]. The increase of biodiesel in the blends lowered CO emissions and reduced air and noise pollution while improving engine sound quality. By contrast, NOx emissions increased, keeping fuel consumption constant. These authors found that cetane number exhibited a stronger effect over noise production than bulk modulus, whereas the reverse effect was found concerning NOx emissions.

In summary, the use of biofuels in alternative internal combustion engines has both advantages and disadvantages in their operation. Even so, the future of biofuels is promising and in a short period of time their use will be an increasingly solid and favourable option.

## 6. Conclusions

At present, the solely actual energetic application of the three olive orchard by-products described in this paper, i.e., olive-tree pruning, olive stones and olive pomaces (including olive pomace from olive mills and olive pomace extracted from extraction plants), is their combustion in boilers either in small industries or households. Among these by-products, the extracted pomace (issued from pomace extraction plants) is the most used at industrial scale. It is regarded as an excellent solid biofuel for is used in industrial boilers and in electric power generation industries.

As previously stated, olive tree prunings and olive stones could be used for bioethanol production. The most promising by-product for bioethanol production could be the olive-pruning debris, since its HHV is low, and thus it is the worst raw material for combustion, and it has the largest percentage of hollocelullose, mainly cellulose. In addition, due to its low lignin content, the pre-treatment, enzymatic hydrolysis and fermentation are easier to be carried out. Nevertheless, second generation bioethanol is not economically feasible nowadays at industrial scale. Reducing the number of stages (as opposed to what can be found in recent literature on laboratory scale researches) and a complete use of sugars from both hemicellulose and cellulose are mandatory. The resulting solid from the raw material at the end of the process, composed mainly by lignin, could be used for pellets production.

Olive stones are nowadays used in small boilers in the own olive mills and in buildings, hotels, shopping centres, etc., around the world. Olive stones have shown some potential for electricity production through gasification. The efficiency could be improved if the syngas is cleaned and then burnt in a turbine. Torrefaction, instead of pyrolysis, would provide a solid with better fuel characteristics for subsequent combustion or gasification with moderate energy expenditure. In spite of their lower hollocelulose content, compared to that of olive-tree pruning, olive stones have provided notable bioethanol yields, so this energetic route should be not discarded in medium term.

Olive pomaces are currently used in Andalusia (Spain) in thermal power stations, and it is envisaged that more plants will be set up in the next years. The biodiesel production from olive pomace is currently being studied at laboratory scale, although an industrial development seems difficult due to its low fat matter content and the high price of olive pomace oil when using directly the oil extracted from olive pomace.

Finally, it should be highlighted that the integral use of wastes and by-products from the Olive Oil Industry must evolve towards the current technological concept of 'Integrated Biorefinery', which means maximum reuse of by-products and wastes through biochemical and thermochemical processes, of greater efficiency and minimal environmental impact, in the production of biofuels (such as bioethanol, biodiesel, biogas and synthetic gas), as well as other products of high added value.

**Author Contributions:** Section 1 (Introduction): J.F.G.M., S.S. and C.-H.F.; Section 2 (Olive-pruning debris): J.F.G.M.; Section 3 (Olive stones): M.C. and J.F.G.M.; Section 4 (Olive pomaces): S.S., P.Á.M., J.F.G.M. and M.C.; Section 5 (Biofuels and engines): J.F.G.M., M.T.G., S.S. and C.-H.F.; Section 6 (Conclusions): J.F.G.M. and S.S. All authors have read and agree to the published version of the manuscript.

**Funding:** This research received no external funding.

**Acknowledgments:** Chao-Hui Feng would like to express her gratitude to University of Seville for the mobility grant (VIPPIT-2018-I.3) awarded under the VI Plan Propio de Investigación y Transferencia of the University of Seville. Chao-Hui Feng also wishes to acknowledge the RIKEN's Special Postdoctoral Researcher Program for funding her research career.

**Conflicts of Interest:** The authors declare no conflict of interest.

# References


1. Feng, C.H.; Wang, W.; Makino, Y.; García-Martín, J.F.; Alvarez-Mateos, P.; Song, X.Y. Evaluation of storage time and temperature on physicochemical properties of immersion vacuum cooled sausages stuffed in the innovative casings modified by surfactants and lactic acid. *J. Food Eng.* **2019**, *257*, 34–43. [CrossRef]

2. Feng, C.H.; Makino, Y. Colour analysis in sausages stuffed in modified casings with different storage days using hyperspectral imaging—A feasibility study. *Food Control* **2020**, *111*, 107047. [CrossRef]

3. Cuevas, M.; Martínez-Cartas, M.L.; Pérez-Villarejo, L.; Hernández, L.; García-Martín, J.F.; Sánchez, S. Drying kinetics and effective water diffusivities in olive stone and olive-tree pruning. *Renew. Energy* **2019**, *132*, 911–920. [CrossRef]

4. Cuevas, M.; Sánchez, S.; Bravo, V.; García, J.F.; Baeza, J.; Parra, C.; Freer, J. Determination of optimal pre-treatment conditions for ethanol production from olive-pruning debris by simultaneous saccharification and fermentation. *Fuel* **2010**, *89*, 2891–2896. [CrossRef]

5. The Food and Agriculture Organization. Crops. Available online: http://www.fao.org/faostat/en/#data/QC/visualize (accessed on 3 February 2020).

6. Sánchez, S.; Moya, A.J.; Moya, M.; Romero, I.; Torrero, R.; Bravo, V.; San Miguel, M.P. Aprovechamiento del residuo de poda del olivar mediante conversión termoquímica. *Ing. Quim.* **2002**, *391*, 184–202.

7. Garcia-Maraver, A.; Rodriguez, M.L.; Serrano-Bernardo, F.; Diaz, L.F.; Zamorano, M. Factors affecting the quality of pellets made from residual biomass of olive trees. *Fuel Process. Technol.* **2015**, *129*, 1–7. [CrossRef]

8. Pérez, A.; Martín-Lara, M.A.; Gálvez-Pérez, A.; Calero, M.; Ronda, A. Kinetic analysis of pyrolysis and combustion of the olive tree pruning by chemical fractionation. *Bioresour. Technol.* **2018**, *249*, 557–566. [CrossRef]

9. Iáñez-Rodríguez, I.; Martín-Lara, M.Á.; Blázquez, G.; Osegueda, Ó.; Calero, M. Thermal analysis of olive tree pruning and the by-products obtained by its gasification and pyrolysis: The effect of some heavy metals on their devolatilization behavior. *J. Energy Chem.* **2019**, *32*, 105–117. [CrossRef]

10. García Martín, J.F.; Sánchez, S.; Bravo, V.; Cruz, N.; Cuevas, M.; Rigal, L.; Doumeng, C. Hidrólisis ácida del residuo de poda de olivo en un reactor de extrusión. *Afinidad* **2008**, *65*, 39–44.

11. García, J.F.; Sánchez, S.; Bravo, V.; Rigal, L.; Cuevas, M. Acid hydrolysis of olive-pruning debris for D-xylose production. *Collect. Czechoslov. Chem. Commun.* **2008**, *73*, 637–648. [CrossRef]

12. Martínez-Patiño, J.C.; Ruiz, E.; Romero, I.; Cara, C.; López-Linares, J.C.; Castro, E. Combined acid/alkaline-peroxide pretreatment of olive tree biomass for bioethanol production. *Bioresour. Technol.* **2017**, *239*, 326–335. [CrossRef] [PubMed]

13. Cara, C.; Ruiz, E.; Ballesteros, M.; Manzanares, P.; Negro, M.J.; Castro, E. Production of fuel ethanol from steam-explosion pretreated olive tree pruning. *Fuel* **2008**, *87*, 692–700. [CrossRef]

14. Peinado, S.; Mateo, S.; Sánchez, S.; Moya, A.J. Effectiveness of sodium borohydride treatment on acid hydrolyzates from olive-tree pruning biomass for bioethanol production. *Bioenergy Res.* **2019**, *12*, 302–311. [CrossRef]

15. Mateo, S.; Puentes, J.G.; Moya, A.J.; Sánchez, S. Ethanol and xylitol production by fermentation of acid hydrolysate from olive pruning with *Candida tropicalis* NBRC 0618. *Bioresour. Technol.* **2015**, *190*, 1–6. [CrossRef]

16. García, J.F.; Sánchez, S.; Bravo, V.; Cuevas, M.; Rigal, L.; Gaset, A. Xylitol production from olive-pruning debris by sulphuric acid hydrolysis and fermentation with *Candida tropicalis*. *Holzforschung* **2011**, *65*, 59–65. [CrossRef]

17. Mata-Sánchez, J.; Pérez-Jiménez, J.A.; Díaz-Villanueva, M.J.; Serrano, A.; Núñez-Sánchez, N.; López-Giménez, F.J. Statistical evaluation of quality parameters of olive stone to predict its heating value. *Fuel* **2013**, *113*, 750–756. [CrossRef]

18. Saleh, M.; Cuevas, M.; García, J.F.; Sánchez, S. Valorization of olive stones for xylitol and ethanol production from dilute acid pretreatment via enzymatic hydrolysis and fermentation by *Pachysolen tannophilus*. *Biochem. Eng. J.* **2014**, *90*, 286–293. [CrossRef]

19. Mata-Sánchez, J.; Pérez-Jiménez, J.A.; Díaz-Villanueva, M.J.; Serrano, A.; Núñez-Sánchez, N.; López-Giménez, F.J. Development of olive stone quality system based on biofuel energetic parameters study. *Renew. Energy* **2014**, *66*, 251–256. [CrossRef]

20. Fernández-Bolaños, J.; Felizón, B.; Heredia, A.; Guillén, R.; Jiménez, A. Characterization of the lignin obtained by alkaline delignification and of the cellulose residue from steam-exploded olive stones. *Bioresour. Technol.* **1999**, *68*, 121–132. [CrossRef]

21. Barreca, F.; Fichera, C.R. Use of olive stone as an additive in cement lime mortar to improve thermal insulation. *Energy Build.* **2013**, *62*, 507–513. [CrossRef]

22. Rodríguez, G.; Lama, A.; Rodríguez, R.; Jiménez, A.; Guillén, R.; Fernández-Bolaños, J. Olive stone an attractive source of bioactive and valuable compounds. *Bioresour. Technol.* **2008**, *99*, 5261–5269. [CrossRef] [PubMed]

23. Cuevas, M.; Saleh, M.; García-Martín, J.F.; Sánchez, S. Acid and enzymatic fractionation of olive stones for ethanol production using *Pachysolen tannophilus*. *Processes* **2020**, *8*, 195. [CrossRef]

24. Martín-Lara, M.A.; Hernáinz, F.; Calero, M.; Blázquez, G.; Tenorio, G. Surface chemistry evaluation of some solid wastes from olive-oil industry used for lead removal from aqueous solutions. *Biochem. Eng. J.* **2009**, *44*, 151–159. [CrossRef]

25. Fernández-Bolaños, J.; Felizón, B.; Heredia, A.; Rodríguez, R.; Guillén, R.; Jiménez, A. Steam-explosion of olive stones: Hemicellulose solubilization and enhancement of enzymatic hydrolysis of cellulose. *Bioresour. Technol.* **2001**, *79*, 53–61. [CrossRef]

26. Lama-Muñoz, A.; Romero-García, J.M.; Cara, C.; Moya, M.; Castro, E. Low energy-demanding recovery of antioxidants and sugars from olive stones as preliminary steps in the biorefinery context. *Ind. Crops Prod.* **2014**, *60*, 30–38. [CrossRef]

27. Pattara, C.; Cappelletti, G.M.; Cichelli, A. Recovery and use of olive stones: Commodity, environmental and economic assessment. *Renew. Sustain. Energy Rev.* **2010**, *14*, 1484–1489. [CrossRef]

28. García, G.B.; Calero De Hoces, M.; Martínez García, C.; Cotes Palomino, M.T.; Gálvez, A.R.; Martín-Lara, M.Á. Characterization and modeling of pyrolysis of the two-phase olive mill solid waste. *Fuel Process. Technol.* **2014**, *126*, 104–111. [CrossRef]

29. Blanco-López, M.C.; Blanco, C.G.; Martínez-Alonso, A.; Tascón, J.M.D. Composition of gases released during olive stones pyrolysis. *J. Anal. Appl. Pyrolysis* **2002**, *65*, 313–322. [CrossRef]

30. Skodras, G.; Grammelis, P.; Basinas, P.; Kakaras, E.; Sakellaropoulos, G. Pyrolysis and combustion characteristics of biomass and waste-derived feedstock. *Ind. Eng. Chem. Res.* **2006**, *45*, 3791–3799. [CrossRef]

31. Vera, D.; Jurado, F.; Margaritis, N.K.; Grammelis, P. Experimental and economic study of a gasification plant fuelled with olive industry wastes. *Energy Sustain. Dev.* **2014**, *23*, 247–257. [CrossRef]

32. García, R.; Pizarro, C.; Lavín, A.G.; Bueno, J.L. Characterization of Spanish biomass wastes for energy use. *Bioresour. Technol.* **2012**, *103*, 249–258. [CrossRef] [PubMed]

33. Skoulou, V.; Koufodimos, G.; Samaras, Z.; Zabaniotou, A. Low temperature gasification of olive kernels in a 5-kW fluidized bed reactor for $H_2$-rich producer gas. *Int. J. Hydrog. Energy* **2008**, *33*, 6515–6524. [CrossRef]

34. Di Giovacchino, L. I sottoprodotti della lavorazione delle olive in oleificio. In *Tecnologie di Lavorazione Delle Olive in Frantoio*; Di Giovacchino, L., Ed.; Tecnica Alimentare: Milano, Italy, 2010; pp. 243–258. ISBN 978-88-481-2554-3.

35. Yang, B.; Boussaid, A.; Mansfield, S.D.; Gregg, D.J.; Saddler, J.N. Fast and efficient alkaline peroxide treatment to enhance the enzymatic digestibility of steam-exploded softwood substrates. *Biotechnol. Bioeng.* **2002**, *77*, 678–684. [CrossRef]

36. Zheng, Y.; Zhao, J.; Xu, F.; Li, Y. Pretreatment of lignocellulosic biomass for enhanced biogas production. *Prog. Energy Combust. Sci.* **2014**, *42*, 35–53. [CrossRef]

37. Nitayavardhana, S.; Shrestha, P.; Rasmussen, M.L.; Lamsal, B.P.; van Leeuwen, J.H.; Khanal, S.K. Ultrasound improved ethanol fermentation from cassava chips in cassava-based ethanol plants. *Bioresour. Technol.* **2010**, *8*, 2741–2747. [CrossRef]

38. Perrone, O.M.; Colombari, F.M.; Rossi, J.S.; Moretti, M.M.S.; Bordignon, S.E.; Nunes, C.D.C.C.; Gomes, E.; Boscolo, M.; Da-Silva, R. Ozonolysis combined with ultrasound as a pretreatment of sugarcane bagasse: Effect on the enzymatic saccharification and the physical and chemical characteristics of the substrate. *Bioresour. Technol.* **2016**, *218*, 69–76. [CrossRef]

39. Martínez-Patiño, J.C.; Gullón, B.; Romero, I.; Ruiz, E.; Brnčić, M.; Žlabur, J.Š.; Castro, E. Optimization of ultrasound-assisted extraction of biomass from olive trees using response surface methodology. *Ultrason. Sonochem.* **2019**, *51*, 487–495. [CrossRef]

40. García, A.; González Alriols, M.; Labidi, J. Evaluation of the effect of ultrasound on organosolv black liquor from olive tree pruning residues. *Bioresour. Technol.* **2012**, *108*, 155–161. [CrossRef]

41. Ponnusamy, V.K.; Nguyen, D.D.; Dharmaraja, J.; Shobana, S.; Banu, J.R.; Saratale, R.G.; Chang, S.W.; Kumar, G. A review on lignin structure, pretreatments, fermentation reactions and biorefinery potential. *Bioresour. Technol.* **2019**, *271*, 462–472. [CrossRef]

42. Peretz, R.; Gerchman, Y.; Mamane, H. Ozonation of tannic acid to model biomass pretreatment for bioethanol production. *Bioresour. Technol.* **2017**, *241*, 1060–1066. [CrossRef]

43. Negro, M.J.; Alvarez, C.; Ballesteros, I.; Romero, I.; Ballesteros, M.; Castro, E.; Manzanares, P.; Moya, M.; Oliva, J.M. Ethanol production from glucose and xylose obtained from steam exploded water-extracted olive tree pruning using phosphoric acid as catalyst. *Bioresour. Technol.* **2014**, *153*, 101–107. [CrossRef] [PubMed]

44. Barbanera, M.; Buratti, C.; Cotana, F.; Foschini, D.; Lascaro, E. Effect of steam explosion pretreatment on sugar production by enzymatic hydrolysis of olive tree pruning. *Energy Procedia* **2015**, *81*, 146–154. [CrossRef]

45. Tortosa, J.F.; Rubio, M.; Gómez, D. Autohidrólisis de tallo de maiz en suspensión acuosa. *Afinidad* **1995**, *52*, 181–188.

46. Overend, R.P.; Chornet, E. Fractionation of lignocellulosics by steam-aqueous pretreatments. *Philos. Trans. R. Soc. London. Ser. A Math. Phys. Sci.* **1987**, *321*, 523–536.

47. García Martín, J.F.; Cuevas, M.; Bravo, V.; Sánchez, S. Ethanol production from olive prunings by autohydrolysis and fermentation with *Candida tropicalis*. *Renew. Energy* **2010**, *35*, 1602–1608. [CrossRef]

48. García, J.F.; Sánchez, S.; Bravo, V.; Cuevas, M. Autohydrolysis and acid post-hydrolysis of olive-pruning debris. *Afinidad* **2010**, *67*, 279–282.

49. Gatt, E.; Rigal, L.; Vandenbossche, V. Biomass pretreatment with reactive extrusion using enzymes: A review. *Ind. Crops Prod.* **2018**, *122*, 329–339. [CrossRef]

50. García Martín, J.F.; Sánchez, S.; Cuevas, M. Evaluation of the effect of the dilute acid hydrolysis on sugars release from olive prunings. *Renew. Energy* **2013**, *51*, 382–387. [CrossRef]

51. Martínez-Patiño, J.C.; Lu-Chau, T.A.; Gullón, B.; Ruiz, E.; Romero, I.; Castro, E.; Lema, J.M. Application of a combined fungal and diluted acid pretreatment on olive tree biomass. *Ind. Crops Prod.* **2018**, *121*, 10–17. [CrossRef]

52. Romero, I.; Moya, M.; Sánchez, S.; Ruiz, E.; Castro, E.; Bravo, V. Ethanolic fermentation of phosphoric acid hydrolysates from olive tree pruning. *Ind. Crops Prod.* **2007**, *25*, 160–168. [CrossRef]

53. Ho, M.C.; Ong, V.Z.; Wu, T.Y. Potential use of alkaline hydrogen peroxide in lignocellulosic biomass pretreatment and valorization – A review. *Renew. Sustain. Energy Rev.* **2019**, *112*, 75–86. [CrossRef]

54. Sun, Y.; Cheng, J. Hydrolysis of lignocellulosic materials for ethanol production: A review. *Bioresour. Technol.* **2002**, *83*, 1–11. [CrossRef]

55. Palmqvist, E.; Grage, H.; Meinander, N.Q.; Hahn-Hägerdal, B. Main and interaction effects of acetic acid, furfural, and p- hydroxybenzoic acid on growth and ethanol productivity of yeasts. *Biotechnol. Bioeng.* **1999**, *63*, 46–55. [CrossRef]

56. García, J.F.; Sánchez, S.; García, J. Ethanol from biomass: Application to the olive-pruning debris. In *Liquid Fuels: Types, Properties and Production*; Carasillo, D.A., Ed.; Nova Science Publishers: Hauppauge, NY, USA, 2012; pp. 239–254. ISBN 9781614704355.

57. Bayrock, D.P.; Ingledew, W.M. Ethanol production in multistage continuous, single stage continuous, Lactobacillus-contaminated continuous, and batch fermentations. *World J. Microbiol. Biotechnol.* **2005**, *21*, 83–88. [CrossRef]

58. Sánchez, S.; Bravo, V.; García, J.F.; Cruz, N.; Cuevas, M. Fermentation of D-glucose and D-xylose mixtures by *Candida tropicalis* NBRC 0618 for xylitol production. *World J. Microbiol. Biotechnol.* **2008**, *24*, 709–716. [CrossRef]

59. Sánchez, S.; Bravo, V.; Castro, E.; Moya, A.J.; Camacho, F. The fermentation of mixtures of D-glucose and D-xylose by *Candida shehatae*, *Pichia stipitis* or *Pachysolen tannophilus* to produce ethanol. *J. Chem. Technol. Biotechnol.* **2002**, *77*, 641–648. [CrossRef]

60. Moya, A.J.; Bravo, V.; Mateo, S.; Sánchez, S. Fermentation of acid hydrolysates from olive-tree pruning debris by *Pachysolen tannophilus*. *Bioprocess Biosyst. Eng.* **2008**, *31*, 611–617. [CrossRef]

61. Costa, P.; Dell'Omo, P.P.; La Froscia, S. Multistage milling and classification for improving both pellet quality and biogas production from hazelnut and olive pruning. *Ann. Chim. Sci. des Mater.* **2018**, *42*, 471–487. [CrossRef]

62. Messineo, A.; Maniscalco, M.P.; Volpe, R. Biomethane recovery from olive mill residues through anaerobic digestion: A review of the state of the art technology. *Sci. Total Environ.* **2020**, *703*, 135508. [CrossRef]

63. Carone, M.T.; Pantaleo, A.; Pellerano, A. Influence of process parameters and biomass characteristics on the durability of pellets from the pruning residues of *Olea europaea* L. *Biomass Bioenergy* **2011**, *35*, 402–410. [CrossRef]

64. Zambon, I.; Colosimo, F.; Monarca, D.; Cecchini, M.; Gallucci, F.; Proto, A.R.; Lord, R.; Colantoni, A. An innovative agro-forestry supply chain for residual biomass: Physicochemical characterisation of biochar from olive and hazelnut pellets. *Energies* **2016**, *9*, 526. [CrossRef]

65. Martín-Lara, M.A.; Ronda, A.; Zamora, M.C.; Calero, M. Torrefaction of olive tree pruning: Effect of operating conditions on solid product properties. *Fuel* **2017**, *202*, 109–117. [CrossRef]

66. Martín-Lara, M.A.; Blázquez, G.; Zamora, M.C.; Calero, M. Kinetic modelling of torrefaction of olive tree pruning. *Appl. Therm. Eng.* **2017**, *113*, 1410–1418. [CrossRef]

67. Zabaniotou, A.; Rovas, D.; Monteleone, M. Management of olive grove pruning and solid waste from olive oil extraction via thermochemical processes. *Waste Biomass Valorization* **2015**, *6*, 831–842. [CrossRef]

68. Calahorro, C.V.; Serrano, V.G.; Alvaro, J.H.; García, A.B. The use of waste matter after olive grove pruning for the preparation of charcoal. The influence of the type of matter, particle size and pyrolysis temperature. *Bioresour. Technol.* **1992**, *40*, 17–22. [CrossRef]

69. Hidayat, A.; Rochmadi; Wijaya, K.; Nurdiawati, A.; Kurniawan, W.; Hinode, H.; Yoshikawa, K.; Budiman, A. Esterification of palm fatty acid distillate with high amount of free fatty acids using coconut shell char based catalyst. *Energy Procedia* **2015**, *75*, 969–974. [CrossRef]

70. Álvarez-Mateos, P.; García-Martín, J.F.; Guerrero-Vacas, F.J.; Naranjo-Calderón, C.; Barrios, C.C.; Pérez Camino, M.D.C.; Barrios, C.C. Valorization of a high-acidity residual oil generated in the waste cooking oils recycling industries. *Grasas Aceites* **2019**, *40*, e335. [CrossRef]

71. Bartoli, M.; Rosi, L.; Giovannelli, A.; Frediani, P.; Frediani, M. Characterization of bio-oil and bio-char produced by low-temperature microwave-assisted pyrolysis of olive pruning residue using various absorbers. *Waste Manag. Res.* **2020**, *38*, 213–225. [CrossRef]

72. Nilsson, S.; Gómez-Barea, A.; Fuentes-Cano, D.; Haro, P.; Pinna-Hernández, G. Gasification of olive tree pruning in fluidized bed: Experiments in a laboratory-scale plant and scale-up to industrial operation. *Energy Fuels* **2017**, *31*, 542–554. [CrossRef]

73. Nilsson, S.; Gómez-Barea, A.; Fuentes-Cano, D.; Campoy, M. Gasification kinetics of char from olive tree pruning in fluidized bed. *Fuel* **2014**, *125*, 192–199. [CrossRef]

74. Gallucci, F.; Longo, L.; Santangelo, E.; Guerriero, E.; Paolini, V.; Carnevale, M.; Colantoni, A.; Tonolo, A. Assessment of syngas produced from gasification of olive tree pruning in a downdraft reactor. In Proceedings of the European Biomass Conference and Exhibition Proceedings, Copenhagen, Denmark, 14–18 May 2018.

75. Vera, D.; De Mena, B.; Jurado, F.; Schories, G. Study of a downdraft gasifier and gas engine fueled with olive oil industry wastes. *Appl. Therm. Eng.* **2013**, *51*, 119–129. [CrossRef]

76. Sánchez, J.M.; Jiménez, J.A.P.; Villanueva, M.J.D.; Serrano, A.; Núñez, N.; Giménez, J.L. New techniques developed to quantify the impurities of olive stone as solid biofuel. *Renew. Energy* **2015**, *78*, 566–572. [CrossRef]

77. Vega-Nieva, D.J.; Ortiz Torres, L.; Míguez Tabares, J.L.; Morán, J. Measuring and predicting the slagging of woody and herbaceous Mediterranean biomass fuels on a domestic pellet boiler. *Energy Fuels* **2016**, *30*, 1085–1095. [CrossRef]

78. Miranda, M.T.; Cabanillas, A.; Rojas, S.; Montero, I.; Ruiz, A. Combined combustion of various phases of olive wastes in a conventional combustor. *Fuel* **2007**, *86*, 367–372. [CrossRef]

79. Gómez-de la Cruz, F.J.; Casanova-Peláez, P.J.; Palomar-Carnicero, J.M.; Cruz-Peragón, F. Drying kinetics of olive stone: A valuable source of biomass obtained in the olive oil extraction. *Energy* **2014**, *75*, 146–152. [CrossRef]

80. Gómez-De La Cruz, F.J.; Palomar-Carnicero, J.M.; Casanova-Peláez, P.J.; Cruz-Peragón, F. Experimental determination of effective moisture diffusivity during the drying of clean olive stone: Dependence of temperature, moisture content and sample thickness. *Fuel Process. Technol.* **2015**, *137*, 320–326. [CrossRef]

81. Coimbra, M.A.; Waldron, K.W.; Selvendran, R.R. Isolation and characterisation of cell wall polymers from the heavily lignified tissues of olive (Olea europaea) seed hull. *Carbohydr. Polym.* **1995**, *27*, 285–294. [CrossRef]

82. Cuevas, M.; Sánchez, S.; Bravo, V.; Cruz, N.; García, J.F. Fermentation of enzymatic hydrolysates from olive stones by *Pachysolen tannophilus*. *J. Chem. Technol. Biotechnol.* **2009**, *84*, 461–467. [CrossRef]

83. Cuevas, M.; García, J.F.; Hodaifa, G.; Sánchez, S. Oligosaccharides and sugars production from olive stones by autohydrolysis and enzymatic hydrolysis. *Ind. Crops Prod.* **2015**, *70*, 100–106. [CrossRef]

84. Cuevas, M.; Sánchez, S.; García, J.F.; Baeza, J.; Parra, C.; Freer, J. Enhanced ethanol production by simultaneous saccharification and fermentation of pretreated olive stones. *Renew. Energy* **2015**, *74*, 839–847. [CrossRef]

85. Cuevas, M.; Saleh, M.; García-Martín, J.F.; Sánchez, S. Influence of solid loading on D-xylose production through dilute sulphuric acid hydrolysis of olive stones. *Grasas Aceites* **2015**, *66*, e084.

86. Antonopoulou, G.; Kampranis, A.; Ntaikou, I.; Lyberatos, G. Enhancement of liquid and gaseous biofuels production from agro-industrial residues after thermochemical and enzymatic pretreatment. *Front. Sustain. Food Syst.* **2019**, *3*, 92. [CrossRef]

87. Montané, D.; Salvadó, J.; Torras, C.; Farriol, X. High-temperature dilute-acid hydrolysis of olive stones for furfural production. *Biomass Bioenergy* **2002**, *22*, 295–304. [CrossRef]

88. Nabarlatz, D.; Ebringerová, A.; Montané, D. Autohydrolysis of agricultural by-products for the production of xylo-oligosaccharides. *Carbohydr. Polym.* **2007**, *69*, 20–28. [CrossRef]

89. Hernández, V.; Romero-García, J.M.; Dávila, J.A.; Castro, E.; Cardona, C.A. Techno-economic and environmental assessment of an olive stone based biorefinery. *Resour. Conserv. Recycl.* **2014**, *92*, 145–150. [CrossRef]

90. Goldfarb, J.L.; Buessing, L.; Gunn, E.; Lever, M.; Billias, A.; Casoliba, E.; Schievano, A.; Adani, F. Novel integrated biorefinery for olive mill waste management: Utilization of secondary waste for water treatment. *ACS Sustain. Chem. Eng.* **2017**, *5*, 876–884. [CrossRef]

91. Acharya, B.; Sule, I.; Dutta, A. A review on advances of torrefaction technologies for biomass processing. *Biomass Convers. Biorefinery* **2012**, *2*, 349–369. [CrossRef]

92. Costa, F.F.; Wang, G.; Costa, M. Combustion kinetics and particle fragmentation of raw and torrified pine shells and olive stones in a drop tube furnace. *Proc. Combust. Inst.* **2015**, *35*, 3591–3599. [CrossRef]

93. Cuevas, M.; Martínez Cartas, M.L.; Sánchez, S. Effect of short-time hydrothermal carbonization on the properties of hydrochars prepared from olive-fruit endocarps. *Energy Fuels* **2019**, *33*, 313–322. [CrossRef]

94. Álvarez-Murillo, A.; Román, S.; Ledesma, B.; Sabio, E. Study of variables in energy densification of olive stone by hydrothermal carbonization. *J. Anal. Appl. Pyrolysis* **2015**, *113*, 307–314. [CrossRef]

95. Sánchez, F.; San Miguel, G. Improved fuel properties of whole table olive stones via pyrolytic processing. *Biomass Bioenergy* **2016**, *92*, 1–11. [CrossRef]

96. Martín-Lara, M.A.; Ronda, A.; Blázquez, G.; Pérez, A.; Calero, M. Pyrolysis kinetics of the lead-impregnated olive stone by non-isothermal thermogravimetry. *Process Saf. Environ. Prot.* **2018**, *113*, 448–458. [CrossRef]

97. Bartocci, P.; D'Amico, M.; Moriconi, N.; Bidini, G.; Fantozzi, F. Pyrolysis of olive stone for energy purposes. *Energy Procedia* **2015**, *82*, 374–380. [CrossRef]

98. Marcilla, A.; García, A.N.; Pastor, M.V.; León, M.; Sánchez, A.J.; Gómez, D.M. Thermal decomposition of the different particles size fractions of almond shells and olive stones. Thermal behaviour changes due to the milling processes. *Thermochim. Acta* **2013**, *564*, 24–33. [CrossRef]

99. Gomez-Martin, A.; Chacartegui, R.; Ramirez-Rico, J.; Martinez-Fernandez, J. Performance improvement in olive stone's combustion from a previous carbonization transformation. *Fuel* **2018**, *228*, 254–262. [CrossRef]

100. Adánez, J.; De Diego, L.F.; García-Labiano, F.; Abad, A.; Abanades, J.C. Determination of biomass char combustion reactivities for FBC applications by a combined method. *Ind. Eng. Chem. Res.* **2001**, *40*, 4317–4323. [CrossRef]

101. Karakaş, C.; Özçimen, D.; İnan, B. Potential use of olive stone biochar as a hydroponic growing medium. *J. Anal. Appl. Pyrolysis* **2017**, *125*, 17–23. [CrossRef]

102. Molina-Sabio, M.; Jesús Sánchez-Montero, M.; Juarez-Galan, J.M.; Salvador, F.; Rodríguez-Reinoso, F.; Salvador, A. Development of porosity in a char during reaction with steam or supercritical water. *J. Phys. Chem. B* **2006**, *110*, 12360–12364. [CrossRef]

103. Alslaibi, T.M.; Abustan, I.; Ahmad, M.A.; Foul, A.A. Cadmium removal from aqueous solution using microwaved olive stone activated carbon. *J. Environ. Chem. Eng.* **2013**, *1*, 589–599. [CrossRef]

104. Hazzaa, R.; Hussein, M. Adsorption of cationic dye from aqueous solution onto activated carbon prepared from olive stones. *Environ. Technol. Innov.* **2015**, *4*, 36–51. [CrossRef]

105. Moreno, N.; Caballero, Á.; Morales, J.; Rodríguez-Castellón, E. Improved performance of electrodes based on carbonized olive stones/S composites by impregnating with mesoporous $TiO_2$ for advanced Li-S batteries. *J. Power Sources* **2016**, *313*, 21–29. [CrossRef]

106. Moreno, N.; Caballero, A.; Hernán, L.; Morales, J. Lithium-sulfur batteries with activated carbons derived from olive stones. *Carbon N. Y.* **2014**, *70*, 241–248. [CrossRef]

107. Gómez-Barea, A.; Arjona, R.; Ollero, P. Pilot-plant gasification of olive stone: A technical assessment. *Energy Fuels* **2005**, *19*, 598–605. [CrossRef]

108. Di Blasi, C.D.; Signorelli, G.; Portoricco, G. Countercurrent fixed-bed gasification of biomass at laboratory scale. *Ind. Eng. Chem. Res.* **1999**, *38*, 2581. [CrossRef]

109. Skoulou, V.; Swiderski, A.; Yang, W.; Zabaniotou, A. Process characteristics and products of olive kernel high temperature steam gasification (HTSG). *Bioresour. Technol.* **2009**, *100*, 2444–2451. [CrossRef]

110. Dogru, M. Experimental results of olive pits gasification in a fixed bed downdraft gasifier system. *Int. J. Green Energy* **2013**, *10*, 348–361. [CrossRef]

111. Margaritis, N.K.; Grammelis, P.; Vera, D.; Jurado, F. Assessment of operational results of a downdraft biomass gasifier coupled with a gas engine. *Procedia—Soc. Behav. Sci.* **2012**, *48*, 857–867. [CrossRef]

112. Rosúa, J.M.; Pasadas, M. Biomass potential in Andalusia, from grapevines, olives, fruit trees and poplar, for providing heating in homes. *Renew. Sustain. Energy Rev.* **2012**, *16*, 4190–4195. [CrossRef]

113. Morán, J.C.; Míguez, J.L.; Porteiro, J.; Patiño, D.; Granada, E. Low-quality fuels for small-scale combustion boilers: An experimental study. *Energy Fuels* **2015**, *29*, 3064–3081. [CrossRef]

114. González, J.F.; González-García, C.M.; Ramiro, A.; González, J.; Sabio, E.; Gañán, J.; Rodríguez, M.A. Combustion optimisation of biomass residue pellets for domestic heating with a mural boiler. *Biomass Bioenergy* **2004**, *27*, 145–154. [CrossRef]

115. Mami, M.A.; Mätzing, H.; Gehrmann, H.J.; Stapf, D.; Bolduan, R.; Lajili, M. Investigation of the olive mill solid wastes pellets combustion in a counter-current fixed bed reactor. *Energies* **2018**, *11*, 1965. [CrossRef]

116. Grammelis, P.; Kakaras, E. Biomass combustion modeling in fluidized beds. *Energy Fuels* **2005**, *19*, 292–297. [CrossRef]

117. Scala, F.; Chirone, R. Characterization and early detection of bed agglomeration during the fluidized bed combustion of olive husk. *Energy Fuels* **2006**, *20*, 120–132. [CrossRef]

118. Abreu, P.; Casaca, C.; Costa, M. Ash deposition during the co-firing of bituminous coal with pine sawdust and olive stones in a laboratory furnace. *Fuel* **2010**, *89*, 4040–4048. [CrossRef]

119. De Moraes Pinheiro, S.M.; Font, A.; Soriano, L.; Tashima, M.M.; Monzó, J.; Borrachero, M.V.; Payá, J. Olive-stone biomass ash (OBA): An alternative alkaline source for the blast furnace slag activation. *Constr. Build. Mater.* **2018**, *178*, 327–338. [CrossRef]

120. Font, A.; Soriano, L.; de Moraes Pinheiro, S.M.; Tashima, M.M.; Monzó, J.; Borrachero, M.V.; Payá, J. Design and properties of 100% waste-based ternary alkali-activated mortars: Blast furnace slag, olive-stone biomass ash and rice husk ash. *J. Clean. Prod.* **2020**, *243*, 118568. [CrossRef]

121. Topal, H.; Taner, T.; Naqvi, S.A.H.; Altınsoy, Y.; Amirabedin, E.; Ozkaymak, M. Exergy analysis of a circulating fluidized bed power plant co-firing with olive pits: A case study of power plant in Turkey. *Energy* **2017**, *140*, 40–46. [CrossRef]

122. Vamvuka, D.; Papas, M.; Galetakis, M.; Sfakiotakis, S. Thermal valorization of an animal sludge for energy recovery via co-combustion with olive kernel in a fluidized bed unit: Optimization of emissions. *Energy Fuels* **2016**, *30*, 5825–5834. [CrossRef]

123. Manzanares, P.; Ballesteros, I.; Negro, M.J.; González, A.; Oliva, J.M.; Ballesteros, M. Processing of extracted olive oil pomace residue by hydrothermal or dilute acid pretreatment and enzymatic hydrolysis in a biorefinery context. *Renew. Energy* **2020**, *145*, 1235–1245. [CrossRef]

124. García-Martín, J.F.; Barrios, C.C.; Alés-Álvarez, F.J.; Dominguez-Sáez, A.; Alvarez-Mateos, P. Biodiesel production from waste cooking oil in an oscillatory flow reactor. Performance as a fuel on a TDI diesel engine. *Renew. Energy* **2018**, *125*, 546–556. [CrossRef]

125. Liang, M.H.; Jiang, J.G. Advancing oleaginous microorganisms to produce lipid via metabolic engineering technology. *Prog. Lipid Res.* **2013**, *52*, 395–408. [CrossRef] [PubMed]

126. Lama-Muñoz, A.; Álvarez-Mateos, P.; Rodríguez-Gutiérrez, G.; Durán-Barrantes, M.M.; Fernández-Bolaños, J. Biodiesel production from olive-pomace oil of steam-treated alperujo. *Biomass Bioenergy* **2014**, *67*, 443–450. [CrossRef]

127. Dodos, G.S.; Tsesmeli, C.E.; Zannikos, F. Evaluation of the antimicrobial activity of synthetic and natural phenolic type antioxidants in biodiesel fuel. *Fuel* **2017**, *201*, 150–161. [CrossRef]

128. Hernández, D.; Astudillo, L.; Gutiérrez, M.; Tenreiro, C.; Retamal, C.; Rojas, C. Biodiesel production from an industrial residue: Alperujo. *Ind. Crops Prod.* **2014**, *52*, 495–498. [CrossRef]

129. Akgün, N.; Iscan, E. Effects of process variables for biodiesel production by transesterification. *Eur. J. Lipid Sci. Technol.* **2007**, *5*, 486–492. [CrossRef]

130. Yücel, Y. Optimization of biocatalytic biodiesel production from pomace oil using response surface methodology. *Fuel Process. Technol.* **2012**, *99*, 97–102. [CrossRef]

131. Ouachab, N.; Tsoutsos, T. Study of the acid pretreatment and biodiesel production from olive pomace oil. *J. Chem. Technol. Biotechnol.* **2013**, *88*, 1175–1181. [CrossRef]

132. Willson, R.M.; Wiesman, Z.; Brenner, A. Analyzing alternative bio-waste feedstocks for potential biodiesel production using time domain (TD)-NMR. *Waste Manag.* **2010**, *30*, 1881–1888. [CrossRef]

133. Alvarez Serafni, M.S.; Tonetto, G.M. Production of fatty acid methyl esters from an olive oil industry waste. *Brazilian J. Chem. Eng.* **2019**, *36*, 285–297. [CrossRef]

134. Ayadi, I.; Belghith, H.; Gargouri, A.; Guerfali, M. Screening of new oleaginous yeasts for single cell oil production, hydrolytic potential exploitation and agro-industrial by-products valorization. *Process Saf. Environ. Prot.* **2018**, *119*, 104–114. [CrossRef]

135. Kamoun, O.; Ayadi, I.; Guerfali, M.; Belghith, H.; Gargouri, A.; Trigui-Lahiani, H. *Fusarium verticillioides* as a single-cell oil source for biodiesel production and dietary supplements. *Process Saf. Environ. Prot.* **2018**, *118*, 68–78. [CrossRef]

136. Kamoun, O.; Muralitharan, G.; Belghith, H.; Gargouri, A.; Trigui-Lahiani, H. Suitable carbon sources selection and ranking for biodiesel production by oleaginous *Mucor circinelloides* using multi-criteria analysis approach. *Fuel* **2019**, *257*, 116117. [CrossRef]

137. Tekin, A.R.; Dalgiç, A.C. Biogas production from olive pomace. *Resour. Conserv. Recycl.* **2000**, *30*, 301–313. [CrossRef]

138. Milanese, M.; De Risi, A.; De Riccardis, A.; Laforgia, D. Numerical study of anaerobic digestion system for olive pomace and mill wastewater. *Energy Procedia* **2014**, *45*, 141–149. [CrossRef]

139. Orive, M.; Cebrián, M.; Zufía, J. Techno-economic anaerobic co-digestion feasibility study for two-phase olive oil mill pomace and pig slurry. *Renew. Energy* **2016**, *97*, 532–540. [CrossRef]

140. Riggio, V.; Comino, E.; Rosso, M. Energy production from anaerobic co-digestion processing of cow slurry, olive pomace and apple pulp. *Renew. Energy* **2015**, *83*, 1043–1049. [CrossRef]

141. El Gnaoui, Y.; Sounni, F.; Bakraoui, M.; Karouach, F.; Benlemlih, M.; Barz, M.; El Bari, H. Anaerobic co-digestion assessment of olive mill wastewater and food waste: Effect of mixture ratio on methane production and process stability. *J. Environ. Chem. Eng.* **2020**, *8*, 103874. [CrossRef]

142. De la Lama, D.; Borja, R.; Rincón, B. Performance evaluation and substrate removal kinetics in the semi-continuous anaerobic digestion of thermally pretreated two-phase olive pomace or "Alperujo". *Process Saf. Environ. Prot.* **2017**, *105*, 288–296. [CrossRef]

143. Elalami, D.; Carrere, H.; Abdelouahdi, K.; Garcia-Bernet, D.; Peydecastaing, J.; Vaca-Medina, G.; Oukarroum, A.; Zeroual, Y.; Barakat, A. Mild microwaves, ultrasonic and alkaline pretreatments for improving methane production: Impact on biochemical and structural properties of olive pomace. *Bioresour. Technol.* **2020**, *299*, 22591. [CrossRef]

144. Barbanera, M.; Lascaro, E.; Stanzione, V.; Esposito, A.; Altieri, R.; Bufacchi, M. Characterization of pellets from mixing olive pomace and olive tree pruning. *Renew. Energy* **2016**, *88*, 185–191. [CrossRef]

145. Hernández, D.; Fernández-Puratich, H.; Rebolledo-Leiva, R.; Tenreiro, C.; Gabriel, D. Evaluation of sustainable manufacturing of pellets combining wastes from olive oil and forestry industries. *Ind. Crops Prod.* **2019**, *134*, 338–346. [CrossRef]
146. Lajili, M.; Guizani, C.; Escudero Sanz, F.J.; Jeguirim, M. Fast pyrolysis and steam gasification of pellets prepared from olive oil mill residues. *Energy* **2018**, *150*, 61–68. [CrossRef]
147. Benavente, V.; Calabuig, E.; Fullana, A. Upgrading of moist agro-industrial wastes by hydrothermal carbonization. *J. Anal. Appl. Pyrolysis* **2015**, *113*, 89–98. [CrossRef]
148. Zribi, M.; Lajili, M.; Escudero-Sanz, F.J. Hydrogen enriched syngas production via gasification of biofuels pellets/powders blended from olive mill solid wastes and pine sawdust under different water steam/nitrogen atmospheres. *Int. J. Hydrog. Energy* **2019**, *44*, 11280–11288. [CrossRef]
149. Schacht, C.; Zetzl, C.; Brunner, G. From plant materials to ethanol by means of supercritical fluid technology. *J. Supercrit. Fluids* **2008**, *46*, 299–321. [CrossRef]
150. Lee, Z.; Park, S. Particulate and gaseous emissions from a direct-injection spark ignition engine fueled with bioethanol and gasoline blends at ultra-high injection pressure. *Renew. Energy* **2020**, *149*, 80–90. [CrossRef]
151. Redel-Macías, M.D.; Pinzi, S.; Leiva, D.; Cubero-Atienza, A.J.; Dorado, M.P. Air and noise pollution of a diesel engine fueled with olive pomace oil methyl ester and petrodiesel blends. *Fuel* **2012**, *95*, 615–625. [CrossRef]
152. López, I.; Pinzi, S.; Leiva-Candia, D.; Dorado, M.P. Multiple response optimization to reduce exhaust emissions and fuel consumption of a diesel engine fueled with olive pomace oil methyl ester/diesel fuel blends. *Energy* **2016**, *117*, 398–404. [CrossRef]
153. Özçelik, A.E.; Acaroğlu, M.; Köse, H. Determination of combustion characteristics of olive pomace biodiesel–Eurodiesel fuel mixtures. *Energy Sources Part A Recover. Util. Environ. Eff.* **2019**, *42*, 1476–1489. [CrossRef]

 *processes*

*Article*

# Determination of Hemicellulose, Cellulose, and Lignin Content in Different Types of Biomasses by Thermogravimetric Analysis and Pseudocomponent Kinetic Model (TGA-PKM Method)

David Díez [1,2,*], Ana Urueña [1,2], Raúl Piñero [1,2], Aitor Barrio [3] and Tarja Tamminen [4]

[1] CARTIF Centre of Technology, Parque Tecnológico de Boecillo, 205, Boecillo, 47151 Valladolid, Spain; anauru@cartif.es (A.U.); raupin@cartif.es (R.P.)

[2] ITAP Institute, University of Valladolid, Paseodel Cauce 59, 47011 Valladolid, Spain

[3] TECNALIA, Basque Research and Technology Alliance (BRTA), Área Anardi 5, E-20730 Azpeitia, Spain; aitor.barrio@tecnalia.com

[4] VTT-Technical Research Centre of Finland, P.O. Box 1000, VTT, FI-02044 Espoo, Finland; tarja.tamminen@vtt.fi

* Correspondence: davdie@cartif.es

Received: 30 July 2020; Accepted: 21 August 2020; Published: 27 August 2020

**Abstract:** The standard method for determining the biomass composition, in terms of main lignocellulosic fraction (hemicellulose, cellulose and lignin) contents, is by chemical method; however, it is a slow and expensive methodology, which requires complex techniques and the use of multiple chemical reagents. The main objective of this article is to provide a new efficient, low-cost and fast method for the determination of the main lignocellulosic fraction contents of different types of biomasses from agricultural by-products to softwoods and hardwoods. The method is based on applying deconvolution techniques on the derivative thermogravimetric (DTG) pyrolysis curves obtained by thermogravimetric analysis (TGA) through a kinetic approach based on a pseudocomponent kinetic model (PKM). As a result, the new method (TGA-PKM) provides additional information regarding the ease of carrying out their degradation in comparison with other biomasses. The results obtained show a good agreement between experimental data from analytical procedures and the TGA-PKM method (±7%). This indicates that the TGA-PKM method can be used to have a good estimation of the content of the main lignocellulosic fractions without the need to carry out complex extraction and purification chemical treatments. In addition, the good quality of the fit obtained between the model and experimental DTG curves ($R^2_{Adj} = 0.99$) allows to obtain the characteristic kinetic parameters of each fraction.

**Keywords:** TGA; hemicellulose; cellulose; lignin; pseudocomponent kinetic model; biomass

## 1. Introduction

The use of biomass resources for energy generation has been of considerable importance in recent years [1]. The global increase in energy demand has been one of the main reasons for their use. Added to this situation, there is also a need for dealing with certain problems, such as the depletion of fossil fuel reserves and the increase in environmental pollution from the use of these energy sources [2].

In this context, biomass has the advantage of being the only renewable resource that can be used in solid, liquid and gaseous forms [3]. Furthermore, biomass has the great capacity of producing by-products of high interest, such as catalytic carbons [4] and bioplastics [5]. However, biomass has a number of features that make it difficult to use, including its moisture content, low-energy density and complex structure. Lignocellulosic biomass is made up of a structure that includes mainly cellulose, hemicellulose and lignin [6,7]. The proportions and distribution of these components in the biomass physical structure is complex and depends on the type of species. The knowledge of this composition is very important for its use in different industrial applications.

Up to now, the determination of biomass composition, in terms of hemicellulose, cellulose and lignin contents, has been made by the chemical method. However, it is a slow and expensive methodology, which requires complex techniques and the use of multiple chemical reagents [8]. This means that it is not a suitable method for use in industrial applications.

Thermogravimetric analysis and, especially, the derivative thermogravimetric (DTG) curve is often used for the preliminary study of various thermochemical processes with biomass, since it allows the determination of the different stages of biomass devolatilization. In general, the process of devolatilization of the biomass in the absence of oxygen usually differentiates four stages corresponding to the loss of moisture and the three lignocellulosic components (hemicellulose, cellulose and lignin) [3]. Numerous articles have been published in which the thermal decomposition intervals of these lignocellulosic components are presented based on the deconvolution of the DTG curves [9–15]. It has been observed that, after moisture removal that takes place up to 150 °C, the decomposition of the three biomass lignocellulosic components takes place: hemicellulose is the first component to decompose between 200–300 °C, followed by cellulose between 250–380 °C. Regarding the thermal decomposition of lignin, it is the component with the most complex structure, and its decomposition range is the widest [16], occurring from 200 °C up to high temperatures such as 1000 °C [17,18].

There are different studies based on determining the lignocellulosic composition by analyzing DTG curves. However, most of these studies are based exclusively on applying deconvolution methods without taking into account their kinetic interpretation of the process [3,19].

On the other hand, kinetic studies on the thermal decomposition of biomass are extensive, in which the use of different kinetic models is analyzed [20,21], providing the kinetic parameters that best fit the experimental data. However, these studies do not focus on finding a method that allows the quantification of the three main lignocellulosic fractions of the biomass.

The use of kinetic analysis to the quantification of the main lignocellulosic fractions allows to include restrictions for a more precise quantification, while a physical interpretation is added to the deconvolution process.

The main objective of this work is to provide a new efficient, low-cost and fast method for the determination of the hemicellulose, cellulose and lignin contents of different types of biomasses, from agricultural by-products to wood. The method is based on applying deconvolution techniques on DTG pyrolysis curves based on a kinetic analysis of the process, and the kinetic model used is based on the assumption that the degradation of each lignocellulosic fraction can be represented by the evolution of a certain number of pseudocomponents.

## 2. Materials and Methods

### 2.1. Biomass Samples

Five raw materials representing different types of biomass have been selected, including agricultural biomass (wheat straw) and forest biomass, both as softwood barks (spruce bark and pine bark) and hardwoods (poplar and willow).

The pine bark originated from Sweden, while the other biomasses (wheat straw, poplar, spruce bark and willow) came from the South of France.

## 2.2. Experimental Method

Each sample was crushed in a mill (Model A 10 basic, IKA-Werke GmbH & Co. KG, Staufen, Germany) and then sieved. The sample sizes were all less than 100 μm in order to minimize the heat transfer resistances and mass transfer diffusion effects.

The TG (thermogravimetric) analysis was performed on a TG-DTA analyzer (Model DTG-60H, SHIMADZU Co. Ltd., Kyoto, Japan). The analyses were carried out using a nitrogen atmosphere with a flow rate of 50 mL min$^{-1}$. The heating rate used was 5 °C min$^{-1}$, from room temperature to a final temperature of 1000 °C. The sample weight was c.a. 10 mg.

To reduce temperature-related errors, the equipment used was calibrated across the entire temperature range. In addition, the actual sample temperature was used directly to solve the kinetic equations and to calculate the actual sample heating rate [22].

The information obtained in these analyses was the weight loss as the temperature and time of analysis increase (TG curve).

## 2.3. Data Treatment

The TG analysis provides the weight loss as a function of temperature over time. The analysis can be used to determine the different fractions of volatiles released as a function of temperature, as well as the solid residue remaining after heat treatment. However, for the determination of kinetics, it is more useful to use the derivative thermogravimetric (DTG) of weight loss as a function of time, because this signal is much more sensitive to small changes.

Before proceeding with its calculation, it is necessary to preprocess the data in order to obtain a curve that depends exclusively on the process variables.

The first step is the normalization of the TG signal. The normalization has been carried out in relation to the initial weight of the sample ($m_0$) and the final weight ($m_\infty$) of the sample. To do this, the weight fraction of the volatiles remaining in the sample has been calculated for each instant of discrete time $i$, as indicated in Equation (1).

$$X_i = \frac{m_i - m_\infty}{m_0 - m_\infty} \tag{1}$$

In this case, $m_\infty$ represents the mass of char obtained at the end of each TG analysis and includes the mass of ash and fixed carbon at the final temperature of the analysis.

## 2.4. DTG Curves

The DTG curve is obtained from the weight over time derivative for each experimental point, i.e.,

$$\frac{dX_i}{dt} = \frac{X_i - X_{i-\Delta}}{t_i - t_{i-\Delta}} \tag{2}$$

where $\Delta$ is the interval of the experimental data taken into account. In this case, $\Delta = 1$ has been used.

## 2.5. Kinetic Model

The thermochemical decomposition of the biomass can be represented by three main kinetics that correspond to the degradation of hemicellulose, cellulose and lignin. The most commonly used model consists of assuming that the process can be represented by the decomposition reactions of each of these compounds [23,24]. In addition, the decomposition of these compounds can be represented by a number of parallel and independent first-order Arrhenius-type reactions, named pseudocomponents.

Thus, for the adjustment of the DTG curve of each biomass, it has been assumed that the process follows the model that consists of the decomposition of hemicellulose, cellulose and lignin independently, so that the overall kinetics can then be expressed as follows:

$$\frac{dX}{dt} = \frac{dX_H}{dt} + \frac{dX_C}{dt} + \frac{dX_L}{dt} \tag{3}$$

where $H$, $C$ and $L$ represent the mass fraction of hemicellulose, cellulose and lignin, respectively.

At the same time, the kinetics of each of these fractions can be represented by a set of parallel reactions, expressed in the form:

$$\frac{dX_H}{dt} = \sum_{j=1}^{m_H} \frac{dX_{H_j}}{dt} = -\sum_{j=1}^{m_H} K_{H_j} exp\left(\frac{-E_{H_j}}{RT}\right) X_{H_j} \tag{4}$$

$$\frac{dX_C}{dt} = \sum_{j=1}^{m_C} \frac{dX_{C_j}}{dt} = -\sum_{j=1}^{m_C} K_{C_j} exp\left(\frac{-E_{C_j}}{RT}\right) X_{C_j} \tag{5}$$

$$\frac{dX_L}{dt} = \sum_{j=1}^{m_L} \frac{dX_{L_j}}{dt} = -\sum_{j=1}^{m_L} K_{L_j} exp\left(\frac{-E_{L_j}}{RT}\right) X_{L_j} \tag{6}$$

where $T$: temperature, in K; $R$: ideal gas constant, $8.314 \times 10^{-3}$ kJ (K mol)$^{-1}$; $j$: number of pseudocomponents of the fractions of hemicellulose, cellulose and lignin, which take the values from 1 to the total number of pseudocomponents of each fraction of hemicellulose; cellulose and lignin ($mH$, $mC$ and $mL$); $K_{Hj}$, $K_{Cj}$ and $K_{Lj}$: pre-exponential factors of the pseudocomponents of the hemicellulose, cellulose and lignin fractions, expressed in s$^{-1}$ and $E_{Hj}$, $E_{Cj}$ and $E_{Lj}$: activation energies of the pseudocomponents of the hemicellulose, cellulose and lignin fractions, expressed in kJ mol$^{-1}$.

In general, the kinetic equation of each pseudocomponent j, corresponding to fraction $F$ ($F = H, C, L$), in a nonisothermal process at constant heating rate $\beta = dT/dt$, is given by

$$\frac{dX_{F_j}}{X_{F_j}} = -\frac{K_{F_j}}{\beta} exp\left(\frac{-E_{F_j}}{RT}\right) dT \tag{7}$$

The integral of the second term can be resolved by using the exponential integral, defined as follows:

$$\int_u^\infty \frac{e^{-u}}{u} du, \ u = \frac{E}{R} \tag{8}$$

Thus, Equation (7), integrated between $To$ and $T$, can be expressed in the form

$$X_{F_{j,i}} = X_{F_{j,0}} . exp\left\{ -\frac{K_{F_j}}{\beta} \left[ T_i . exp\left(\frac{-E_{F_j}}{RT_i}\right) - \int_{E_{F_j}/RT_i}^\infty \frac{exp\left(\frac{-E_{F_j}}{RT}\right)}{T} dT \right] \right\} \tag{9}$$

Therefore, the kinetics of each pseudocomponent depends on three variables: the pre-exponential factor, the activation energy and the initial concentration of the pseudocomponent in the biomass ($X_{Fj,0}$).

A restriction that the system must satisfy is that the sum of the mass fractions of all the pseudocomponents must be equal to the mass fraction of all volatiles generated for each instant of time $t = i$.

$$X_i = X_{H_i} + X_{C_i} + X_{L_i} = \sum_{j=1}^{m_H} X_{H_{j,i}} + \sum_{j=1}^{m_C} X_{C_{j,i}} + \sum_{j=1}^{m_L} X_{L_{j,i}} \tag{10}$$

Combining Equations (9) and (10) for each instant of discrete time $i$ gives a system of equations with $3 \times (m_H + m_C + m_L) - 1$ unknowns, which needs to be solved.

### 2.6. Calculation Procedure

For the calculation of unknown variables, an optimization method based on the minimization by least squares has been used. As an objective function (OF), the square of the errors between the values of the experimental curve and the model has been used for each instant of time $i$, in which the model has been evaluated.

$$O.F. = \sum_{i=1}^{n}\left[\left(\frac{dX}{dt}\right)_{i,exp} - \left(\frac{dX}{dt}\right)_{i,model}\right]^2 \tag{11}$$

The solution has been made with MATLAB using the *lsqcurvefit* command to find the constants that best fit the system of equations. The final solution was obtained when the percentage variation of the OF was less than 0.01% during five consecutive cycles of 200 iterations each ($\Delta OF_5 < 0.01\%$).

The obtained quality of fit (QOF) between the simulated and experimental curves was evaluated with the expression (12).

$$QOF\ (\%) = 100\ x\ \sum_{i=1}^{n}\frac{\sqrt{\left[\left(\frac{dX}{dt}\right)_{i,exp} - \left(\frac{dX}{dt}\right)_{i,model}\right]^2/n}}{\max\left[\left(\frac{dX}{dt}\right)_{i,exp}\right]} \tag{12}$$

where $n$ is the number of experimental points employed (967).

Additionally, the goodness of fit was evaluated by the adjusted R-squared, $R^2_{Adj}$, which represents the response that is explained by the model and was calculated as the ratio between the sum of square of the residuals (SSE) and the total sum of squares (SST) as follows [25]:

$$R^2_{adj} = 1 - \frac{(n-1)xSSE}{(n-(k+1))xSST} = 1 - \frac{(n-1)x\sum_{i=1}^{n}\left[\left(\frac{dX}{dt}\right)_{i,exp} - \left(\frac{dX}{dt}\right)_{i,model}\right]^2}{(n-(k+1))x\sum_{i=1}^{n}\left[\left(\frac{dX}{dt}\right)_{i,exp} - \overline{\left(\frac{dX}{dt}\right)_{i,exp}}\right]^2} \tag{13}$$

where $k$ is the number of variables.

The initial values of the constants were taken after an initial analysis of the kinetics, using as initial seed values the restrictions on the concentrations of the hemicellulose, cellulose and lignin fractions obtained from the literature review (Table 1).

**Table 1.** Literature references of the main lignocellulosic fraction compositions related to used biomasses.

| Biomass | Ref. | Hemicellulose, wt.% | Cellulose, wt.% | Lignin, wt.% | Extractives, wt.% | Ash, wt.% |
|---|---|---|---|---|---|---|
| Pine bark | [26] [a] | 25.0 | 19.0 | 38.0 | 18.0 | |
| Spruce bark | [24] [a] | 27.0 | 42.0 | 26.0 | | |
| | [27] [a] | 24.3 | 41.0 | 30.0 | | |
| | [28] [a] | 21.2 | 50.8 | 27.5 | | |
| | [26] [a] | 28.0 | 22.0 | 31.0 | 19.0 | |
| Poplar | [17] [b] | 26.0 | 50.0 | 24.0 | | |
| | [19] [a] | 28.0 | 43.0 | 25.0 | 5 | |
| | [29] [a] | 18.0–26.6 | 46.5–52.0 | 16.0–25.9 | | |
| | [3] [b] | 22.0 | 49.0 | 28.0 | | |
| | [30] [a] | 24.0 | 49.0 | 20.0 | 5.9 | 1.0 |
| Willow | [30] [a] | 16.7 | 41.7 | 29.3 | 9.7 | 2.5 |
| Wheat straw | [30] [a,c] | 24.6 | 39.2 | 17.0 | | |
| | [28] [a] | 29.0 | 38.0 | 15.0 | | |
| | [28] [a] | 39.1 | 28.8 | 18.6 | | |
| | [30] [a] | 25.0 | 37.5 | 20.2 | 4.0 | 3.7 |

[a] By chemical methods, [b] by thermogravimetric analysis (TGA) and [c] cellulose as glucan and hemicellulose as xylan.

The decision tree of the calculation process is as Figure 1:

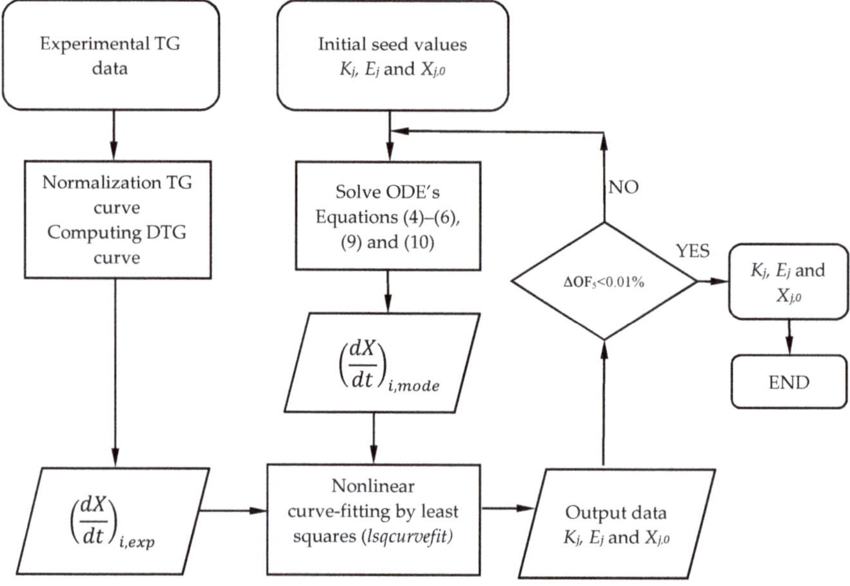


**Figure 1.** Decision tree of the calculation procedure.

## 3. Results and Discussion

### 3.1. Analytical Method

The lignocellulosic biomass wt.% composition was determined by chemical methods by the VTT and TECNALIA laboratories; the detailed procedure was described in [31]. Biomasses were previously sampled and prepared through TAPPI T257 and then conditioned through TAPPI 264. Table 2 includes the analytical results obtained.

**Table 2.** Composition by chemical methods for the raw biomasses (wt.%, dry basis).

| Biomass Component | Analysis Method | Pine Bark | Spruce Bark | Poplar | Willow | Wheat Straw |
|---|---|---|---|---|---|---|
| Hemicellulose | TAPPI T249 | 18.30 | 13.90 | 21.70 | 22.60 | 23.80 |
| Cellulose | TAPPI T249 | 21.90 | 29.70 | 42.70 | 44.30 | 37.50 |
| Lignin | TAPPI T222 | 40.70 | 45.10 | 26.90 | 25.10 | 20.50 |
| Extractives | Internal Method TAPPI 204 | 15.20 | 4.90 | 4.40 | 8.00 | 15.70 |
| Ash | XP CEN/TS 14775 TAPPI 211 | 2.80 | 5.22 | 2.80 | 2.30 | 8.30 |

The results obtained in Table 2 are in-line with the results obtained by other researchers [32]. According to the literature, the softwood bark composition corresponds to a cellulose content of 18–38%, the hemicellulose content is 15–33% and the lignin content is 30–60%. For hardwood biomasses, the cellulose content is 43–47%, the hemicellulose content is 25–35% and the lignin content is 16–24%. Finally, the composition of herbaceous biomass, such as cereal straw, is 33–38% cellulose, 26–32% hemicellulose and 17–19% lignin.

Therefore, according to the literature review [32,33] and the analyses carried out (Table 2), softwood bark has higher lignin content than hardwood and agricultural biomasses. On the other hand, hardwood has a higher cellulose content than the rest of the biomasses analyzed.

It also should be noted that, during the thermogravimetric analysis (TGA), it is possible to differentiate the biomass into its three main lignocellulosic fractions, but it is not possible to distinguish the extractives from the other fractions. Extractives are a group of compounds that can be obtained from the biomass using organic solvents, such as benzene, alcohol or water [34]. The main components of the lipophilic extracts are triglycerides, fatty acids, resin acids, sterile esters and sterols and of hydrophilic extracts are lignin [35]. The extractives thermally degrade in the temperature range of 200–400 °C, which falls within the range in which hemicellulose and cellulose and, also, lignin is degraded. For this reason, in order to get comparable results with those obtained by the TGA method, the analytical data are expressed in weight % on a dry and ash and extractives-free basis.

### 3.2. Devolatilization Behavior

The performance of the DTG curves shows similar behavior (Figure 2). At first sight, two large peaks can be observed in all of them: the first one appears from room temperature to about 150 °C and corresponds to the loss of moisture. At temperatures exceeding 150 °C, degradation of lignocellulosic compounds begins [30,32,36,37]. The second large peak is located in the range of temperature between 250 and 380 °C and corresponds to the degradation of cellulose. Two other peaks, which are more or less perceptible depending on the type of biomass, can be seen overlapping the cellulose peak. Thus, at temperatures between 200 and 300 °C, the degradation of hemicellulose occurs, which proves a deformation of the cellulose peak in that temperature range. Finally, lignin is the component with the most complex structure, and its decomposition range is the widest, occurring from 200 °C to the final temperature of the analysis. The degradation of lignin is more significant near the 400 °C zone, where a small peak can be observed that overlaps with the end of the cellulose degradation.

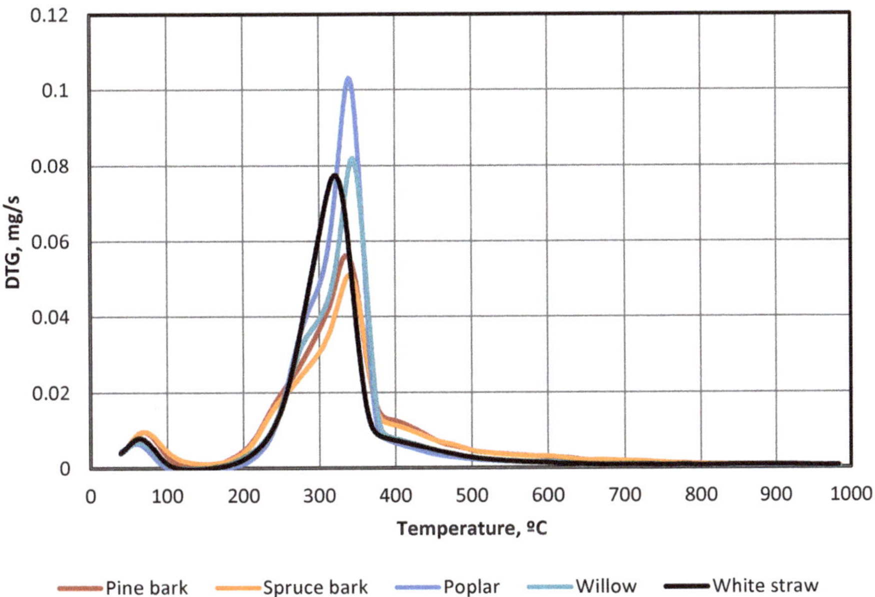


**Figure 2.** DTG curves comparison.

In relation to the development of each type of biomass, it is observed that pine bark and spruce bark have very similar development patterns. Both barks, as compared to the rest of the biomasses (poplar, willow and white straw), have a higher peak near 400 °C corresponding to the degradation of lignin and a lower peak height corresponding to the degradation of cellulose (~350 °C) and hemicellulose (~300 °C). Therefore, these softwood barks have a higher lignin content and lower cellulose and hemicellulose contents, as compared to other biomasses (Table 2). It is also observed that these two biomasses have the lowest DTG area, so they are the ones that release the least amounts of total volatiles.

On the other hand, willow and poplar show very similar behaviors, which indicates that their compositions will be very similar. Both biomasses present a greater generation of volatiles in the cellulose degradation zone. This is in agreement with the fact that both biomasses have higher cellulose contents and lower lignin contents compared to the rest of the biomasses analyzed (Table 2).

Finally, wheat straw presents a single peak in the degradation zone of hemicellulose and cellulose and is slightly displaced to the low temperature zone. This suggests a higher hemicellulose content, while the evolution of the lignin content is very similar to that of poplar and willow.

### 3.3. TGA-PKM Method

The first step was to determine the minimum number of pseudocomponents needed to adequately represent the evolution of each of the three main lignocellulosic fractions and all volatiles generated during the thermal degradation process.

This analysis was carried out by means of an initial kinetic analysis, in which a division of the DTG was established according to the degradation temperatures of the three main constituents of the biomass (hemicellulose, cellulose and lignin), in addition to water. Each of these regions was initially attributed a single pseudocomponent; then, the number of pseudocomponents was gradually increased, until an adequate performance of the evolution of the volatiles was achieved. The minimum numbers of pseudocomponents necessary for the quantifications of each fraction are shown in the Table 3. The use of a larger number of pseudocomponents could induce overfitting.

**Table 3.** Minimum number of components for each biomass fraction.

| Component | Temperature Range, °C | Number of Pseudocomponents |
| --- | --- | --- |
| Water | 25–150 | 1 |
| Hemicellulose | 200–350 | 2 |
| Cellulose | 250–400 | 1 |
| Lignin | 150–1000 | 3 |

The next step was to determine the minimum number of heating rates needed to achieve the objective of quantifying the main biomass fractions. The use of three or more heating rates while reducing the effect of kinetic compensation and improving the accuracy of kinetic parameters requires the use of significantly different heating rates, which involves, in practice, the use of higher heating rates. However, higher heating rates worsen the separation of lignocellulosic fractions, making their identification more difficult. Additionally, the use of various heating rates for the quantification of the lignocellulosic fractions is more time-consuming.

Therefore, a low heating rate achieves a better separation of the degraded compounds and is less time-consuming. This is the reason why a single heating rate of 5 °C min$^{-1}$ has been employed in the determination of the main lignocellulosic fractions. However, a validation of the method using three heating rates has been carried out and is reported in Section 3.4.

To improve the accuracy of the kinetic parameters, it was found that the use of upper and lower limits of the kinetic parameters (Tables 4 and 5) was necessary, not only to ensure adequate values of the pre-exponential and activation energy but, also, to provide adequate seed values for the determination of the hemicellulose, cellulose and lignin fractions.

**Table 4.** Upper bonds of the pseudocomponents (PC).

| Kinetic Parameters | PC 2 | PC 3 | PC 4 | PC 5 | PC 6 | PC 7 |
|---|---|---|---|---|---|---|
| $K$ (s$^{-1}$) | $1.00 \times 10^9$ | $1.50 \times 10^5$ | $2.40 \times 10^{15}$ | $5.00 \times 10^1$ | $3.00$ | $1.80$ |
| $E$ (kJ mol$^{-1}$) | 120.00 | 80.00 | 240.00 | 60.00 | 60.00 | 68.00 |
| $X_{j,0}$ (wt.%) | 50.00 | 50.00 | 60.00 | 60.00 | 20.00 | - |

**Table 5.** Lower bonds of the pseudocomponents (PC).

| Kinetic Parameters | PC 2 | PC 3 | PC 4 | PC 5 | PC 6 | PC 7 |
|---|---|---|---|---|---|---|
| $K$ (s$^{-1}$) | $7.00 \times 10^8$ | $1.40 \times 10^5$ | $1.50 \times 10^{15}$ | $4.00 \times 10^7$ | $2.30$ | $1.00 \times 10^{-1}$ |
| $E$ (kJ mol$^{-1}$) | 100.00 | 70.00 | 160.00 | 55.00 | 45.00 | 50.00 |
| $X_{j,0}$ (wt.%) | 0.1 | 1 | 5.00 | 15.00 | 0.10 | - |

Finally, taking into account the above procedure, the values of the kinetic parameters of each pseudocomponent were calculated by the TGA-PKM method and are summarized in Table 6.

Overall, the results obtained in this study are in reasonable ranges when compared to the results corresponding to the kinetics of other biomasses published, as can be seen in Table 7.

**Table 6.** Kinetic parameters of the pseudocomponents.

| Biomass | Kinetic Parameters | Water PC 1 | Hemicellulose PC 2 | Hemicellulose PC 3 | Cellulose PC 4 | Cellulose PC 5 | Lignin PC 6 | Lignin PC 7 |
|---|---|---|---|---|---|---|---|---|
| Pine bark | $K$ (s$^{-1}$) | $9.14 \times 10^4$ | $6.00 \times 10^8$ | $1.50 \times 10^5$ | $1.69 \times 10^{15}$ | $5.00 \times 10$ | $2.44$ | $9.83 \times 10^{-1}$ |
| | $E$ (kJ mol$^{-1}$) | 48.58 | 120.00 | 75.60 | 204.74 | 55.00 | 50.20 | 61.31 |
| | $X_{j,0}$ (wt.%) | 6.81 | 14.71 | 7.00 | 24.66 | 27.68 | 13.16 | 5.98 |
| Spruce bark | $K$ (s$^{-1}$) | $4.52 \times 10^3$ | $7.00 \times 10^8$ | $1.42 \times 10^5$ | $1.51 \times 10^{15}$ | $5.00 \times 10$ | $2.33$ | $1.11$ |
| | $E$ (kJ mol$^{-1}$) | 40.74 | 119.99 | 74.93 | 205.17 | 55.00 | 48.79 | 60.95 |
| | $X_{j,0}$ (wt.%) | 8.25 | 14.04 | 6.56 | 24.53 | 24.61 | 14.18 | 7.82 |
| Poplar | $K$ (s$^{-1}$) | $7.22 \times 10^5$ | $6.00 \times 10^8$ | $1.40 \times 10^5$ | $2.30 \times 10^{15}$ | $4.81 \times 10$ | $2.79$ | $5.07 \times 10^{-1}$ |
| | $E$ (kJ mol$^{-1}$) | 52.44 | 120.00 | 80.00 | 207.39 | 55.00 | 53.22 | 59.71 |
| | $X_{j,0}$ (wt.%) | 3.81 | 21.72 | 1.00 | 51.85 | 15.34 | 4.24 | 2.04 |
| Willow | $K$ (s$^{-1}$) | $2.98 \times 10^5$ | $6.00 \times 10^8$ | $1.47 \times 10^5$ | $1.94 \times 10^{15}$ | $5.00 \times 10$ | $2.60$ | $8.77 \times 10^{-1}$ |
| | $E$ (kJ mol$^{-1}$) | 50.69 | 120.00 | 74.88 | 208.03 | 55.00 | 49.43 | 62.61 |
| | $X_{j,0}$ (wt.%) | 4.69 | 21.91 | 1.91 | 44.33 | 15.00 | 8.25 | 3.91 |
| Wheat straw | $K$ (s$^{-1}$) | $8.23 \times 10^5$ | $6.00 \times 10^8$ | $1.50 \times 10^5$ | $1.51 \times 10^{15}$ | $5.00 \times 10$ | $3.00$ | $1.62 \times 10^{-1}$ |
| | $E$ (kJ mol$^{-1}$) | 53.65 | 120.00 | 71.38 | 200.64 | 55.00 | 51.08 | 52.16 |
| | $X_{j,0}$ (wt.%) | 5.18 | 24.16 | 1.00 | 39.51 | 19.73 | 6.90 | 3.51 |

**Table 7.** Kinetic parameters from other studies.

| Component | Temperature, °C | E, kJ mol$^{-1}$ | K, min$^{-1}$ | Reference |
|---|---|---|---|---|
| Hemicellulose | 200–350 | 127.00 | $9.5 \times 10^{10}$ | [38] |
| | | 83.20–96.40 | $4.55 \times 10^6$–$1.57 \times 10^8$ | [39] |
| Cellulose | 300–340 | 227.02 | $3.36 \times 10^{18}$ | [37] |
| | | 239.70–325.00 | $16.30 \times 10^{19}$–$3.62 \times 10^{26}$ | [39] |
| Lignin | 220–380 | 7.80 | $2.96 \times 10^{-3}$ | [37] |
| | 25–900 | 47.90–54.50 | $6.80 \times 10^2$–$6.60 \times 10^4$ | [17] |
| | 160–680 | 25.20 | $4.70 \times 10^2$ | [18] |
| | | 20.00–29.10 | $5.35 \times 10$–3.18 | [39] |

As shown in Tables 6 and 7, cellulose is the compound with the highest activation energies. This is attributed to the fact that the cellulose is a very long polymer of glucose units without any branches [18], while hemicellulose has a random branched amorphous structure that gives a lower activation energy; this is the reason why hemicellulose decomposes more easily in a lower temperature range [38].

Lignin has a very complex structure composed of three kinds of heavily crosslinked phenylpropane structures [18]. Additionally, it is observed that the activation energy is lower than for hemicellulose and cellulose, which indicates that its thermal degradation is easier. However, it presents much lower values of pre-exponential factors that cause a lower reaction rate; this fact is reflected in the wide range of temperatures in which its degradation takes place and in the high temperature required to reach a complete degradation.

In addition, Figures 3–7 show the fit of the model to the DTG experimental data, as well as the contribution of the different pseudocomponents to the model. In all the figures, it can be seen that a good fit is achieved between the global model, obtained as the envelope resulting from the sum of the seven pseudocomponents, and the experimental DTG curve.

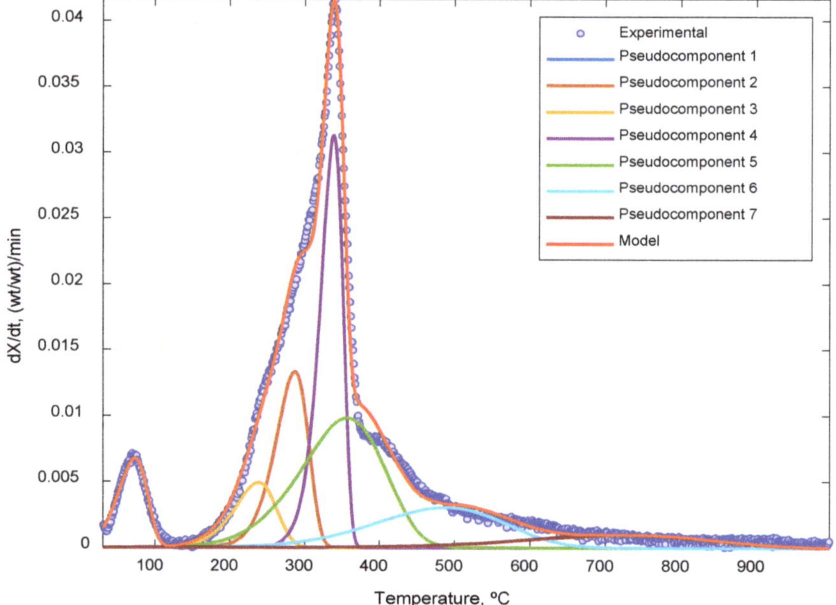


**Figure 3.** Model fitted to the experimental pine bark DTG curve.

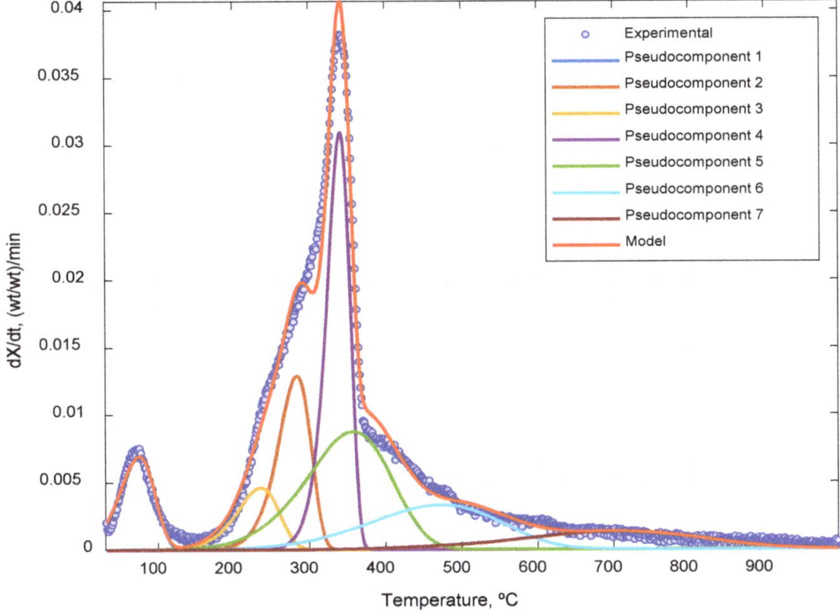

**Figure 4.** Model fitted to the experimental spruce bark DTG curve.

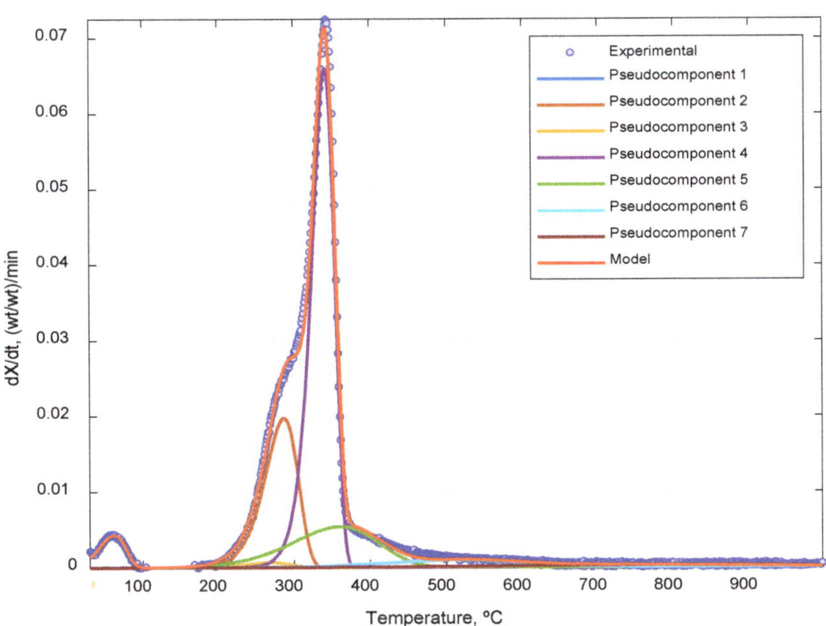

**Figure 5.** Model fitted to the experimental poplar DTG curve.

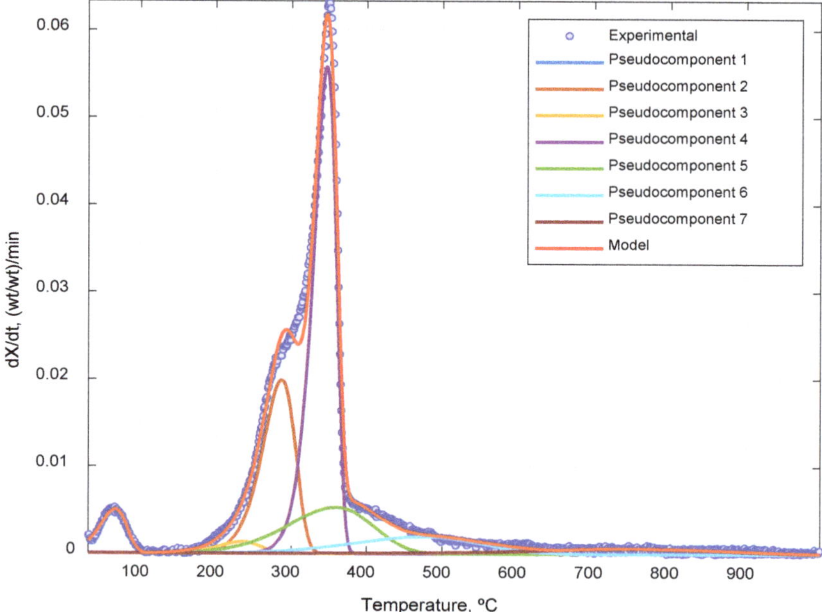

**Figure 6.** Model fitted to the experimental willow DTG curve.

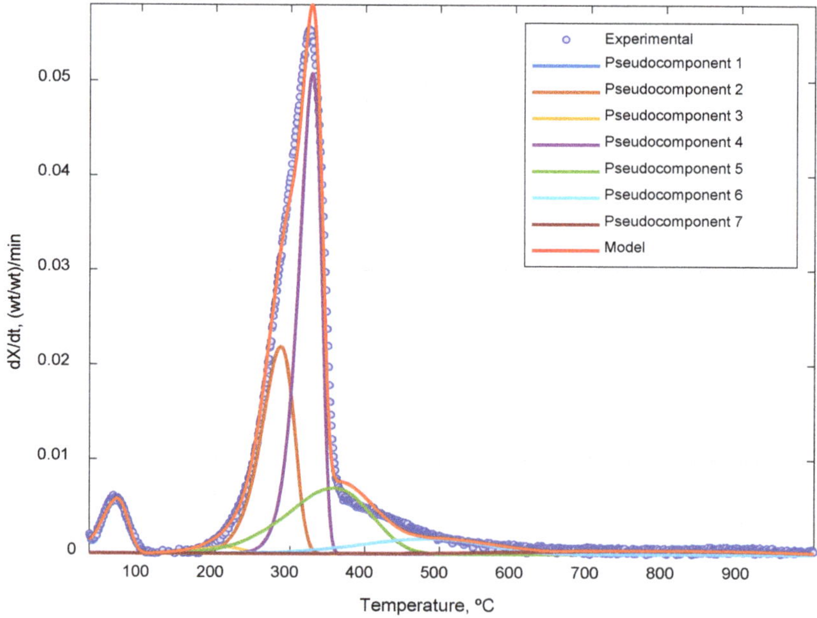

**Figure 7.** Model fitted to the experimental wheat straw DTG curve.

By comparison, between the kinetic constants in Table 6 and Figures 3–7, it can be seen that low activation energy leads to a reaction in the low temperature zone and vice versa. With respect to the pre-exponential factor, low values cause the reaction rate to be slower and to take place over a wider temperature range, which is characteristic of the lignin pseudocomponents. On the contrary, high values of the pre-exponential factor increase the reaction rate, leading to a narrower temperature range, which is characteristic of cellulose, for example.

On the other hand, at the same activation energy, a higher pre-exponential factor causes the reaction to take place in the high temperature zone. For example, there are lignin pseudocomponents with a similar activation energy as hemicellulose pseudocomponents (Table 6) but with much lower pre-exponential factors, which cause the reaction to take place at higher temperatures.

The quality of the fit expressed as $R^2$ and QOF% can be observed in Table 8.

**Table 8.** Quality of the fit expressed as $R^2_{Adj}$ and QOF%.

| Biomass | QOF% | $R^2_{Adj}$ |
|---|---|---|
| Pine bark | 1.51 | 0.9939 |
| Spruce bark | 1.94 | 0.9905 |
| Poplar | 1.09 | 0.9960 |
| Willow | 1.42 | 0.9933 |
| Wheat straw | 1.79 | 0.9921 |

In addition, Figures 8–12 show the fit of the global model to the TG experimental data. The TG curve model has been obtained simultaneously with the DTG curve model by solving Equations (9) and (10). As can be seen, the TG curve model achieves good results not only with respect to the model fitting to the experimental TG curve along the operating temperature but, also, with respect to the final value.

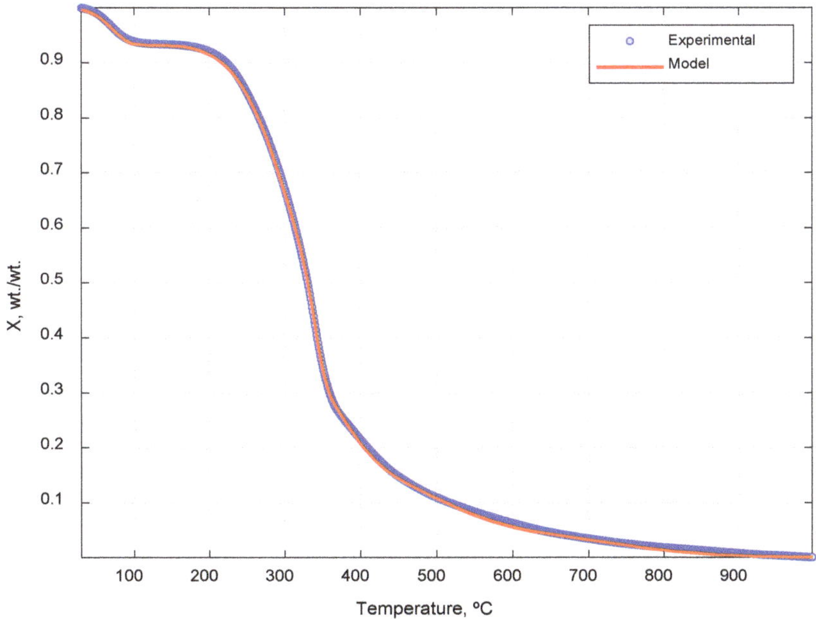


**Figure 8.** Model fitted to the experimental pine bark TG curve.

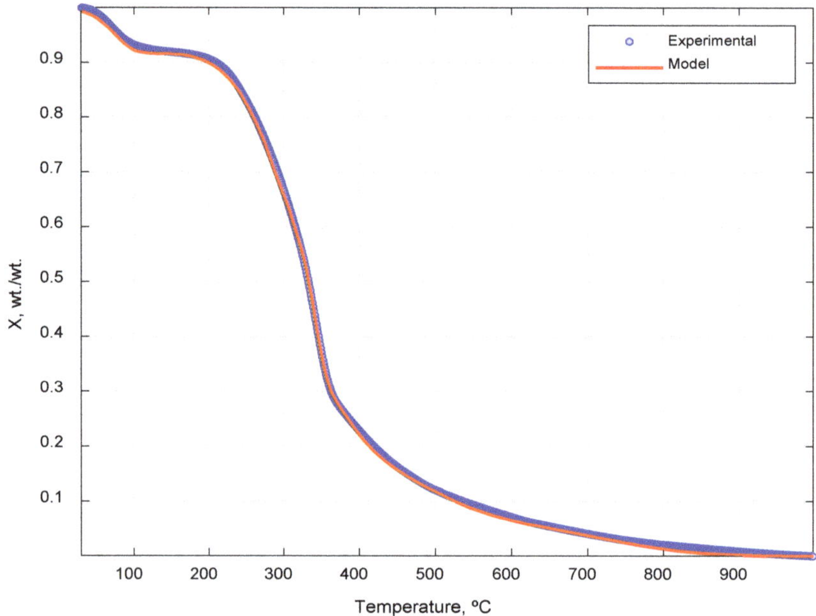

**Figure 9.** Model fitted to the experimental spruce bark TG curve.

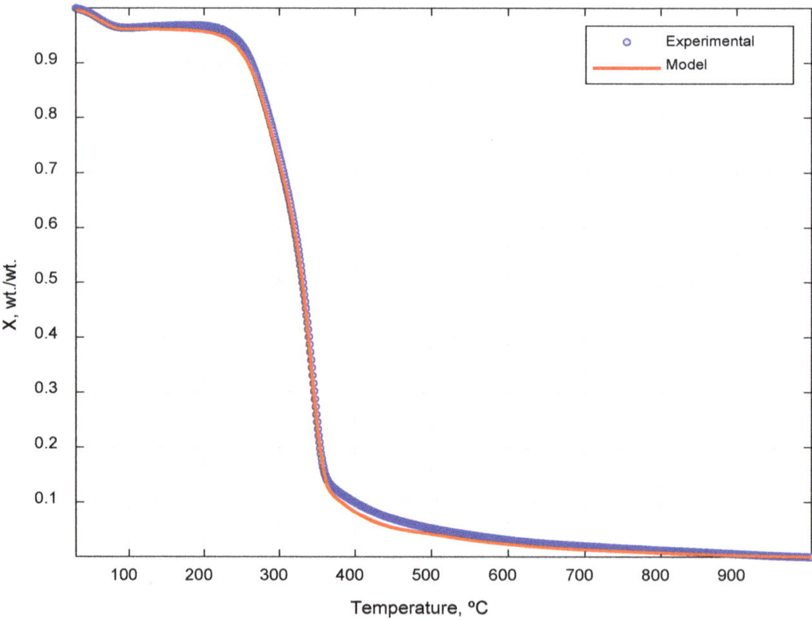

**Figure 10.** Model fitted to the experimental poplar TG curve.

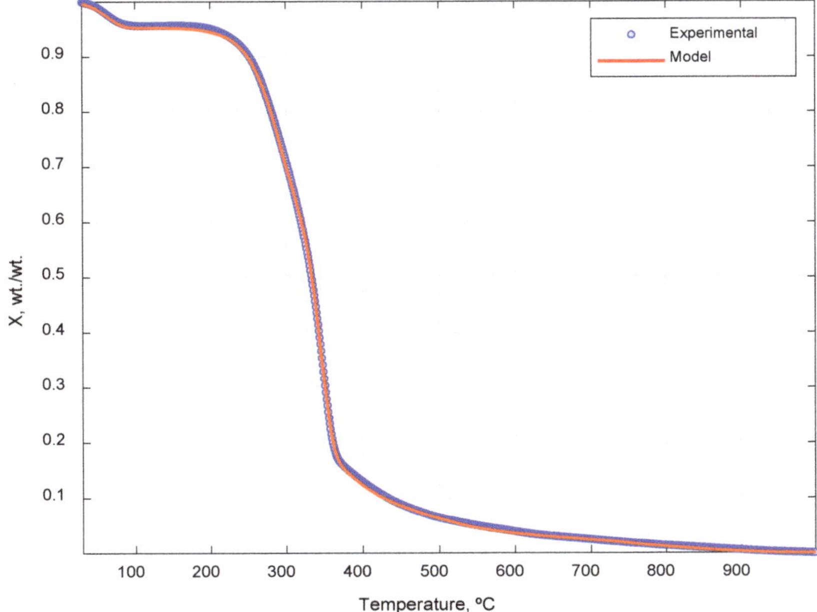

**Figure 11.** Model fitted to the experimental willow TG curve.

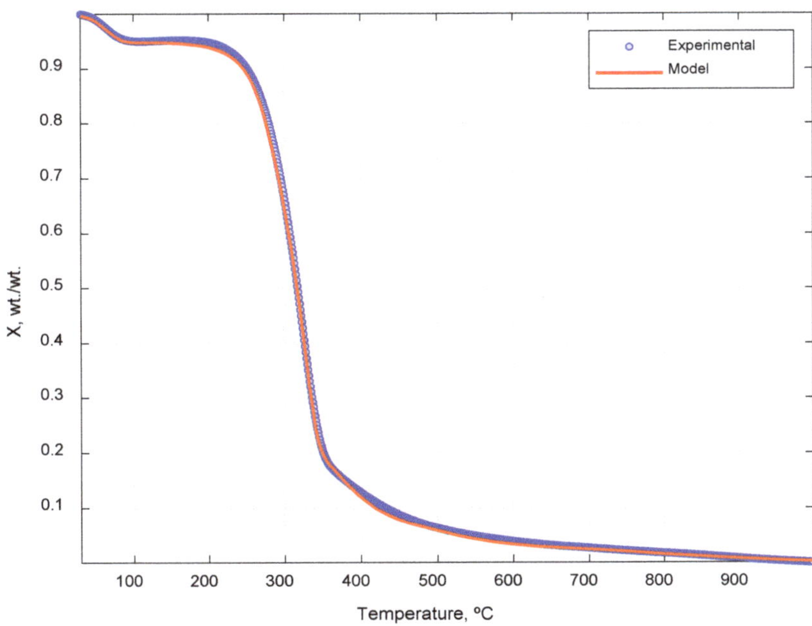

**Figure 12.** Model fitted to the experimental wheat straw TG curve.

Table 9 shows the comparison between the analytical composition and the data obtained with the TGA-PKM method. As can be seen, there is a good agreement between the data obtained through analytical procedures and the TGA-PKM model. This indicates that the new method can be used to have a good estimation of the content of the main lignocellulosic fractions of the analyzed biomasses without the need to carry out complex extraction and purification chemical treatments.

**Table 9.** Comparison between the analytical and thermogravimetric analysis-pseudocomponent kinetic model (TGA-PKM) results.

| Biomass | Component | Analytical Method wt.%, Dry, Ash and Extractives-Free Basis | TGA-PKM Method wt.%, Dry, Ash and Extractives-Free Basis | Error, wt.% |
|---|---|---|---|---|
| Poplar | Hemicellulose | 23.77 | 23.62 | −0.15 |
| | Cellulose | 46.77 | 53.90 | 7.14 |
| | Lignin | 29.46 | 22.48 | −6.99 |
| Willow | Hemicellulose | 24.57 | 25.00 | 0.43 |
| | Cellulose | 48.15 | 46.51 | −1.64 |
| | Lignin | 27.28 | 28.49 | 1.21 |
| Wheat straw | Hemicellulose | 29.10 | 26.54 | −2.56 |
| | Cellulose | 45.84 | 41.67 | −4.18 |
| | Lignin | 25.06 | 31.80 | 6.73 |
| Spruce Bark | Hemicellulose | 15.67 | 22.45 | 6.78 |
| | Cellulose | 33.48 | 26.74 | −6.74 |
| | Lignin | 50.85 | 50.81 | −0.04 |
| Pine bark | Hemicellulose | 22.62 | 23.30 | 0.68 |
| | Cellulose | 27.07 | 26.47 | −0.61 |
| | Lignin | 50.31 | 50.24 | −0.07 |

The following error ranges are obtained between the values measured analytically and those measured by the TGA-PKM method for each of the main lignocellulosic fractions: hemicellulose (−2.56–6.78), cellulose (−6.74–7.14) and lignin (−6.99–6.73). The level of accuracy achieved is considered suitable, taking into account that it is within the error range of the chemical methods. For example, Korpinen et al. found that the determination of lignin by different chemical methods can be as high as 10 wt.% [39]; Ioelovich [40] also determined a difference of 4 wt.% between the TAPPI and NERL methods in determination of the cellulose content. In this way, the TGA-PKM method allows to obtain a fast estimation of the contents of the main lignocellulosic fractions within the ranges that would be obtained by a chemical analysis.

*3.4. Validation of the TGA-PKM Method*

In order to check the validity of the method, an additional fit of the poplar biomass devolatilization was performed using, simultaneously, three heating rates: 3, 5 and 10 °C min$^{-1}$ datasets.

Figure 13 shows the graphical results by fitting the model to the DTG and TG curves for each heating rate.

Additionally, the quality of the fit achieved for each heating rate and for the global fit are summarized in Table 10, where QOF% and $R^2_{Adj}$ are shown for each heating rate dataset and for the three heating rates simultaneously.

**Table 10.** Quality of the fit for each dataset.

| Quality of the Fit | 3 °C min$^{-1}$ | 5 °C min$^{-1}$ | 10 °C min$^{-1}$ | Global |
|---|---|---|---|---|
| QOF% | 1.16 | 1.58 | 0.96 | 1.35 |
| $R^2_{Adj}$ | 0.9957 | 0.9917 | 0.9971 | 0.9959 |

The results obtained (Figure 13 and Table 10) indicate that the quality of the fit obtained is very satisfactory, since the model is capable of representing the evolution of the devolatilization process when different heating rates are used.

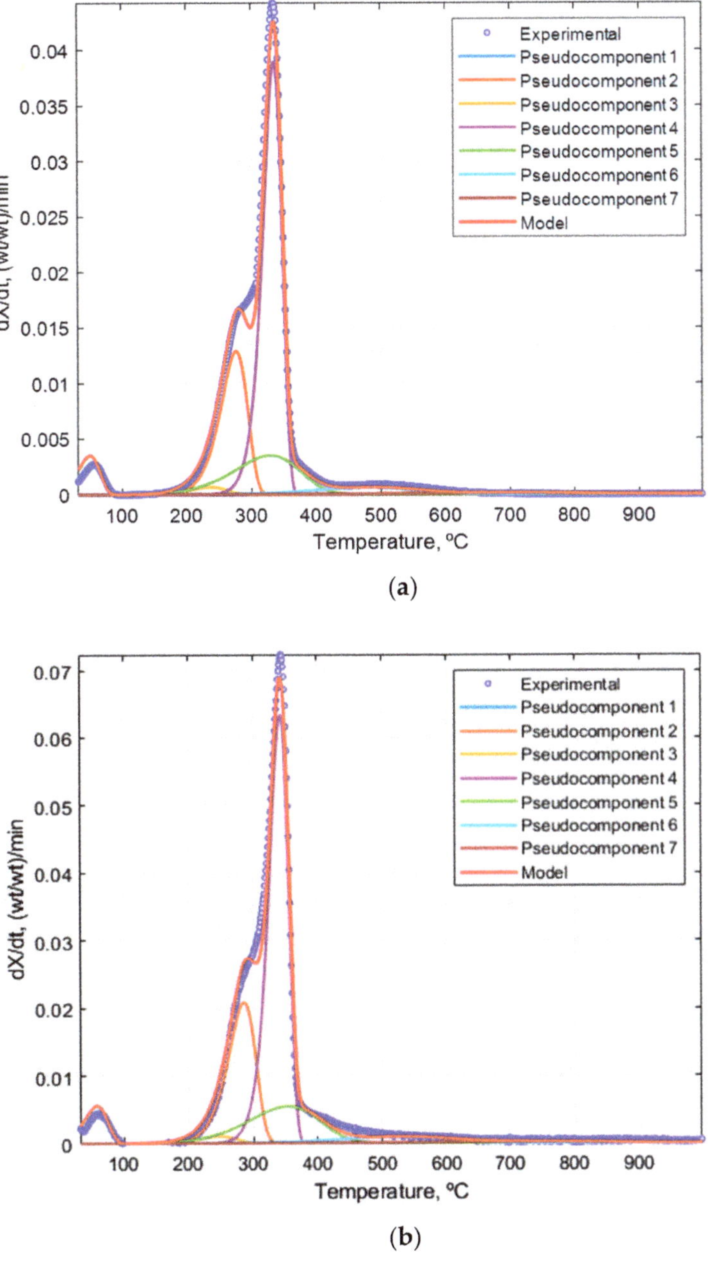

(a)

(b)

**Figure 13.** *Cont.*

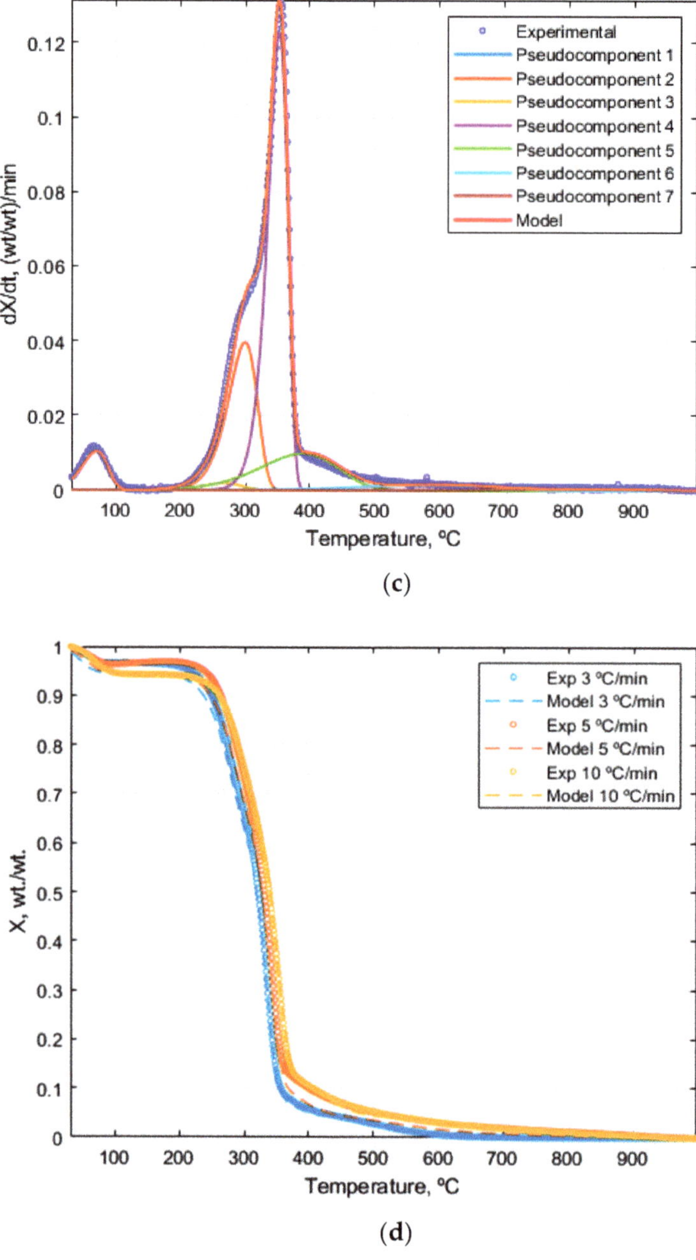

**(c)**

**(d)**

**Figure 13.** Model fitted to the experimental poplar DTG and TG curves: (**a**) DTG at 3 °C min−1, (**b**) 5 °C min−1 and (**c**) 10 °C min−1. (**d**) TG at the three heating rates.

Table 11 shows the kinetic parameters by fitting the model using a single heating rate and three heating rates simultaneously. The obtained results by both datasets are very similar. For example, the activation energy obtained is identical for almost all the pseudocomponents, and only pseudocomponents 3 and 7 have a relative standard deviation of 4%.

**Table 11.** Kinetic parameters of each pseudocomponents calculated using a single heating rate and three simultaneous heating rates.

| | | Hemicellulose | | Cellulose | | Lignin | |
|---|---|---|---|---|---|---|---|
| **Number of Heating Rates** | **Kinetic Parameters** | **PC 2** | **PC 3** | **PC 4** | **PC 5** | **PC 6** | **PC 7** |
| Single heating rate | $K \ (s^{-1})$ | $6.00 \times 10^8$ | $1.40 \times 10^5$ | $2.30 \times 10^{15}$ | $4.81 \times 10^1$ | $2.79 \times 10^0$ | $5.07 \times 10^{-1}$ |
| | $E \ (kJ \ mol^{-1})$ | 120.00 | 80.00 | 207.39 | 55.00 | 53.22 | 59.71 |
| | $X_{j,0} \ (wt.\%)$ | 21.72 | 1.00 | 51.85 | 15.34 | 4.24 | 2.04 |
| Three simultaneous heating rates | $K \ (s^{-1})$ | $6.56 \times 10^8$ | $1.50 \times 10^5$ | $2.23 \times 10^{15}$ | $5.49 \times 10^1$ | $2.79 \times 10^0$ | $5.57 \times 10^{-1}$ |
| | $E \ (kJ \ mol^{-1})$ | 119.98 | 77.18 | 207.48 | 55.00 | 53.99 | 57.48 |
| | $X_{j,0} \ (wt.\%)$ | 22.78 | 1.46 | 49.97 | 15.08 | 3.88 | 1.51 |

Finally, the results obtained in the determination of the lignocellulosic fractions are shown in Table 12. The results obtained with a single heating rate are comparable to those obtained with three heating rates, because the deviation between the results calculated by the model and by the analytical method are of the same order when a single heating rate or three heating rates are considered. However, slightly better results are achieved if a single heating rate of 5 °C min$^{-1}$ is used, but mainly, it requires considerably less analysis time, which justifies the use of a single heating rate.

**Table 12.** Comparison between TGA-PKM results using three simultaneous heating rates and the analytical method.

| Biomass | Component | **Analytical Method Wt.%, Dry, Ash and Extractives-Free Basis** | **TGA-PKM Method (Three Simultaneous Heating Rates) wt.%, Dry, Ash and Extractives-Free Basis** | **Error, wt.%** |
|---|---|---|---|---|
| Poplar | Hemicellulose | 23.77 | 25.60 | 1.83 |
| | Cellulose | 46.77 | 52.78 | 6.01 |
| | Lignin | 29.46 | 21.62 | −7.85 |

## 4. Conclusions

Five lignocellulosic samples have been characterized by the TGA-PKM experimental protocol, covering different types of woody and herbaceous biomasses from both forest and agricultural origins (spruce bark, pine bark, poplar, willow and wheat straw).

The TGA-PKM method developed allows the determination of the main lignocellulosic fractions of biomasses without the need to use long and complex chemical methods; e.g., TAPPI methods T222 and T249 require several long successive steps (hydrolysis, extraction, filtration, neutralization, reduction, etc.) [41], which may require several days of work in the laboratory, while the new method may be performed in a few hours. Thus, it would be possible to reduce the cost of analysis and processing time by 80–90%.

The accuracy of the TGA-PKM method was tested and proved to be significantly good and consistent within the order of magnitude of the standard analytical methods to determine the contents of the main lignocellulosic fractions.

**Author Contributions:** Conceptualization, D.D. and A.U.; methodology, D.D. and A.U.; software, D.D. and A.U.; validation, D.D., A.U., R.P.; formal analysis, D.D., A.U., A.B. and T.T.; investigation, D.D. and A.U.; resources, A.B., T.T. and R.P.; data curation, D.D., A.U.; writing (original draft preparation), D.D., A.U.; writing (review and editing), R.P, A.B. and T.T.; visualization, D.D., A.U., R.P., A.B. and T.T.; supervision, D.D., A.U., R.P., A.B. and T.T.; project administration, A.B; funding acquisition, A.B., T.T. and R.P. All authors have read and agreed to the published version of the manuscript.

**Funding:** This research was funded by the European Union's Horizon 2020 research and innovation programme under grant agreement No 723670, with the title "Systemic approach to reduce energy demand and CO$_2$ emissions of processes that transform agroforestry waste into high added value products (REHAP)".

**Acknowledgments:** The authors would like to thank María González Martínez from IMT Mines Albi (Université de Tolousse) for her technical contribution and support.

**Conflicts of Interest:** The authors declare no conflict of interest.

## References


1. Kim, S.; Dale, B.E. All biomass is local: The cost, volume produced, and global warming impact of cellulosic biofuels depend strongly on logistics and local conditions. *Biofuels Bioprod. Biorefining* **2015**, *9*, 422–434. [CrossRef]
2. McKendry, P. Energy production from biomass (part 1): Overview of biomass. *Bioresour. Technol.* **2002**, *83*, 37–46. [CrossRef]
3. Rego, F.; Dias, A.P.S.; Casquilho, M.; Rosa, F.C.; Rodrigues, A. Fast determination of lignocellulosic composition of poplar biomass by thermogravimetry. *Biomass Bioenergy* **2019**, *122*, 375–380. [CrossRef]
4. Álvarez-Mateos, P.; Alés-Álvarez, F.J.; García-Martín, J.F. Phytoremediation of highly contaminated mining soils by *Jatropha curcas* L. and production of catalytic carbons from the generated biomass. *J. Environ. Manag.* **2019**, *231*, 886–895. [CrossRef] [PubMed]
5. Snell, K.D.; Peoples, O.P. PHA bioplastic: A value-added coproduct for biomass biorefineries. *Biofuels Bioprod. Biorefining Innov. Sustain. Econ.* **2009**, *3*, 456–467. [CrossRef]
6. Pereira, B.L.C.; Carneiro, A.D.C.O.; Carvalho, A.M.M.L.; Colodette, J.L.; Oliveira, A.C.; Fontes, M.P.F. Influence of chemical composition of Eucalyptus wood on gravimetric yield and charcoal properties. *BioResources* **2013**, *8*, 4574–4592. [CrossRef]
7. Melzer, M.; Blin, J.; Bensakhria, A.; Valette, J.; Broust, F. Pyrolysis of extractive rich agroindustrial residues. *J. Anal. Appl. Pyrolysis* **2013**, *104*, 448–460. [CrossRef]
8. Park, J.I.; Liu, L.; Ye, X.P.; Jeong, M.K.; Jeong, Y.S. Improved prediction of biomass composition for switchgrass using reproducing kernel methods with wavelet compressed FT-NIR spectra. *Expert Syst. Appl.* **2012**, *39*, 1555–1564. [CrossRef]
9. Yu, J.; Paterson, N.; Blamey, J.; Millan, M. Cellulose, xylan and lignin interactions during pyrolysis of lignocellulosic biomass. *Fuel* **2017**, *191*, 140–149. [CrossRef]
10. Shen, D.; Xiao, R.; Gu, S.; Luo, K. The pyrolytic behavior of cellulose in lignocellulosic biomass: A review. *RSC Adv.* **2011**, *1*, 1641–1660. [CrossRef]
11. Shen, D.K.; Gu, S.; Bridgwater, A.V. The thermal performance of the polysaccharides extracted from hardwood: Cellulose and hemicellulose. *Carbohydr. Polym.* **2010**, *82*, 39–45. [CrossRef]
12. Shen, D.K.; Gu, S. The mechanism for thermal decomposition of cellulose and its main products. *Bioresour. Technol.* **2009**, *100*, 6496–6504. [CrossRef] [PubMed]
13. Li, S.; Lyons-Hart, J.; Banyasz, J.; Shafer, K. Real-time evolved gas analysis by FTIR method: An experimental study of cellulose pyrolysis. *Fuel* **2001**, *80*, 1809–1817. [CrossRef]
14. Qiao, Y.; Wang, B.; Ji, Y.; Xu, F.; Zong, P.; Zhang, J.; Tian, Y. Thermal decomposition of castor oil, corn starch, soy protein, lignin, xylan, and cellulose during fast pyrolysis. *Bioresour. Technol.* **2019**, *278*, 287–295. [CrossRef]
15. Wang, S.; Lin, H.; Ru, B.; Sun, W.; Wang, Y.; Luo, Z. Comparison of the pyrolysis behavior of pyrolytic lignin and milled wood lignin by using TG–FTIR analysis. *J. Anal. Appl. Pyrolysis* **2014**, *108*, 78–85. [CrossRef]
16. Di Blasi, C. Modeling chemical and physical processes of wood and biomass pyrolysis. *Prog. Energy Combust. Sci.* **2008**, *34*, 47–90. [CrossRef]
17. Zhou, H.; Long, Y.; Meng, A.; Li, Q.; Zhang, Y. The pyrolysis simulation of five biomass species by hemi-cellulose, cellulose and lignin based on thermogravimetric curves. *Thermochim. Acta* **2013**, *566*, 36–43. [CrossRef]
18. Skreiberg, A.; Skreiberg, Ø.; Sandquist, J.; Sørum, L. TGA and macro-TGA characterisation of biomass fuels and fuel mixtures. *Fuel* **2011**, *90*, 2182–2197. [CrossRef]
19. Carrier, M.; Loppinet-Serani, A.; Denux, D.; Lasnier, J.M.; Ham-Pichavant, F.; Cansell, F.; Aymonier, C. Thermogravimetric analysis as a new method to determine the lignocellulosic composition of biomass. *Biomass Bioenergy* **2011**, *35*, 298–307. [CrossRef]
20. Hu, S.; Jess, A.; Xu, M. Kinetic study of Chinese biomass slow pyrolysis: Comparison of different kinetic models. *Fuel* **2007**, *86*, 2778–2788. [CrossRef]

21. Aboyade, A.O.; Carrier, M.; Meyer, E.L.; Knoetze, J.H.; Görgens, J.F. Model fitting kinetic analysis and characterisation of the devolatilization of coal blends with corn and sugarcane residues. *Thermochim. Acta* **2012**, *530*, 95–106. [CrossRef]

22. Vyazovkin, S.; Burnham, A.K.; Criado, J.M.; Pérez-Maqueda, L.A.; Popescu, C.; Sbirrazzuoli, N. ICTAC Kinetics Committee recommendations for performing kinetic computations on thermal analysis data. *Thermochim. Acta* **2011**, *520*, 1–19. [CrossRef]

23. Manya, J.J.; Velo, E.; Puigjaner, L. Kinetics of biomass pyrolysis: A reformulated three-parallel-reactions model. *Ind. Eng. Chem. Res.* **2003**, *42*, 434–441. [CrossRef]

24. Burhenne, L.; Messmer, J.; Aicher, T.; Laborie, M.P. The effect of the biomass components lignin, cellulose and hemicellulose on TGA and fixed bed pyrolysis. *J. Anal. Appl. Pyrolysis* **2013**, *101*, 177–184. [CrossRef]

25. O'Brien, C.M. Statistical Applications for Environmental Analysis and Risk Assessment by Joseph Ofungwu. *Int. Stat. Rev.* **2014**, *82*, 487–488. [CrossRef]

26. Raitanen, J.E.; Järvenpää, E.; Korpinen, R.; Mäkinen, S.; Hellström, J.; Kilpeläinen, P.; Jaana Liimatainen, J.; Ora, A.; Tupasela, T.; Jyske, T. Tannins of Conifer Bark as Nordic Piquancy—Sustainable Preservative and Aroma? *Molecules* **2020**, *25*, 567. [CrossRef]

27. Hofbauer, H.; Kaltschmitt, M.; Nussbaumer, T. *Energie aus Biomasse–Grundlagen, Techniken, Verfahren*; Springer: Berlin, Germany, 2009.

28. Demirbaş, A. Calculation of higher heating values of biomass fuels. *Fuel* **1997**, *76*, 431–434. [CrossRef]

29. Rowell, R.M.; Pettersen, R.; Han, J.S.; Rowell, J.S.; Tshabalala, M.A. Cell wall chemistry. In *Handbook of Wood Chemistry and Wood Composites*; CRC Press: Boca Ratón, FL, USA, 2005; Volume 2.

30. Wang, S.; Dai, G.; Yang, H.; Luo, Z. Lignocellulosic biomass pyrolysis mechanism: A state-of-the-art review. *Prog. Energy Combust. Sci.* **2017**, *62*, 33–86. [CrossRef]

31. Pasangulapati, V.; Ramachandriya, K.D.; Kumar, A.; Wilkins, M.R.; Jones, C.L.; Huhnke, R.L. Effects of cellulose, hemicellulose and lignin on thermochemical conversion characteristics of the selected biomass. *Bioresour. Technol.* **2012**, *114*, 663–669. [CrossRef]

32. González Martinez, M. Woody and Agricultural Biomass Torrefaction: Experimental Study and Modelling of Solid Conversion and Volatile Species Release Based on Biomass Extracted Macromolecular Components. Ph.D. Thesis, University of Toulouse, Toulouse, France, 2018.

33. Saini, J.K.; Saini, R.; Tewari, L. Lignocellulosic agriculture wastes as biomass feedstocks for second-generation bioethanol production: Concepts and recent developments. *3 Biotech* **2015**, *5*, 337–353. [CrossRef]

34. González Martínez, M.; Dupont, C.; da Silva Perez, D.; Mortha, G.; Thiéry, S.; Meyer, X.M.; Gourdon, C. Understanding the torrefaction of woody and agricultural biomasses through their extracted macromolecular components. Part 1: Experimental thermogravimetric solid mass loss. *Energy* **2020**, *205*, 118067.

35. Guo, X.J.; Wang, S.R.; Wang, K.G.; Qian, L.I.U.; Luo, Z.Y. Influence of extractives on mechanism of biomass pyrolysis. *J. Fuel Chem. Technol.* **2010**, *38*, 42–46. [CrossRef]

36. Ebringerová, A.; Hromádková, Z.; Heinze, T.; Hemicellulose, T.H. Polysaccharides I. *Adv. Polym. Sci.* **2005**, *186*, 67.

37. Yang, H.; Yan, R.; Chen, H.; Lee, D.H.; Zheng, C. Characteristics of hemicellulose, cellulose and lignin pyrolysis. *Fuel* **2007**, *86*, 1781–1788. [CrossRef]

38. Chen, W.H.; Wang, C.W.; Ong, H.C.; Show, P.L.; Hsieh, T.H. Torrefaction, pyrolysis and two-stage thermodegradation of hemicellulose, cellulose and lignin. *Fuel* **2019**, *258*, 116168. [CrossRef]

39. Yeo, J.Y.; Chin, B.L.F.; Tan, J.K.; Loh, Y.S. Comparative studies on the pyrolysis of cellulose, hemicellulose, and lignin based on combined kinetics. *J. Energy Inst.* **2019**, *92*, 27–37. [CrossRef]

40. Korpinen, R.; Kallioinen, M.; Hemming, J.; Pranovich, A.; Mänttäri, M.; Willför, S. Comparative evaluation of various lignin determination methods on hemicellulose-rich fractions of spruce and birch obtained by pressurized hot-water extraction (PHWE) and subsequent ultrafiltration (UF). *Holzforschung* **2014**, *68*, 971–979. [CrossRef]

41. Ioelovich, M. Methods for determination of chemical composition of plant biomass. *J. SITA* **2015**, *17*, 208–214.

*Article*

# Acid and Enzymatic Fractionation of Olive Stones for Ethanol Production Using *Pachysolen tannophilus*

Manuel Cuevas [1,2,*], Marwa Saleh [1], Juan F. García-Martín [2,3,*] and Sebastián Sánchez [1,2]

[1]  Department of Chemical, Environmental and Materials Engineering, University of Jaén,
    Campus 'Las Lagunillas', 23071 Jaén, Spain; marwah.saleh@gmail.com (M.S.); ssanchez@ujaen.es (S.S.)
[2]  Center for Advanced Studies in Olive Grove and Olive Oils, Science and Technology Park GEOLIT,
    23620 Mengibar, Spain
[3]  Departamento de Ingeniería Química, Facultad de Química, Universidad de Sevilla,
    C/Profesor García González, 1, 41012 Seville, Spain
*   Correspondence: mcuevas@ujaen.es (M.C.); jfgarmar@us.es (J.F.G.-M.)

Received: 29 December 2019; Accepted: 31 January 2020; Published: 6 February 2020

**Abstract:** Olive stones are an abundant lignocellulose material in the countries of the Mediterranean basin that could be transformed to bioethanol by biochemical pathways. In this work, olive stones were subjected to fractionation by means of a high-temperature dilute-acid pretreatment followed by enzymatic hydrolysis of the pretreated solids. The hydrolysates obtained in these steps were separately subjected to fermentation with the yeast *Pachysolen tannophilus* ATCC 32691. Response surface methodology with two independent variables (temperature and reaction time) was applied for optimizing D-xylose production from the raw material by dilute acid pretreatment with 0.01 M sulfuric acid. The highest D-xylose yield in the liquid fraction was obtained in the pretreatment at 201 °C for 5.2 min. The inclusion of a detoxification step of the acid prehydrolysate, by vacuum distillation, allowed the fermentation of the sugars into ethanol and xylitol. The enzymatic hydrolysis of the pretreated solids was solely effective when using high enzyme loadings, thus leading to easily fermentable hydrolysates into ethanol. The mass macroscopic balances of the overall process illustrated that the amount of inoculum used in the fermentation of the acid prehydrolysates strongly affected the ethanol and xylitol yields.

**Keywords:** bioethanol; dilute acid pretreatment; enzymatic hydrolysis; olive stones; *Pachysolen tannophilus*; response surface methodology

---

## 1. Introduction

Global warming is a problem that could be mitigated by replacing fossil energy sources by renewable energy sources, such as green biomass. [1]. In recent decades numerous research papers have addressed the use of lignocellulose materials to obtain ethanol through biochemical routes [2–4], describing bioprocesses that are mainly composed of three major stages: Biomass pretreatment, cellulose hydrolysis, and sugars fermentation. Acid prehydrolysis at high temperature (around 200 °C) is one of the most efficient pretreatments to remove hemicelluloses and extracts present in lignocellulose materials, obtaining a cellulose and lignin-rich pretreated solid [5,6]. This type of pretreatment can be carried out with low concentrations of acid and short reaction times (few minutes), being able to generate liquid prehydrolysates with high concentrations of hemicellulose sugars, provided that suitable operating conditions are used. Thus, for a fixed acid concentration, there will be optimal temperature and reaction time conditions that will lead to complete hydrolysis of the hemicellulose and will maximize the concentration of hemicellulose sugars in the liquid prehydrolysate. The search for these conditions could be approached using response surface methodology as previously described elsewhere [7,8]. In relation to the type of acid used, sulfuric acid is generally used instead of hydrochloric acid due to its low volatility, lesser equipment corrosion and lower cost per mole of protons [9], so it is frequently used at this stage of the bioprocess.

To improve the biomass fractionation, pretreated solids could be subjected to enzymatic hydrolysis with cellulases, which would allow a selective conversion of cellulose into D-glucose under mild operating conditions (pH 4.8, 50 °C temperature). It is of great importance to assess the effect of enzyme loading in the enzymatic hydrolysis, since a low concentration of the same would lead to low monosaccharide yields, while a high concentration would excessively increase the operating costs.

Concerning sugars fermentation, the use of non-traditional yeasts, such as *Pachysolen tannophilus*, would allow converting both hexoses and pentoses into ethanol [10], provided that fermentative inhibitors concentrations are cut down. In this sense, the levels of furfural and acetic acid in the acid prehydrolysates could be reduced by vacuum distillation because of the higher vapor pressures of these compounds with respect to those of monosaccharides. A variable of great interest in the fermentative stage is the inoculum concentration since this is responsible, among other factors, for the bioprocess rate.

Olive stones are a lignocellulose material that is obtained in great quantities in the olive oil mills [11]. Therefore, this biomass is a potential source of biofuels in countries with large olive production, such as Spain, Italy, Greece, or Portugal. The biochemical conversion of olive stones into ethanol, using hydrothermal or acid pretreatments, has been partially studied by different authors. Thus, Miranda et al. [12] applied a hydrothermal pretreatment to olive stones (130 °C, 30 min) achieving a good enzymatic digestibility of the cellulose fraction. However, the further use of liquid prehydrolysates is usually not addressed. The liquid prehydrolysates are rich in D-xylose and xylooligosaccharides as described in other works in which water was used as a hydrolytic agent [13,14]. On the other hand, Romero-García et al. [15] applied 2% (wt.) sulfuric acid (130 °C, 60 min) to olive stones, achieving a high production of hemicellulose sugars (mainly D-xylose) that were fermented to ethanol after undergoing detoxification by overliming. Notwithstanding, this work did not address the use of the pretreated solids, thus remaining as waste. Saleh et al. [16] applied dilute sulfuric acid (0.025 M $H_2SO_4$) to olive stones (195 °C, 5 min) to hydrolyze the hemicellulose and obtain a D-xylose-rich prehydrolysate, which was subsequently detoxified and fermented. The pretreated solids were subjected to enzymatic hydrolysis, but the obtained hydrolysates were not fermented despite their high sugars content.

The present study aims to develop a complete scheme of ethanol production from olive stones by applying two consecutive hydrolytic stages (acidic and enzymatic) followed by the subsequent stage of sugars fermentation. The acid pretreatment was optimized to maximize the recovery of hemicellulose sugars, while the effect of enzyme loading on D-glucose production in the subsequent stage of enzymatic hydrolysis was studied. The non-traditional yeast *Pachysolen tannophilus* was used for the fermentation of both the acid prehydrolysates and the enzymatic hydrolysates, and the effect of inoculum concentration on the fermentation performance was assessed. The fractionation applied to the biomass was envisaged by determining the mass macroscopic balances at the different stages of the process. In this way, a rapid description of the conversion of olive stones into ethanol and other bioproducts was achieved.

## 2. Materials and Methods

### 2.1. Raw Material

Olive stones (fragmented endocarps) were supplied by the olive mill S.C.A. San Juan (Jaén, Spain, UTM coordinates: 37°47′58.52″ N, 3°47′09.42″ W). Once at laboratory, the raw material was washed and then dried at room temperature for three weeks. Afterwards, the olive stones were screened using a vibratory screen (Restch, Mod. Vibro). The solids used in this research had diameters lower than 3 mm while those of diameter lower than 0.85 mm represented less than 3% of the total sample weight. Finally, the dry solids were stored in sealed plastic bags at room temperature until used.

## 2.2. Dilute Acid Pretreatment

Dilute acid hydrolysis was carried out in a 2 dm$^3$ Parr reactor, Series 4522 (Moline, IL, USA). Fo experiments, this reactor was loaded with 50 g of dry solids and 300 cm$^3$ of 0.010 M sulfuric-acid solution. The chosen H$_2$SO$_4$ concentration was significantly lower than that used in a previous work published by us (0.025 M, [16]) to reduce the corrosion that the acid causes on the reactor. The mixture was stirred at 250 rpm and quickly heated up to work temperature, which was maintained over the selected reaction time (see Section 2.3). Subsequently, the reactor was cooled to room temperature (in less than 10 min) by circulating cold water through an internal coil. The liquid prehydrolysate was separated from the pretreated solid by vacuum filtration. The latter one was water-washed and dried at room temperature. The water used for washing the pretreated solid was added to the prehydrolysate until reaching a final volume of 2 dm$^3$. The two separate phases were stored for later characterization and used in the following stages of the scheme.

## 2.3. Response Surface Methodology (RSM)

Response surface methodology (RSM) was applied to data to optimize the acid pretreatment conditions by multiple regression analysis. To do this, the Modde 7.0 (Umetrics AB, Umeå, Sweden) software was used. The $2^2$ central composite circumscribed design (CCCD) with two independent variables (temperature and time) at two different levels, four star (axial) points and three central points (total 11 runs, Table 1) was assayed to find linear, quadratic and interaction effects of the independent variables (operational parameters) on the experimental responses (experimental results). The temperature (180–220 °C) and time (2–8 min) ranges were selected from the results obtained in preliminary experiments (data not shown) with the aim of achieving as D-xylose recovery as possible in the liquid prehydrolysates. The statistical validation was carried out by one-way ANOVA test (95% confidence), and the optimal conditions values were determined from the response surfaces using the SIMPLEX method.

**Table 1.** Operational conditions assayed as dimensional and dimensionless independent variables.

| Run | Temperature (°C) | | Time (min) | |
|---|---|---|---|---|
| | Actual Values ($T$) | Coded Values ($X_1$) | Actual Values ($t$) | Coded Values ($X_2$) |
| 1 | 172 | 0 | 5.00 | −1.414 |
| 2 | 180 | −1 | 2.00 | −1 |
| 3 | 180 | 1 | 8.00 | −1 |
| 4 | 200 | −1.414 | 0.76 | 0 |
| 5 | 200 | 0 | 5.00 | 0 |
| 6 | 200 | 0 | 5.00 | 0 |
| 7 | 200 | 0 | 5.00 | 0 |
| 8 | 200 | 1.414 | 9.24 | 0 |
| 9 | 220 | −1 | 2.00 | 1 |
| 10 | 220 | 1 | 8.00 | 1 |
| 11 | 228 | 0 | 5.00 | 1.414 |

## 2.4. Detoxification and Fermentation of Acid Prehydrolysates

The acid prehydrolysates were detoxified using a Buchi R-114 rotary evaporator (BÜCHI Labortechnik AG, Flawil, Switzerland) at 40 °C, thus removing inhibitor compounds (acetic acid, furfural, and 5-hydroxymethylfurfural) and achieving a D-xylose concentration close to 15 g/dm$^3$.

Fermentations of acid prehydrolysates were carried out using the yeast *P. tannophilus* ATCC 32,691, supplied by the American Type Culture Collection. The microorganism was stored in a cold room (5–10 °C) in 100-cm$^3$ test tubes on a sterilized solid culture medium with the following composition: 3 g/dm$^3$ yeast extract (Becton Dickinson Co), 3 g/dm$^3$ malt extract (Merck), 5 g/dm$^3$ peptone from casein (Merck), 10 g/dm$^3$ D-xylose (≥99% purity, Panreac), and 20 g/dm$^3$ agar–agar (Panreac). For pre-inocula, cells were transferred to a sterile medium with the same aforementioned composition and kept in an incubator at 30 °C for 60 h in order to obtain cells at the same growth stage at the beginning of each fermentation. Afterwards, the pre-inocula were transferred to 250-cm$^3$ Erlenmeyer flasks along with 100 cm$^3$ of sterile liquid culture made of 4 g/dm$^3$ yeast extract (Becton Dickinson Co), 3.6 g/dm$^3$ peptone from casein (Merck), 3 g/dm$^3$ (NH$_4$)$_2$SO$_4$ (99% purity, Panreac), 2 g/dm$^3$ MgSO$_4$·7H$_2$O (99,5% purity, Carlo Erba), 2 g/dm$^3$ KH$_2$PO$_4$ (99% purity, Panreac), and 25 g/dm$^3$ D-xylose (≥99% purity, Panreac). Cultures were incubated at 30 °C for 24 h in an orbital shaker (150 rpm). Then, yeast cells were recovered by centrifugation at 7000 rpm for 10 min, washed with a dilute NaCl solution, and suspended in the fermentation medium to obtain initial inoculum concentrations of 0.5, 1.0, 2.0, and 4.0 g/dm$^3$. These concentrations ($x$, kg/m$^3$) were calculated from the absorbances of the cultures at 620 nm using a previously obtained absorbance versus dry-weight calibration line [10].

Fermentations were carried out with 30 cm$^3$ of prehydrolysate inside 100 cm$^3$ Erlenmeyer flasks. The prehydrolysates were supplemented with 2 g/dm$^3$ yeast extract (Becton Dickinson Co), 1.8 g/dm$^3$ peptone from casein (Merck), 1.5 g/dm$^3$ (NH$_4$)$_2$SO$_4$ (99% purity, Panreac), 1 g/dm$^3$ MgSO$_4$·7H$_2$O (99,5% purity, Carlo Erba), and 1 g/dm$^3$ KH$_2$PO$_4$ (99% purity, Panreac). The resulting cultures were sterilized using a glasswool pre-filter and a 0.2-μm pore-size cellulose nitrate filter. Temperature (30 °C) and pH (4.5) were chosen according to previous works [10,17] and kept constant over fermentations. The aeration was only supplied by the stirring vortex (microaerobic conditions). The cultures were sampled at fixed intervals to analyze the biomass, D-glucose, D-xylose, acetic acid, ethanol, and xylitol concentrations. Two replicas of each fermentation were performed.

### 2.5. Enzymatic Hydrolysis

The washed and dried water-insoluble solids obtained in the dilute acid hydrolysis of olive stones were submitted to enzymatic hydrolysis in order to obtain D-glucose from the cellulose. To do this, 3 g of dry solids were suspended in 30 cm$^3$ of 0.05 M citrate buffer solution (pH 4.8) inside 125 cm$^3$ Erlenmeyer flasks. Enzymatic hydrolyses were carried out at 50 °C for 72 h on an orbital shaker (150 rpm). A commercial preparation of *Trichoderma reesei* cellulases (Celluclast 1.5L, Novo Nordisk Bioindustrial, Madrid, Spain) was used throughout this research. Enzyme loadings of 10, 20, 40 and 60 FPU per g dried solid were added to the Erlenmeyer flasks. 1-cm$^3$ samples were withdrawn from the reaction media at 4, 10, 24, 48, 72, and 120 h to analyse the D-glucose concentration. The D-glucose yield was calculated as g of D-glucose per 100 g of initial pretreated solid. All the enzymatic hydrolyses were performed in duplicate.

### 2.6. Fermentation of Enzymatic Hydrolysates

The fermentation of the enzymatic hydrolysates was performed with the yeast *P. tannophilus* ATCC 32691 following the procedure described in Section 2.4 but with two modifications: Enzymatic hydrolysates were not detoxified, and the initial inoculum concentrations assayed were 0.5, 1.0, 1.5, and 3.0 g/dm$^3$. The cultures were sampled at fixed intervals to analyse the biomass, D-glucose and ethanol concentrations. Two replicas of each fermentation were performed.

### 2.7. Analytical Methods

The raw material as well as the solids resulting from the acid treatments and from enzymatic hydrolyses were characterized according their contents in moisture (TAPPI T257 standard), hemicellulose and cellulose [18], and insoluble acid lignin (TAPPI T222 os-74 standard). Besides, ash (TAPPI T211 standard), extractives (ASTM D 1107 84 standard), and soluble acid lignin [19] were additionally analyzed in the raw material.

The Puls method [20], with a modification described elsewhere [16], was used to determine the percentage of xylans and acetyl groups in the raw material. The concentrations of D-glucose, D-xylose, L-arabinose, D-galactose and 5-hydroxymethyl-furfural in prehydrolysates, enzymatic hydrolysates and cultures were analyzed by high-performance liquid ionic chromatography (HPLIC). The HPLIC system (Dionex ICS 3000, Sunnyvale, CA, USA) was equipped with a CARBOPAD PA20 analytical column (3 mm × 150 mm) combined with a CARBOPAD PA20 guard column (3 mm × 30 mm), and a pulsed amperometer detector (gold electrode). Elution took place at 30 °C, the eluent being 1 cm$^3$/min 0.002 M NaOH. After dilution, the samples were filtered through a 0.2 μm nylon membrane (Sartorius). Finally, ethanol, xylitol and acetic acid concentrations in liquid samples were quantified using enzymatic methods [21–23], using test-combination kits purchased from R-Biopharm AG (Darmstadt, Germany). All the analytical determinations were performed in duplicate.

## 3. Results

### 3.1. Dilute Acid Pretreatment: Experimental Results

Table 2 shows the characterization of the raw material as well as the pretreated solids and the liquid prehydrolysates obtained by dilute acid hydrolysis of the olive stones. The increase in temperature and reaction time led to a continuous decrease in the total gravimetric recovery of pretreated solid (TGR), a parameter that reached values between 56.54% and 86.53% for the most and less severe pretreatment conditions, respectively (228 °C—5 min and 172 °C—5 min). The loss of solid is due to the hydrolysis of different components of the raw material. Thus, hemicellulose began to depolymerize from the lowest temperature tested (172 °C) and practically was removed from the pretreated solids at 200 °C—5 min. Under these conditions, extractives (6.0% of the raw material) and the soluble acid lignin (2.1% of the raw material) were removed along with the hemicellulose (28.1% of the raw material), so the sum of the three fractions led to a TGR of 63.8%, theoretical value close to the average experimental value of TGR (62.78 ± 0.67%) obtained in experiments 5, 6, and 7 (Table 2). The loss of hemicellulose, extractives and acid insoluble lignin (AIL) caused the increase in the percentage of cellulose in the acid-pretreated solid, reaching a maximum of 38.62% for the experiment carried out at 200 °C—9.24 min, while the highest temperatures assayed (220 °C and 228 °C) led to the decrease of this percentage, which would indicate a partial hydrolysis of cellulose. In relation to the AIL percentage in acid-pretreated solids, this continuously increased, from 32.01% to 48.64%, with the increase of the temperature and pretreatment time (Table 2).

**Table 2.** Total gravimetric recovery and composition of acid-pretreated solids, and products yields (as g/100 g dry raw material) in the prehydrolysates obtained at different acid hydrolysis conditions.

| Run Number | Acid-Treated Solids [a] | | | | | Prehydrolysates | | | | | |
| | T | t | TGR | Hem | Cel | AIL | Product Yield (as g/100 g Dry Raw Material) | | | | |
| | (°C) | (min) | (%) | (%) | (%) | (%) | D-Xylose | L-Arabinose | D-Galactose | D-Glucose | AA | 5-HMF |
|---|---|---|---|---|---|---|---|---|---|---|---|---|
| 1 | 172 | 5 | 86.53 | 23.05 | 30.38 | 32.01 | 1.07 | 0.92 | 0.11 | <0.01 | 2.47 | <0.01 |
| 2 | 180 | 2 | 82.96 | 16.72 | 33.35 | 32.87 | 1.45 | 1.11 | 0.14 | <0.01 | 1.45 | <0.01 |
| 3 | 180 | 8 | 73.55 | 9.42 | 35.56 | 35.44 | 5.53 | 0.77 | 0.25 | <0.01 | 2.71 | <0.01 |
| 4 | 200 | 0.76 | 67.56 | 2.31 | 37.72 | 35.97 | 10.38 | 1.04 | 0.34 | <0.01 | 3.43 | <0.01 |
| 5 | 200 | 5 | 62.47 | 0.00 | 37.76 | 40.04 | 18.99 | 0.87 | 0.49 | <0.01 | 3.44 | <0.01 |
| 6 | 200 | 5 | 63.55 | 1.03 | 35.97 | 38.96 | 21.83 | 0.97 | 0.62 | <0.01 | 3.34 | <0.01 |
| 7 | 200 | 5 | 62.32 | 0.00 | 36.93 | 40.98 | 19.31 | 0.85 | 0.50 | <0.01 | 3.77 | <0.01 |
| 8 | 200 | 9.24 | 60.52 | 0.00 | 38.62 | 42.07 | 13.90 | 0.32 | 0.42 | <0.01 | 4.12 | 0.03 |
| 9 | 220 | 2 | 58.45 | 0.00 | 33.20 | 44.20 | 8.17 | 0.43 | 0.24 | 0.24 | 4.59 | 0.12 |
| 10 | 220 | 8 | 56.78 | 0.00 | 32.02 | 45.87 | 4.57 | 0.21 | 0.23 | 1.46 | 5.77 | 0.80 |
| 11 | 228 | 5 | 56.54 | 0.00 | 25.10 | 48.64 | 2.81 | 0.12 | 0.17 | 1.63 | 5.29 | 1.07 |

[a] Chemical composition of 100 g of olive stones: 29.9 ± 1.1 g cellulose, 28.1 ± 1.7 g hemicellulose (of which 20.6 ± 1.1 g xylans and 4.0 ± 0.2 g acetyl groups), 27.7 ± 2.1 g acid-insoluble lignin, 2.1 ± 0.3 g acid-soluble lignin, 6.0 ± 0.3 g extractives, and 0.7 ± 0.0 g ash. TGR: total gravimetric recovery; Hem: hemicelluloses; Cel: cellulose; AIL: acid insoluble lignin; AA: acetic acid; 5-HMF: 5-hydroxymethylfurfural.

In relation to liquid prehydrolysates, product yields were strongly influenced by reaction conditions. D-xylose was the most abundant monosaccharide in the liquid phase, reaching a maximum experimental yield of 20.04 ± 1.56 g/100 g dry raw material under the conditions of 200 °C—5 min (Table 2), which represents 85.6% of the potential D-xylose in the biomass. The decrease in performance for more severe conditions would be explained by the thermal degradation of the monosaccharide. The maximum recoveries of L-arabinose (1.11 g/100 g dry raw material) and D-galactose (0.537 ± 0.072 g/100 g dry raw material) were achieved at low severities, while the maximum yield of D-glucose (1.63 g/100 g dry raw material) was reached at the maximum temperature assayed (228 °C, run 11), in which there was an intense hydrolysis of the cellulosic fraction. Apart from carbohydrates, certain compounds that can act as inhibitors in fermentation processes, such as acetic acid and HMF, were found in the liquid prehydrolysates. The maximum yields of acetic acid (5.77 g/100 g dry raw material) and 5-HMF (1.07 g/100 g dry raw material) were achieved in the experiments carried out at the highest temperatures; in the first case, from the hydrolysis of acetyl groups of the hemicellulose and, in the second one, as a consequence of the thermal degradation of D-glucose.

### 3.2. Dilute Acid Pretreatment: Modelling and Optimization

The application of response surface methodology (RSM) can lead to mathematical models that describe the modification of the composition of the solid residue and the liquid hydrolysate according to the studied independent variables: Temperature ($X_1$) and reaction time ($X_2$). With this aim, the mathematical principles described in Section 2.3 were applied to the experimental data (Table 2), obtaining the values included in Tables 3 and 4 in terms of normalized values. Data on percentages of total solids solubilization ($x_{total\ solids}$), hemicellulose conversion ($x_{hemicellulose}$) and cellulose conversion ($x_{cellulose}$) were obtained by relating the weight of each material removed during the acid pretreatment (total solids, hemicellulose and cellulose, respectively) to the weight of each material available in the raw material, and multiplying the result by one hundred. All equations were validated using the ANOVA test using the MODDE software. The $R^2$ values obtained for the seven equations varied between 0.925 and 0.999 (Tables 3 and 4), indicating that the models explain between 92.5% and 99.9% of the variability contained in the responses.

**Table 3.** Acid-pretreated solids: Estimated effects (EE), standard deviations (SD), and significance level ($p$) for the models representing total solids solubilisation ($x_{total\ solids}$), hemicellulose conversion ($x_{hemicellulose}$), cellulose conversion ($x_{cellulose}$), and acid insoluble lignin percentages (*AIL*).

| Response Variable | Coefficient | EE | SD | $p$-Value (Prob > F) | $R^2$ | $R_{adjust}^2$ |
|---|---|---|---|---|---|---|
| $x_{total\ solids}$, % | Constant | 37.220 | ±0.290 | $5.460 \times 10^{-10}$ | 0.999 | 0.996 |
| | $X_1$ | 10.462 | ±0.178 | $2.674 \times 10^{-8}$ | | |
| | $X_2$ | 2.630 | ±0.178 | $2.547 \times 10^{-5}$ | | |
| | $X_1 \cdot X_1$ | −4.415 | ±0.212 | $4.673 \times 10^{-6}$ | | |
| | $X_2 \cdot X_2$ | −0.667 | ±0.212 | $2.532 \times 10^{-2}$ | | |
| | $X_1 \cdot X_2$ | −1.935 | ±0.251 | $5.891 \times 10^{-4}$ | | |
| $x_{hemicellulose}$, % | Constant | 98.430 | ±1.964 | $4.238 \times 10^{-9}$ | 0.978 | 0.964 |
| | $X_1$ | 21.829 | ±1.653 | $1.166 \times 10^{-5}$ | | |
| | $X_2$ | 4.075 | ±1.653 | $4.882 \times 10^{-2}$ | | |
| | $X_1 \cdot X_1$ | −16.978 | ±1.881 | $1.036 \times 10^{-4}$ | | |
| | $X_1 \cdot X_2$ | −6.183 | ±2.338 | $3.831 \times 10^{-2}$ | | |
| $x_{cellulose}$, % | Constant | 22.557 | ±0.770 | $1.051 \times 10^{-7}$ | 0.994 | 0.990 |
| | $X_1$ | 13.938 | ±0.472 | $9.964 \times 10^{-8}$ | | |
| | $X_2$ | 2.391 | ±0.472 | $2.289 \times 10^{-3}$ | | |
| | $X_1 \cdot X_1$ | 4.445 | ±0.562 | $2.159 \times 10^{-4}$ | | |
| | $X_2 \cdot X_2$ | −2.565 | ±0.562 | $3.819 \times 10^{-3}$ | | |
| *AIL*, % | Constant | 39.732 | ±0.283 | $7.366 \times 10^{-15}$ | 0.975 | 0.969 |
| | $X_1$ | 5.660 | ±0.332 | $1.411 \times 10^{-7}$ | | |
| | $X_2$ | 1.608 | ±0.332 | $1.272 \times 10^{-3}$ | | |

$X_1$: temperature (in coded form), $X_2$: time (in coded form). Significance level was defined as $p < 0.05$.

**Table 4.** Acid prehydrolysates: Estimated effects (EE), standard deviations (SD), and significance level ($p$) for the models representing D-xylose ($Y_{xyl}$), L-arabinose ($Y_{ara}$) and acetic acid ($Y_{AA}$) yields.

| Responses | | Factors | | | Regression | | |
|---|---|---|---|---|---|---|---|
| | | EE | SD | $p$-Value (Prob > F) | $R^2$ | $R_{adjust}^2$ | $p$-Value |
| $Y_{xyl}$, % | Constant | 20.044 | ±1.083 | $8.478 \times 10^{-6}$ | 0.970 | 0.939 | 0.001 |
| | $X_1$ | 1.028 | ±0.663 | $1.820 \times 10^{-1}$ | | | |
| | $X_2$ | 0.682 | ±0.663 | $3.509 \times 10^{-1}$ | | | |
| | $X_1 \cdot X_1$ | −9.581 | ±0.790 | $6.717 \times 10^{-5}$ | | | |
| | $X_2 \cdot X_2$ | −4.480 | ±0.790 | $2.368 \times 10^{-3}$ | | | |
| | $X_1 \cdot X_2$ | −1.920 | ±0.938 | $9.602 \times 10^{-2}$ | | | |
| $Y_{ara}$, % | Constant | 0.897 | ±0.048 | $1.512 \times 10^{-6}$ | 0.967 | 0.945 | 0.000 |
| | $X_1$ | −0.296 | ±0.029 | $5.496 \times 10^{-5}$ | | | |
| | $X_2$ | −0.197 | ±0.029 | $5.294 \times 10^{-4}$ | | | |
| | $X_1 \cdot X_1$ | −0.181 | ±0.035 | $2.069 \times 10^{-3}$ | | | |
| | $X_2 \cdot X_2$ | −0.101 | ±0.035 | $2.791 \times 10^{-2}$ | | | |
| $Y_{AA}$, % | Constant | 3.671 | ±0.115 | $1.029 \times 10^{-9}$ | 0.925 | 0.906 | 0.000 |
| | $X_1$ | 1.274 | ±0.135 | $1.321 \times 10^{-5}$ | | | |
| | $X_2$ | 0.427 | ±0.135 | $1.341 \times 10^{-2}$ | | | |

$X_1$: temperature (in coded form), $X_2$: time (in coded form). Significance level was defined as $p < 0.05$.

Figure 1 shows the response surfaces and corresponding contour plots built from data in Tables 3 and 4. In general, mathematical models show that total solids solubilization, hemicellulose, and cellulose conversions, as well as the percentage of AIL in the pretreated solids, increased with the increase in the severity of pretreatment (temperature and reaction time, Figure 1). Notwithstanding, for $x_{total\ solids}$ and, mainly, for $x_{hemicellulose}$ a stabilization of the conversions was observed at relatively high temperatures and reaction times. Thus, Figure 1B shows that total conversion of the hemicellulose fraction was achieved at around 200 °C, the temperature exerting an effect on the response greater than that of the reaction time.

In relation to the main obtained products in the acid prehydrolysate, the response surfaces for D-xylose yield (Figure 1E) and L-arabinose yield (Figure 1F) presented maximum values within the region studied. Partial differentiation of the multivariate functions $Y_{xyl} = f(X_1, X_2)$ and $Y_{ara} = f(X_1, X_2)$ was carried out to determinate the values of temperature and time that provide these maximums. The predicted values were 201 °C and 5.2 min, with a response corresponded to 20.1 ± 2.8 g D-xylose per 100 g dry raw material (85.9% D-xylose extraction), and 183.6 °C and 2.08 min, with a response of 1.11 g L-arabinose per 100 g dry raw material. The D-xylose extraction was similar to that achieved in rice straw (80.8%) by other authors, the most suitable conditions to depolymerize xylans into xylose being 201 °C, 10 min retention time and 0.5% sulfuric acid concentration [24]. Figure 1G shows that the highest temperature and reaction time assayed (220 °C and 8 min, respectively) were the best conditions for acetic acid recovery.

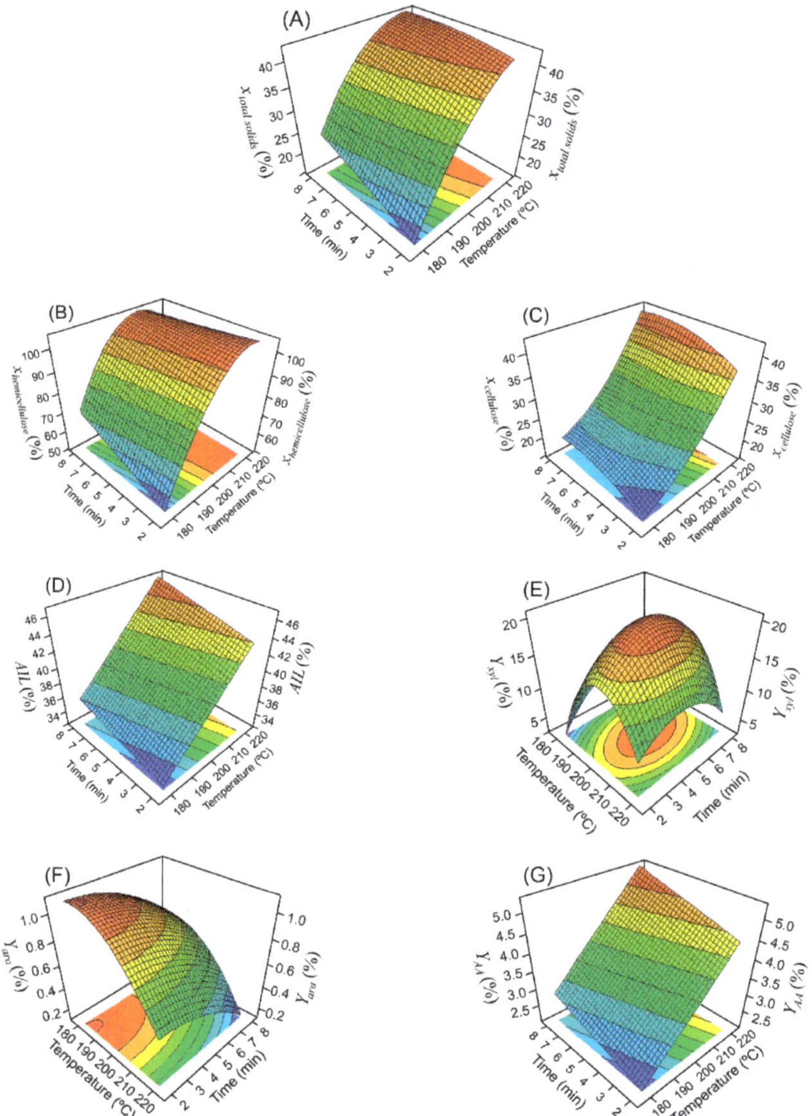

**Figure 1.** Response surfaces and contour plots for (**A**) total solids solubilization, (**B**) hemicellulose conversion, (**C**) cellulose conversion, (**D**) acid-insoluble lignin percentage, (**E**) D-xylose yield, (**F**) L-arabinose yield, and (**G**) acetic acid yield as a function of reaction temperature (°C) and reaction time (min) at fixed acid concentration of 0.010 M.

### 3.3. Fermentation of Acid Prehydrolysates

Figure 2 and Table 5 show the effect of the inoculum concentration on the fermentation of the acid prehydrolysate obtained under the conditions that maximized D-xylose recovery (201 °C—5.2 min), which was previously subjected to vacuum distillation until achieving the following composition (g/dm$^3$): 14.45 D-xylose, 0.73 L-arabinose, 0.28 D-galactose, and 2.23 acetic acid. D-glucose was not detected at the beginning of the fermentation stage. It is worth noting that *P. tannophilus* yeast completely uptook both D-xylose and acetic acid, although uptake rates depended strongly on the inoculum concentration. Thus, D-xylose was almost depleted in prehydrolysates after 144 h and 48 h for fermentations carried out with initial biomass concentrations of 0.5 g/dm$^3$ and 4.0 g/dm$^3$, respectively (Figure 2). In relation to acetic acid, this compound was completely uptaken within 72 h in all the fermentations. This demonstrate the low inhibition exerted by the medium on *P. tannophilus*, which is also evident when analyzing the biomass growth data over fermentations. In this sense, there was only a lag phase (about 10 h) in the bioreactor with the lowest initial concentration of microorganism (Figure 2A). The increase in the initial inoculum concentration from 0.5 g/dm$^3$ to 4.0 g/dm$^3$ caused a continuous decrease (from 4.8 g/dm$^3$ to 1.9 g/dm$^3$) in the net biomass production (Table 5). Regarding the production of ethanol and xylitol, it was observed that the latter compound was the main product of cellular metabolism except for fermentations performed with initial inoculum of 4.0 g/dm$^3$. Thus, the maximum concentrations of ethanol and xylitol achieved were 0.25 g/dm$^3$ and 4.81 g/dm$^3$, respectively, for the inoculum of 0.5 g/dm$^3$, and 1.8 g/dm$^3$ and 1.5 g/dm$^3$, respectively, for the inoculum of 4.0 g/dm$^3$ (Table 5). Therefore, the fermentations with the lowest inoculum concentrations were those that led to the highest yields and volumetric xylitol productivity (0.42 g/g and 0.07 g/dm$^3$·h), respectively), which were obtained at 72 h, while the inoculum of 4 g/dm$^3$ led to the highest yield of ethanol (0.17 g/g D-xylose). When comparing these results with those obtained by Saleh et al. [16] it can be pointed out that the decrease in the concentration of sulfuric acid from 0.025 M to 0.010 M in the pretreatment stage results in prehydrolysates with a lower capacity to inhibit *P. tannophilus*. These sugar media allow reaching, for a fixed inoculum concentration, higher concentrations of ethanol and lower biomass productions. With regards to xylitol, both prehydrolysates reached similar maximum yields (0.42 g/g in this work; 0.44 g/g in the research of Saleh et al. [16]).

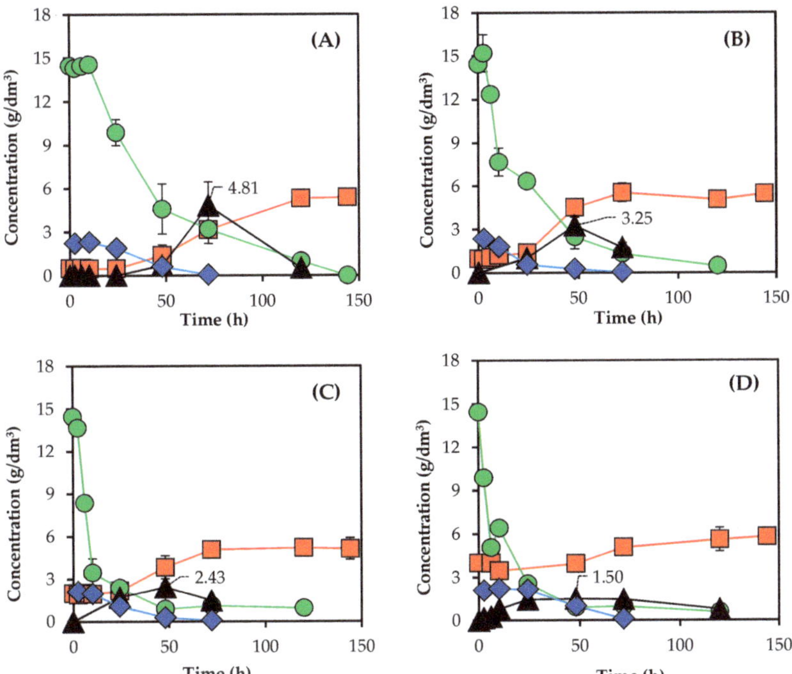

**Figure 2.** Acid prehydrolysate: Effect of inoculum concentration ((**A**), 0.5 g/dm$^3$; (**B**), 1.0 g/dm$^3$; (**C**), 2.0 g/dm$^3$; (**D**), 4.0 g/dm$^3$) on D-xylose (•) and acetic acid (♦) consumption, and biomass (■) and xylitol (▲) production by *P. tannophilus* at 30 °C and pH 4.5.

**Table 5.** Maximum parameters of xylitol, ethanol and biomass production by *P. tannophilus* from hemicellulose hydrolysates obtained by sulfuric-acid hydrolysis of olive stones. Effect of inoculum concentrations.

| Starting Inoculum Concentration (g/dm$^3$) | 0.5 | 1.0 | 2.0 | 4.0 |
| --- | --- | --- | --- | --- |
| Net biomass concentration (g/dm$^3$) | 4.8 ± 0.2 (120 h) [1] | 4.6 ± 0.7 (72 h) | 3.2 ± 0.1 (72 h) | 1.9 ± 0.2 (72 h) |
| Ethanol concentration (g/dm$^3$) | 0.25 ± 0.00 (10 h) | 0.41 ± 0.04 (10 h) | 0.97 ± 0.10 (10 h) | 1.8 ± 0.1 (10 h) |
| Xylitol concentration (g/dm$^3$) | 4.81 ± 1.63 (72 h) | 3.25 ± 0.59 (48 h) | 2.43 ± 0.30 (48 h) | 1.50 ± 0.05 (48 h) |
| Xylitol yield [2] (g/g) | 0.42 ± 0.08 (72 h) | 0.26 ± 0.08 (48 h) | 0.18 ± 0.01 (48 h) | 0.14 ± 0.02 (48 h) |
| Xylitol volumetric productivity (g/dm$^3$·h) | 0.07 ± 0.01 (72 h) | 0.07 ± 0.02 (48 h) | 0.05 ± 0.01 (48 h) | 0.04 ± 0.00 (48 h) |

[1] Culture time, at which the parameter was calculated, is shown in brackets. [2] Based on consumed D-xylose.

### 3.4. Enzymatic Hydrolysis of Pretreated Solids

The acid pretreatment carried out at 201 °C—5.2 min on the olive stones led to a solid without hemicellulose and rich in cellulose (35.2%) and insoluble acid lignin (40.0%). To study the enzymatic digestibility of pretreated cellulose, enzymatic hydrolyses were carried out with Celluclast 1.5 L using the following enzyme loadings: 10, 20, 40, and 60 FPU/g pretreated solid. The yield in D-glucose, expressed as grams of monosaccharide generated per gram of pretreated solid, over time is shown in Figure 3, showing that the increase in enzyme loading led to the increase in D-glucose yield. Thus, the yields of D-glucose for enzyme loadings of 10, 20, 40, and 60 FPU/g solid were 0.131, 0.137, 0.316, and 0.342 g D-glucose per gram of solid, respectively, at 120 h of reaction, which are equivalent to values of 0.335, 0.350, 0.808, and 0.875 g D-glucose per gram of potential D-glucose in pretreated cellulose. The above data illustrate the capacity of Celluclast 1.5L to hydrolyze above 80% of pretreated cellulose, although high cellulases loadings are required for this. These data could prove that the pretreatment is capable of considerably increasing the porosity of the solid and, therefore, the accessibility of the enzyme to the pretreated cellulose, although high catalyst loadings are necessary to compensate for the losses caused by the adsorption of protein on the pretreated lignin. Fernandez et al. achieved 83% glucan conversion from extracted olive pomace that was previously subjected to autohydrolysis at 230 °C [25].

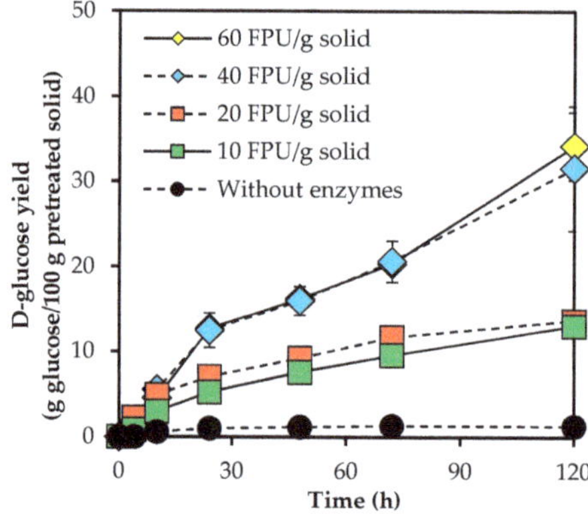


**Figure 3.** Enzymatic digestibility of acid-pretreated olive stones (201 °C, 5.2 min) at different enzyme loadings.

### 3.5. Fermentation of Enzymatic Hydrolysates

The enzymatic hydrolysate obtained in Section 3.4, using an enzyme loading of 40 FPU/g pretreated solid, was diluted with water up to achieve a D-glucose concentration of 20.6 g/dm$^3$ in order to ferment it with *P. tannophilus*. The evolution over time of the fermentations carried out with four inoculum levels (0.5, 1.0, 1.5, and 3.0 g/dm$^3$) is shown in Figure 4. The absence of fermentative inhibitors caused D-glucose uptake to be completed within 24 first hours for the inoculum of 0.5 g/dm$^3$, and in around 10 h for the rest of inocula. In these fermentations, the yeast generated ethanol as the main product along with a low biomass production. Thus, for initial yeast concentrations of 0.5, 1.0, 1.5, and 3.0 g/dm$^3$, the final biomass concentration were 1.51, 1.85, 2.75, and 4.18 g/dm$^3$, respectively. The maximum ethanol concentrations detected for the inocula 0.5, 1.0, 1.5, and 3.0 g/dm$^3$ were 9.6, 10.1, 10.8, and 9.7 g/dm$^3$, respectively, resulting in ethanol yields of 0.464, 0.491, 0.523, and 0.472 g ethanol per g D-glucose, respectively, i.e., values close to the stoichiometric maximum.

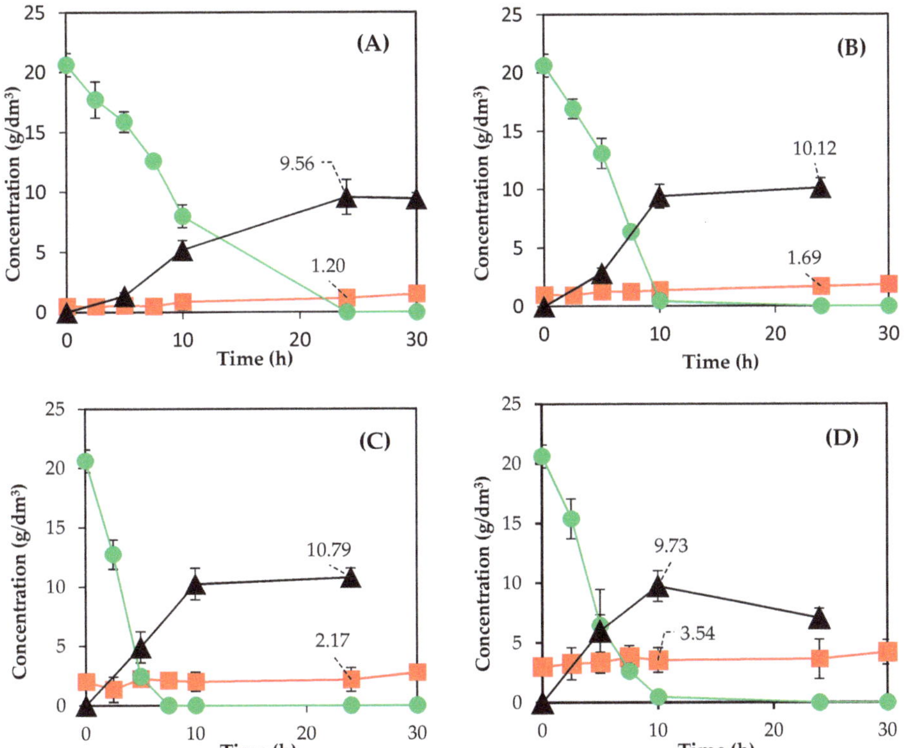

**Figure 4.** Effect of inoculum concentration ((**A**), 0.5 g/dm$^3$; (**B**), 1.0 g/dm$^3$; (**C**), 1.5 g/dm$^3$; (**D**), 3.0 g/dm$^3$) on D-glucose consumption (•), and biomass (■) and ethanol (▲) production by *P. tannophilus* at 30 °C and pH 4.5.

### 3.6. Mass Macroscopic Balance for Complete Process

Figure 5 shows the mass balance for the complete ethanol production process developed in this work. When 100 g of olive stones were pretreated at 201 °C—5.2 min with 0.010 M sulfuric acid, a liquid prehydrolysate was obtained with the maximum recovery of D-xylose achieved in this work (20.0 g, equivalent to 85.6% monosaccharide recovery) along with a hemicellulose-free solid residue, rich in acid insoluble lignin and cellulose. The enzymatic hydrolysis of the pretreated solid led to high amounts of D-glucose (19.8 g), which were easily metabolized by *P. tannophilus*, rendering 9.7 g ethanol. With regards to the liquid prehydrolysate, the previous vacuum distillation to concentrate fermentable sugars allowed *P. tannophilus* to ferment them into ethanol or xylitol. This fermentative stage was strongly influenced, both on its duration and on the production of ethanol and xylitol, by the initial yeast concentration so that two alternative schemes could be considered. In the first scheme (option A, Figure 5), using an initial inoculum concentration of 4 g/dm$^3$, similar amounts of ethanol (2.50 g) and xylitol (2.09 g) would be obtained after 48 h fermentation. In the scheme B an inoculum of 0.5 g/dm$^3$ would be used, and a much richer medium in xylitol (6.69 g) than in ethanol (0.35 g) would be obtained after 72 h fermentation, 4.46 g D-xylose remaining in the fermentation culture. Although the first scheme would generate a total of 12.2 g ethanol per 100 of olive stones, in the second scheme the lower production of ethanol (10.1 g/100 g olive stones) could be compensated with an important production of xylitol, which could reach 8.42 g if the whole D-xylose present in the fermentation medium were used.

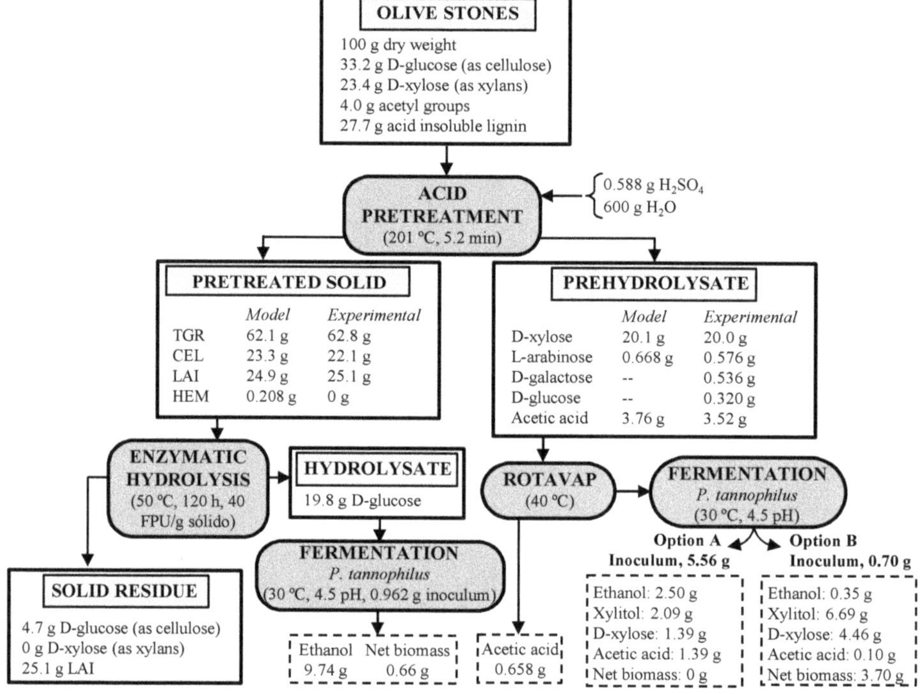


**Figure 5.** Mass macroscopic balance for the ethanol production flowsheet proposed: Acid pretreatment of olive stones, enzymatic hydrolysis of pretreated solids, detoxification with rotary evaporator and fermentation of hydrolysates using *P. tannophilus*.

## 4. Conclusions

The proposed flowsheet for the fractionation of the olive stones led to a suitable valorization of their hemicellulose fraction. In this sense, the application of a response surface methodology to the acid hydrolysis stage led to high D-xylose recovery into the liquid prehydrolysate, which could be fermented into ethanol and xylitol using the non-traditional yeast *P. tannophilus*. The production of these compounds was strongly influenced by the initial concentration of inoculum in the fermentation stage, and fermentation conditions that led to high xylitol production were found. To be specific, starting from a yeast concentration of 0.5 g/dm$^3$ each gram of D-xylose consumed by *P. tannophilus* was transformed into 0.42 g of xylitol. In relation to the pretreated solids, these materials led to hydrolysates rich in D-glucose (35 g/dm$^3$) when high loadings of cellulases were used, i.e., 40 FPU/g pretreated solid. Therefore, the enzymatic hydrolysis stage still remains to be upgraded in order to reduce operating costs and thus enhance the feasibility of the overall process.

**Author Contributions:** M.C. performed some experiments and the analysis of the data, and contributed to the aspects related to the design of figures and writing the initial draft paper; M.S. performed some experiments; J.F.G.-M. performed the English translation, text and figures formatting and revision of the paper; S.S. provided the funding, performed the experimental design, and contributed to the revision of the paper. All authors have read and agreed to the published version of the manuscript.

**Funding:** This work was funded the projects 01272/2005 and AGR/6509 granted by Andalusia Regional Government (Spain).

**Acknowledgments:** Marwa Saleh wish to thank her grant from the AECI (Spain). The authors also thank to Novo Nordisk Bioindustrial (Madrid, Spain) and 'S.C.A. San Juan' (Jaén, Spain) for providing the enzyme preparation and the olive stones, respectively, used throughout this research.

**Conflicts of Interest:** The authors declare no conflict of interest.

## References

1. Dhillon, R.S.; Wuehlisch, G.V. Mitigation of global warming through renewable biomass. *Biomass Bioenergy* **2013**, *48*, 75–89. [CrossRef]
2. Balat, M. Production of bioethanol from lignocellulosic materials via the biochemical pathway: A review. *Energy Convers. Manage.* **2011**, *52*, 858–875. [CrossRef]
3. Prasad, R.K.; Chatterjee, S.; Mazumder, P.B.; Gupta, S.K.; Sharma, S.; Vairale, M.G.; Datta, S.; Dwivedi, S.K.; Gupta, D.K. Bioethanol production from waste lignocelluloses: A review on microbial degradation potential. *Chemosphere* **2019**, *231*, 588–606. [CrossRef] [PubMed]
4. Bhatia, S.K.; Jagtap, S.S.; Bedekar, A.A.; Bhatia, R.K.; Patel, A.K.; Pant, D.; Banu, R.; Rao, C.V.; Kim, Y.-G.; Yang, Y.-H. Recent development in pretreatment technologies on lignocellulosic biomass: Effect of key parameters, technological improvements, and challenges. *Bioresour. Technol* **2020**, 121936. [CrossRef] [PubMed]
5. Kumar, P.; Barret, D.M.; Delwiche, M.J.; Stroeve, P. Methods for pretreatment of lignocellulosic biomass for efficient hydrolysis and biofuel production. *Ind. Eng. Chem. Res.* **2009**, *48*, 3713–3729. [CrossRef]
6. Cuevas, M.; García, J.F.; Sánchez, S. Enhanced enzymatic hydrolysis of pretreated almond-tree prunings for sugar production. *Carbohydr. Polym.* **2014**, *99*, 791–799. [CrossRef] [PubMed]
7. Cao, L.; Chen, H.; Tsang, D.C.W.; Luo, G.; Hao, S.; Zhang, S.; Chen, J. Optimizing xylose production from pinewood sawdust through dilute-phosphoric-acid hydrolysis by response surface methodology. *J. Clean. Prod.* **2018**, *178*, 572–579. [CrossRef]
8. Jang, S.-K.; Kim, J.-H.; Jeong, H.; Choi, J.-H.; Lee, S.-M.; Choi, I.-G. Investigation of conditions for dilute acid pretreatment for improving xylose solubilization and glucose production by supercritical water hydrolysis from *Quercus mongolica*. *Renew. Energy* **2018**, *117*, 150–156. [CrossRef]
9. García-Martín, J.F.; Sánchez, S.; Cuevas, M. Evaluation of the effect of the dilute acid hydrolysis on sugars release from olive prunings. *Renew. Energy* **2013**, *51*, 382–387. [CrossRef]
10. Sánchez, S.; Bravo, V.; Moya, A.J.; Castro, E.; Camacho, F. Influence of temperature on the fermentation of D-xylose by *Pachysolen tannophilus* to produce ethanol and xylitol. *Process Biochem.* **2004**, *39*, 673–679. [CrossRef]
11. Cuevas, M.; Sánchez, S.; García, J.F. Thermochemical and biochemical conversion of olives stones. In *Agricultural Wastes: Characteristics, Types and Management*; Foster, C.N., Ed.; Nova Science Publishers: New York, NY, USA, 2015; pp. 61–86.
12. Miranda, I.; Simões, R.; Medeiros, B.; Nampoothiri, K.M.; Sukumaran, R.K.; Rajan, D.; Pereira, H.; Ferreira-Dias, S. Valorization of lignocellulosic residues from the olive oil industry by production of lignin, glucose and functional sugars. *Bioresour. Technol.* **2019**, *292*, 121936. [CrossRef] [PubMed]
13. Cuevas, M.; Sánchez, S.; Bravo, V.; Cruz, N.; García, J.F. Fermentation of enzymatic hydrolysates from olives stones by *Pachysolen tannophilus*. *J. Chem. Technol. Biotechnol.* **2009**, *84*, 461–467. [CrossRef]

14. Cuevas, M.; García, J.F.; Hodaifa, G.; Sánchez, S. Oligosaccharides and sugars production from olive stones by autohydrolysis and enzymatic hydrolysis. *Ind. Crops Prod.* **2015**, *70*, 100–106. [CrossRef]

15. Romero-García, J.M.; Martínez-Patiño, C.; Ruiz, E.; Romero, I.; Castro, E. Ethanol production from olive stone hydrolysates by xylose fermenting microorganisms. *Bioethanol* **2016**, *2*, 51–65. [CrossRef]

16. Saleh, M.; Cuevas, M.; García, J.F.; Sánchez, S. Valorization of olive stones for xylitol and ethanol production from dilute acid pretreatment via enzymatic hydrolysis and fermentation by *Pachysolen tannophilus*. *Biochem. Eng. J.* **2014**, *90*, 286–293. [CrossRef]

17. Sánchez, S.; Bravo, V.; Castro, E.; Moya, A.J.; Camacho, F. The fermentation of mixtures of D-glucose and D-xylose by *Candida shehatae*, *Pichia stipitis* or *Pachysolen tannophilus* to produce ethanol. *J. Chem. Technol. Biotechnol.* **2002**, *77*, 641–648. [CrossRef]

18. Van Soest, P.J.; Wine, R.H. Use of detergent in the analysis of fibrous feeds. IV. Determination of plant cell-wall constituents. *J. Assoc. Off. Anal. Chem.* **1967**, *50*, 50–55.

19. Sluiter, A.; Hames, B.; Ruiz, R.; Scarlata, C.; Sluiter, J.; Templeton, D.; Crocker, D. Determination of Structural Carbohydrates and Lignin in Biomass. In *Laboratory Analytical Procedure (LAP)*; NREL/TP-510-42618; National Renewable Energy Laboratory: Golden, CO, USA, 2012.

20. Puls, J.; Poutanen, K.; Körner, H.; Viikari, L. Biotechnical utilization of wood carbohydrates after steaming pretreatment. *Appl. Microbiol. Biotechnol.* **1985**, *22*, 416–423. [CrossRef]

21. Beutler, H.O. Ethanol. In *Methods of Enzymatic Analysis*; Bergmeyer, H., Ed.; Academic Press: New York, NY, USA, 1974; pp. 594–606.

22. Beutler, H.O.; Becker, J. Enzymatische bestimmung von D-sorbit und xylit in lebensmitteln. *Dtsch. Lebensmitt. Rundsch.* **1977**, *6*, 182–187.

23. Bergmeyer, H.; Möllering, H. Acetic acid. In *Methods of Enzymatic Analysis*; Bergmeyer, H., Ed.; Academic Press: New York, NY, USA, 1974; pp. 1520–1528.

24. Karimi, K.; Kheradmandinia, S.; Taherzadeh, M.J. Conversion of rice straw to sugars by dilute-acid hydrolysis. *Biomass Bioenergy* **2006**, *30*, 247–253. [CrossRef]

25. Fernandes, M.C.; Brás, T.S.; Lourenço, P.M.L.; Duarte, L.C.; Carvalheiro, F.; Bernardo, P.; Marques, S.; Neves, L.A. The effect of autohydrolysis pretreatment on the production of second generation bioethanol from extracted olive pomace. In Proceedings of the 4th International Conference on Engineering for Waste and Biomass Valorisation, Porto, Portugal, 10–13 September 2012.

*Article*

# Esterification of Free Fatty Acids with Glycerol within the Biodiesel Production Framework

**Juan Francisco García Martín** [1,*], **Javier Carrión Ruiz** [1], **Miguel Torres García** [2],
**Chao-Hui Feng** [3] **and Paloma Álvarez Mateos** [1]

1    Departamento de Ingeniería Química, Facultad de Química, Universidad de Sevilla, C/Profesor García
     González, 1, 41012 Seville, Spain; carriparadas@hotmail.com (J.C.R.); palvarez@us.es (P.Á.M.)
2    Departamento de Ingeniería Energética, E.T.S. de Ingeniería, Universidad de Sevilla, Camino de los
     Descubrimientos, s/n, 41092 Seville, Spain; migueltorres@us.es
3    RIKEN Centre for Advanced Photonics, RIKEN, 519-1399 Aramaki-Aoba, Aoba-ku, Sendai 980-0845, Japan;
     chaohui.feng@riken.jp
*    Correspondence: jfgarmar@us.es; Tel.: +34-954-55-71-83

Received: 18 October 2019; Accepted: 5 November 2019; Published: 8 November 2019

**Abstract:** Companies in the field of the collection and treatment of waste cooking oils (WCO) for subsequent biodiesel production usually have to cope with high acidity oils, which cannot be directly transformed into fatty acid methyl esters due to soap production. Since glycerine is the main byproduct of biodiesel production, these high acidity oils could be esterified with the glycerine surplus to transform the free fatty acids (FFA) into triglycerides before performing the transesterification. In this work, commercial glycerol was esterified with commercial fatty acids and commercial fatty acid/lampante olive oil mixtures over tin (II) chloride. In the first set of experiments, the esterification of linoleic acid with glycerol excess from 20 to 80% molar over the stoichiometric was performed. From 20% glycerol excess, there was no improvement in FFA reduction. Using 20% glycerol excess, the performance of a biochar obtained from heavy metal-contaminated plant roots was compared to that of $SnCl_2$. Then, the effect of the initial FFA content was assessed using different oleic acid/lampante olive oil mixtures. The results illustrated that glycerolysis was impeded at initial FFA contents lower than 10%. Finally, the glycerolysis of a WCO with 9.94% FFA was assayed, without success.

**Keywords:** biodiesel; esterification; free fatty acids; glycerol; waste cooking oil

---

## 1. Introduction

Vegetable oils undergo numerous physical and chemical alterations during frying due the exposure to high temperatures (160–200 °C) and the presence of oxygen and water. One of these alterations is the increase in acidity in the resulting waste cooking oil (WCO), provoked by the release of free fatty acids (FFA) from the partial hydrolysis of triglycerides.

Waste cooking oils are regarded as an alternative to raw vegetable oils for biodiesel production due to the high cost of the vegetable oils and the threat to food security [1,2]. This biodiesel is typically obtained by transesterification, which involves the reaction of triglycerides with a short chain alcohol (generally methanol because of its low cost) in the presence of an alkaline catalyst (mainly sodium hydroxide) to render fatty acid methyl esters (FAME), which are ultimately biodiesel [2].

Companies responsible for the collection, storage, and treatment of WCO usually have to deal with oils with high acidity, which cannot be used as raw material for subsequent biodiesel production [3,4]. This is because free fatty acids react with the catalyst of the transesterification reaction, rendering soaps (saponification reaction, Equation (1)). The formation of soap significantly reduces the process efficiency, resulting in low biodiesel yields, hence the need to overcome this problem.

$$R - COOH + NaOH \rightarrow R - COONa + H_2O. \tag{1}$$

One proposed solution is the previous esterification of the FFA with methanol to obtain FAME [5], i.e., biodiesel, using an acid catalyst (usually sulfuric acid), as shown in Equation (2). This reaction is reversible, so an excess of water can displace the equilibrium towards the formation of FFA [6], thus water should be removed during the process. The most suitable conditions found by several authors for this procedure when treating oils with high acidity were 5% $H_2SO_4$ (wt.) as catalyst [4,7,8], 15:1 methanol to WCO molar ratio [4], 160 °C [4,8] and 2 h reaction time [4,8]. Under these conditions, the acidity of a WCO was reduced from 60.5% to 1% FFA. Notwithstanding this, the problem is the further transesterification of the triglycerides, which is not only affected by the presence of water (hydrolysis of triglycerides) but is also inhibited by the presence of high amounts of the product of this reaction, i.e., FAME. As result, the previous esterification drastically reduces the efficiency of the subsequent transesterification, leading to incomplete triglycerides conversion [4].

$$CH_3 - OH + R - COOH \leftrightarrow R - COO - CH_3 + H_2O. \tag{2}$$

Another alternative consists of firstly hydrolyzing the triglycerides with an enzyme (lipase) in aqueous medium to obtain FFA (Equation (3)), and then esterifying the resulting FFA with methanol, in the presence of an acid catalyst, into FAME. Lipases from the yeast *Candida rugose* [4] and from the fungus *Rhizopus microsporus* [6] have been assayed in the hydrolysis step, none of them achieving high FFA yields.

$$\tag{3}$$

On the other hand, the increasing production of biodiesel has resulted in an oversupply of glycerine as a byproduct [9]. It is estimated that 1 kg of glycerine (crude glycerol) is produced from every 10 kg of biodiesel produced by transesterification [10]. Since 42 billion liters of biodiesel will be produced in 2020, according to projections, roughly 4.2 billion liters of glycerine will be available that year [10]. Glycerine from biodiesel production is composed of glycerol (40–70% wt.), methanol, water, salts, and soap, as well as traces of mono-, di-, and triglycerides that have not entirely reacted. Hence, one last and promising alternative to take advantage of high acidity WCO is to esterify first the FFA with glycerol (glycerolysis) using a Lewis acid catalyst to render triglycerides according to Equation (4), and then perform the transesterification [11]. This alternative is within the framework of circular economy and the biorefinery concept. Nevertheless, a previous purification stage of the glycerine is required to obtain high-purity glycerol for the glycerolysis. At present, the best glycerine purification method is a sequential physicochemical scheme based on acidification driven phase separation with phosphoric acid, glycerol extraction with propanol, and subsequent adsorption with activated carbon, the resulting glycerol purity ranging between technical grade [9] and USP [10].

$$\text{(structural reaction scheme)} \tag{4}$$

Glycerolysis has been reported to be effective solely when the free fatty acid content is high (5–60% FFA) [9]. This reaction is carried out in excess of glycerol to shift the equilibrium towards the products. The presence of water is detrimental since the glycerolysis is reversible: water can hydrolyze the formed mono-, di-, and triglycerides, rendering glycerol again [12,13].

Different inorganic catalysts have been assayed for the esterification of FFA with glycerol, including $AlC_2 \cdot 6H_2O$, $Al_2O_3$, $CdCl_2 \cdot 2H_2O$, $FeCl_3 \cdot 6H_2O$, FeO, $HgCl_2$, $MgCl_2 \cdot 6H_2O$, MgO, $MnCl_2 \cdot H_2O$, $MnO_2$, NaOH, Ni, $NiCl_2 \cdot 2H_2O$, $PbCl_2$, PbO, $SbCl_2$, $SnCl_2 \cdot 2H_2O$, $SnCl_4 \cdot 5H_2O$, $SnO_2$, $ZnCl_2$, and ZnO [14]. According to literature, tin (II) chloride dihydrate ($SnCl_2 \cdot 2H_2O$) leads to the highest glycerolysis yields, so this commercial catalyst is the most widely used [13–17]. In this sense, the FFA content of a WCO was reduced from 4.2% to 1.5% over 0.1% $SnCl_2 \cdot 2H_2O$ (relative to WCO) at 160 °C in 1 h, using a 0.2 mass ratio of crude glycerol to WCO [13]. Furthermore, the FFA content of soapstocks decreased from 50 to 5% after 3 h of glycerolysis at 200 °C [17].

Temperatures between 120 °C and 200 °C are used for glycerolysis, according to literature [15]. In spite of accelerating the reaction rate, the use of high temperatures can lead to the formation of acrolein from glycerol at 167 °C. This gaseous compound (boiling point = 53 °C) represents a serious health hazard, so it is preferable to use lower temperatures.

Finally, aiming for a greener process, the inorganic catalyst tin (II) chloride should be replaced with a natural catalyst that has not been submitted to chemical treatment. In this sense, the resulting biochars from the pyrolysis of heavy metal-contaminated roots of the plant *Jatropha curcas* L. [18] have shown a graphite-like structure and a great performance in the catalysis of similar reactions, such as glycerol esterification with acetic acid or acetic anhydride to obtain oxygenated fuel additives [10], or the esterification of FFA with methanol to render FAME [4].

The aim of this work was to study the reaction of glycerolysis over $SnCl_2$ at a relatively low temperature (160 °C) using commercial glycerol. In the first set of experiments, the ratio of FFA to glycerol was assessed, as well as the performance of a biocatalyst. Secondly, the effect of the initial FFA concentration was studied. Finally, the glycerolysis of a WCO was performed.

## 2. Materials and Methods

### 2.1. Esterification Reaction

Esterification of commercial glycerine (99.5% glycerol) from Panreac Química S.A.U. (Barcelona, Spain) with FFA was performed in a 250-cm$^3$ bath reactor equipped with a temperature controller and a water-cooled condenser, the stirring speed being set at 500 rpm. Reaction times between 60 and 120 min and a temperature of 160 °C were selected based on the findings of Yeom and Go [13]. As sources of FFA to react with glycerol, commercial linoleic acid (55% linoleic acid, 35% oleic acid) from Sigma-Aldrich Química S.L. (Madrid, Spain), commercial oleic acid (65–88% purity) from Panreac Química S.A.U. (Barcelona, Spain), a lampante virgin oil supplied by Agroalimentaria Virgen del Rocío S.C.A. (Almonte, Spain), and a waste cooking oil supplied by Grupo BIOSEL (Aznalcóllar, Spain) were assayed. Two catalysts were assayed: commercial $SnCl_2 \cdot 2H_2O$ from Panreac Química S.A.U. (Barcelona, Spain) and a biocatalyst obtained from the pyrolysis of heavy metal-contaminated roots of *Jatropha curcas* L. at 550 °C [18]. This biocatalyst has shown a similar performance in esterification reactions to commercial catalyst Amberlyst-15 [4,10]. Its BET surface, average pore diameter, and total pore volume were 346 $m^2 \cdot g^{-1}$, 4.3 nm, and 0.0446 $cm^3 \cdot g^{-1}$, respectively. The amount of catalyst added to the reaction medium was 1.08 g (2.4% wt. to linoleic acid).

Two sets of experiments were performed. First, the effect of the excess of glycerol (over the stoichiometric) over $SnCl_2 \cdot H_2O$ was assessed and a comparison between the performance of commercial catalyst and that of the biocatalyst was performed under the most favorable conditions (Table 1). For these experiments, linoleic acid was used as the source of FFA and the temperature was fixed at 160 °C.

**Table 1.** Components and incubation time of the esterification reactions of commercial linoleic acid with different amounts of glycerol at 160 °C.

| Run | Linoleic Acid (g) | Glycerol (g) | Glycerol Excess (%) | Catalyst | t (min) |
|-----|-------------------|--------------|---------------------|----------|---------|
| 1 | 45 | 4.48 | 0 | $SnCl_2 \cdot 2H_2O$ | 80 |
| 2 | 45 | 5.37 | 20 | $SnCl_2 \cdot 2H_2O$ | 80 |
| 3 | 45 | 6.27 | 40 | $SnCl_2 \cdot 2H_2O$ | 80 |
| 4 | 45 | 8.06 | 80 | $SnCl_2 \cdot 2H_2O$ | 80 |
| 5 | 45 | 5.37 | 20 | $SnCl_2 \cdot 2H_2O$ | 120 |
| 6 | 45 | 5.37 | 20 | Biocatalyst | 120 |

In the second set of experiments, different mixtures of commercial oleic acid and the lampante olive oil were used as the source of FFA for the esterification of glycerol over commercial $SnCl_2$ (Table 2). For these experiments, the temperature and reaction time were fixed to 160 °C and 60 min, respectively.

**Table 2.** Esterifications of mixtures of commercial oleic acid and lampante olive oil over $SnCl_2 \cdot 2H_2O$ at 160 °C for 60 min (mean ± SD, $n$ = 2).

| Run | Oleic Acid (%) | Lampante Oil (%) | FFA (%) |
|-----|----------------|------------------|---------|
| 7 | 100 | 0 | 99.9 ± 4.5 |
| 8 | 80 | 20 | 79.6 ± 3.8 |
| 9 | 60 | 40 | 60.0 ± 2.9 |
| 10 | 40 | 60 | 37.1 ± 2.1 |
| 11 | 20 | 80 | 20.8 ± 1.9 |
| 12 | 0 | 100 | 1.3 ± 0.5 |

Finally, the esterification of the FFA of a WCO (9.94 ± 0.13% FFA) with 20% glycerol excess over $SnCl_2 \cdot 2H_2O$ at 160 °C for 60 min was assayed. All the experiments were performed in duplicate.

### 2.2. Analytical Methods

Free fatty acids in the different sources of FFA were quantified following the UNE-EN 140140 standard, and expressed as oleic acid percentage. Briefly, 20 g WCO was placed into 250-$cm^3$ wide-mouth Erlenmeyer flasks, along with 100 $cm^3$ ethanol:diethyl ether solution (1:1 v/v) and a few drops of phenolphthalein, and then neutralized with 0.1 N KOH, previously standardized with benzoic acid. The titration ended when a reddish-brown color change was observed. Determinations were performed in duplicate.

The percentage of acidity of the oil was calculated according to the following equation:

$$\mathrm{FFA}(\%) = \frac{V \times 0.1 \mathrm{N(KOH)} \times 0.282}{m} \times 100, \tag{5}$$

where V is the spent volume of KOH in mL, 0.1 N stands for the KOH normality, 0.282 is the equivalent weight of oleic acid in meq, and m is the mass of the source of FFA in g.

The glyceride composition of the samples, before and after the esterification reactions, was analyzed by high performance size exclusion chromatography (HPSEC). This technique makes it possible to separate the compounds according to their molecular size. The elution order was as follows: polymers of triglycerides (trimers and dimers), triglycerides (TG), diglycerides (DG), monoglycerides (MG), and finally FFA. For their quantification, a liquid chromatograph Hewlett Packard 1050 working with an isocratic flow rate of 0.7 $cm^3 \cdot min^{-1}$ of tetrahydrofuran (THF) was used. The equipment was provided with a manual rheodyne injector with 20 μL loop, a column Agilent PL-gel 3 μm (size of pore 100 Å), and it was connected to a refractive index detector Merck L-7490. The sample concentration was 50 $mg \cdot cm^{-3}$ THF. The data were processed using the 32 Karat program (Beckman Coulter, Inc.). The total time of the chromatographic analysis was 15 min.

## 3. Results

### 3.1. Effect of Mass Ratio of Crude Glycerol to FFA

In the first set of experiments, esterification of glycerine with FFA was performed at 160 °C for 80 min using $SnCl_2 \cdot 2H_2O$ as the catalyst (runs 1 to 4, Table 1). This salt was selected because it is widely used as a catalyst for glycerol esterification [13,15,16]. In their study, Yeom and Go reported that the optimum conditions for glycerol esterification were 0.1% catalyst concentration, 0.2 mass ratio of crude glycerol to waste cooking oil (4.2% FFA), 160 °C, and 1 h of reaction time [13]. We selected the same optimal conditions, but extended the reaction time to 80 min (Table 1). This was because commercial linoleic acid was used as the source of FFA in our study. The content of FFA of this commercial linoleic acid was 94.4 ± 0.4%, according to our analytical determination (UNE-EN 140140 standard), far higher than the waste cooking oil assayed by Yeom and Go [13]. This led us to assess the ratio FFA to glycerol as well.

The results revealed that there was no difference in esterification efficiency for the different assayed excess glycerol amounts (Figure 1), so 20% excess glycerol was selected for further experiments. This result is similar to that found by other authors in the esterification of soapstock with 50% FFA at 220 °C [17]. As can be seen in Figure 1, the reaction mixture took about 20 min to reach the desired temperature, during which glycerol esterification took place to some extent, so the initial FFA content at time zero (beginning of the reaction at 160 °C) was lower than the original FFA content of the source of FFA (Figure 1). Obviously, the reaction rates at temperatures lower than 160 °C (between –20 and 0 min) were lower than the reaction rate at 160 °C, as is illustrated in Figure 1.

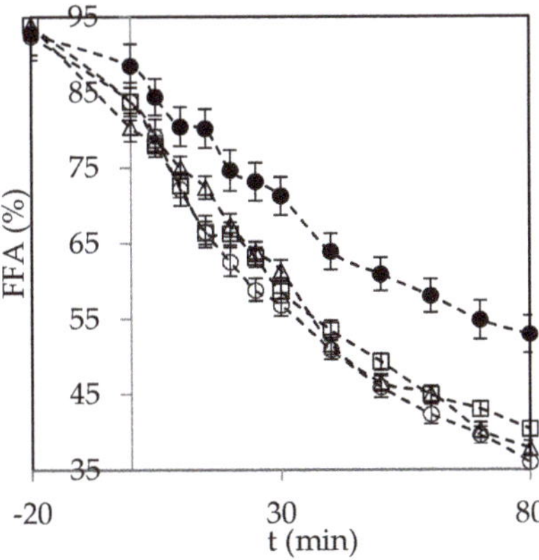

**Figure 1.** Time course of esterification of linoleic acid with stoichiometric (●), 20% excess (Δ), 40% excess (□), and 80% excess (○) glycerol (mean ± SD, $n$ = 2).

From Figure 1 it cannot be stated that 80 min was enough to reach the completion of the reaction, so a new experiment was performed with 20% excess glycerol but for 120 min (run 5, Table 1). The results obtained from this experiment revealed that the end of the reaction was close to 80 min

(Figure 2). From this time on, the FFA concentration remained roughly constant, being the maximum FFA reduction attained close to 60%.

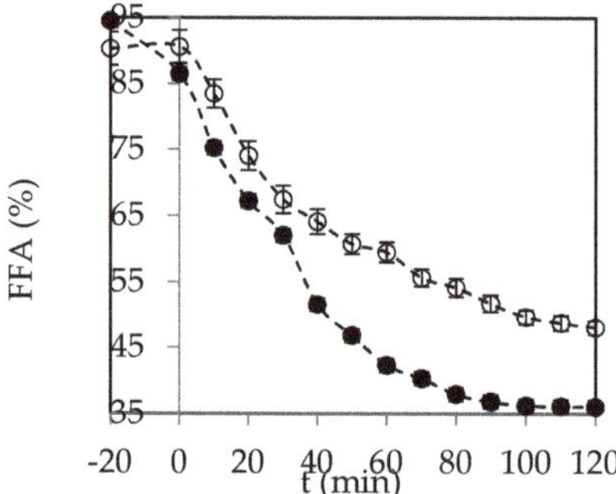


**Figure 2.** Esterification of linoleic acid with 20% excess glycerol over $SnCl_2 \cdot 2H_2O$ (●) or biocatalyst (○) at 160 °C for 120 min (mean ± SD, $n = 2$).

On the other hand, the performance of the biochar obtained from the pyrolysis at 550 °C of heavy metal-contaminated roots of *Jatropha curcas* L. as a catalyst in this reaction was assayed (run 6, Table 1). This biochar was previously demonstrated to achieve an excellent performance in other esterification reactions [4,10]. Notwithstanding this, the performance of this biocatalyst was significantly lower than that of the commercial $SnCl_2 \cdot 2H_2O$ (Figure 2), so it was eliminated from further experiments.

The initial and resulting reaction mixtures of the two experiments with linoleic acid and 20% excess glycerol at 160 °C for 120 min were analyzed by HPSEC (Table 3). The percentages of FFA obtained by HPSEC were in agreement with those obtained by the titration method. The HPSEC results verified that glycerol esterification with FFA occurred, rendering triglycerides (TG), diglycerides (DG), and monoglycerides (MG). The results obtained by HPSEC also illustrated the higher esterification performance of the commercial catalyst, since $SnCl_2$ led to both higher formation of TG and higher conversion of FFA into glyceride compounds (Table 3).

**Table 3.** Free fatty acid (FFA) conversion yield and triglyceride (TG), diglyceride (DG), monoglyceride (MG) and free fatty acid (FFA) content obtained by high performance size exclusion chromatography (HPSEC) at the beginning and at the end of the esterifications over $SnCl_2 \cdot 2H_2O$ and the biocatalyst (mean ± SD, n = 2).

| Run | Catalyst | t (min) | TG (%) | DG (%) | MG (%) | FFA (%) | Yield (%) |
|---|---|---|---|---|---|---|---|
| 5 | $SnCl_2 \cdot 2H_2O$ | 0 | 2.1 ± 0.03 | 3.9 ± 0.09 | - | 94.0 ± 3.9 | 61.1 ± 1.8 |
| | | 120 | 15.4 ± 0.91 | 37.4 ± 1.38 | 10.6 ± 0.71 | 36.6 ± 2.3 | |
| 6 | Biocatalyst | 0 | 2.3 ± 0.06 | 4.7 ± 0.05 | - | 93.0 ± 3.6 | 49.3 ± 0.7 |
| | | 120 | 8.3 ± 0.37 | 25.7 ± 0.23 | 18.4 ± 0.55 | 47.7 ± 2.7 | |

### 3.2. Trials with Different FFA Concentrations

In this set of experiments, we covered roughly the full range of initial FFA concentrations (Table 2) by adding a commercial fatty acid (to be specific, oleic acid) to a lampante olive oil. An excess of 20% of commercial glycerol over the stoichiometric one was used. The acidities of the commercial oleic acid and the lampante oil were 99.9 ± 0.1% and 1.3 ± 0.1%, respectively, based on the titration method.

The results revealed that glycerolysis yield decreased when decreasing the initial FFA content. Figure 3 illustrates the change in the content of free fatty acids, this content measured by the titration method.

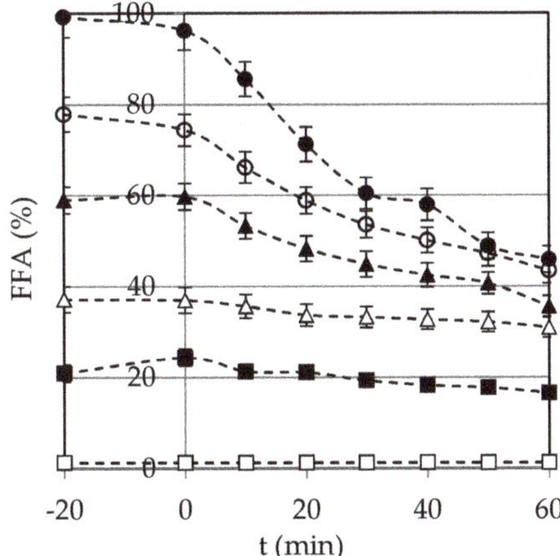


**Figure 3.** Glycerolysis of mixtures of linoleic acid with 0% (●), 20% (○), 40% (▲), 60% (Δ), 80% (■), and 100% (□) lampante olive oil (mean ± SD, n = 2).

The analysis performed by HPSEC provided additional information. When esterifying pure oleic acid (run 7, Table 4), large quantities of mono-, di-, and triglycerides were formed, with DG being the main compound found. One could think that this was be due to steric hindrance to the formation of TG. Notwithstanding this, the TG concentration decreased in the glycerolysis of mixtures of oleic acid and lampante olive (runs 8 to 11, Table 4), with the final concentrations being lower than the initial ones. This can be attributed to triglyceride (and diglyceride) hydrolysis by the water produced in the esterification reaction. As other authors have emphasized, low FFA contents are not suitable for glycerol esterification, at least at the assayed temperature (160 °C), because the water generated in this reaction would shift the equilibrium of transesterification to the reverse reaction [8]. As a result, the esterification with 20% excess glycerol of lampante olive oil failed (run 12, Table 4), and it only slightly reduced the FFA content of the mixture 20% oleic acid/80% lampante olive oil (run 11, Table 4).

**Table 4.** Triglyceride (TG), diglyceride (DG), monoglyceride (MG) and free fatty acid (FFA) content of the mixtures of oleic acid and lampante olive oil mixtures obtained by HPSEC at the beginning and at the end of the esterification reactions, and conversion yields (mean ± SD, $n$ = 2).

| Run | Oleic Acid (%) | Lampante Oil (%) | t (min) | TG (%) | DG (%) | MG (%) | FFA (%) | Yield (%) |
|---|---|---|---|---|---|---|---|---|
| 7 | 100 | 0 | 0 | 0.65 ± 0.01 | 1.75 ± 0.06 | - | 97.57 ± 3.81 | 60.0 ± 1.6 |
|  |  |  | 120 | 10.23 ± 0.38 | 37.28 ± 0.82 | 13.48 ± 0.90 | 39.01 ± 2.10 |  |
| 8 | 80 | 60 | 0 | 20.76 ± 0.15 | 1.85 ± 0.08 | - | 77.39 ± 2.02 | 44.6 ± 0.8 |
|  |  |  | 120 | 17.26 ± 0.01 | 28.33 ± 0.75 | 11.51 ± 0.88 | 42.90 ± 2.63 |  |
| 9 | 60 | 40 | 0 | 40.67 ± 0.25 | 1.71 ± 0.03 | - | 57.62 ± 2.31 | 44.5 ± 0.8 |
|  |  |  | 120 | 33.32 ± 0.15 | 25.55 ± 0.38 | 9.16 ± 0.75 | 31.97 ± 1.80 |  |
| 10 | 40 | 60 | 0 | 59.34 ± 0.35 | 2.14 ± 0.04 | - | 38.52 ± 1.71 | 23.8 ± 0.6 |
|  |  |  | 120 | 52.64 ± 0.38 | 13.83 ± 0.50 | 4.17 ± 0.06 | 29.36 ± 1.51 |  |
| 11 | 20 | 80 | 0 | 80.10 ± 2.98 | - | - | 19.90 ± 0.92 | 23.6 ± 1.0 |
|  |  |  | 120 | 64.60 ± 1.52 | 13.85 ± 0.55 | 6.35 ± 0.07 | 15.20 ± 0.81 |  |
| 12 | 0 | 100 | 0 | 98.69 ± 3.55 | - | - | 1.31 ± 0.03 | 0.0 ± 0.0 |
|  |  |  | 120 | 98.66 ± 3.56 | - | - | 1.34 ± 0.03 |  |

### 3.3. Glycerolysis of a Waste Cooking Oil (WCO)

Finally, the glycerolysis of a WCO over SnCl$_2$·2H$_2$O using 20% excess glycerol was assayed at 160 °C for 60 min. The FFA content of this WCO was 9.94% ± 0.13%, based on the titration method, and only slightly changed over the course of glycerolysis.

When analyzing the WCO at the beginning and at the end of the glycerolysis by HPSEC, it was found that not only did esterification not occur, but also triglyceride hydrolysis occurred to some extent. The HPSEC analysis illustrated that the WCO was mainly composed of glycerides dimers (7.8%), TG (81%), and FFA (10.4%), with negligible amounts of DG and MG (Figure 4a). In the HPSEC chromatogram of one of the replica of this experiment (Figure 4b), an increase in the areas in the retention times of DG and MG can be observed. A tentative quantification of these areas gave the following composition: 7.3% glyceride dimers, 69.4% TG, 11% DG, 1.8% MG, and 10.1% FFA, which accounts for the hydrolysis of TG into DG and MG, probably due to water contained into the WCO (not measured) and in the catalyst. These results are in contrast to those found by Yeom and Go [13], who stated that the FFA content of a WCO decreased from 4.2% to 1.5% in 1 h when the glycerolysis of this WCO was performed at 160 °C for 1 h over 0.1% SnCl$_2$·2H$_2$O relative to WCO.

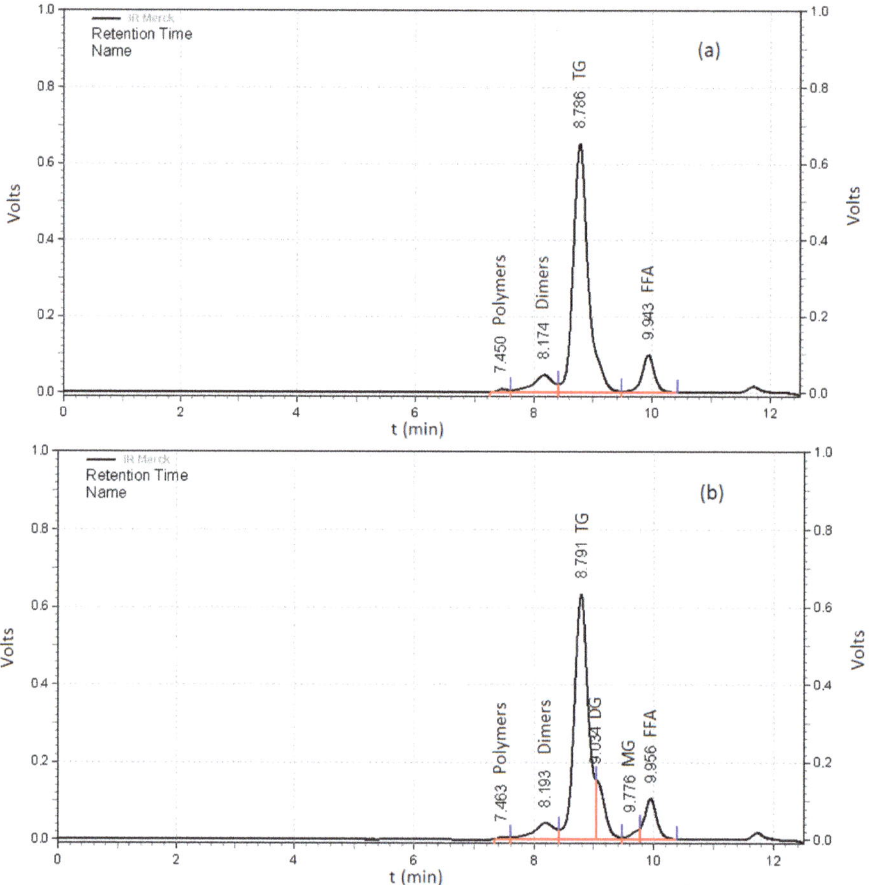

**Figure 4.** HPSEC chromatograms of the waste cooking oil before (**a**) and after (**b**) glycerolysis.

## 4. Conclusions

The results presented in this work show that glycerolysis might be used as a first stage to reduce the FFA content in high-acidic and low-cost feedstocks for biodiesel production. The results illustrated that an excess of glycerol over the stoichiometric is required to increase the FFA conversion, with 20% being the most suitable glycerol excess. In the esterification of a commercial fatty acid (linoleic acid) with 20% glycerol excess at 160 °C, it was found that 90% of the maximum FFA conversion was reached within 60 min of glycerolysis. The performance of the biochar obtained from heavy metal-contaminated roots of the plant *Jatropha curcas* L. was lower than that of the commercial catalyst tin (II) chloride dihydrate.

Thus, the FFA conversion yield was 49.3% for the biocatalyst, whereas that for $SnCl_2 \cdot 2H_2O$ was 61.1%. In the set of experiments with mixtures of oleic acid (99.9% FFA) and a lampante olive oil (1.3% FFA), a competition was observed between glycerolysis (formation of mono-, di- and triglycerides) and hydrolysis (release of FFA from mono-, di-, and triglycerides). As a result, the main reaction product was diglycerides. Due to this competition, glycerolysis is not suitable for oils with relative low acidity. In this sense, the esterification of a waste cooking oil (9.94% FFA) under the most suitable conditions in this study was unsuccessful.

**Author Contributions:** Conceptualization, P.Á.M.; methodology, P.Á.M.; formal analysis, J.F.G.M., C.-H.F. and M.T.G.; investigation, J.C.R.; data curation, J.C.R., J.F.G.M. and P.Á.M.; writing—original draft preparation, J.F.G.M.; writing—review and editing, J.F.G.M.; supervision, P.Á.M. and J.F.G.M.; project administration, P.Á.M.; funding acquisition, P.Á.M. and M.T.G.

**Funding:** This research was funded by European Union under grant LIFE 13-Bioseville ENV/1113.

**Acknowledgments:** Chao-Hui Feng would like to express her gratitude to University of Seville for the mobility grant (VIPPIT-2018-I.3) awarded under the VI Plan Propio de Investigación y Transferencia of the University of Seville.



## References

1. Sánchez, S.; Cuevas, M.; García-Martín, J.F. Bioethanol production: Corn v/s lignocellulose biomass from olive oil industry and the potential role in ensuring food security. *J. Fundam. Renew. Energy Appl.* **2015**, *05*, 18.
2. García-Martín, J.F.; Barrios, C.C.; Alés-Álvarez, F.J.; Dominguez-Sáez, A.; Alvarez-Mateos, P. Biodiesel production from waste cooking oil in an oscillatory flow reactor. Performance as a fuel on a TDI diesel engine. *Renew. Energy.* **2018**, *125*, 546–556. [CrossRef]
3. García Martín, J.F.; del Carmen Barrera, M.L.; Torres García, M.; Zhang, Q.A.; Álvarez Mateos, P. Determination of the acidity of waste cooking oils by near infrared spectroscopy. *Processes* **2019**, *7*, 1–7. [CrossRef]
4. Álvarez-Mateos, P.; García-Martín, J.F.; Guerrero-Vacas, F.J.; Naranjo-Calderón, C.; Barrios, C.C.; Pérez-Camino, M.C. Valorization of a high-acidity residual oil generated in the waste cooking oils recycling industries. *Grasas y Aceites* **2019**, *40*, 335. [CrossRef]
5. Pereda Marín, J.; Barriga Mateos, F.; Álvarez Mateos, P. Aprovechamiento de las oleinas residuales procedentes del proceso de refinado de los aceites vegetales comestibles para la fabricación de biodiesel. *Grasas y Aceites* **2003**, *54*, 130–137.
6. Botton, V.; Piovan, L.; Meier, H.F.; Mitchell, D.A.; Cordova, J.; Krieger, N. Optimization of biodiesel synthesis by esterification using a fermented solid produced by Rhizopus microsporus on sugarcane bagasse. *Bioprocess Biosyst. Eng.* **2018**, *41*, 573–583. [CrossRef] [PubMed]
7. Marchetti, J.M.; Errazu, A.F. Esterification of free fatty acids using sulfuric acid as catalyst in the presence of triglycerides. *Biomass Bioenergy* **2008**, *32*, 892–895. [CrossRef]
8. Chai, M.; Tu, Q.; Lu, M.; Yang, Y.J. Esterification pretreatment of free fatty acid in biodiesel production, from laboratory to industry. *Fuel Process. Technol.* **2014**, *125*, 106–113. [CrossRef]
9. Manosak, C.R.; Limpattayanate, S.; Hunsom, M. Sequential-refining of crude glycerol derived from waste used-oil methyl ester plant via a combined process of chemical and adsorption. *Fuel Process. Technol.* **2011**, *92*, 92–99. [CrossRef]
10. García-Martín, J.F.; Alés-Álvarez, F.J.; Torres-García, M.; Feng, C.-H.; Álvarez-Mateos, P. Production of Oxygenated Fuel Additives from Residual Glycerine Using Biocatalysts Obtained from Heavy-Metal-Contaminated *Jatropha curcas* L. Roots. *Energies* **2019**, *12*, 740. [CrossRef]
11. Vitiello, R.; Li, C.; Russo, V.; Tesser, R.; Turco, R.; Di Serio, M. Catalysis for esterification reactions: A key step in the biodiesel production from waste oils. *Rend. Lincei* **2017**, *28*, 117–123. [CrossRef]
12. Cai, Z.Z.; Wang, Y.; Teng, Y.L.; Chong, K.M.; Wang, J.W.; Zhang, J.W.; Yang, D.-P. A two-step biodiesel production process from waste cooking oil via recycling crude glycerol esterification catalyzed by alkali catalyst. *Fuel Process. Technol.* **2015**, *137*, 186–193. [CrossRef]

13.  Yeom, S.H.; Go, Y.W. Optimization of a Novel Two-step Process Comprising Re-esterification and Transesterification in a Single Reactor for Biodiesel Production Using Waste Cooking Oil. *Biotechnol. Bioprocess Eng.* **2018**, *23*, 432–441. [CrossRef]

14.  Albisu Aguado, M.; Fernández Gil, P. Aceites y grasas industriales. In *Bases la Aliment Humana*, 2nd ed.; Reverté: Barcelona, Spain, 2008; pp. 103–116.

15.  Kombe, G.G.; Temu, A.K.; Rajabu, H.M.; Mrema, G.D.; Kansedo, J.; Lee, K.T. Pre-Treatment of High Free Fatty Acids Oils by Chemical Re-Esterification for Biodiesel Production—A Review. *Adv. Chem. Eng. Sci.* **2013**, *3*, 242–247. [CrossRef]

16.  Cardoso, A.; Neves, S.; Da Silva, M. Esterification of Oleic Acid for Biodiesel Production Catalyzed by SnCl$_2$: A Kinetic Investigation. *Energies* **2008**, *1*, 79–92. [CrossRef]

17.  Felizardo, P.; MacHado, J.; Vergueiro, D.; Correia, M.J.N.; Gomes, J.P.; Bordado, J.M. Study on the glycerolysis reaction of high free fatty acid oils for use as biodiesel feedstock. *Fuel Process. Technol.* **2011**, *92*, 1225–1229. [CrossRef]

18.  Álvarez-Mateos, P.; Alés-Álvarez, F.-J.; García-Martín, J.F. Phytoremediation of highly contaminated mining soils by *Jatropha curcas* L. and production of catalytic carbons from the generated biomass. *J. Environ. Manag.* **2019**, *231*, 886–895. [CrossRef] [PubMed]

*Article*

# Influences of Water Content in Feedstock Oil on Burning Characteristics of Fatty Acid Methyl Esters

**Cherng-Yuan Lin * and Lei Ma**

Department of Marine Engineering, National Taiwan Ocean University, Keelung 202, Taiwan;
awpcsawp@yahoo.com.tw
* Correspondence: Lin7108@ntou.edu.tw

Received: 15 August 2020; Accepted: 8 September 2020; Published: 10 September 2020

**Abstract:** Strong alkaline-catalyst transesterification with short-chain alcohol is generally used for biodiesel production due to its dominant advantages of shorter reaction time and higher conversion rate over other reactions. The existence of excess water content in the feedstock oil might retard the transesterification rate and in turn deteriorate the fuel characteristics of the fatty acid methyl esters. Hence, optimum water content in the raw oil, aimed towards both lower production cost and superior fuel properties, becomes significant for biodiesel research and industrial practices. Previous studies only concerned the influences of water contents on the yield or conversion rate of fatty acid methyl esters through transesterification of triglycerides. The effects of added water in the reactant mixture on burning characteristics of fatty acid methyl esters are thus first investigated in this study. Raw palm oil was added with preset water content before being transesterified. The experimental results show that the biodiesel produced from the raw palm oil containing a 0.05 wt.% added water content had the highest content of saturated fatty acids and total fatty acid methyl esters (FAME), while that containing 0.11 wt.% water content had the lowest content of total FAME and fatty acids of longer carbon chains than C16 among the biodiesel products. Regarding burning characteristics, palm-oil biodiesel made from raw oil with a 0.05 wt.% added water content among those biodiesels was found to have the highest distillation temperatures, flash point, and ignition point, which implies higher safety extents during handling and storage of the fuel. The added water content 0.05 wt.% in raw oil was considered the optimum to produce palm-oil biodiesel with superior fuel structure of fatty acids and burning characteristics. Higher or lower water content than 0.05 wt.% would cause slower nucleophilic substitution reaction and thus a lower conversion rate from raw oil and deteriorated burning characteristics in turn.

**Keywords:** burning characteristics; fatty acid methyl ester; added water content; fuel structure; distillation temperature

---

## 1. Introduction

Biodiesel is composed of mono-alkyl esters of long-chain fatty acids primarily produced through transesterification of vegetable oils, animal fats or microalgae lipids with short chain alcohols by virtue of nucleophilic substitution. Biodiesel has been considered a superior alternative fuel to petro-diesel due to its dominant advantages including superior biodegradability, being free of SOx emissions and acid rain, having enhanced combustion due to its higher oxygen content, exhibiting excellent lubricity, containing no carcinogenic PAHs (polycyclic aromatic hydrocarbons), etc. [1,2]. The application of biodiesel fuel could alleviate the emission of greenhouse gas $CO_2$ owing to the lower carbon content of biodiesel by about 10 wt.% compared to petro-diesel. However, in comparison with diesel fuel, biodiesel has a higher kinematic viscosity and inferior low-temperature fluidity. Heating or adding adequate antifreeze would improve these characteristics of biodiesel [3].

International fuel specifications for biodiesel, such as ASTM D6751 and EN 14214, have been drafted to regulate fuel properties in order to protect users' equipment. Water content is a significant fuel characteristic of biodiesel. Higher water content in biodiesel will accelerate the corrosion rate of metallic engine parts [4]. Partial emulsion may be formed from accumulation of water content with liquid fuel to block the fuel feeding system [5]. During the production process of biodiesel, the water content has a dominant influence on the conversion rate of feedstock and the appearance of the saponification phenomenon.

Bitonto and Pastore [6] found that water content, acid value, and free fatty acids (FFA) of feedstock oils should be lower than 0.06 wt.%, 1 mg KOH/g, and 0.5 wt.%, respectively, to prevent negative effects on the biodiesel product. Hakimi et al. [7] even suggested all reactants should be substantially anhydrous during alkali-catalyzed transesterification. Yasar [8] studied the effect of water content of the feedstock on the ester content of biodiesel. Chen et al. [9] further indicated that the upper limit of water content in raw oil is 0.05 wt.%, for which the conversion rate of transesterification could reach above 90%. The conversion rate is only 5.6% if the added water in feedstock oil is 5 wt.%. Shi et al. [10], after investigating the effects of water content in rapeseed oil on transesterification, found that the addition of 2.5 wt.% to the feedstock oil achieved the highest conversion rate of transesterification. They considered that the addition of an adequate amount of water enhances the hydrolysis of fatty acids. However, the free fatty acids formed from such a hydrolysis process facilitate a transesterification reaction towards biodiesel production [11].

The effects of added water contents on the types of reaction and the yields of methyl esters in transesterification of triglycerides have been widely studied. They inferred that water presence in biodiesel might cause ester hydrolysis, leading to hydrolytic and oxidative degradation and rapid growth of microorganisms. In addition, the engine performance and emission characteristics of emulsion of water-in-biodiesel were investigated previously. Zhang et al. [12] studied the effects of water addition in biodiesel emulsion on spray, combustion, and emission characteristics of a diesel engine. They found water in the emulsion might enhance micro-explosion, resulting in improving fuel-air mixing and reduction of NOx and CO emissions. Rao and Anad [13] prepared biodiesel emulsion added with 5 to 10 wt.% water and observed lower brake thermal efficiency and higher NO emission for the emulsion than those for neat diesel. The effects of water addition in the corn-oil biodiesel on engine performance were studied by Sudalaimuthu et al. [14]. Zakaria et al. [15] experimentally found that the water contents in palm-oil biodiesels increased with the increase of storage temperatures and storage time, leading to degradation of fuel properties. Lawen et al. [16] observed that occurrence of intensive microbial activity in biodiesel might cause the increase of its water content. Delfino et al. [17] developed an alternative method of electrochemical impedance spectroscopy to determine water content in biodiesel. Although the fuel properties might be influenced by added water contents of feedstock oil, the water effects on burning characteristics of fatty acid methyl esters have not been investigated as yet in the literature [18–21]. The optimum water content for achieving superior burning characteristics of biodiesel have not been studied either. Therefore, the effects of the added water content in palm oil feedstock on the burning characteristics of a biodiesel product including the profile of fatty acid compounds, heating value, flash point, etc., were first experimentally investigated in this study. The results of this study could provide valuable references to possible audience for adopting adequate process of water removing from or adding into feedstock oils during transesterification reaction.

## 2. Experimental Details

### 2.1. Preparation of Biodiesel from Palm Oil with Various Water Contents Added

Palm oil, with water contents ranging from 0.02 wt.% to 0.12 wt.%, was added and stirred by a mechanical homogenizer (Model Ultra-Turrax T50, IKA Inc., Staufen, Germany). The properties of the palm oil that were provided by the vender (Formosa Oilseed Processing Ltd. in Taichung

City, Taiwan) are shown in Table 1. The palm oil and water mixtures were then preheated to 60 °C. Methanol was mixed with the alkaline catalyst NaOH using a mechanical homogenizer. The molar ratio of methanol to palm oil was set at 6. The alkaline catalyst NaOH was weighted to be 1 wt.% of the palm oil. The premixed methanol and catalyst NaOH solution was slowly added into the preheated palm oil and water mixture and stirred using a mechanical homogenizer at a speed of 6000 rpm to undergo transesterification for 30 min. After the completion of the transesterification reaction, adequate amounts of glacial acetic acid were added to the product mixture to neutralize the pH value, and this was stirred for 1 min. The product settled and separated to create an upper biodiesel layer and a lower glycerol layer. The biodiesel, after being removed from the glycerol layer, was heated to 70 °C for 30 min to vaporize any volatile impurity, such as methanol, away from the biodiesel product. The biodiesel was then water-washed with 10 wt.% de-ionized water and settled for 15 min to remove the lower-layer liquid. The biodiesel was then distilled at 110 °C for 30 min to separate from the residual water and methanol to complete the production process.

**Table 1.** Properties of palm oil feedstock.

| Item | Property |
| --- | --- |
| Water content (wt.%) | 0.029 |
| Acid value (mg KOH/g) | 0.16 |
| Peroxide value (meq/kg) | 0.53 |
| Lovibond Tintometer | R1.5 Y15 |
| Melting point (°C) | 23.01 |
| Specific gravity | 0.907 |
| Cold filter plugging Point (°C) | 16 |

*2.2. Analysis of Burning Characteristics of Fatty Acid Methyl Esters from Palm Oil with Various Water Contents*

The burning characteristics of biodiesel produced through a transesterification reaction from palm oil with various added water contents were analyzed. An optical microscope (Model BX-60, Olympus Inc., Tokyo, Japan) along with a charged-couple device, Image-Pro Plus version 4.1 analysis software (Media Cybernetics Inc., Rockville, MD, USA), and an image analyzer (Model TK-C1380, JVC Inc., Yokohama, Japan) were utilized to observe the added water droplets within the palm oil layer. The weight proportions of the fatty acids of biodiesel produced from palm oil with various water contents added were analyzed by a gas chromatograph (GC) analyzer (Model GC14A, Shimadzu Inc., Kyoto, Japan) accompanied with a Flame Ionization Detector (FID) and a chromatograph data management system (Avantech Inc., Taipei, Taiwan). The fused silica capillary column (Model Zebron ZB-5HT Inferon Column, Phenomenex Inc., Torrance, CA, USA) used in the GC analyzer was 30 min length, 0.32 mm in inside diameter, and 0.25 μm in film thickness. Adequate type of capillary column is significant to identify fatty acid compounds. The compound of heptadecanoic acid methyl ester of 99% purity was used as the internal standard to mix with the biodiesel sample. The temperature of the injector and FID was set at 250 °C. Nitrogen gas at 20~100 mL/min flow rate was used as the carrier gas. The retention times and elution order were used to chromatographically resolved into the types of methyl esters appeared in the biodiesel samples. The weight fraction of the corresponding fatty ester i ($C_i$) could be determined by the following formula:

$$C_i = \frac{A_i}{A_{EI}} \left[ \frac{C_{EI} \times V_{EI}}{m} \right] \tag{1}$$

where $A_i$ is the peak area of the corresponding fatty acid, $A_{EI}$ is the peak area of heptadecanoic acid methyl ester, $C_{EI}$ and $V_{EI}$ are the concentration and volume of the internal standard, and m is the mass of the sample. The weight fraction of total fatty acid methyl esters can be calculated based on the following formula:

$$C = \frac{(\sum A) - A_{EI}}{A_{EI}} \times \frac{C_{EI} \times V_{EI}}{m} \times 100\% \tag{2}$$

where $\sum A$ is the integrated peak areas of the fatty acid methyl esters identified in the biodiesel sample [22]. The weight percentage of longer carbon-chain fatty acids than C16 was calculated by summing up the weight percentages of those fatty acids longer than C16.

The heating value, in units of cal/g or MJ/kg, is defined as the amount of heat released after the complete burning of a tested fuel. An oxygen bomb calorimeter (Model 1261 automatically adiabatic, Parr Inc., Demopolis, AL, USA) was used to analyze the heating value of the biodiesel sample. The specific gravity (sg) of the fuel sample at 15 °C was measured with a hydrometer (Model 0709, Ho Yu Inc., Taoyuan City, Taiwan) placed in a graduated cylinder. The flash point and ignition point, which are two important safety indicators during fuel storage and transportation, were measured with a Pensky-Marten closed-cup flash point tester based on the ISO 3679:2015 standard method [23]. When a fuel sample is heated at some temperature to accumulate its vaporized gas concentration, a flame holder is swept over the gas environment to cause an instantaneous spark and then be distinguished. Such a temperature is termed a flash point. If the tested fuel sample is heated at some temperature to accumulate vaporizing fuel gas, the burning of the fuel sample could occur and last continuously for at least 5 sec; the ignition point was recorded for that temperature.

The distillation temperatures of the tested samples were analyzed by a distillation temperature analyzer (Model HAD-620, Petroleum Analyzer Inc., Houston, TX, USA). An ASTM D86 curve of liquid fuel for comparison can be plotted using the data for the distillation temperatures corresponding to various volumetric percentages of distilled and condensed fuel. The distillation temperature at 50 vol.% liquid fuel distilled, condensed, and collected is denoted as $T_{50}$. The specific gravity (sg) together with the $T_{50}$ of the sample fuel can be used to calculate the cetane index (CI) of the liquid fuel [24], which indicates the time delay of compression-ignition of the sample fuel:

$$CI = -420.34 + 0.016\ API^2 + 0.192\ (\log T_{50}) + 65.01\ (\log T_{50})^2 - 0.0001809\ T_{50} \tag{3}$$

where

$$API = 141.5/sg - 131.5 \tag{4}$$

## 3. Results and Discussion

The effects of the added water content in palm oil on the burning characteristics of fatty acid methyl esters were experimentally investigated in this study. The mean values of the experimental data were recorded after at least three repetitions. The experimental uncertainties of the results were estimated based on the method by Holman [25]. The experimental uncertainties of the flash point, specific gravity, ignition point, distillation temperature, and the heating value were ±1.27%, ±3.16%, ±2.93%, ±3.52%, and ±1.83%, respectively. The experimental results were described and discussed in the following.

### 3.1. Micrograph of Water in Palm Oil and Fatty Acid Methyl Esters (FAME)

Palm oil with various water contents added, ranging from 0.02 wt.% to 0.12 wt.%, was used as the raw oil to undergo a methanol assisted transesterification reaction with strong alkaline catalyst NaOH. A micrograph of the added water droplets of 0.05 wt.% distributed within the palm oil layer captured by an optical electron microscope in conjunction with a charged-couple device is shown in Figure 1. A rather even distribution of micrometer-sized water droplets within the palm oil layer at 50× magnification was observed. The mean diameter of the water droplets was 0.229 µm. Mechanical stirring using a homogenizer was employed to mix the added water of 0.05 wt.% with the palm oil without adding any surfactant before observing and capturing the results using optical microscope equipment. Micro-explosion might occur after the μm-sized water droplets absorbed sufficient surrounding heat to explode outwards through enveloping oil layer [26], leading to much increase of contacting surface among the reactants and in turn a larger extent of chemical reaction. In addition, the even distribution of μm-sized water droplets in palm oil might increase the homogeneous mixing extent with hydrophilic methanol, leading to enhancement of alkali-catalyzed reaction and formation of fatty acid methyl esters.

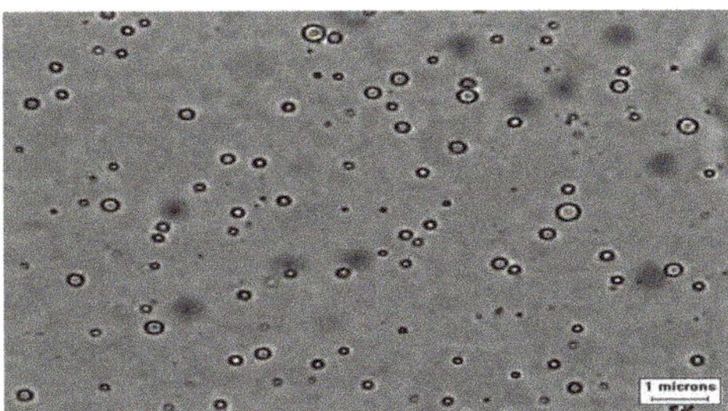

**Figure 1.** Photograph of physical structure of water droplets distributed within palm oil layer when 0.05 wt.% water was added.

Fatty acid methyl esters (FAME) were produced from the transesterification of triglyceride-rich vegetable oil or animal fats with short-chain alcohol particularly methanol. The FAME content is available to determine the extent of transesterification. Higher FAME amount indicates higher purity of the biodiesel product [27]. On the contrary, inferior fuel properties exist for a biodiesel with lower FAME content. The Gas chromatograph (GC) method was used to analyze the fatty acid compositions of the biodiesel produced from palm oil with seven different water contents added. The results of the fatty acid compositions, analyzed by GC equipment, are shown in Table 2. Biodiesel is excellent alternative fuel to petro-diesel due to their similar carbon-chain structure and fuel characteristics. The fatty acid compositions of biodiesel made from vegetable oil or animal fat are mostly in the similar range between C14 and C18 as those carbon-chains of petro-diesel. This can be justified that the total contents of FAME in the range between C14 and C18 for those seven biodiesel samples are only from 81.1 wt.% to 82.1 wt.%. In addition, the FAME contents of carbon chains longer than C16 of those seven biodiesel samples are at least 80.9 wt.% in Table 2. The FAME produced from palm oil added with 0.05 wt.% water content was observed to have the highest saturated fatty acids, which amounted to 46.3 wt.%. Carbon chains of fatty acids ranging from C14 to C24 are frequently identified in biodiesel samples made from various feedstocks. The fatty acid compositions in Table 2 are similar to those of biodiesel structures in previous studies [28,29]. Hence, the biodiesels in this study were successfully produced. The total contents of the palmitic acid (C16:0), stearic acid (C18:0), and oleic acid (C18:1) of those biodiesels accounted for more than 70 wt.% in Table 2, which agrees well with the results of Pinzi et al. [30].

**Table 2.** Comparison of fatty acid compositions of the biodiesel produced from palm oil added with various water content through transesterification.

| Types of Fatty Acids | Added Water Contents (wt.%) | | | | | | |
|---|---|---|---|---|---|---|---|
| | 0.02 | 0.03 | 0.05 | 0.07 | 0.09 | 0.11 | 0.12 |
| C14:0 | 0.8 | 0.6 | 0.8 | 0.6 | 0.6 | 0.6 | 0.8 |
| C16:0 | 33.3 | 31.1 | 34.2 | 27.6 | 27.3 | 28.3 | 34.1 |
| C18:0 | 43.4 | 14.1 | 11.2 | 14.6 | 15.5 | 48.9 | 43.2 |
| C18:1 | | 32.4 | 31.7 | 35.4 | 34.4 | | |

Table 2. Cont.

| Types of Fatty Acids | Added Water Contents (wt.%) | | | | | | |
|---|---|---|---|---|---|---|---|
| | 0.02 | 0.03 | 0.05 | 0.07 | 0.09 | 0.11 | 0.12 |
| C18:2 | 3.6 | 3.9 | 3.8 | 3.4 | 3.4 | 3.3 | 3.2 |
| C18:3 | 0.4 | | | 0.1 | 0.1 | | |
| C24:0 | 0.1 | 0.1 | 0.1 | 0.1 | 0.1 | 0.1 | 0.1 |
| C24:1 | 0.3 | 0.3 | 0.2 | 0.3 | 0.3 | 0.3 | 0.3 |
| Saturated fatty acids | - | 45.9 | 46.3 | 42.9 | 43.5 | - | - |
| Longer carbon-chain fatty acids than C16 | 81.1 | 81.9 | 81.2 | 81.5 | 81.1 | 80.9 | 80.9 |
| Total FAME | 93.5 | 95 | 97.3 | 95.3 | 94.1 | 93.3 | 93.7 |

The variations in the total fatty acid compositions with water contents added to the palm oil are shown in Figure 2. The highest content of fatty acid methyl esters, which amounted to 97.3 wt.%, was produced from palm oil with 0.05 wt.% water added. This is probably owing to the enhancement of the dissociation of OH⁻ radicals from the water to conjugate with the long carbon-chain fatty acids. Although Wu et al. [31] suggested that a water content that is as low as possible in raw oil is required to result in a more complete transesterification reaction, insufficient or excessive amounts of OH⁻ radicals dissociated from the water might be ineffective to move forward the transesterification reaction. Hence, adequate water content would facilitate the conversion reaction, and this postulate agrees well with Nguyen et al. [32]. Less or larger than 0.05 wt.% water added to the palm oil caused less extent of transesterification and thus lower production of total fatty acid methyl esters (FAME) in Figure 2. Hence, the lower FAME formation appeared when the biodiesel produced from palm oil added with 0.02 wt.% or 0.11 wt.% water contents. Sun et al. [33] found that water content was negative to algae dissolution and [Bmim] [HSO₄] catalyzed in-situ transesterification. The biodiesel production from wet algae was thus reduced. Arumugam and Ponnusami [34] observed that the highest conversion rate of triglycerides (92.5%) was produced from waste sardine oil at a water content of 10 vol.% for transesterification reaction catalyzed by enzymes. Excess water in reactants favors hydrolysis and thus decreases biodiesel production.

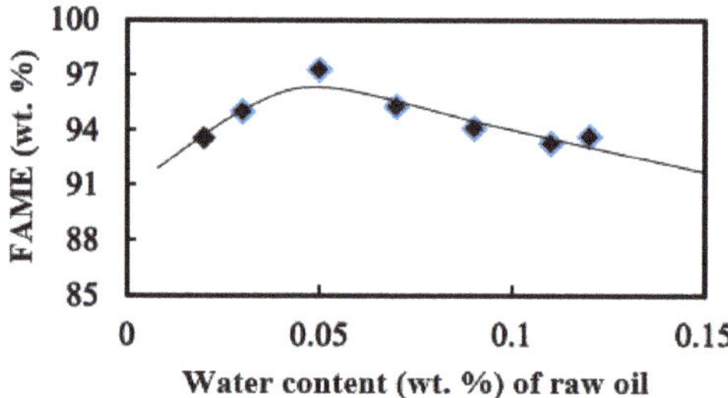


**Figure 2.** Effects of added water content in palm oil on the fatty acid methyl esters.

Table 2 reveals the analytic results of fatty acid methyl esters from the biodiesel produced from palm oil with various water contents added. The biodiesel produced from palm oil with 0.05 wt.% water added was primarily composed of palmitic acid (C16:0), oleic acid (C18:1), and stearic acid (C18:0), which accounted for 34.2 wt.%, 31.7 wt.%, and 11.2 wt.%, respectively. This implies that the biodiesel is relatively oxidatively stable, and thus, fuel properties are not prone to deterioration. In contrast, the biodiesels produced from palm oil added with 0.02 wt.% and 0.12 wt.% water were shown to have much less formation of total fatty acid methyl esters. The content of fatty acids from C16 to C18 amounted to 80.7 wt.% and 80.5 wt.% for the biodiesel made from palm oil added with 0.02 wt.% and 0.12 wt.% water, respectively. This implied that although the biodiesel made from palm oil added with water contents from 0.02 wt.% to 0.12 wt.% resulted in various extents of transesterification and different amounts of fatty acid methyl esters, all the biodiesel produced were composed of almost carbon-chained compounds from C16 to C18. Hence, the biodiesel products are adequate alternative fuel to petro-diesel due to similar carbon-chain chemical structure. Moreover, palm oil is a competitive and abundant feedstock oil source for biodiesel production.

Free fatty acids might be produced through the hydrolysis of fatty acids with water [35]. Excessive water content in reactant mixture of esterification reaction might cause frequent attack of lipids by water. Fatty acids of longer carbon chain lengths would be hydrolyzed, resulting in the formation of free fatty acids and shorter carbon-chain fatty acids [36]. The chemical composition of the biofuel is changed accordingly, resulting in worsened fuel properties. The significant phenomena arising are the occurrence of odor, viscosity increase, and color change, which is the so-called rancidity of the lipid [37].

*3.2. Heating Value*

The heating value is defined as the amount of heat released from the complete burning of fuel. Fuel with a higher heating value requires only lower fuel consumption to attain the same power output. The heating value of biodiesel is lower than petro-diesel by around 10% [38]. The heating values of biodiesels made from palm oil were in the range of 39.5 MJ/kg to 40.9 MJ/kg and were shown to increase with the increase in added water content to the palm oil, as seen in Figure 3. Biodiesel produced from palm oil with 0.12 wt.% water added was found to have the highest heating value, while that with 0.02 wt.% water added had the lowest heating value among those biodiesels, as shown in Figure 3. Shi et al. [10] found that water content that was too low might cause a low extent of hydrolysis of lipid towards reformation of fatty acid methyl esters during transesterification, resulting in a low conversion rate from raw oil and thus, a reduced heating value. In contrast, high water content might render continuous hydrolysis of lipid to form $H^+$ and $OH^-$ radicals and in turn biodiesel, as observed by He et al. [39]. Caetano et al. [40] inferred that water presence might lead to the enhancement of catalytic activity of lipase because the alcohol removed the hydration layer of the enzyme. Adequate water existence thus facilitates both transesterification and hydrolysis. Therefore, higher water content in the raw palm oil appeared to have a higher heating value in the biodiesel product.

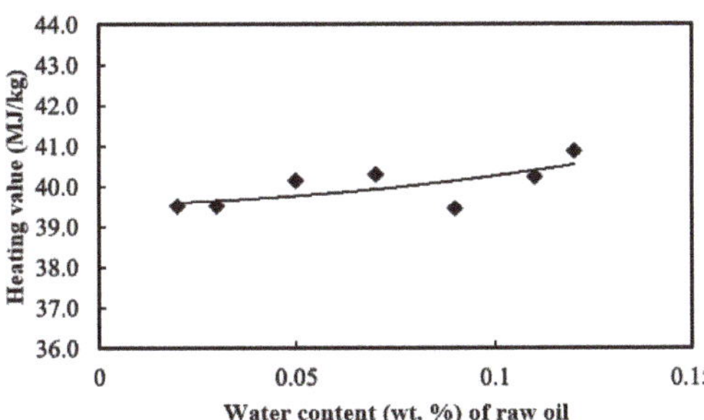


**Figure 3.** Effects of added water content in palm oil on the heating value of the biodiesel product.

Elsanusi et al. [41] investigated the effects of water concentrations in biodiesel emulsions on fuel characteristics and engine performance. They found that the brake thermal efficiency (BTE) increased with the increase of water content in the biodiesel emulsions. This implies that a larger amount of heat was released from burning the biodiesel emulsion with larger water content to result in higher BTE.

*3.3. Specific Gravity*

Specific gravity is defined as the ratio of density of some liquid to that of water at 4 °C. The specific gravity of biodiesel is in the range of 0.86 to 0.9 based on the EN 14,214 standard. The highest and lowest specific gravities were observed for the biodiesel made from palm oil with 0.09 and 0.02 wt.% water added, as shown in Figure 4. The curve trend of specific gravity shown in Figure 4 almost totally agrees with that of the fatty acid methyl esters of the biodiesel product shown in Figure 2. Hence, a lower specific gravity corresponds to a lower total FAME content of the biodiesel. In addition, the type of fatty acid compositions influences the specific gravity of the biodiesel. For example, a larger content of longer carbon-chain fatty acids appeared to create a larger specific gravity of the biodiesel, as found by Hajilar and Shafei [42].

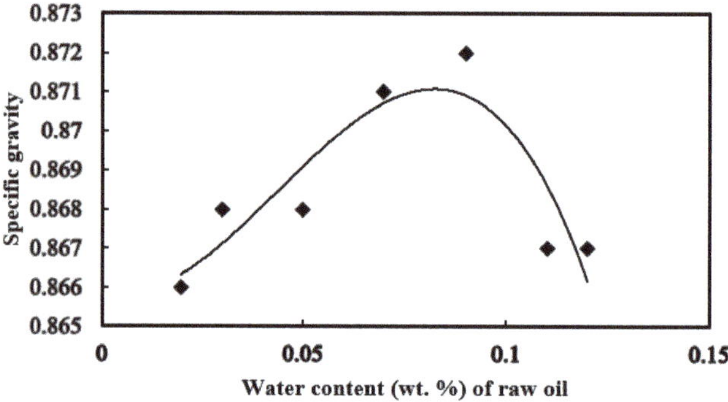

**Figure 4.** Effects of added water content in palm oil on the specific gravity of the biodiesel product.

Ramírez-Verduzco et al. [43] presented characterization of two biodiesel samples through their FAME profiles and derived empirical equations to correlate biodiesel properties with their fatty acid structures. They found that the specific gravity of biodiesel increased as molecular weight decreased and degree of unsaturation increased. The larger specific gravity of the biodiesel made from palm oil added with 0.09 wt.% might thus be ascribed to its larger content of unsaturated fatty acids, as shown in Table 2. Refaat [44] and Folayan et al. [45] also confirmed that specific gravity of biodiesel increases with the increase of unsaturated fatty acids and the decrease of chain length.

*3.4. Flash Point and Ignition Point*

The flash point is one major safety indicator during storage and transportation of liquid fuel. The temperature at which liquid fuel is heated to form and accumulate fuel vapor to a certain concentration, where an instantaneous spark is flashed after a flame crosses over the fuel vapor, is defined as the flash point. The temperature at which the fuel vapor is formed to cause the spark and further continuous burning is denoted as the ignition point of liquid fuel. The flash points of biodiesel were found to range from 160 to 176 °C and peaked corresponding to the 0.05 wt.% water content added to the palm oil, as shown in Figure 5. The curve trend of the flash point shown in Figure 5 conformed to that of FAME profile in Figure 2. The highest FAME content in biodiesel rendered the highest flash point when 0.05 wt.% water was added to raw palm oil. The peak flash point could also be observed from Table 2, where the total carbon-chain fatty acids for the biodiesel produced from palm oil with 0.05 wt.% water content added reached the highest 97.3 wt.% among all the cases of added water contents. Marlina et al. [46] also found that biodiesel composed of a greater content of longer carbon-chain fatty acids tended to have a higher flash point. Too high or low added water to raw oil caused a slower nucleophilic substitution reaction, as proposed by Paula et al. [47]. A lower conversion rate thus occurred, resulting in a lower FAME and in turn a lower flash point in those cases. The flash point was decreased with the increase of the content of unsaturated fatty acid methyl esters in Figure 5—a result that agreed with that of Ayoola [48]. Su et al. [49] proposed a correlation equation of flash point with chain length and unsaturation of biodiesel. Rao et al. [50] also derived a correlation equation to relate flash point of biodiesel linearly with its specific gravity. Hence, similar curve trends between those of specific gravity and flash point could be observed in Figures 4 and 5. Flash point was also observed to influence higher heating values of biodiesel [51].

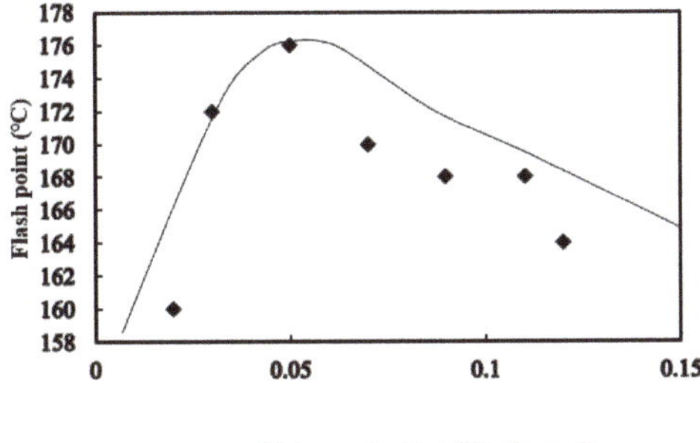

**Figure 5.** Effects of added water content in palm oil on the flash point of the biodiesel product.

The highest ignition point of biodiesel was found to be made from palm oil with 0.05 wt.% water content added, as shown in Figure 6. This can probably be ascribed to the highest FAME formation among the biodiesels from palm oil with various water contents added, shown in Figure 2. In comparison with Figure 2, the curve trend of the ignition points of the biodiesel with respect to the added water content in Figure 6 was observed to agree with that of FAME contents in the biodiesel products. This implies that higher FAME content in the biodiesel product increased the ignition point. In addition, the higher specific gravity of the biodiesel was shown to have a higher ignition point in comparison to Figures 4 and 6; this inference agrees well with the findings of Kumar and Bansal [52] and Rao et al. [50]. The increase of ignition point might also be ascribed to the increase of saturated fatty acid methyl esters, as observed by Ayoola [48]. Bukkarapu et al. [53] found that the increase of kinematic viscosity caused the increase of ignition point of the biodiesel.

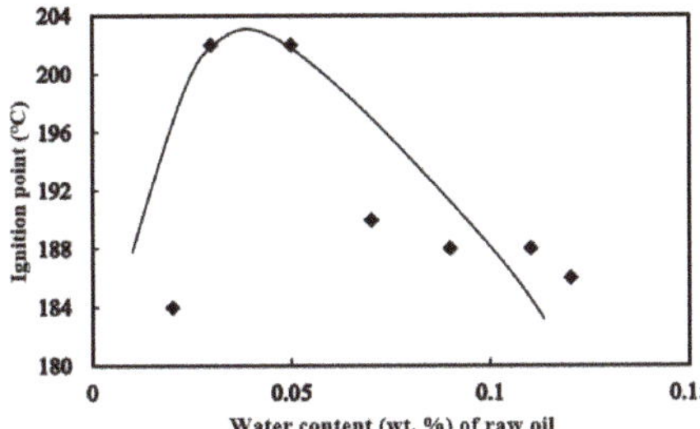

**Figure 6.** Effects of added water content in palm oil on the ignition point of the biodiesel product.

*3.5. Distillation Temperature and Cetane Index*

Distillation temperature is one of the significant indicators of volatility and combustion characteristics of liquid fuel. The tendency of forming smoke and soot can be indicated by distillation temperature as well. In contrast to the distillation temperatures of petro-diesel, biodiesel has much narrower range of boiling points due to mostly alkyl esters in biodiesel [54]. A distillation temperature curve based on ASTM D86 is prepared to reveal the range of boiling points of various compounds in liquid fuel. The curve of the distillation temperatures can be used to determine the distribution from light to heavy compounds. The temperature at which a liquid drop is vaporized, condensed, and collected is referred to as $T_{IBP}$. Similarly, $T_{50}$ is the temperature for a 50 vol.% liquid fuel, and $T_{EP}$ is the highest temperature corresponding to the final liquid drop that is vaporized, condensed, and collected.

The ASTM D86 distillation temperature curve for biodiesel made from palm oil with various water contents added, ranging from 0.03 wt.% to 0.12 wt.%, is shown in Figure 7. Biodiesel made from palm oil with 0.05 wt.% water added was found to have the highest distillation temperature, while that with 0.12 wt.% water added had the lowest distillation temperatures among those three biodiesels. For example, the $T_{EP}$ of the biodiesel made from palm oil added with 0.05 wt.% and 0.12 wt.% water contents were 354 and 342 °C, respectively. This is ascribed to the fact that the addition of 0.05 wt.% water content caused the highest formation of fatty acid methyl esters (FAME), while that of 0.12 wt.% water formed the lowest FAME, as shown in Figure 2. In addition, Yao et al. [55] considered that $T_{90}$ is an indicator for the content of heavier compounds in liquid fuel. A higher $T_{90}$ implies a larger amount of heavier compounds and greater viscosity of a liquid fuel, which might result in deteriorated atomization, slower vaporization, and in turn, incomplete burning. The biodiesel made from palm oil with 0.05 wt.% water added was shown to have the highest $T_{90}$, which thus implies production of a higher extent of pollutants from burning such biodiesel. Distillation temperature influences the combustion and emission characteristics of biodiesel. Lower distillation temperature results in higher volatility and enhances homogeneity of reactant mixture [56].

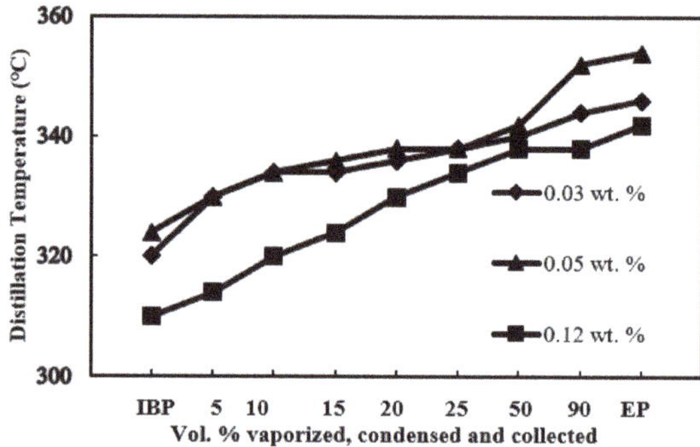


**Figure 7.** Effects of added water content in palm oil on the distillation temperature curve of the biodiesel product.

The cetane number (CN) is used to indicate the compression-ignition quality of liquid fuel in a diesel engine. Fuel bearing a higher cetane number would shorten the period of ignition delay in a diesel engine and thus reduce the burning time and residence period of the peak flame within the engine cylinder, resulting in a lower occurrence of engine knocking and NOx formation. The cetane index, which is an alternative to the cetane number, is obtained by calculations using the data from $T_{50}$ and API gravity based on Equation (3). The lowest cetane index is found for biodiesel made from palm oil with 0.09 wt.% water added, shown in Figure 8. This is probably due to its larger $T_{50}$ and the specific gravity of the biodiesel, as shown in Figure 4. A higher cetane index existed when water content that was either lower or higher than 0.09 wt.% was added to palm oil for manufacturing the biodiesel. Cetane number of biodiesel was determined by its fatty acid composition, number of double bonds, degree of unsaturation, chain length, and molecular weight [57]. A few correlation equations which relate cetane number with those physicochemical properties of biodiesel have been proposed for CN prediction [58,59]. In addition, Mishra et al. [60] and Moser [61] found that the cetane number of the biodiesel increased with the increased amount of long carbon-chain fatty acids or saturated fatty acids. It was found that the increase of number of double bonds leads to the decrease of cetane number of biodiesel [62]. Higher or lower water content than 0.09 wt.% might cause an increase in saturated fatty acids and heating value in turn. Hence, those two curve trends between the cetane index and heating value agree well with each other in comparison with Figures 3 and 8.

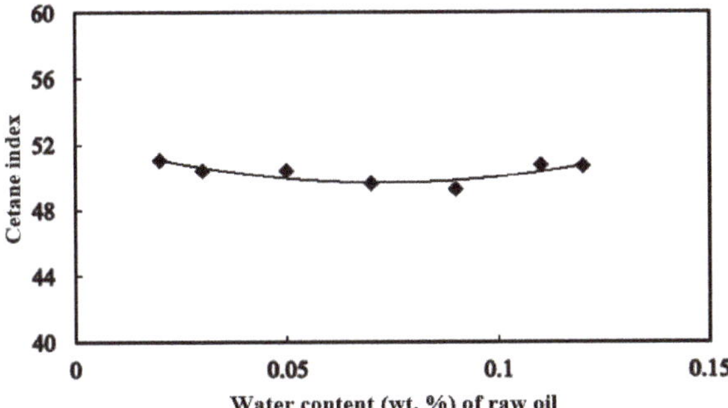

**Figure 8.** Effects of added water content in palm oil on the cetane index of the biodiesel product.

## 4. Conclusions

Various water contents were added to palm oil to undergo strong alkaline-catalyst transesterification for biodiesel production. The burning characteristics of those fatty acid methyl esters, such as flash point and heating value, were analyzed. Major experimental results are summarized below.

A rather even distribution of water droplets with the mean diameter of 0.229 µm within the palm oil layer was produced when 0.05 wt.% water was added to the raw palm oil and stirred using a mechanical stirrer. The fatty acid methyl esters produced from the palm oil with the water added ranging from 0.02 wt.% to 0.12 wt.% were composed of over 70 wt.% of palmitic acid (C16:0), stearic acid (C18:0), and oleic acid (C18:1). The biodiesel produced from palm oil with 0.05 wt.% water added through strong alkaline-catalyst transesterification were found to form the highest total fatty acid methyl esters (FAME) and saturated fatty acids, which amounted to 97.3 wt.% and 46.3 wt.%, respectively among those seven biodiesel samples. Moreover, the fatty acid methyl esters produced from palm oil with 0.05 wt.% water added appeared to have the highest flash point, ignition point, and distillation temperature and thus, the highest safety level during storage and transportation of the biodiesel. The total carbon-chain fatty acids longer than C16 reached as high as 81.2 wt.% in such biodiesel. The raw palm oil with 0.05 wt.% water content was found to produce a biodiesel with superior fatty acid composition and fuel characteristics.

In contrast, the biodiesel produced from palm oil with 0.02 wt.% water added was found to have the lowest heating value and specific gravity. The lowest distillation temperature and formation of fatty acid methyl esters were found to be in the biodiesel made from palm oil with 0.12 wt.% water added. In addition, biodiesel made from palm oil with 0.09 wt.% water added was observed to have the highest specific gravity along with the lowest cetane index. Hence, added water content higher or lower than 0.05 wt.%, such as 0.02 wt.% and 0.12 wt.%, caused poorer fatty acids compositions and deteriorated fuel properties.

**Author Contributions:** Conceptualization, principal investigation, experimental design, funding acquisition, writing and editing, and supervision, C.-Y.L.; carrying out experiment and data acquisition, L.M. All authors have read and agreed to the published version of the manuscript.

**Funding:** This research was funded by the Ministry of Science and Technology, Taiwan, ROC under grant number: MOST 107-2221-E-019-056-MY2 and MOST 105-2221-E-019-066 and the APC was funded by National Taiwan Ocean University, Taiwan, ROC.

**Conflicts of Interest:** The authors declare no conflict of interest.

## References

1. Østerstrøm, F.F.; Anderson, J.E.; Mueller, S.A.; Collings, T.; Ball, J.C.; Wallington, T.J. Oxidation stability of rapeseed biodiesel/petroleum diesel blends. *Energy Fuels* **2016**, *30*, 344–351. [CrossRef]
2. Othman, M.F.; Adam, A.; Najafi, G.; Mamat, R. Green fuel as alternative fuel for diesel engine: A review. *Renew. Sustain. Energy Rev.* **2017**, *80*, 694–709. [CrossRef]

3.  Kandasamy, S.; Samudrala, S.P.; Bhattacharya, S. The route towards sustainable production of ethylene glycol from a renewable resource, biodiesel waste: A review. *Cat. Sci. Tec.* **2019**, *9*, 567–577. [CrossRef]

4.  Ahmmad, M.S.; Haji Hassan, M.B.; Kalam, M.A. Comparative corrosion characteristics of automotive materials in Jatropha biodiesel. *Int. J. Green Energy* **2018**, *15*, 393–399. [CrossRef]

5.  Santos, D.; da Rocha, E.C.; Santos, R.L.; Cancelas, A.J.; Franceschi, E.; Santos, A.F.; Fortuny, M.; Dariva, C. Demulsification of water-in-crude oil emulsions using single mode and multimode microwave irradiation. *Sep. Purif. Technol.* **2017**, *189*, 347–356. [CrossRef]

6.  di Bitonto, L.; Pastore, C. Metal hydrated-salts as efficient and reusable catalysts for pre-treating waste cooking oils and animal fats for an effective production of biodiesel. *Renew. Energy* **2019**, *143*, 1193–1200. [CrossRef]

7.  Hakimi, M.I.; Goembira, F.; Ilham, Z. Engine-compatible biodiesel from Leucaena leucocephala seed oil. *J. Soc. Automot. Eng. Malays.* **2017**, *1*, 86–93.

8.  Yaşar, F. Comparision of fuel properties of biodiesel fuels produced from different oils to determine the most suitable feedstock type. *Fuel* **2020**, *264*, 116817. [CrossRef]

9.  Chen, J.; Li, J.; Dong, W.; Zhang, X.; Tyagi, R.D.; Drogui, P.; Surampalli, R.Y. The potential of microalgae in biodiesel production. *Renew. Sustain. Energy Rev.* **2018**, *90*, 336–346. [CrossRef]

10.  Shi, W.J.; Liu, H.Q.; Feng, S.B.; Zheng, E.L. Study on preparation of biodiesel with low acid value rapeseed oil. *Renew. Energy Resour.* **2009**, *27*, 37–39.

11.  Jia, W.; Xu, G.; Liu, X.; Zhou, F.; Ma, H.; Zhang, Y.; Fu, Y. Direct Selective Hydrogenation of Fatty Acids and Jatropha Oil to Fatty Alcohols over Cobalt-Based Catalysts in Water. *Energy Fuels* **2018**, *32*, 8438–8446. [CrossRef]

12.  Zhang, Z.; Jiaqiang, E.; Chen, J.; Zhu, H.; Zhao, X.; Han, D.; Yin, Z. Effects of low-level water addition on spray, combustion and emission characteristics of a medium speed diesel engine fueled with biodiesel fuel. *Fuel* **2019**, *239*, 245–262. [CrossRef]

13.  Rao, M.S.; Anand, R.B. Performance and emission characteristics improvement studies on a biodiesel fuelled DICI engine using water and AlO (OH) nanoparticles. *Appl. Therm. Eng.* **2016**, *98*, 636–645.

14.  Sudalaimuthu, G.; Rathinam, S.; Munuswamy, D.B.; Thirugnanasambandam, A.; Devarajan, Y. Testing and evaluation of performance and emissions characteristics of water-biodiesel aspirated research engine. *J. Test. Eval.* **2020**, *48*, 20180306. [CrossRef]

15.  Zakaria, H.; Khalid, A.; Sies, M.F.; Mustaffa, N.; Manshoor, B. Effect of storage temperature and storage duration on biodiesel properties and characteristics. *Appl. Mech. Mater.* **2014**, *465*, 316–321. [CrossRef]

16.  Lawan, I.; Zhou, W.; Idris, A.L.; Jiang, Y.; Zhang, M.; Wang, L.; Yuan, Z. Synthesis, properties and effects of a multi-functional biodiesel fuel additive. *Fuel Process. Technol.* **2020**, *198*, 106228. [CrossRef]

17.  Delfino, J.R.; Pereira, T.C.; Viegas, H.D.C.; Marques, E.P.; Ferreira, A.A.P.; Zhang, L.; Marques, A.L.B. A simple and fast method to determine water content in biodiesel by electrochemical impedance spectroscopy. *Talanta* **2018**, *179*, 753–759. [CrossRef]

18.  Kusdiana, D.; Saka, S. Effects of water on biodiesel fuel production by supercritical methanol treatment. *Bioresour. Technol.* **2004**, *91*, 289–295. [CrossRef]

19.  Tan, K.T.; Lee, K.T.; Mohamed, A.R. Effects of free fatty acids, water content and co-solvent on biodiesel production by supercritical methanol reaction. *J. Supercrit. Fluids* **2010**, *53*, 88–91. [CrossRef]

20.  Srimhan, P.; Kongnum, K.; Taweerodjanakarn, S.; Hongpattarakere, T. Selection of lipase producing yeasts for methanol-tolerant biocatalyst as whole cell application for palm-oil transesterification. *Enzyme Microb. Technol.* **2011**, *48*, 293–298. [CrossRef]

21.  Yan, S.; Salley, S.O.; Ng, K.Y.S. Simultaneous transesterification and esterification of unrefined or waste oils over ZnO-La2O3 catalysts. *Appl. Catal. A.* **2009**, *353*, 203–212. [CrossRef]

22.  Shimadzu Corporation. *GC-2014 Gas Chromatograph Instruction Manual*; Shimadzu Corporation: Kyoto, Japan, 2014.

23.  International Organization for Standardization. *Determination of Flash Point-Rapid Equilibrium Closed Cup Method*; American National Standard Institute: Washington, DC, USA, 2015; ISO 3679:2015.

24.  Kalargaris, I.; Tian, G.; Gu, S. Experimental evaluation of a diesel engine fuelled by pyrolysis oils produced from low-density polyethylene and ethylene–vinyl acetate plastics. *Fuel Process. Technol.* **2017**, *161*, 125–131. [CrossRef]

25. Holman, J.P. Analysis of experimental data. In *Experimental Methods for Engineers*, 8th ed.; McGraw Hill Inc.: Singapore, 2012; pp. 60–77.

26. Moussa, O.; Tarlet, D.; Massoli, P.; Bellettre, J. Parametric study of the micro-explosion occurrence of W/O emulsions. *Int. J. Therm. Sci.* **2018**, *133*, 90–97. [CrossRef]

27. Pangestu, T.; Kurniawan, Y.; Soetaredjo, F.E.; Santoso, S.P.; Irawaty, W.; Yuliana, M.; Ismadji, S. The synthesis of biodiesel using copper based metal-organic framework as a catalyst. *J. Environ. Chem. Eng.* **2019**, *7*, 103277. [CrossRef]

28. Hoekman, S.K.; Broch, A.; Robbins, C.; Ceniceros, E.; Natarajan, M. Review of biodiesel composition, properties, and specifications. *Renew. Sustain. Energy Rev.* **2012**, *16*, 143–169. [CrossRef]

29. Singh, D.; Sharma, D.; Soni, S.L.; Sharma, S.; Kumari, D. Chemical compositions, properties, and standards for different generation biodiesels: A review. *Fuel* **2019**, *253*, 60–71. [CrossRef]

30. Pinzi, S.; Leiva, D.; Arzamendi, G.; Gandia, L.M.; Dorado, M.P. Multiple response optimization of vegetable oils fatty acid composition to improve biodiesel physical properties. *Bioresour. Technol.* **2011**, *102*, 7280–7288. [CrossRef]

31. Wu, L.; Wei, T.Y.; Tong, Z.F.; Zou, Y.; Lin, Z.J.; Sun, J.H. Bentonite-enhanced biodiesel production by NaOH-catalyzed transesterification of soybean oil with methanol. *Fuel Process. Technol.* **2016**, *144*, 334–340. [CrossRef]

32. Nguyen, H.C.; Nguyen, M.L.; Wang, F.M.; Juan, H.Y.; Su, C.H. Biodiesel production by direct transesterification of wet spent coffee grounds using switchable solvent as a catalyst and solvent. *Bioresour. Technol.* **2020**, *296*, 122334. [CrossRef]

33. Sun, Y.; Xu, C.; Igou, T.; Liu, P.; Hu, Z.; Van Ginkel, S.W.; Chen, Y. Effect of water content on [Bmim][HSO4] assisted in-situ transesterification of wet Nannochloropsis oceanica. *Appl. Energy* **2018**, *226*, 461–468. [CrossRef]

34. Arumugam, A.; Ponnusami, V. Production of biodiesel by enzymatic transesterification of waste sardine oil and evaluation of its engine performance. *Heliyon* **2017**, *3*, e00486. [CrossRef]

35. Oliverira, E.D.C.; Silva, P.R.D.; Ramos, A.P.; Aranda, D.A.G.; Freire, D.M.G. Study of soybean oil hydrolysis catalyzed by Thermomyces Ianuginosus lipase and its application to biodiesel production via hydroesterification. *Enzyme Res.* **2011**, *2011*, 1–8. [CrossRef] [PubMed]

36. Edeh, I.; Overton, T.; Bowra, S. Optimization of subcritical water-mediated lipid extraction from activated sludge for biodiesel production. *Biofuels* **2019**, 1–7. [CrossRef]

37. Nejad, A.S.; Zahedi, A.R. Optimization of biodiesel production as a clean fuel for thermal power plants using renewable energy source. *Renew. Energy* **2018**, *119*, 365–374. [CrossRef]

38. Yamin, J.A.; Sheet, E.A.E.; Hdaib, I. Exergy analysis of biodiesel fueled direct injection CI engines. *Energy Sources Part A* **2018**, *40*, 1351–1358. [CrossRef]

39. He, Y.; Wu, T.; Wang, X.; Chen, B.; Chen, F. Cost-effective biodiesel production from wet microalgal biomass by a novel two-step enzymatic process. *Bioresour. Technol.* **2018**, *268*, 583–591. [CrossRef]

40. Caetano, N.S.; Caldeira, D.; Martins, A.A.; Mata, T.M. Valorisation of spent coffee grounds: Production of biodiesel via enzymatic catalysis with ethanol and a co-solvent. *Waste Biomass Valori.* **2017**, *8*, 1981–1994. [CrossRef]

41. Elsanusi, O.A.; Roy, M.M.; Sidhu, M.S. Experimental investigation on a diesel engine fueled by diesel-biodiesel blends and their emulsions at various engine operating conditions. *Appl. Energy* **2017**, *203*, 582–593. [CrossRef]

42. Hajilar, S.; Shafei, B. Thermal transport properties at interface of fatty acid esters enhanced with carbon-based nanoadditives. *Int. J. Heat Mass Tran.* **2019**, *145*, 118762. [CrossRef]

43. Ramírez-Verduzco, L.F.; Rodríguez-Rodríguez, J.E.; del Rayo Jaramillo-Jacob, A. Predicting cetane number, kinematic viscosity, density and higher heating value of biodiesel from its fatty acid methyl ester composition. *Fuel* **2012**, *91*, 102–111. [CrossRef]

44. Refaat, A.A. Correlation between the chemical structure of biodiesel and its physical properties. *Int. J. Environ. Sci. Technol.* **2009**, *6*, 677–694.

45. Folayan, A.J.; Anawe, P.A.L.; Aladejare, A.E.; Ayeni, A.O. Experimental investigation of the effect of fatty acids configuration, chain length, branching and degree of unsaturation on biodiesel fuel properties obtained from lauric oils, high-oleic and high-linoleic vegetable oil biomass. *Energ. Rep.* **2019**, *5*, 793–806. [CrossRef]

46. Marlina, E.; Wijayanti, W.; Yuliati, L.; Wardana, I.N.G. The role of pole and molecular geometry of fatty acids in vegetable oils droplet on ignition and boiling characteristics. *Renew. Energy* **2020**, *145*, 596–603. [CrossRef]

47. Paula, A.J.; Stéfani, D.; Filho, A.G.S.; Kim, Y.A.; Endo, M.; Alves, O.L. Surface chemistry in the process of coating mesoporous $SiO_2$ onto carbon nanotubes driven by the formation of Si-O-C Bonds. *Chem.-Eur. J.* **2011**, *17*, 3228–3237. [CrossRef] [PubMed]

48. Ayoola, A.A.; Anawe, P.A.L.; Ojewumi, M.E.; Amaraibi, R.J. Comparison of the properties of palm oil and palm kerneloil biodiesel in relation to the degree of unsaturation of their oil feedstocks. *Int. J. App. Nat. Sci.* **2016**, *5*, 1–8.

49. Su, Y.C.; Liu, Y.A.; Diaz Tovar, C.A.; Gani, R. Selection of prediction methods for thermophysical properties for process modeling and product design of biodiesel manufacturing. *Ind. Eng. Chem. Res.* **2011**, *50*, 6809. [CrossRef]

50. Rao, G.L.N.; Ramadhas, A.S.; Nallusamy, N.; Sakthivel, P. Relationships among the physical properties of biodiesel and engine fuel system design requirement. *Int. J. Energ. Environ.* **2010**, *1*, 919–926.

51. Sivaramakrishnan, K.; Ravikumar, P. Determination of cetane number of biodiesel and its influence on physical properties. *ARPN J. Eng. Appl. Sci.* **2012**, *7*, 205–211.

52. Kumar, J.; Bansal, A. Application of artificial neural network to predict properties of diesel—biodiesel blends. *Kathman Univ. J. Sci. Eng. Technol.* **2010**, *6*, 98–103. [CrossRef]

53. Bukkarapu, K.R.; Rahul, T.S.; Kundla, S.; Vardhan, G.V. Effects of blending on the properties of diesel and palm biodiesel. In *IOP Conference Series: Materials Science and Engineering*; IOP Publishing Ltd.: Hyderabad, India, 2018; Volume 330, pp. 1–15. [CrossRef]

54. Yao, C.; Dou, Z.; Wang, B.; Liu, M.; Lu, H.; Feng, J.; Feng, L. Experimental study of the effect of heavy aromatics on the characteristics of combustion and ultrafine particle in DISI engine. *Fuel* **2017**, *203*, 290–297. [CrossRef]

55. Phan, A.N.; Phan, T.M. Biodiesel production from waste cooking oils. *Fuel* **2008**, *87*, 3490–3496. [CrossRef]

56. Chen, H.; Xie, B.; Ma, J.; Chen, Y. NOx emission of biodiesel compared to diesel: Higher or lower? *Appl. Therm. Eng.* **2018**, *137*, 584–593. [CrossRef]

57. Giakoumis, E.G.; Sarakatsanis, C.K. A comparative assessment of biodiesel cetane number predictive correlations based on fatty acid composition. *Energies* **2019**, *12*, 422. [CrossRef]

58. de Oliveira, F.M.; de Carvalho, L.S.; Teixeira, L.S.; Fontes, C.H.; Lima, K.M.; Câmara, A.B.; Sales, R.V. Predicting cetane index, flash point, and content sulfur of diesel–biodiesel blend using an artificial neural network model. *Energy Fuels* **2017**, *31*, 3913–3920. [CrossRef]

59. Bemani, A.; Xiong, Q.; Baghban, A.; Habibzadeh, S.; Mohammadi, A.H.; Doranehgard, M.H. Modeling of cetane number of biodiesel from fatty acid methyl ester (FAME) information using GA-, PSO-, and HGAPSO-LSSVM models. *Renew. Energy* **2020**, *150*, 924–934. [CrossRef]

60. Mishra, S.; Anand, K.; Mehta, P.S. Predicting the cetane number of biodiesel fuels from their fatty acid methyl ester composition. *Energy Fuels* **2016**, *30*, 10425–10434. [CrossRef]

61. Moser, B.R. Biodiesel production, Properties and Feedstocks. In *Biofuels*; Springer: Singapore, 2011; pp. 285–347.

62. Giakoumis, E.G.; Sarakatsanis, C.K. Estimation of biodiesel cetane number, density, kinematic viscosity and heating values from its fatty acid weight composition. *Fuel* **2018**, *222*, 574–585. [CrossRef]

 *processes*

*Article*

# Environmental Assessment of Olive Mill Solid Waste Valorization via Anaerobic Digestion Versus Olive Pomace Oil Extraction

Bernabé Alonso-Fariñas [1], Armando Oliva [2,3], Mónica Rodríguez-Galán [1], Giovanni Esposito [4], Juan Francisco García-Martín [5], Guillermo Rodríguez-Gutiérrez [6], Antonio Serrano [6,7] and Fernando G. Fermoso [6,*]

[1] Departamento de Ingeniería Química y Ambiental, Escuela Técnica Superior de Ingeniería, Universidad de Sevilla, Camino de los Descubrimientos s/n. 41092 Seville, Spain; bernabeaf@us.es (B.A.-F.); mrgmonica@us.es (M.R.-G.)

[2] Department of Civil and Mechanical Engineering, University of Cassino and Southern Lazio, Via Di Biasio 43, 03043 Cassino, Italy; a.oliva1@nuigalway.ie

[3] Departmnet of Microbiology, National University of Ireland Galway, University Road, H91 TK33 Galway, Ireland

[4] Department of Civil, Architectural and Environmental Engineering, University of Napoli Federico II, via Claudio 21, 80125 Napoli, Italy; gioespos@unina.it

[5] Departamento de Ingeniería Química, Facultad de Química, Universidad de Sevilla, C/ Profesor García González, 1, 41012 Seville, Spain; jfgarmar@us.es

[6] Instituto de Grasa, Spanish National Research Council (CSIC), Ctra. de Utrera, km. 1, 41013 Seville, Spain; guirogu@ig.csic.es (G.R.-G.); antonio.serrano@ig.csic.es (A.S.)

[7] School of Civil Engineering, The University of Queensland, Campus St. Lucia - AEB Ed 49, St Lucia 4067, QLD, Australia

* Correspondence: fgfermoso@ig.csic.es; Tel.: +34-954611550

Received: 20 April 2020; Accepted: 21 May 2020; Published: 23 May 2020

**Abstract:** Anaerobic digestion is a promising alternative to valorize agrifood wastes, which is gaining interest under an environmental sustainability overview. The present research aimed to compare anaerobic digestion with olive pomace oil extraction, by using life cycle assessment, as alternatives for the valorization of the olive mill solid waste generated in the centrifugation process with a two-outlet decanter from oil mills. In the case of olive pomace oil extraction, two cases were defined depending on the type of fuel used for drying the wet pomace before the extraction: natural gas or a fraction of the generated extracted pomace. The anaerobic digestion alternative consisted of the production of biogas from the olive mill solid waste, heat and electricity cogeneration by the combustion of the generated biogas, and composting of the anaerobic digestate. The life cycle assessment showed that anaerobic digestion was the best alternative, with a global environmental impact reduction of 88.1 and 85.9% respect to crude olive pomace oil extraction using natural gas and extracted pomace, respectively, as fuel.

**Keywords:** biogas; environmental impact; life cycle assessment; olive pomace; sustainability

## 1. Introduction

The olive oil industry represents one of the fastest-growing industrial sectors worldwide, being of great importance in the economy of countries, such as Spain, Greece, and Italy, and becoming an important industry in countries, such as Chile, South Africa, or Argentina. The volume of processed olives in the main olive oil producer countries, such as Spain, leads to the generation of circa 4–5 million metric tons of annual waste. The olive mill solid waste is the main waste produced in olive mills that uses the two-outlet decanter, the most used system for olive oil extraction. Olive mill solid waste is a semi-solid with a high degree of humidity and high organic load [1].

In general, the olive mill solid waste obtained from the two-outlet decanters is transported to the pomace extraction plants to extract the crude pomace oil from them, mainly by extraction with organic solvents (technical hexane) [2]. Before extraction, a drying phase is necessary to reduce the moisture and volatile matter of the olive mill solid waste (between 65 and 75%) to less than 8%. This drying phase involves the highest energy consumption of the whole process of pomace oil extraction, and it is normally fed by natural gas or by the resulting extracted pomace from pomace oil extraction. Of note is that this extracted pomace, once dried, is regarded as an excellent solid biofuel that is currently used in industrial boilers and electric power generation industries [3], thus decreasing the demand for natural gas. In Spain, this energy production at an industrial scale has been possible, thanks to government incentives for the production of electricity from biomass. Notwithstanding, such incentives have been drastically reduced for running plants and have been canceled for new plants. As a result, the economic feasibility of the plants that use extracted pomace as the thermal source for the generation of electrical energy has decreased. Other challenges that pomace oil producers are facing are the fluctuation of olive mill solid waste generation and, mainly, the low commercial value of crude pomace oil [3].

In this context, anaerobic digestion (AD) has been shown to offer a possible solution for the management of the olive mill solid waste [4,5]. AD of olive mill solid waste produces mainly two streams, i.e., methane, as a source of bioenergy, and a stabilized digestate for use in agriculture as fertilizer, avoiding the need to resort to a drying process.

This study aimed to compare the environmental impacts of the two alternatives considered for olive pomace valorization, i.e., (a) AD of the olive mill solid waste, combustion of the generated biogas for heat and electricity production, and dewatering of the digestate for subsequent composting, and (b) extraction of crude pomace oil after drying with natural gas or extracted pomace.

## 2. Materials and Methods

In this study, a comparative attributional life cycle assessment (LCA) was carried out according to the ISO 14040/44 standards [6,7]. The goal and scope, the inventory data, and the impact assessment method used in this study are described in the following sections.

### 2.1. Goal and Scope of the Study

The main goal was to estimate and compare the life cycle environmental impacts of two alternatives for olive mill solid waste valorization: AD and crude olive pomace oil extraction (OPOE).

The scope of the study was from 'gate to gate'. Figure 1 shows all the foreground and background processes included in the system boundaries for each alternative—AD and OPOE. For the sake of clarity, foreground processes were framed with a dashed line box per each valorization alternative, including (1) for AD: biogas generation in an AD reactor, combustion of the biogas in a cogeneration engine for the production of heat and electricity, dewatering of the digestate in a decanter, and composting of the solid phase from the decanter, and (2) for OPOE: drying of the olive mill solid waste, extraction of the oil from the dried waste, and refining of the crude olive pomace oil. Hence, in the function of the energy pricing policies, producers might prefer burning natural gas and selling the extracted pomace from an economic point of view; two cases for OPOE were considered depending on the source of energy employed for olive mill solid waste drying: natural gas (OPOE-A) and extracted pomace (OPOE-B). The construction and decommissioning of the treatment plants were excluded under the hypothesis that the lifespan of these infrastructures is long enough to assume that the impacts of these stages per functional unit can be considered negligible.

The functional unit was defined as the valorization of 1 metric ton of olive pomace.

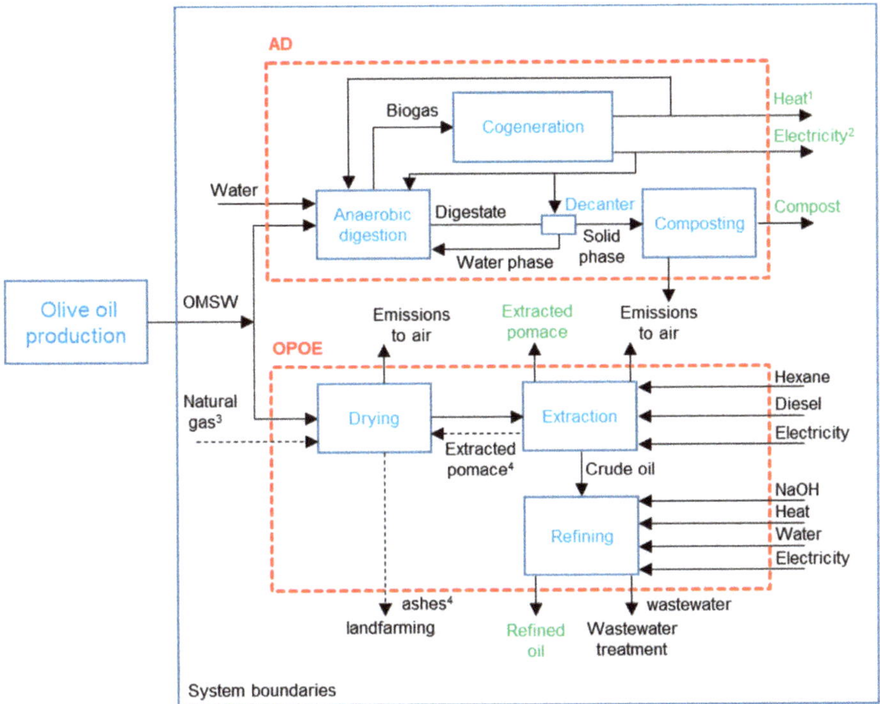



[1] Net heat production after subtracting the total heat consumed from the heat produced.
[2] Net electricity production after subtracting the total electricity consumed from the electricity produced.
[3] Only for case OPOE-A (olive pomace oil extraction, drying with natural gas).
[4] Only for case OPOE-B (olive pomace oil extraction, drying with extracted pomace).

**Figure 1.** System boundaries.

*2.2. Description of the Systems*

As shown in Figure 1, the system under study consisted of two alternative pathways for olive mill solid waste valorization. Each of the processes included in the foreground is described below. The reasons for choosing each background process are justified in Section 2.3, concerning inventory data. As a common practice, it was assumed that the olive husks were removed from the olive mill solid waste in the olive mill. The main characteristics of the olive mill solid waste are summarized in Table 1 [1].

**Table 1.** Olive mill solid waste characterization.

| | |
|---|---|
| **Total solids (g/L)** | 266 ± 4 |
| **Volatile solids (g/L)** | 250 ± 4 |
| **pH** | 4.97 ± 0.01 |
| **Alkalinity (mg CaCO$_3$/L)** | 6559 ± 5 |

2.2.1. Anaerobic Digestion (AD)

The first stage of this alternative was the production of biogas via AD of the olive mill solid waste stream in the anaerobic reactor. The AD conditions for olive mill solid waste were based on experimental results obtained in previous research works [1,8]. Water was consumed for dilution in AD as to reduce the total solids concentration until 9%wt before feeding the reactor [9]. The heat was necessary for keeping the temperature of the digester around 30 °C for mesophilic conditions, and electricity was used for pumping and stirring.

The digestate was dewatered using a decanter. The solid phase was valorized by composting, while the liquid phase was recirculated to the AD reactor to reduce the water consumption for diluting the olive mill solid waste. During composting, moisture was reduced from 65% from the solid phase to 35% in the compost, and $NH_3$, $N_2O$, and $CH_4$ were emitted to air. The compost was sold for its use as organic fertilizer. Direct application of digestate to the soil was not considered since it is a practice that is increasingly limited in the legislation, forcing the implementation of stabilization processes, such as composting, before the reuse of the anaerobic digestate [10].

The generated biogas was combusted in a cogeneration engine in which heat and electricity were simultaneously generated. Both heat and electricity were enough to cover the energy requirement of the rest of the stages of the system—AD (heat and electricity) and decanter (electricity). The surplus energy was sold and fed to the grid.

### 2.2.2. Olive Pomace Oil Extraction (OPOE)

Firstly, the olive mill solid waste was dried to reduce its humidity in a rotary dryer. As aforementioned, two options were considered depending on the fuel selected to supply the energy required for drying: natural gas (OPOE-A) or extracted pomace resulting from the crude pomace oil extraction stage (OPOE-B). Flue gas was emitted due to fuel combustion. When extracted pomace was used as fuel, ashes were generated and used in landfarming.

The dried olive mill solid waste, or olive pomace, was then subjected to the oil extraction phase. The extraction with an organic solvent, namely, technical hexane, was chosen for this study as it is widely used at industrial scale in the extraction plants. The extracted phase was distilled to remove the solvent from the pomace oil. The recovered solvent was then recirculated to the extraction process. The main intakes for this process were considered in the study: technical hexane, electricity, and diesel for heat production [11]. The emission from diesel combustion and hexane losses were also included within the boundary limits.

Due to its high acidity, the crude pomace oil must be refined. The most employed method for olive pomace oil refining is the chemical refining, in which the crude pomace oil reacts with an alkali solution to neutralize the free fatty acids [12]. Caustic soda was added, forming soap stock by reacting with the fatty acids. A centrifuge was used for separating the oil/soap mixture and, subsequently, the oil from the soap was clarified by filtration. The generated wastewater stream was sent to appropriate treatment. Heat, electricity, water, and sodium hydroxide were considered in the study as needed supplies. Soap stock is not considered as a by-product according to the results reported in [13].

### 2.3. Inventory Data

The life cycle inventory (LCI) data for both alternatives for olive mill solid waste valorization are detailed in Table 2. All background data were sourced from Ecoinvent v3.3 [14]. The figures shown in Table 2 were calculated based on the data and assumptions summarized below.

**Table 2.** Inventory data referred to 1 metric ton of olive mill solid waste. na; not applied.

| Category | Unit Per Metric Ton of Olive Mill Solid Waste | Anaerobic Digestion (AD) | Olive Pomace Oil Extraction (OPOE-A) [3] | Olive Pomace Oil Extraction (OPOE-B) [4] |
|---|---|---|---|---|
| Electricity | kWh | −215.06 [1] | 2.42 | 2.42 |
| Heat | MJ | −850.97 [2] | 0.06 | 0.06 |
| Diesel | kg | na | 1.07 | 1.07 |
| Natural gas | $m^3$ | na | 61.9 | na |
| Compost | kg | 221.03 | na | na |
| Refined olive pomace oil | kg | na | 22.17 | 22.17 |
| Extracted olive pomace | kg | na | 224.39 | 82.46 |
| Water | kg | 157.75 | 3652 | 3652 |
| Technical hexane | kg | na | 0.03 | 0.03 |
| NaOH solution | kg | na | 0.11 | 0.11 |
| Emissions to air | | | | |
| $CO_2$ (fossil) | kg | na | 143.07 | 3.37 |
| $CH_4$ | kg | 1.15 | 0.91 | 0.91 |
| $N_2O$ | kg | 0.05 | 0.05 | 0.05 |
| $NH_3$ | kg | 0.25 | na | na |
| Technical hexane | | na | 0.02 | 0.02 |
| Olive mill solid waste transportation | tkm | na | 100.00 | 100.00 |
| Wastewater to treatment | $m^3$ | na | 3.10 | 3.10 |
| Ashes to landfarming | kg | na | na | 3.78 |

[1] Net electricity production after subtracting the total electricity consumed from the electricity produced. [2] Net heat production after subtracting the total heat consumed from the heat produced. [3] Natural gas employed as fuel for drying. [4] A fraction of the extracted olive pomace is employed as fuel for drying.

### 2.3.1. Anaerobic Digestion (AD)

Anaerobic digester. The ultimate methane production ($G_{max}$) obtained from olive mill solid waste in previous work by using biomethane potential tests was 216 $cm^3$ $CH_4$/g volatile solids (VS) [1]. The biomethane production was obtained then by applying a scale-up factor of 0.85 to this experimental $G_{max}$ value (216 $cm^3$ $CH_4$/g VS) [9].

Decanter. The electricity consumed by the decanter for dewatering was 3.5 kW/h per metric ton of digestate [15].

Composting. Emission to air during composting was calculated according to average reported values [16].

Cogeneration engine and energy integration. The energy generation efficiency in a cogeneration biogas engine was considered 33% for electricity and 55% for thermal energy (30% in hot water and 25% in exhausted gas) [17]. The 200 kJ of thermal energy per kg of waste fed to the AD was consumed to keep the operating temperature of the reactor [18]. The electricity consumption in the AD section reached 15% of the electricity generated by the co-generation biogas [19].

### 2.3.2. Olive Pomace Oil Extraction (OPOE)

Drying and extraction. Data from the literature [11] were adapted to consider both fuel options for drying. Olive mill solid waste needed to be dried until 10%wt humidity. The energy requirement for drying was 2176 MJ/t wet olive mill solid waste [20]. Lower heating value (LHV) of natural gas and extracted oil pomace was 42.4 MJ/kg and 15.33 MJ/kg, respectively [20]. Emissions to air from fuel combustion were calculated using emission factors from the Intergovernmental Panel on Climate Change [21]. Background data from Ecoinvent was used for the use of the ashes in landfarming.

Refining. Inventory data from the Ecoinvent database for the chemical refining of crude vegetable oil were adapted by using an average acidity of 10% for the crude olive pomace oil [12]. Background inventory data for a specific treatment for wastewater from vegetable oil refinery were also included.

### 2.3.3. Transport

Transport background from the Ecoinvent database was assumed for all materials except for the olive mill solid waste. The AD facility was considered to be located in the same area as the olive mill. In this sense, the transport of the olive mill solid waste to the AD reactor could be despised. The distance from the olive mill to the extraction and refining plant was assumed to be 100 km.

### 2.3.4. System Expansion Approach

To compare both alternatives for the valorization of olive mill solid waste from the 2-outlet decanter, for which the obtained products were different, system expansion was applied. Each alternative was credited for avoiding the production of products that could be substituted by the different valuable outcomes. The credits were equal to the impacts of the production, by current production processes, of the substituted products. The data for these avoided production systems were sourced from Ecoinvent. Table 3 summarizes the credits associated with substituted products for each valorization alternative.

**Table 3.** Credits associated with avoided products for each valorization alternative.

| Anaerobic digestion | | |
|---|---|---|
| **Outcomes** | **Credits for avoided products** | **Equivalence ratio** |
| Electricity | Medium voltage-Spanish mix | 1:1 (kwh) |
| Heat | Industrial heat from natural gas | 1:1 (MJ) |
| Compost | Peat [1] | 1:1 (kg) [1] |
| **Olive pomace oil extraction** | | |
| **Outcomes** | **Credits for avoided products** | **Equivalence ratio** |
| Refined olive pomace oil | Refined vegetable oil | 1:1 (kg) |
| Extracted olive pomace | Natural gas | 1:1 (MJ) [2] |

[1] According to [16]. [2] LHV (lower heating value): natural gas = 42.4 MJ/kg; extracted olive pomace = 15.33 MJ/kg [20].

### 2.4. Impact Assessment

SimaPro v.8.3. software from Pré Consultants B.V. (Amersfoort, The Netherlands) was used to model the life cycle. The latest available version of CML 2001 (Centrum voor Milieuwetenschappen, January 2016 version) impact assessment method was used to calculate the environmental impacts [22]. The eleven impact categories included in the CML 2 method were assessed: abiotic depletion potential of elements (ADe), abiotic depletion potential of fossil fuel resources (ADf), global warming potential (GWP), ozone depletion potential (ODP), human toxicity potential (HTP), freshwater aquatic ecotoxicity potential (FWEP), marine aquatic ecotoxicity potential (MWEP), terrestrial ecotoxicity potential (TEP), photochemical oxidants creation potential (POP), acidification potential (AP), and eutrophication potential (EP).

Despite ISO standards do not require normalization and weighting, they are frequently applied in practice to identify important impact categories or to solve tradeoffs between results [23]. In this study, normalization was included to obtain a single score per alternative by using the reference values included in the CML 2001 method, as well as a default weighting factor of one.

## 3. Results

The environmental impacts of the alternatives considered for olive mill solid waste valorization are shown in Figure 2. A thorough discussion determining the main contributors to each impact category has been addressed in the following section for a better understanding of the environmental differences between both alternatives. To illustrate the origin of the environmental impacts, Figure 3, Figure 4, and Figure 5 for AD, OPOE-A, and OPOE-B, respectively, show the percentage contribution of the different concepts included in the inventory to each impact category, distinguishing between positive and negative (credits) impacts. Additionally, Figure 6 shows the percentage of contributions to the impacts for the refining of the crude pomace oil (OPOE-A and OPOE-B). In general terms, it is worth noting that most of the credits for AD came from avoiding the external production of electricity. This was in contrast to OPOE, in which the main contributors to the credits depended on the impact category.

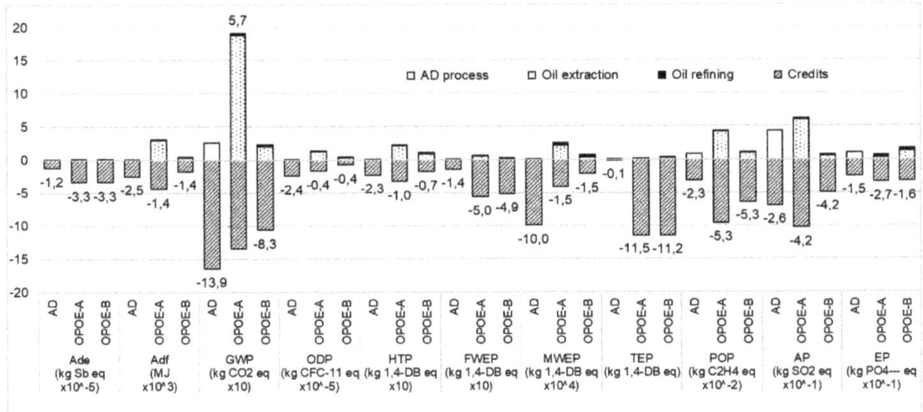

**Figure 2.** Life cycle environmental impacts of olive mill solid waste valorization via anaerobic digestion and olive pomace oil extraction. AD: anaerobic digestion; OPOE-A: crude olive pomace oil extraction with natural gas as fuel for olive pomace drying; OPOE-B: crude olive pomace oil extraction with extracted pomace as fuel for olive pomace drying. The values shown on top of each bar represent the total impact after the system credits have been applied. Some impacts have been scaled to fit. To obtain the original values, multiply by the factor shown on the x-axis for the relevant impacts.

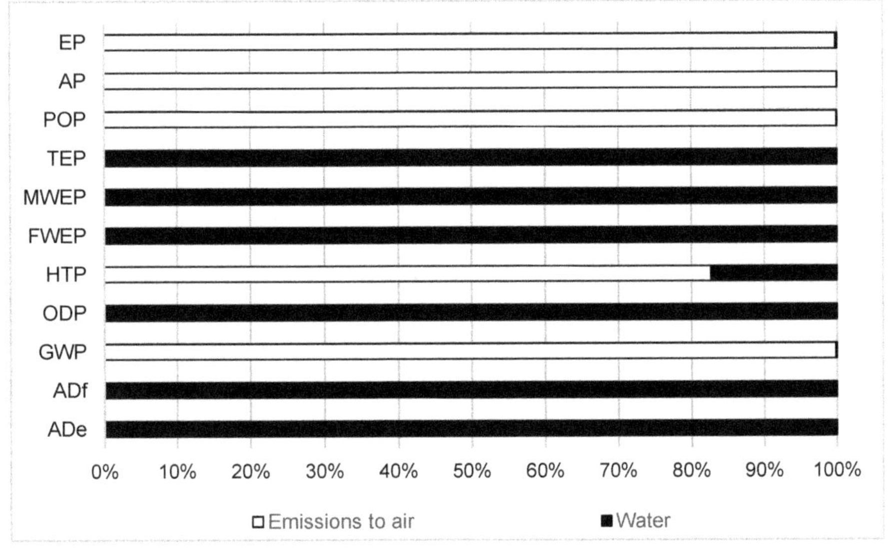

(a)

**Figure 3.** *Cont.*

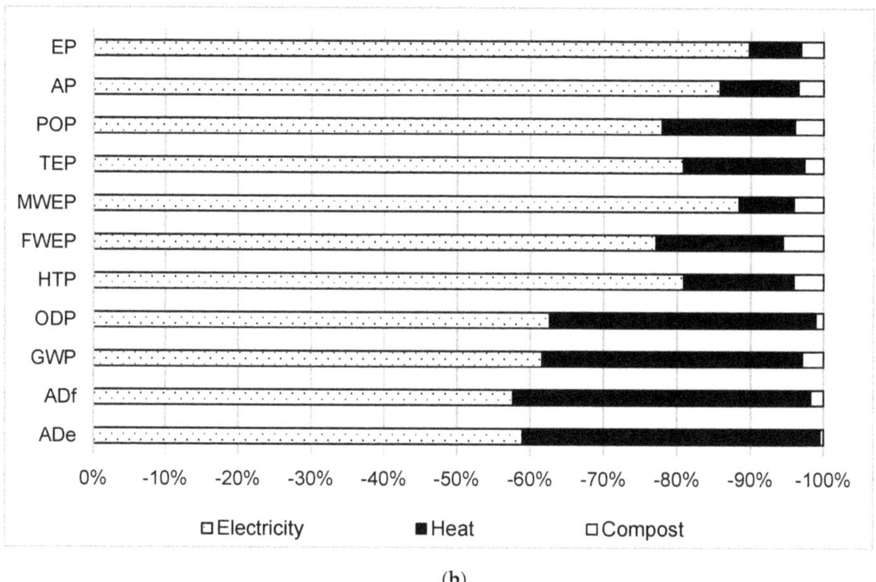



(**b**)

**Figure 3.** Percentage contribution to the impacts for anaerobic digestion scheme (AD): (**a**) positive contribution; (**b**) credits.

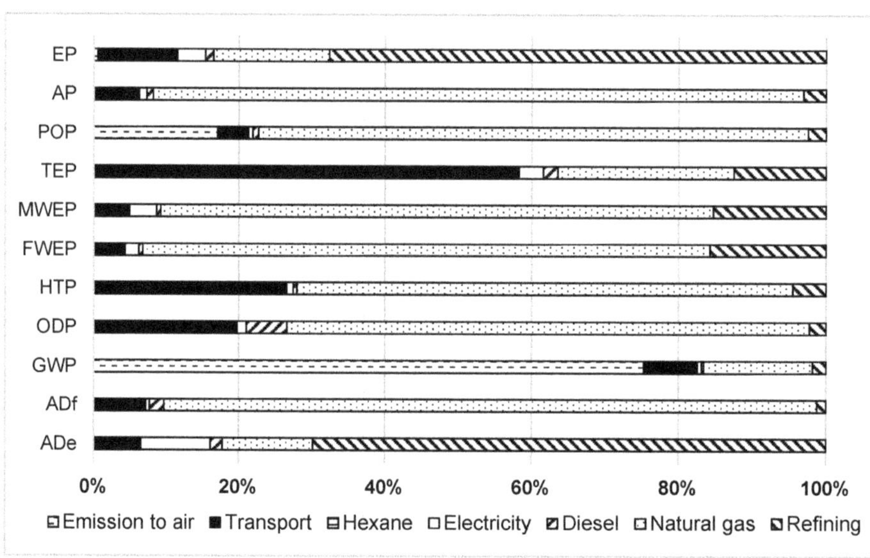



(**a**)

**Figure 4.** *Cont.*

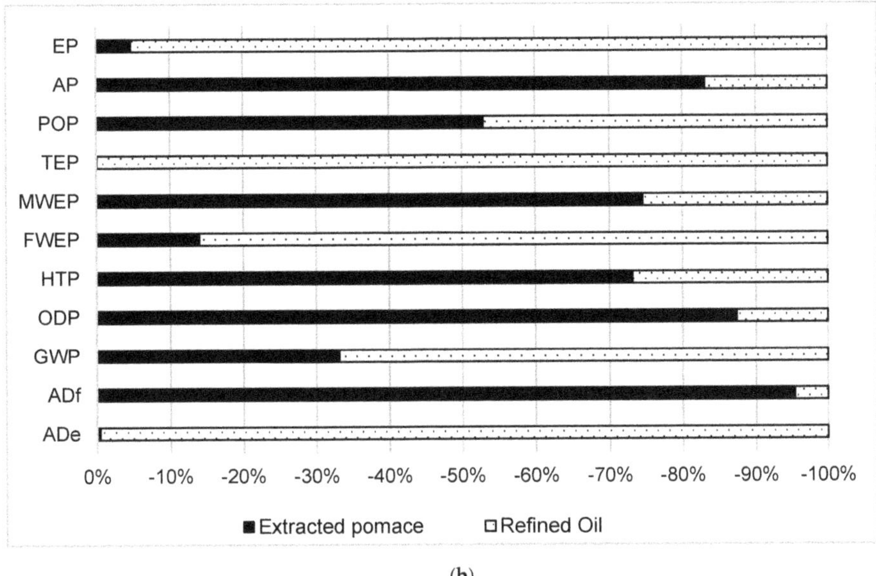



(**b**)

**Figure 4.** Percentage contribution to the impacts for crude olive pomace oil extraction, burning natural gas (OPOE-A): (**a**) positive contribution; (**b**) credits.

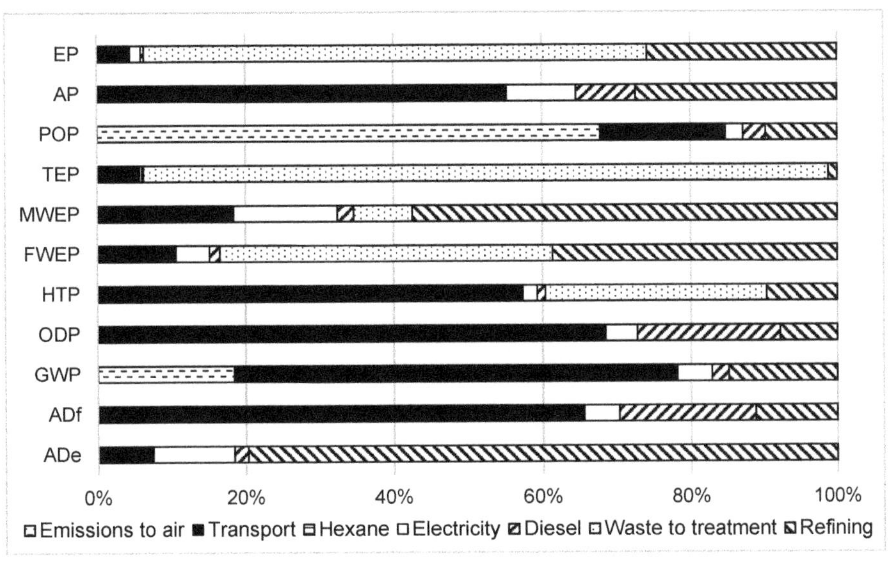



(**a**)

**Figure 5.** *Cont.*

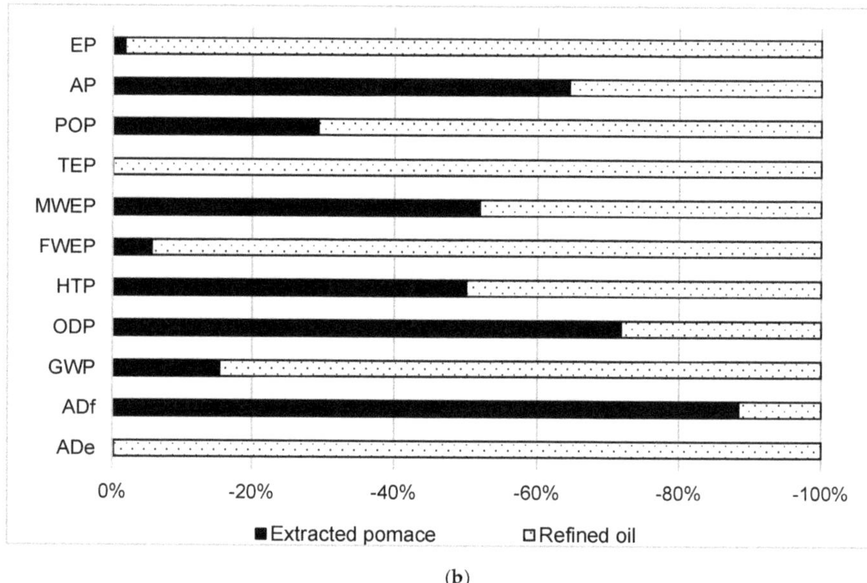

(b)

**Figure 5.** Percentage contribution to the impacts for crude olive pomace oil extraction, burning a fraction of the extracted olive pomace (OPOE-B): (**a**) positive contribution; (**b**) credits.

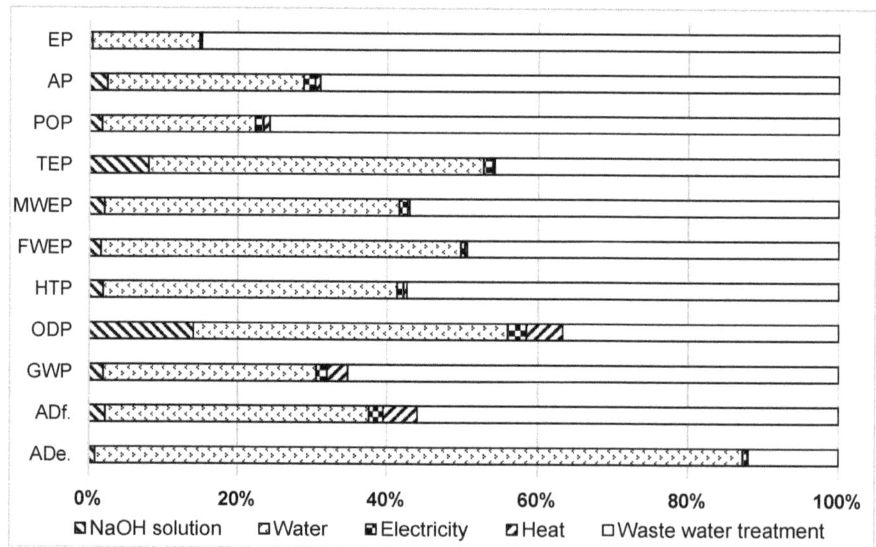

**Figure 6.** Percentage contribution to the impacts of the crude pomace oil refining (OPOE-A and OPOE-B).

Since AD of olive mill solid waste was still a process at the developing stage and due to the relevance of the credits obtained for this alternative by avoiding electricity production, a sensitivity analysis was performed, varying the amount of biogas generated per kg of treated olive mill solid waste. Relative results obtained for AD, related to OPOE-A and OPOE-B environmental impacts and expressed as percentage increment (+ values) or decrement (− values) are represented in Figure 7, including cases with a reduction on biogas production of 20, 35, and 50% respect to the experimental value used as a reference.

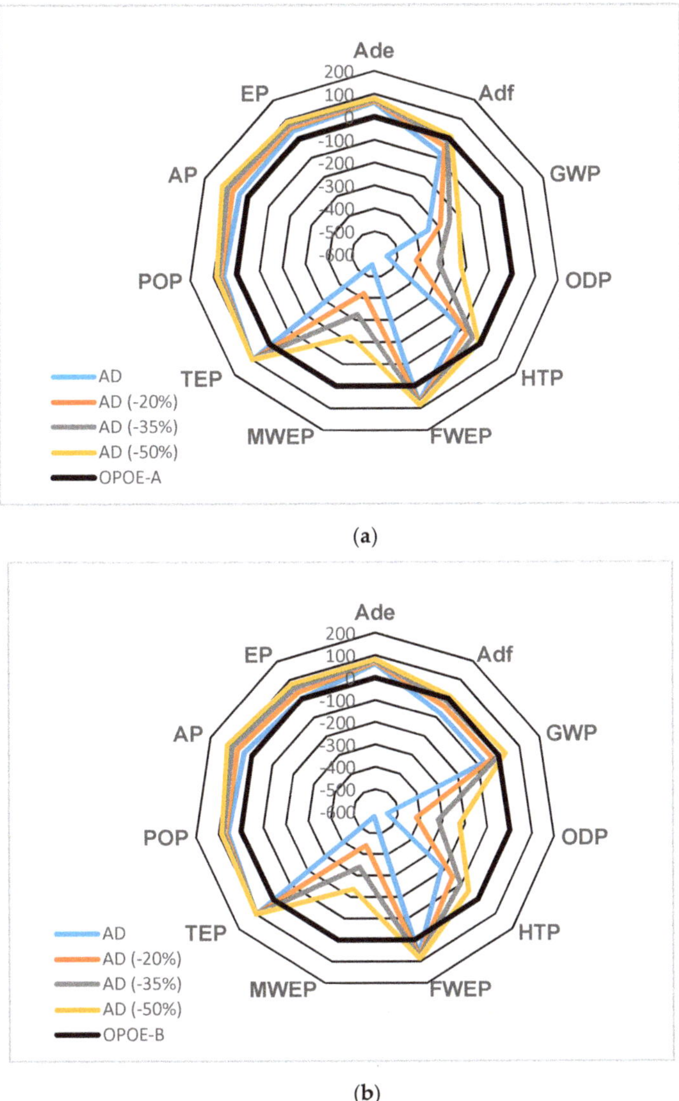



**Figure 7.** Influence of the reduction of the biogas production from olive mill solid waste in the anaerobic digestion (AD) on the environmental LCA (life cycle assessment) comparison with OPOE employing natural gas (**a**) and a fraction of the extracted olive pomace (**b**) as fuel for drying.

Finally, Table 4 illustrates the normalized environmental results and the simple scores, these latter ones by applying a weighting factor of 1.

**Table 4.** Normalized environmental impacts and single scores per kg of olive mill solid waste, where ADe, abiotic depletion potential of elements; ADf, abiotic depletion potential of fossil fuel resources; GWP, global warming potential; ODP, ozone depletion potential; HTP, human toxicity potential; FWEP, freshwater aquatic ecotoxicity potential; MWEP, marine aquatic ecotoxicity potential; TEP, terrestrial ecotoxicity potential; POP, photochemical oxidants creation potential; AP, acidification potential; and EP, eutrophication potential (EP).

| Impact Category | Anaerobic Digestion (AD) | Olive Pomace Oil Extraction (OMOE-A) | Olive Pomace Oil Extraction (OMOE-B) |
|---|---|---|---|
| ADe | $-1.40 \times 10^{-13}$ | $-3.89 \times 10^{-13}$ | $-3.89 \times 10^{-13}$ |
| ADf | $-7.86 \times 10^{-11}$ | $-4.53 \times 10^{-11}$ | $-4.53 \times 10^{-11}$ |
| GWP | $-2.76 \times 10^{-11}$ | $1.13 \times 10^{-11}$ | $-1.65 \times 10^{-11}$ |
| ODP | $-2.71 \times 10^{-13}$ | $-4.19 \times 10^{-14}$ | $-4.19 \times 10^{-14}$ |
| HTP | $-2.92 \times 10^{-12}$ | $-1.33 \times 10^{-12}$ | $-9.27 \times 10^{-13}$ |
| FWEP | $-2.79 \times 10^{-11}$ | $-9.66 \times 10^{-11}$ | $-9.43 \times 10^{-11}$ |
| MWEP | $-8.55 \times 10^{-10}$ | $-1.31 \times 10^{-10}$ | $-1.26 \times 10^{-10}$ |
| TEP | $-2.22 \times 10^{-12}$ | $-2.36 \times 10^{-10}$ | $-2.31 \times 10^{-10}$ |
| POP | $-2.71 \times 10^{-12}$ | $-6.30 \times 10^{-12}$ | $-6.30 \times 10^{-12}$ |
| AP | $-9.36 \times 10^{-12}$ | $-1.50 \times 10^{-11}$ | $-1.50 \times 10^{-11}$ |
| EP | $-1.17 \times 10^{-11}$ | $-2.07 \times 10^{-11}$ | $-1.21 \times 10^{-11}$ |
| Single score [1] | $-1.02 \times 10^{-9}$ | $-5.41 \times 10^{-10}$ | $-5.48 \times 10^{-10}$ |

[1] Weighting factor = 1.

## 4. Discussion

### 4.1. Global Warming

As expected, due to the $CO_2$ of fossil origin emitted with the flue gas during the drying stage, the valorization of olive mill solid waste via AD could involve a great reduction in GWP (345%) with respect to the pomace oil extraction when natural gas is used as fuel (OPOE-A). This reduction was still significant (67%) when a fraction of the extracted pomace was burned (OPOE-B). Without considering the credits, AD and OPOE-B had similar GWP (23.5–25.5 kg $CO_2$ eq /t). The difference in GWP was mainly due to the greater credits obtained in AD (165 kg $CO_2$ eq /t) with respect to OPOE-B (106 kg $CO_2$ eq /t). The main contributors to these credits were the electricity and the olive pomace oil for oil extraction, respectively. The emissions during composting were the main contributors for AD, whereas the transport of olive mill solid waste to the extraction plant was the main contributor for OPOE-B. As expected, most of the GWP came from the combustion of natural gas for OPOE-A. Even though credits in OPOE-A were higher than those obtained in OPOE-B (135 kg $CO_2$ eq./t), this could not compensate for the emissions from natural gas combustion. As aforementioned, the GWP of OPOE-A was the only positive impact (57 kg $CO_2$ eq /t) of all the categories under study.

Assuming a production volume of olive oil in Spain of 1,250,000 metric tons in one regular-season [24] and a ratio of olive mill solid waste to an olive oil of 819:176 kg/kg [25], the application of AD to olive mill solid waste could save the emission of 808 kt $CO_2$ eq. In this sense, AD could be considered environmentally friendlier, in terms of GWP, than OPOE with biomass combustion (saving just 483 kt $CO_2$ eq.) and OPOE when natural gas is burned (almost 330 kt $CO_2$ eq. emission).

This meant that if the olive oil production process could release up to 2.5 kg $CO_2$eq./dm$^3$ olive oil [26], the valorization of the olive mill solid waste by AD could compensate around 28% of the greenhouse gas emissions from olive oil production, which was far higher than the 17% compensated in the same scenario by applying the crude olive pomace oil extraction (OPOE-B).

### 4.2. Abiotic Depletion of Elements

Regarding the abiotic depletion of elements (ADe), the contribution of the processes to the impact was negligible compared with the credits for all studied cases. The fuel used for drying did not affect the impact of the pomace olive oil extraction since almost 100% of the credits were coming from avoiding the production of vegetable oil. Concretely, most of these credits were coming from avoiding the manufacturing of pesticides used in vegetable crops for oil production. These credits made OPOE the best option according to this impact category, being the impact value obtained for AD 64% higher than for OPOE.

### 4.3. Abiotic Depletion of Fossil Resources and Ozone Layer Depletion

Concerning abiotic depletion of fossil resources (ADf), AD was the best option with a low contribution to the process, limited to the water consumption, and the obtained credits due to the avoiding of the production of electricity from fossil fuels. AD had a 74% lower ADf than OPOE. Both cases for OPOE had the same value of ADf due to the energy equivalence applied between the extracted pomace and the avoided natural gas. The amount of extracted pomace consumed in OPOE-B was equivalent to the natural gas consumed in OPOE-A, reducing in the same quantity the credits obtained in OPOE-B due to the sale of extracted pomace to be used as fuel. The results obtained for ozone layer depletion (ODP) were similar to those achieved for ADf. In this case, the impact of AD was 547% lower than for OPOE.

### 4.4. Human Toxicity

As aforementioned, AD was the best option for this category. Another remarkable result that requires deeper analysis is that using extracted pomace as fuel was a worse alternative than using natural gas in terms of HTP. Without counting the credits, OPOE-A had a higher impact than OPOE-B due to the contribution of natural gas production. However, in this case, the reduction of the credits in OPOE-B was not equivalent to the reduction in the impacts, as shown for ADf, due to the contribution of the treatment in a landfill of the ash from the extracted pomace combustion.

### 4.5. Ecotoxicity

OPOE-A had the lowest burdens for freshwater ecotoxicity (FWEP) and terrestrial ecotoxicity (TEP), followed by OPOE-B. The credits from avoiding the production of vegetable oil were responsible for these results. Nevertheless, OPOE had the highest impact on marine water ecotoxicity (MWEP) regardless of the fuel used for drying due to the higher credits assigned to AD from electricity production.

### 4.6. Photochemical Oxidation (POP) and Acidification (AP)

OPOE was the best alternative for photochemical oxidation and acidification. The value of the impact did not depend on the fuel chosen for drying for the same reason explained in Section 4.3 for ADf and ODP.

For POP, without credits, the process for OPOE had always a higher impact than the AD due to the contribution of the production of natural gas (OPOE-A) or the treatment of the ashes from the extracted pomace combustion (OPOE-B). Nevertheless, these impacts were mainly compensated by the credits obtained from avoiding the use of pesticides in vegetable crops for oil production. Conversely, concerning AP, AD had higher burdens than OPOE-B when credits were not considered. The contribution of the emissions to air from composting could be responsible for this fact.

### 4.7. Eutrophication

OPOE-A showed the lowest burdens for eutrophication potential (EP), 42% lower than OPOE-B, which was the second-best option. The higher impacts of OPOE-B and AD were caused by the ash treatment and the emissions to air during composting, respectively.

### 4.8. Sensitivity Analysis

As could be seen in Figure 7, for a reduction of 50% in the production of biogas from olive mill solid waste, remarkable changes were observed in ADf and GWP. On the one hand, for ADf, OPOE turned to be a better option than AD regardless of the fuel used for drying. OPOE-B could be regarded as a better option than AD in terms of GWP.

### 4.9. Normalization

According to the normalized results and applying a weighting factor of 1, AD was the best alternative with a global environmental impact reduction of 85.9–88.1% with respect to both OPOE options. Within a circular economy approach, the use of natural gas (OPOE-A) was the worst option, but the use of a fraction of the extracted pomace as fuel for drying (OPOE-B) offered a reduction of only 1.2% of the environmental impact, expressed as a single score, with respect to the case of using natural gas (Table 4).

## 5. Conclusions

Evaluating each category separately, AD was shown as the best alternative for GWP and the other four categories, including ADf, ODP, HTP, and MWTP. The use of extracted pomace as fuel (OPOE-B) instead of natural gas (OPOE-A) could strongly reduce GWP but, conversely, increase the impact in the other two categories, i.e., HTP and EP, leaving the rest without change or with a negligible increment (FWTP, TEP). More specifically, the refining process of crude oil had a very low contribution compared to the extraction process. After evaluating the three alternatives with normalized results and applying a weighting factor of 1, AD showed a global environmental impact reduction of 88.1 and 85.9% with respect to crude olive pomace oil extraction using natural gas and extracted pomace as fuel, respectively.

**Author Contributions:** Conceptualization, B.A.-F. and M.R.-G.; methodology, B.A.-F. and A.O.; formal analysis, G.E.; resources, G.R.-G. and A.S.; writing—original draft preparation, B.A.-F.; writing—review and editing, J.F.G.-M., A.S., and F.G.F.; supervision, G.E.; funding acquisition, F.G.F. All authors have read and agreed to the published version of the manuscript.

**Funding:** This research was funded by the Spanish Ministry of Economy and Competitiveness by Project CTM2014-55095-R.

**Acknowledgments:** The authors thank Maria Victoria Ruiz Méndez and Rafael Borja Padilla, research members of the Instituto de la Grasa, Spanish National Research Council (CSIC), for their kind help in the inventory data collection of this work.



## References

1. Serrano, A.; Fermoso, F.G.; Alonso-Fariñas, B.; Rodríguez-Gutierrez, G.; Fernandez-Bolaños, J.; Borja, R. Olive mill solid waste biorefinery: High-temperature thermal pre-treatment for phenol recovery and biomethanization. *J. Clean. Prod.* **2017**, *148*, 314–323. [CrossRef]
2. García Martín, J.F.; Cuevas Aranda, M.; Feng, C.-H.; Álvarez Mateos, P.; Torres García, M.; Sánchez Villasclaras, S. Energetic valorisation of olive biomass: Olive-tree pruning, olive stones and pomaces. *Processes* **2020**, *8*, 511. [CrossRef]
3. Olimerca. Cunde el Pesimismo Entre Los Orujeros. Available online: https://www.olimerca.com/noticiadet/cunde-el-pesimismo-entre-los-orujeros-/e49e04531bc5c2f12effc6f6c2f30ed8 (accessed on 15 October 2019).
4. Rincón, B.; Fermoso, F.G.; Borja, R. Olive Oil Mill Waste Treatment: Improving the Sustainability of the Olive Oil Industry with Anaerobic Digestion Technology. In *Olive Oil—Constituents, Quality, Health Properties and Bioconversions*; Boskou, D., Ed.; InTech: London, UK, 2012.
5. Serrano, A.; Fermoso, F.G.; Alonso-Fariñas, B.; Rodríguez-Gutiérrez, G.; López, S.; Fernandez-Bolaños, J.; Borja, R. Performance evaluation of mesophilic semi-continuous anaerobic digestion of high-temperature thermally pre-treated olive mill solid waste. *Waste Manag.* **2019**, *87*, 250–257. [CrossRef] [PubMed]

6. ISO 14040. *Environmental Management—Life Cycle Assessment: Principles and Framework*; International Organization for Standardization: Geneva, Switzerland, 2006.

7. ISO 14044. *Environmental Management—Life Cycle Assessment: Requirements and Guidelines*; International Standardization Organization: Geneva, Switzerland, 2006.

8. Borja, R.; Rincón, B.; Raposo, F. Anaerobic biodegradation of two-phase olive mill solid wastes and liquid effluents: Kinetic studies and process performance. *J. Chem. Technol. Biotechnol.* **2006**, *81*, 1450–1462. [CrossRef]

9. Keith, F.; Tchobanoglous, G. *Handbook of Solid Waste Management*, 2nd ed.; McGraw-Hill: New York, NY, USA, 2002; ISBN 0071500340.

10. Hudcová, H.; Vymazal, J.; Rozkošný, M. Present restrictions of sewage sludge application in agriculture within the European Union. *Soil Water Res.* **2019**, *14*, 104–120. [CrossRef]

11. Figueiredo, F.; Marques, P.; Castanheira, É.G.; Kulay, L.; Freire, F. Greenhouse gas assessment of olive oil in Portugal addressing the valorization of olive mill waste. In Proceedings of the Symbiosis International Conference, Athens, Greece, 19–21 June 2014.

12. Sánchez Moral, P.; Ruiz Méndez, M.V. Production of pomace olive oil. *Grasas y Aceites* **2006**, *57*, 47–55. [CrossRef]

13. Schneider, L.; Finkbeiner, M. Life Cycle Assessment of EU Oilseed Crushing and Vegetable Oil Refining. *Tech. Rep.* **2013**, *2013*, 1–59.

14. Moreno Ruiz, E.; Lévová, T.; Reinhard, J.; Valsasina, L.; Bourgault, G.; Wernet, G. *Documentation of Changes Implemented in Ecoinvent Database 3.3*; Ecoinvent: Zurich, Switzerland, 2016.

15. Tampio, E.; Marttinen, S.; Rintala, J. Liquid fertilizer products from anaerobic digestion of food waste: Mass, nutrient and energy balance of four digestate liquid treatment systems. *J. Clean. Prod.* **2016**, *125*, 22–32. [CrossRef]

16. Saer, A.; Lansing, S.; Davitt, N.H.; Graves, R.E. Life cycle assessment of a food waste composting system: Environmental impact hotspots. *J. Clean. Prod.* **2013**, *52*, 234–244. [CrossRef]

17. Cano, R.; Nielfa, A.; Fdz-Polanco, M. Thermal hydrolysis integration in the anaerobic digestion process of different solid wastes: Energy and economic feasibility study. *Bioresour. Technol.* **2014**, *168*, 14–22. [CrossRef]

18. Gonzalez, A. Environmental, Energetic and Economic Fesiability of the Biomethanization of Wastes from the Agri-Food Industry in Extremadura. Ph.D. Thesis, Universidad de Extremadura, Badajoz, Spain, 2014.

19. Angelidaki, I.; Schmidt, J.E.; Karakashev, D.B. *A Sustainable Solution for Pig Manure Treatment: Environmental Compliance with the Integrated Pollution Prevention and Control Directive*; CORDIS: Nicosia, Cyprus, 2006.

20. *Potencial Energético de los Subproductos de la Industria Olivarera en Andalucía*; Regional Government of Andalusia: Seville, Spain, 2010.

21. Intergovernmental Panel on Climate Change. Emissions Factors Database. Available online: https://www.ipcc-nggip.iges.or.jp/EFDB/main.php (accessed on 7 August 2019).

22. Guinée, J.B.; Gorrée, M.; Heijungs, R.; Huppes, G.; Kleijn, R.; Wegener Sleeswijk, A.; Udo De Haes, H.A.; de Bruijn, J.A.; van Duin, R.; Huijbregts, M.A.J. *Life Cycle Assessment: An Operational Guide to the ISO Standards*; Kluwer Academic Publishers: Dordrecht, The Netherlands, 2001; ISBN 1-4020-0228-9.

23. Pizzol, M.; Laurent, A.; Sala, S.; Weidema, B.; Verones, F.; Koffler, C. Normalisation and weighting in life cycle assessment: Quo vadis? *Int. J. Life Cycle Assess.* **2017**, *22*, 853–866. [CrossRef]

24. Mercacei. Cunde el Pesimismo Entre Los Orujeros. Available online: https://www.mercacei.com/noticia/47472/actualidad (accessed on 5 July 2019).

25. Fernández, J. *Los Residuos de las Agroindustrias Como Biocombustibles Sólidos*; Vida Rural: Madrid, Spain, 2006; Volume 233, pp. 14–18.

26. Olivar del Sur. El Olivar, Importante Aliado del Medio Ambiente. Available online: https://olivadelsur.com/es/blog/el-olivar-importante-aliado-del-medio-ambiente-b100.html (accessed on 5 July 2019).

*processes*

*Article*

# Investigation of Ni–Fe–Cu-Layered Double Hydroxide Catalysts in Steam Reforming of Toluene as a Model Compound of Biomass Tar

David Díez Rodríguez [1,2,*] , Ana Urueña Leal [1,2] and Gregorio Antolín Giraldo [1,2]

[1] CARTIF Centre of Technology, Parque Tecnológico de Boecillo, 205, Boecillo, 47151 Valladolid, Spain; anauru@cartif.es (A.U.L.); great@cartif.es (G.A.G.)
[2] ITAP Institute, University of Valladolid, P del Cauce 59, 47011 Valladolid, Spain
* Correspondence: davdie@cartif.es

**Abstract:** This work focused on the synthesis of a catalyst based on layered double hydroxides with a molar cation concentration Ni/Cu/Fe/Mg/Al of 30/5/5/40/20 and its performance in the steam reforming of toluene as a model compound of biomass tar. Its performance at different temperatures (500, 600, 700, 800, and 900 °C) and steam/carbon molar ratios (S/C ratios) (1, 2, 4, 6, 8) was studied. The contact time used was 0.32 g h mol$^{-1}$. The catalyst obtained allowed us to reach 98–99.87% gas conversion of toluene with a low carbon deposition on catalyst surface (1.4 wt %) at 800 °C and S/C = 4. In addition, conversions in the range of 600–700 °C were higher than 80% and 90%, respectively, and the type of carbon deposited on the catalyst was found to be filamentous, which did not significantly reduce the performance of the catalyst.

**Keywords:** layered double hydroxide; toluene steam reforming; tar; gasification; Ni-based catalyst; hydrotalcite; hydrogen production

**Citation:** Rodríguez, D.D.; Leal, A.U.; Giraldo, G.A. Investigation of Ni–Fe–Cu-Layered Double Hydroxide Catalysts in Steam Reforming of Toluene as a Model Compound of Biomass Tar. *Processes* **2021**, *9*, 76. https://doi.org/10.3390/pr9010076

Received: 30 September 2020
Accepted: 28 December 2020
Published: 31 December 2020

**Publisher's Note:** MDPI stays neutral with regard to jurisdictional claims in published maps and institutional affiliations.

## 1. Introduction

The increase in demand for energy from fossil fuels that has taken place in recent decades is mainly due to the increase in population and living standards. In order to mitigate this situation, efforts are being made to encourage the substitution of fossil fuels by other renewable energy sources.

Among these alternatives, gasification is presented as a promising technology, in which syngas can be obtained from biomass as a renewable energy source. Gasification of biomass into syngas is expected to be used in various fields such as power generation and the production of hydrogen and liquid fuels (e.g., methanol and Fischer–Tropsch fuel) [1].

However, this syngas also contains some impurities, such as organic tars, which need to be removed before its application. Tars are the main contaminants in the syngas, and its content varies with the type of raw material and gasifier from 5 to 100 g/Nm$^3$. Their maximum allowable content depends on the end use, usually being 5 mg/Nm$^3$ in gas turbines and 100 mg/Nm$^3$ in internal combustion engines [2]. Tars are a complex mixture of aromatic and oxygenated hydrocarbons that may cause several operational problems. Within the compounds that constitute the tar, toluene is among the most important, besides naphthalene, indene, and acenaphthylene [3]. The requirement to remove tars introduces significant cost and complexity into the overall gasification process. Among the different strategies to remove tars from the gas, catalytic steam reforming is a useful technic.

By the other hand, layered double hydroxides (LDHs) are anionic clays that represent a class of layered materials with chemical composition expressed by the general formula

$$\left[\left(M^{2+}\right)_{1-x}\left(M^{3+}\right)_x(OH)_2\right]^{x+}\left[\left(A^{n-}\right)_{x/n}\cdot mH_2O\right] \tag{1}$$

119

where $[(M^{2+})_{1-x}\,(M^{3+})_x\,(OH)_2]^{x+}$ represents the metal layer and $[(A^{n-})_{x/n} \cdot mH_2O]$, the interlayer region [4]; $M^{2+}$ represents a divalent cation (i.e., $Ca^{2+}$, $Mg^{2+}$, $Zn^{2+}$, $Co^{2+}$, $Ni^{2+}$, $Cu^{2+}$, $Mn^{2+}$); and $M^{3+}$ represents a trivalent cation (i.e., $Al^{3+}$, $Cr^{3+}$, $Fe^{3+}$, $Co^{3+}$, $Ni^{3+}$, $Mn^{3+}$); whereas $A^{n-}$ is the interlayer anions such as inorganic anions (i.e., $Cl^-$, $NO^{3-}$, $ClO_4^-$, $CO_3^{2-}$, $SO_4^{2-}$) or complex organic molecules [5]. Moreover, $x$ represents the fraction of the trivalent cation and at the same time determines the electrostatic charge of the sheets. Its value is given by the molar ratio of the valence III and II metals, as follows:

$$x = \frac{\sum M^{3+}}{\sum M^{3+} + \sum M^{2+}} \tag{2}$$

Some natural minerals contain a fixed value of $x = 0.33$, although the generally accepted range as suitable for the synthesis of LDH compounds is $0.2 \leq x \leq 0.33$, i.e., an $M^{2+}/M^{3+}$ ratio between 2:1 and 4:1, in which the structure may be stable [6,7].

LDHs have received special attention due to their relatively simple synthesis, low cost, and excellent catalytic properties. LDHs can be calcined to give a stable and homogeneous mixture of oxides with a very small crystal size. These materials when reduced favor a high dispersion of metals, which contributes to avoid the deactivation of the catalyst [8,9].

On the other hand, nickel-based catalysts have been widely used in tar reforming because of is very effective in reforming of tars [10], whose performance can be improved by the addition of promoters such as Fe, Ce, Cu, Pd, Zn, and Ca [1,5,6,8,9].

In this context, the objective of this work is to study steam reforming of toluene as a tar model compound over nickel-based LDH catalysts.

## 2. Materials and Methods

### 2.1. Catalyst Preparation

Ni–Fe–Cu/LDH was prepared by coprecipitation from aqueous solutions at room temperature. Two solutions were prepared—solution A containing metal precursors, and prepared from their respective nitrates: $Ni(NO_3)_2\,6H_2O$, $Cu(NO_3)_2\,3H_2O$, $Fe(NO_3)_3\,9H_2O$, $Mg(NO_3)_2\,6\,H_2O$, and $Al(NO_3)_3\,9\,H_2O$. Solution A was 4M $NO_3^{-2}$, and was prepared to get the following molar cation concentration: Ni/Fe/Cu/Mg/Al: 30/5/5/40/20. Solution B contained $Na_2CO_3$ (0.22 M) and NaOH (3.56 M). Then, 100 mL of solution A and B was slowly dropped by HPLC pumps under vigorous stirring over 100 mL deionized water. To ensure a good mix and dispersion, we used a homogenizer (Model IKA magic, IKA-Werke GmbH & Co. KG, Staufen, Germany) as the reactor. The temperature of the solution was kept constant at 60 °C by means a thermostatic bath. The pH of the solution was also maintained constant during all test by means of a control system that added solution NaOH 1 M to obtain the pH at 10 to ensure the co-precipitation of the metallic salts.

The gel formed was aged for 20 h at 100 °C under reflux in order to improve its crystalline characteristics. The solid obtained was then filtered and washed with distilled water at 90 °C until the solution was a conductivity bellow 30 µS cm$^{-1}$ [11]. The LDH formed was dried at 105 °C overnight, and then finally crushed and calcined using a heating rate of 5 °C min$^{-1}$, from room temperature to 800 °C and keeping at this temperature for 3 h.

In order to obtain a greater understanding of the structure of the catalyst, we synthesized three additional LDHs following the same procedure. Table 1 shows the composition of all synthesized LDHs.

**Table 1.** Molar cation concentration of the synthesized layered double hydroxides (LDHs).

| LDH | Molar Cation Concentration Ni/Fe/Cu/Mg/Al |
|---|---|
| Ni−Fe−Cu/LDH | 30/5/5/40/20 |
| Ni−Fe/LDH | 35/5/0/40/20 |
| Ni−Cu/LDH | 35/0/5/40/20 |
| Ni/LDH | 40/0/0/40/20 |

## 2.2. Catalyst Characterization

### 2.2.1. Chemical Element Analysis

The chemical element analysis of as-synthesized catalysts was performed using the inductively coupled plasma-optical emission spectroscopy (ICP-OES) on the Optima 4300 DV (Perkin Elmer, Waltham, MA, USA). Prior to the analysis, the solid (0.05 g) was dissolved in 4 mL $HNO_3$ and then was heated to 100 °C for 5 min, to 170 for 15 min, and to 240 °C for 10 min; kept in isothermal condition for 15 min; and finally filtered and diluted.

### 2.2.2. Characterization of the Calcination Process

In order to analyze the calcination, process, we conducted a thermogravimetric analysis (TGA).

Thermogravimetric analysis was performed on a DTG-60H TG-DTA analyzer (Shimadzu Co. Ltd., Kyoto, Japan). The analyses were carried out using air with a flow rate of 50 mL $min^{-1}$, and a heating rate of 10 °C $min^{-1}$, from room temperature to 900 °C. The amount of sample analyzed was ≈3 mg. This analysis allowed us to determine the weight loss of a sample as a function of temperature and time (TG curve). In addition, the derivative of the TG signal (DTG curve) helped to determine the number of main thermal processes that took place, as well as their temperature intervals.

### 2.2.3. Temperature-Programmed Reduction (TPR) Analysis

TPR analyses were performed on Auto Chem II 2920 analyzers (Micromeritics Instrument Corporation, Norcross, GA, USA) with a thermal conductivity detector (TCD) used to quantify the amount of $H_2$ consumed during the analysis.

TPR was used to determine the reducible species and the corresponding reduction temperatures. Before measurement, a 0.20 g sample was dried by passing 50 mL $min^{-1}$ of Ar up to 150 °C at a heating rate of 10 °C $min^{-1}$, and holding time under these conditions 30 min. Then, the reduction of the sample was carried out by means of 50 mL $min^{-1}$ of 5% $H_2$/Ar to 950 °C $min^{-1}$, heating rate 10 °C $min^{-1}$, and holding time of 30 min.

### 2.2.4. X-Ray Diffraction (XRD) Analysis

Powder X-ray diffraction (XRD) measurements were performed on a D8 Discover (Bruker, Billerica, MA, USA) using Cu Kα radiation (λ = 0.154 nm) generated at 40 kV and 40 mA. The identification of the diffraction patterns was performed using the Joint Committee of Powder Diffraction Standards (JCPDS) database.

XRD analyses are suitable for determining lattice parameters and particle diameter. Thus, for example, LDHs are a hexagonal system whose most significant lattice parameters are $a$ and $c$—$a$ represents the average cation-cation distance within the layers and $c$ represents the distance between the brucite-like layer and the inter-layer [12,13]. The relationship between lattice parameter and interplanar spacing $d_{hkl}$ is given by

$$1/d_{hkl}^2 = \frac{4}{3}\left(\frac{h^2 + hk + k^2}{a^2}\right) + \frac{l^2}{c^2} \tag{3}$$

where $(h, k, l)$ are Miller indices, $a$ and $c$ are the lattice parameters, and $d$ is interplanar spacing [14]. Lattice parameter $a$ and $c$ are calculated from the (110) and (003) plane, respectively obtained from Equations (4) and (5).

$$a = 2d_{110} \tag{4}$$

$$c = 3d_{003} \tag{5}$$

The interplanar spacing of the (110) and (003) crystal planes was estimated by Bragg's law, which can be represented by Equation (6).

$$2\,d_{hkl}sin(\theta) = n \tag{6}$$

where $n$, $\lambda$, and $\theta$ are the order, X-ray wavelength, and diffraction angle, respectively.

Using Equations (4)–(6) and the diffraction angle obtained from the XRD pattern, we can obtain the lattice parameters $a$ and $c$.

The crystallite size ($D$) was calculated by Scherrer Equation (7)

$$D = \frac{K}{\beta \cos(\theta)} \tag{7}$$

where $K$ is the shape factor ($K = 0.89$), $\beta$ is the full width at half maximum (FWHM) of the analyzed diffraction peak in radians, and $\theta$ is the Bragg angle.

A similar procedure was used to determine the lattice parameters of calcined and reduced LDHs, but using Equation (8), which is characteristic of a face-centered cubic (fcc) structure such as that of NiO and metallic Ni.

$$a = d_{hkl} \sqrt{h^2 + k^2 + l^2} \tag{8}$$

### 2.2.5. Textural Properties

Specific surface area (BET) pore volume and average pore diameter of both as-synthesized and calcined catalysts were measured via $N_2$ adsorption at 198 °C using a surface area analyzer Autosorb-1C (Quantachrome Instruments, Boynton Beach, FL, USA). Surface area ($S_{BET}$) was analyzed using Brunauer–Emmett–Teller (BET) theory, and average pore diameter by Barrett–Joiner–Halenda (BJH) procedure.

### 2.2.6. Scanning Electron Microscopy (SEM)

A high-resolution scanning electron microscope (SEM) FEI ESEM Quanta 200 (FELMI-ZFE, Graz, Austria) was used to analyze the surface morphology of the as-synthetized and used catalyst.

### 2.2.7. Type and Amount of Carbonaceous Species Deposited on the Catalyst

The type of carbonaceous material deposited on the surface of the catalyst at the end of the steam reforming reaction was performed on a DTG-60H TG-DTA analyzer (Shimadzu Co. Ltd., Kyoto, Japan). The analyses were carried out using air with a flow rate of 50 mL min$^{-1}$, and a heating rate of 20 °C min$^{-1}$, from room temperature to 1000 °C. The amount of sample analyzed was ≈3 mg. This method was used to determine the type of carbon deposited by its relationship to its oxidation temperature [15].

The amount of carbon formed on the surface of the catalyst after reaction was measured using an elemental TruSpec CHN analyzer (Leco, St. Joseph, MI, USA).

### 2.3. Catalytic Tests

Steam reforming of toluene used as a model compound of tar was carried out using a fixed-bed reactor made of Hastelloy alloy steel, with an inner diameter of 10 mm and a length of 500 mm (Figure 1). The reactor was heated by a tubular electric furnace coupled to the reactor. The process temperature was measured by a K-type thermocouple placed in the center of the bed and controlled using a controller in closed loop (DEMEDE, Spain).

In each test, 250 mg of calcined catalyst was placed in the middle of the reactor on a 316 stainless steel sintered porous metal filter disc with an internal pore size of 40 μm. Quartz wool was used above and below the bed to help catalyst inside the reactor.

Prior to each test, the system was purged with 50 mL min$^{-1}$ $N_2$ for 15 min, and then the catalyst was reduced with a flow of $H_2$ and $N_2$ ($H_2/N_2 = 50/20$ mL min$^{-1}$) at 900 °C for 2 h to transform Mg(Ni,Al)O periclase species of $Ni^{2+}$ into well-dispersed Ni particles [16]. After reduction, the system was purged again with 50 mL min$^{-1}$ $N_2$ for 15 min while the temperature was kept at the desired temperature of the steam reforming 500–900 °C.

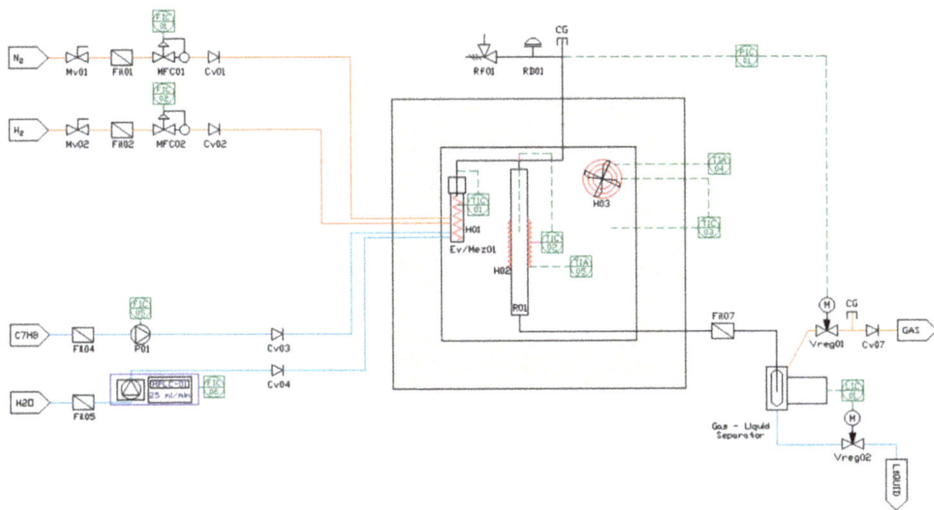

**Figure 1.** Schematic diagram of the experimental system.

Once the system was purged, the steam reforming reaction was started, with the duration of each test being 1 h. To do this, the water and the toluene were fed into a preheater by means of a HPLC pump and syringe pump, respectively. Moreover, to ensure a concentration of toluene in the gas phase of 1.50 vol %, we fed a current of $N_2$ into the preheater through a mass flow controller. This preheater was maintained at 250 °C in order to ensure a complete vaporization of the feedstock before entering the fixed bed reactor containing the catalyst. To prevent condensation and to ensure that feedstock was maintained in a homogeneous vapor phase, we placed the preheater and the reactor in a chamber that was kept warm at 200 °C.

The reaction products were then passed through a cold trap to condense unreacted toluene and water in the effluent stream. All the non-condensed gases ($H_2$, $N_2$, CO, $CO_2$, and $CH_4$) were analyzed online by gas chromatography using a chromatograph MicroGC CP4900 (Varian Inc., Palo Alto, CA, USA) with two channels, one equipped with a molecular sieve column and another one with a Polar Plot Q column, and both coupled to a thermal conductivity detector (TCD). The unreacted toluene and benzene produced were collected at the end of each test and analyzed offline by gas chromatography, using a 6890 N gas chromatograph coupled with mass spectrometer 5973 Network (Agilent, Santa Clara, CA, USA).

Different tests were developed using reaction temperatures of 500, 600, 700, 800, and 900 °C and steam/carbon molar ratios (S/C ratios) of 1, 2, 4, 6, and 8. The total flow rate of the product gases was calculated by means of a mass balance to the nitrogen stream. The contact time in the steam reforming reaction was calculated in all tests to be W/F = 0.32 g h mol$^{-1}$. F is the total molar flow of the feedstock (toluene, water, and nitrogen) and W represents the weight of the catalyst.

Table 2 summarizes the operating conditions for each S/C ratio. As can be seen in Table 2, the toluene flow rate was kept at 0.2 mmol min$^{-1}$, the water flow rate was adjusted to obtain the desired S/C, and the nitrogen flow rate was adjusted to ensure the same concentration of toluene in the feedstock and the same contact time.

**Table 2.** Operating conditions for each steam/carbon molar ratio (S/C ratio) and for $W/F = 0.32$ g h mol$^{-1}$.

| S/C | Reactant | Flow Rate, mmol min$^{-1}$ | Feedstock Concentration, vol % |
|---|---|---|---|
| 1 | Toluene | 0.20 | 1.5% |
|  | Water | 1.38 | 10.5% |
|  | Nitrogen | 11.59 | 88.0% |
| 2 | Toluene | 0.20 | 1.5% |
|  | Water | 2.77 | 21.0% |
|  | Nitrogen | 10.21 | 77.5% |
| 4 | Toluene | 0.20 | 1.5% |
|  | Water | 5.53 | 42.0% |
|  | Nitrogen | 7.45 | 56.5% |
| 6 | Toluene | 0.20 | 1.5% |
|  | Water | 8.30 | 63.0% |
|  | Nitrogen | 4.68 | 35.5% |
| 8 | Toluene | 0.20 | 1.5% |
|  | Water | 11.07 | 84.0% |
|  | Nitrogen | 1.91 | 14.5% |

Benzene yield was evaluated by a carbon balance between the carbon from the benzene formed and the carbon from the toluene at the feed, as shown in Equation (9).

$$Y_b(\%) = \frac{6 \times n_{C_6H_6}}{7 \times n_{C_7H_8}} \times 100 \tag{9}$$

In the same way, the conversion of toluene to gases was calculated by a carbon balance between the carbon from the gaseous products formed (CO, CO$_2$, and CH$_4$), and the carbon from the toluene at the feed.

$$X_g(\%) = \frac{n_{CO} + n_{CO_2} + n_{CH_4}}{7 \times n_{C_7H_8}} \times 100 \tag{10}$$

where $n_{CO}$, $n_{CO2}$, $n_{CH4}$, and $n_{C7H8}$ represent the molar flow rate of CO, CO$_2$, CH$_4$, and toluene, respectively.

The molar gas composition of each gas (H$_2$, CO, CO$_2$, and CH$_4$) was calculated as Equation (11).

$$Y_i(\%) = \frac{n_{gas\ i}}{n_{H2} + n_{CO} + n_{CO_2} + n_{CH_4}} \times 100 \tag{11}$$

## 3. Results and Discussion

### 3.1. Catalyst Characterization

### 3.1.1. Chemical Element Analysis Results

The chemical element analysis of as-synthesized and calcined catalysts is showed in Table 3.

**Table 3.** Chemical element analysis of as-synthesized and calcined Ni–Fe–Cu/LDH.

| LDH | Al | Cu | Mg | Fe | Ni |
|---|---|---|---|---|---|
| As-synthesized Ni–Fe–Cu/LDH | 18.86% | 5.87% | 38.70% | 4.75% | 31.82% |
| Calcined Ni–Fe–Cu/LDH | 18.90% | 5.92% | 38.54% | 4.70% | 31.94% |

The results showed that the LDH obtained corresponded to the one initially synthesized, although a slightly higher substitution of Mg$^{2+}$ by M$^{2+}$ cations (Ni$^{2+}$ and Cu$^{2+}$) was observed. These results also showed that the precipitation of the compounds took place almost in its totality, without important loss of reagent. After calcination at 800 °C, it was also observed that the composition remained almost constant.

### 3.1.2. TGA of the As-Synthesized Ni–Fe–Cu/LDH

The calcination process of the as-synthesized Ni–Fe–Cu/LDH was studied by performing TGA analysis (Figure 2).

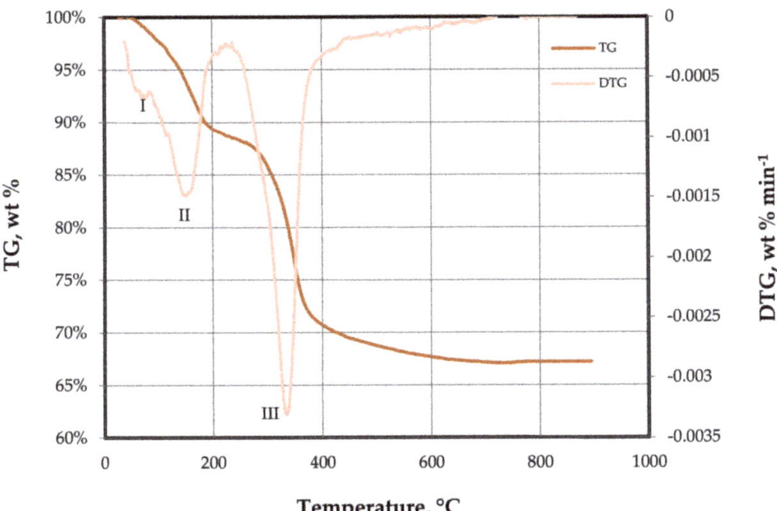

**Figure 2.** Thermogravimetric analysis (TGA) and differential thermal analysis (DTG) from as-synthesized Ni–Fe–Cu/LDH.

The TG curve (Figure 2) showed three main regions of mass loss. The first region was observed between ambient temperature up to 210 °C, which was related with the first two peaks in the DTG signal with maximum values at 100 and 180 °C. The first peak of the DTG curve (peak I) was related with a mass loss of 2.2% from the TG curve and occurring from ambient temperature up to around 100 °C, and was associated with the elimination of water, physically adsorbed on the external surface of the particle. The peak II at about 180 °C presented a mass of 8.7% in the interval 100–210 °C of the TG curve, and may be attributed to the removal of –OH and water from the brucite-like layer [17]. The second region in the TG curve was located between 210 and 400 °C and it corresponded to the second DTG peak, with maximum temperature of 360 °C. The average mass loss in this second region was around 18.5% and was attributed to the dehydroxylation of the brucite type layers, with decomposition of carbonate anions (decarbonization). From this temperature, called the collapse of the layers up to 800 °C, occurred the rearrangement of the structure with the formation of aluminum oxides, magnesia (periclase phase), and the formation of other metal and mixed oxides reaching the structure a great thermal stability, characteristic of oxides [17,18].

Moreover, TGA analysis allowed us to establish the temperature of 800 °C as the temperature from which the sample was thermally stabilized. This temperature has been widely used by other authors to obtain catalysts from LDH [12,15,19].

### 3.1.3. TPR Analysis Results

The reducibility of the calcined LDHs was investigated by TPR measurement; the obtained profiles are shown in Figure 3.

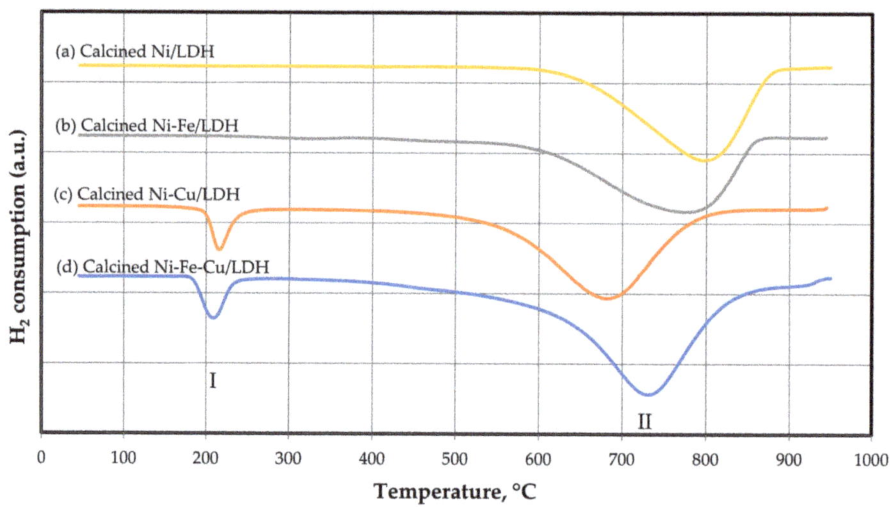



**Figure 3.** Temperature-programmed reduction (TPR) profile for calcined LDHs.

The TPR analysis of the different LDHs showed the presence of two peaks depending on whether the sample contained Cu in its composition or not. Thus, the Ni–Cu/LDH and Ni–Fe–Cu/LDH samples both showed one peak above 210 °C (peak I), which was associated to the reduction of bulk CuO [20].

In addition, Figure 3 shows a second peak (II) in the range of 680–800 °C, which was attributed to the reduction of $Ni^{2+}$ species in thermally stable phases such as Mg(Ni, Al)O solid solution [12,21].

The addition of precursors reduces the maximum reduction temperature, promoting the reduction of $Ni^{+2}$ species by weakening the bond between the Ni species and the support [21]. In the same way, the substitution of Mg by Ni helps to reduce the NiO–MgO interaction, leading to a decrease in the reduction temperature and a low dispersion of the metals [12].

In this way, Figure 3 shows a displacement of the maximum of peak II with respect to Ni–LDH (797 °C) towards lower temperatures when Fe or Cu was added. Thus, it can be seen how the addition of Fe caused a displacement of the maximum up to 776 °C, while the addition of Cu caused a more pronounced displacement up to 680 °C. This meant that the addition of Cu caused a further weakening of the Ni species in the support.

The total $H_2$ consumption during TPR analysis of the different calcined LDHs is shown in Table 4. It can be seen that the substitution of 5 mol % of Ni for Fe or Cu proved an increase in $H_2$ consumption. This increase was more pronounced in the case of Fe. As a result, calcined Ni–Fe–Cu/LDH was the HDL with the highest consumption of $H_2$.

**Table 4.** Total consumption of $H_2$.

| LDH Calcined | Mmol $H_2$/g |
| --- | --- |
| Ni/LDH | 6.68 |
| Ni–Fe/LDH | 7.30 |
| Ni–Cu/LDH | 7.22 |
| Ni–Fe–Cu/LDH | 7.84 |

On the other hand, in order to ensure an adequate reduction of the calcined LDH, we chose a reduction temperature of 900 °C [8] since, as shown in Figure 3, at that temperature there is practically no more consumption of $H_2$.

### 3.1.4. SEM Analysis Results

Figure 4 show SEM micrographs at different magnifications of the as-synthetized Ni–Fe–Cu/LDH. The synthesized LDH showed a homogeneous structure formed by small spheres of about 680 nm in diameter.

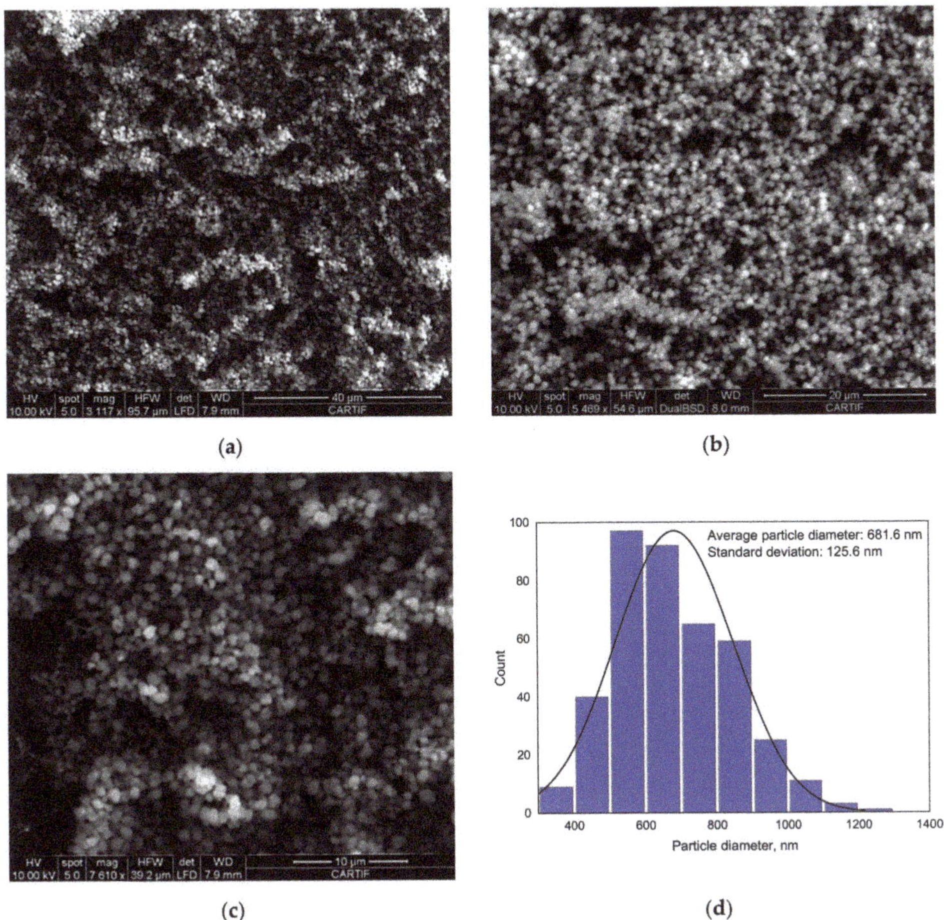

**Figure 4.** Scanning electron microscopy (SEM) micrographs at different magnifications of the as-synthetized Ni–Fe–Cu/LDH at different horizontal field widths (HFWs): (**a**) 95.7 μm, (**b**) 54.6 μm, (**c**) 39.2 μm, and (**d**) histogram.

As seen in Figure 4, as the magnification became larger, it became possible to observe how the synthesized catalyst was formed by small spherical and homogeneous particles, which suggests that the reaction device used was quite suitable in terms of carrying out the synthesis of LDHs.

The determination of the particle diameter was performed using the image analysis program ImageJ, which allowed us to calculate the histogram of the particle size distribution observed by SEM.

The average diameter of the as-synthetized LDH particles observed by SEM was 618.1 nm. After the calcination process, the catalyst lost its hydrotalcite structure, which caused a reduction in size of the particles to 414.2 nm (Figure 5).

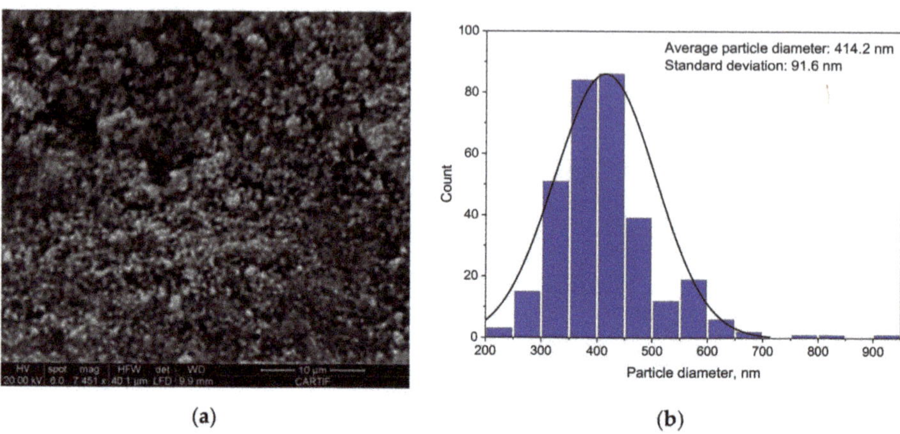

**Figure 5.** SEM micrograph (**a**) and histogram (**b**) of the calcined Ni–Fe–Cu/LDH at 40.1 μm (HFW).

### 3.1.5. XRD Analysis Results

The XRD patterns of as-synthesized, calcined, and hydrogenated Ni–Fe–Cu/LDH are reported in the Figure 6a–c, respectively.

XRD pattern of as-synthesized Ni–Fe–Cu/LDH (Figure 6a) showed several diffraction peaks at 11.5°, 23.0°, 34.6, 38.8°, 46.2, 60.5, 61.8°, and 65.5° indexed to (003), (006), (012), (015), (018), (110), (113), and (116) reflections, respectively. All of them were related to hydrotalcite (JCPDS 22e0700), indicating the formation of a pure hydrotalcite phase. This structure was in agreement with the results obtained by other authors who have studied LDHs [22–24]. No diffraction peaks were observed for any species of nickel, copper, or iron, which may have been due to their integration into the structure or their high dispersion in the LDH [25].

The smaller the size of the crystalline particles, the more defects they have and the wider the diffraction peaks [26]. Therefore, the wide peak (003) indicates the formation of highly crystalline material with a small particle size.

Lattice parameters $a$ and $c$ were calculated following the Equations (4)–(6) and the diffraction angle obtained from the XRD pattern (Figure 6a). Moreover, crystallite size ($D$) was calculated by Equation (7) on the basis of (003) reflections at 11.5°. These parameters are summarized in Table 5.

**Table 5.** Crystallite size and lattice parameters of as-synthesized and calcinated Ni–Fe–Cu/LDH.

| LDH | Crystallite Size | Lattice Parameters | |
|---|---|---|---|
| | $D$, nm | $a$, Å | $c$, Å |
| As-synthesized Ni–Fe–Cu/LDH | 12.83 | 3.06 | 23.06 |
| Calcined Ni–Fe–Cu/LDH | 4.32 | 4.20 | |

The values of as-synthesized Ni–Fe–Cu/LDH were similar to those calculated by other authors that have synthesized Ni-based catalyst from LDHs. For example, Zhou et al. [27] synthesized a Ni–Fe/LDH, finding a crystallite size of 9.1 and lattice parameter values of $a = 3.06$ and $c = 23.2$.

Lattice parameter $c$ is related to the electrostatic interactions between the brucite-like sheet and the interlayer, and is strongly dependent on the $M^{2+}/M^{3+}$ ratio that has a value of 3 in the Ni–Fe–Cu/LDH synthetized [16]. As as-synthesized Ni–Fe–Cu/LDH was calcined up to 900 °C, and the loss of the hydrotalcite structure progressively took place, resulting in a mixture of metal oxides that can be observed in Figure 6b.

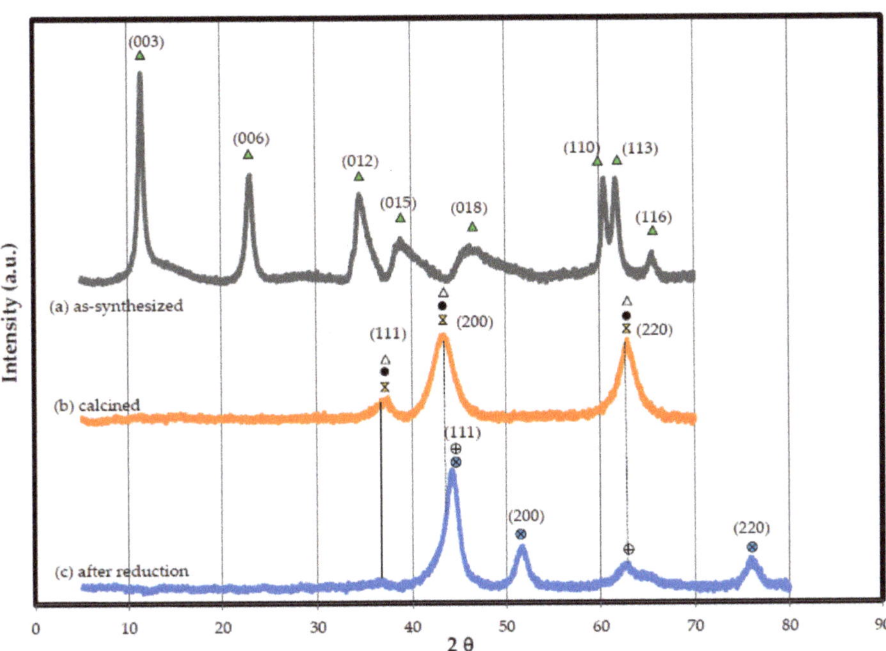

**Figure 6.** Ni-Fe-Cu/LDH XRD profiles (**a**) as-synthesized, (**b**) calcined, and (**c**) after reduction: ($\triangle$) hydrotalcite, ($\overline{\text{X}}$) MgO-like phase, ($\bullet$) NiO, ($\Delta$) Ni(Fe)O$_x$, ($\otimes$) Ni metal, ($\oplus$) Ni–Fe alloy.

As seen in Figure 6b, the XRD pattern for calcined sample showed reflections at around 37°, 43.5°, and 62.5°. These reflections were related to the planes (111), (200), and (220), respectively, which could correspond to the cubic structure of MgO and NiO. Ni$^{2+}$ has an ionic radius similar to that of Mg$^{2+}$, and thus it is likely that it was incorporated into the structure, forming the mixed oxide Mg(Ni,Al)O, as described by several authors [22,28,29], without significant modifications in the network parameter. On the other hand, Ni(Fe)O$_x$ also has a similar cubic structure to NiO, and thus it is possible that Ni(Fe)O$_x$ was also present due to Fe$^{3+}$ substitution in the nickel oxide lattice [27]. In relation to Cu, no patterns were observed regarding this metal, which may have been because it is highly dispersed or forms an amorphous phase. Table 5 also includes the crystallite size (*D*) of particles calculated by Equation (7) on the basis of (200) reflections and lattice parameters by Equation (8). A reduction of the particle size was obtained after calcination, which improved the characteristics of the catalyst.

After the reduction with H$_2$, the reduced Ni–Fe–Cu/LDH was obtained, whose XRD pattern can be seen in Figure 6c. The nickel particles were mostly found as Ni metal, as shown by the reflections at 44.2°, 51.7°, and 75.8°, corresponding to the crystallographic planes (111), (200), and (220). To a lesser extent, NiFe alloy and NiO reflections can also be observed.

### 3.1.6. Textural Properties Analysis Results

Specific surface area (BET), pore volume, and average pore diameter (BJH) of both as-synthesized and calcined catalysts were obtained, and their results are summarized in Table 6.

**Table 6.** Textural characteristics of the as-synthesized and calcined samples.

| Samples | $S_{BET}$ | $V_{pore}$ | Average Pore Diameter |
|---|---|---|---|
| | $m^2\ g^{-1}$ | $cm^3\ g^{-1}$ | nm |
| As-synthesized Ni-Fe-Cu/LDH | 131.10 | 0.509 | 3.90 |
| Calcined Ni-Fe-Cu/LDH | 136.70 | 0.556 | 13.57 |

After calcination, an increase in surface area and pore volume was observed, which may have been due to the removal of carbonate anions in the form of $CO_2$ as the sample was calcined. Moreover, calcination results in an increase in pore volume and pore diameter, which allowed for improved textural properties of the catalyst. Comparing the textural properties obtained from calcined Ni–Fe–Cu/LDH with respect to commercial catalysts, we could see, for example, that calcined catalyst showed a $S_{BET}$ of 136.70 $m^2\ g^{-1}$ larger than the commercial Raney nickel catalyst that had an average Ni surface area of 100 $m^2\ g^{-1}$ [30].

Figure 7 shows the pore diameter distribution curves of as-synthesized and calcined catalyst, obtained from nitrogen adsorption–desorption isotherm measured at 77 K.

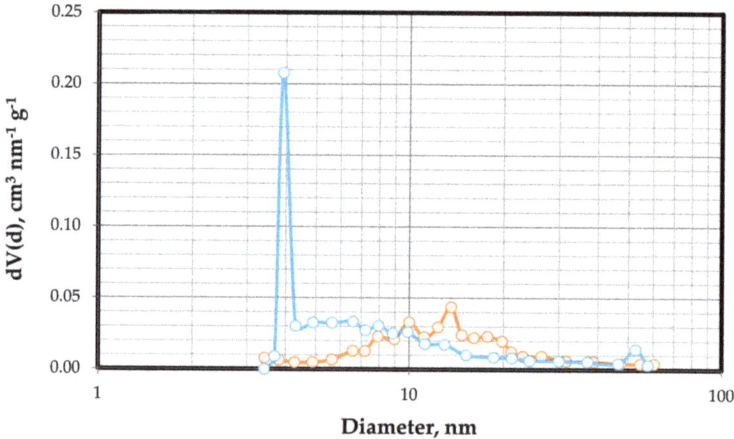


**Figure 7.** Pore size distributions of the catalyst (-○-) as-synthesized; (-○-) calcined.

For as-synthesized catalyst, we observed that the pore size distribution was very narrow, presenting a very sharp maximum over 3 nm, and most of the pore diameters were between 3 and 10 nm (mesopores).

After calcination of fresh LDH at 800 °C for 4 h, the calcined catalyst showed a wider pore size distribution than as-synthetized catalyst. The most part of the pores were between 3 and 21 nm, with a maximum over 13 nm. This may have been due to the fact that the calcination at high temperatures improved the textural properties, especially in the case where the catalyst contained Fe, as demonstrated by Zhou et al. [27].

*3.2. Effect of Temperature and S/C Ratio on Catalytic Performance of Ni–Fe–Cu/LDH Catalyst over Steam Reforming of Toluene*

The main reactions that take place in the steam reforming of toluene are as follows [15,31,32]:

Steam reforming:

$$C_7H_8 + 7H_2O \rightleftharpoons 7CO + 11H_2 \qquad \Delta H = 868.23\ kJ \qquad (12)$$

$$C_7H_8 + 14\ H_2O \rightleftharpoons 7CO_2 + 18H_2 \qquad \Delta H = 578.78\ kJ \qquad (13)$$

Water gas shift:

$$CO + H_2O \rightleftharpoons CO_2 + H_2 \qquad \Delta H = -41.35 \text{ kJ} \qquad (14)$$

Dry reforming:

$$C_7H_8 + 7CO_2 \rightleftharpoons 14CO + 4H_2 \qquad \Delta H = 1157.68 \text{ kJ} \qquad (15)$$

Hydroalkylation:

$$C_7H_8 + H_2 \rightleftharpoons C_6H_6 + CH_4 \qquad \Delta H = -41.91 \text{ kJ} \qquad (16)$$

Benzene hydrocracking:

$$C_6H_6 + 9H_2 \rightleftharpoons 6CH_4 \qquad \Delta H = -532.14 \text{ kJ} \qquad (17)$$

Benzene steam reforming:

$$C_6H_6 + 6H_2O \rightleftharpoons 6CO + 9H_2 \qquad \Delta H = 704.1 \text{ kJ} \qquad (18)$$

Methane steam reforming:

$$CH_4 + H_2O \rightleftharpoons CO + 3H_2 \qquad \Delta H = 206.04 \text{ kJ} \qquad (19)$$

Boudouard reaction:

$$2CO \rightleftharpoons CO_2 + C \qquad \Delta H = -172.53 \text{ kJ} \qquad (20)$$

Most of the reactions involved in toluene steam reforming are endothermic, and thus the process is favored both thermodynamically and kinetically at high temperatures.

Steam reforming of toluene can occur through reactions (12) and (13), depending on the amount of water available, producing $H_2$, and CO and $CO_2$, respectively.

Water gas shift (WGS) reaction (14) is an equilibrium reaction that is thermodynamically favored at low temperatures and kinetically favored at high temperatures [33].

In addition, toluene can react with $CO_2$ through the dry reforming reaction (15), forming CO and $H_2$, and it also can react through the reaction of hydroalkylation (16) with $H_2$ to produce benzene and methane, although this reaction, because it is exothermic, will be less favorable at high temperatures. On the other hand, benzene can react with $H_2$ through a hydrocracking reaction (17) to produce methane. Methane in turn can produce CO and $H_2$ through a steam reforming reaction (19). The Boudouard reaction (20) is an exothermic reaction that transforms CO into $CO_2$ and C.

Finally, it can be observed that in all the reactions except (18) and (20), there was an increase in the volume from reactants to products, and thus the process was favored at low pressure.

### 3.2.1. Effect of Temperature on Gas Yield and Gas Composition

Figure 8 shows the conversion of toluene to gases as a function of temperature and S/C ratio.

It can be noticed how as the temperature increased, the conversion increased. At temperatures below 500 °C, the steam reforming reaction did not take place. However, from 500 °C, the conversion increased sharply with temperature until temperature in the range 700–800 °C was reached, where conversion was highest. For temperatures above 800 °C, the conversion remained constant and no longer increased with temperature. The highest conversion (99.87%) was achieved for an S/C of 4 at a temperature of 800 °C. This could have been due to the fact that for higher S/C values, the water could compete for the active centers of the catalyst.

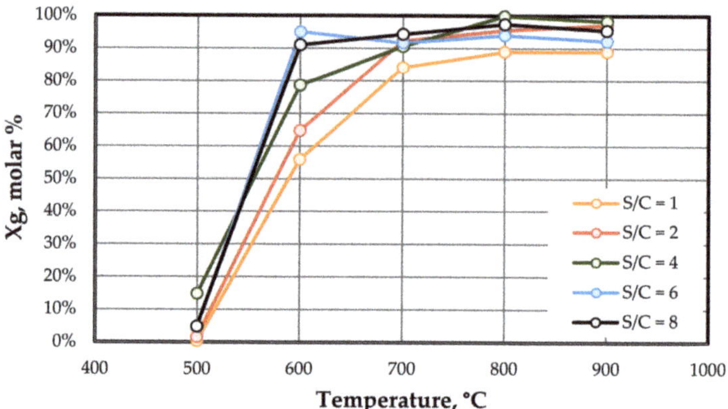

**Figure 8.** Conversion of toluene to gases as a function of temperature and S/C ratio.

Regarding the evolution of CO with temperature (Figure 9a), as the temperature increased, the concentration of CO increased, progressively reaching the maximum value at 900 °C, while the concentration of $CO_2$ (Figure 9b) increased until reaching its maximum value at 600 °C and then decreased slightly. This development may have been due to the fact that the WGS reaction is an exothermic reaction, and thus as the temperature increased, the equilibrium progressively shifted towards the formation of CO [34].

The concentration of $H_2$ (Figure 9c) was the largest during steam reforming with toluene. The concentration of hydrogen decreased strongly at temperatures above 500 °C; from this temperature, a slight decrease was observed as the temperature increased. This behavior has been observed by different authors [35] and can be explained again by the endothermic nature of the WGS reverse reaction, which reduces the concentration of $H_2$ as the temperature increases.

The hydroalkylation (12) and benzene hydrocracking (17) are exothermic reactions and therefore not favored by increasing temperature. As a consequence, the concentration of $CH_4$ (Figure 9d), after reaching a maximum at 600 °C, decreased sharply with increasing temperature. This means that the concentration of toluene occurred preferably through the reaction (12) as the temperature increased.

In the same way, the formation of benzene (Figure 9e) through reaction (16) decreased as the temperature increased due to its exothermic behavior. Benzene yield reached a maximum value at 500 °C of around 8–10%; from this temperature, the concentration of benzene decreased very quickly. This may also have been due to the fact that the catalyst can promote the steam reforming of benzene (18) in the range 600–900 °C.

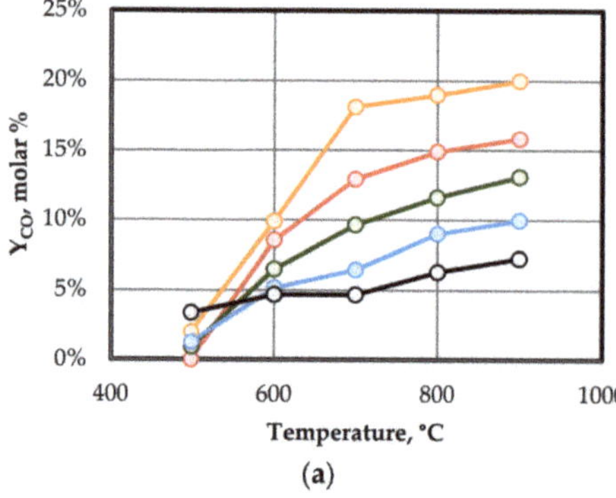

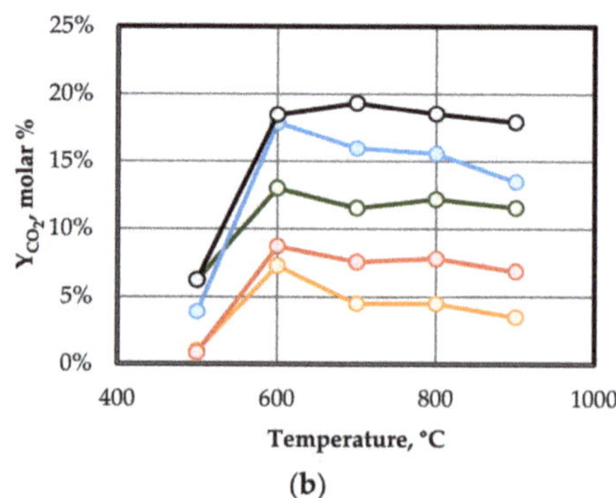

**Figure 9.** *Cont.*

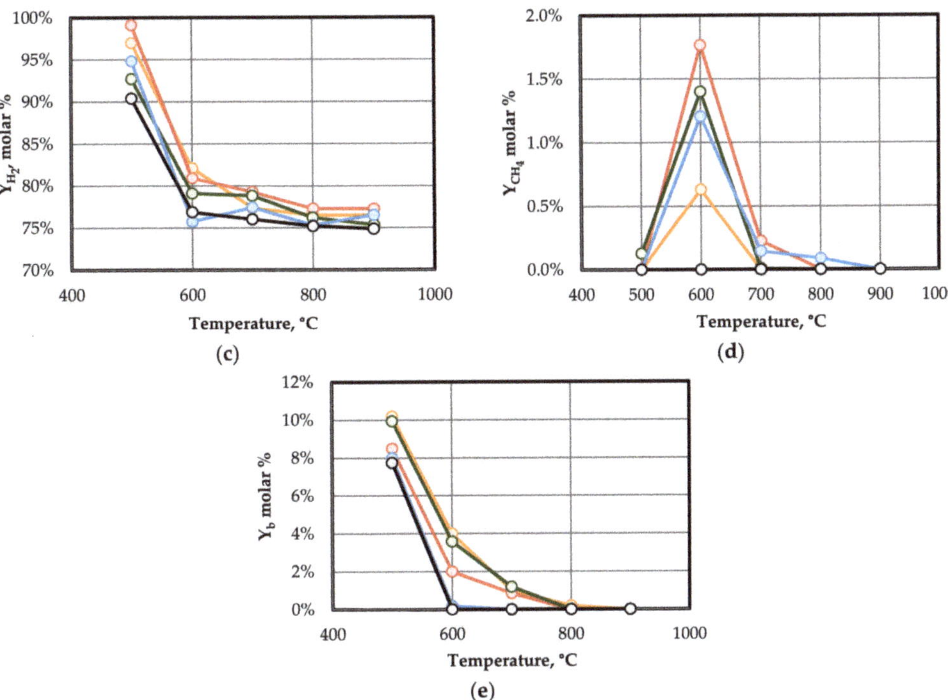

**Figure 9.** Molar gas composition as function of temperature and S/C ratio: (**a**) CO, (**b**) $CO_2$, (**c**) $H_2$, (**d**) $CH_4$. (**e**) Benzene yield. S/C = 1 (–○–), S/C = 2 (–○–), S/C = 4(–○–), S/C = 6 (–○–), S/C = 8 (–○–).

### 3.2.2. Effect of S/C Ratio on Gas Yield and Gas Composition

Regarding the influence of the S/C ratio on conversion of toluene to gases (Figure 8), we observed that the maximum differences were reached in the range of temperatures 500–700 °C. The main tendency was that as the S/C ratio increased, the conversion increased up to S/C ratios of 6. S/C values higher than 6 did not encourage an increase in conversion. This fact was especially noticed at 600 °C, where conversions of 56% were reached for S/C values of 1, while for S/C ratios higher than 6, the conversion was higher than 95%.

With respect to the evolution of the concentration of CO (Figure 9a) and $CO_2$ (Figure 9b), both concentrations were very dependent on the value of S/C. Thus, for example, the concentration of CO was observed to increase when the S/C ratio decreased, and thus the maximum values (17%) were reached for S/C = 1 at 900 °C, while for the same temperature of 900 °C and an S/C = 8, the concentration of CO was around 7%. The evolution of the concentration of $CO_2$ moved in an inverse way; it increased when the S/C ratio increased, reaching maximum values at S/C = 8. This was evidence that increasing the S/C ratio promotes reaction (13).

In respect to the influence of the S/C ratio on the concentration of $H_2$ (Figure 9c), the S/C had much less influence than temperature. Moreover, it was possible to observe that at low temperature, the concentration of $H_2$ decreased as the S/C ratio increased; this effect may have been due to an increase in the basicity. Thus, for example, Ahmed et al. [36] developed a Ni–Fe–Mg/zeolite catalyst and studied its performance in steam reforming reaction of toluene, finding that for S/C ratios higher than 1:1, the amount of $H_2$ decreased, which they related to a decrease in the basicity of the support. This effect occurred mainly at low temperatures, where the degree of conversion was low in these conditions an increase in the S/C ratio can cause an oxidation of the catalyst. At higher temperatures, there was no clear trend because there was very little variation in the concentration of $H_2$ with S/C.

The $CH_4$ concentration (Figure 9d) decreased as the S/C ratio increased, in line with the reaction (19). As shown in Figure 9e, benzene yield was not greatly affected by the S/C ratio; however, as the S/C ratio increased, the amount of benzene generated decreased.

### 3.3. Catalyst Characterization after Reaction

To investigate the characteristics of the catalyst after the steam reforming of toluene, we used an experience using S/C = 4 at 800 °C for 2 h as a reference.

The concentration of the gases remained relatively constant during the development of the test (Figure 10a), reaching an average conversion to gases of 98% (Figure 10b). Throughout the development of the test, no deactivation of the catalyst was observed, and thus the concentration of the gas was quite constant with the following average molar concentration: 76.6% $H_2$, 12.1% CO, and 11.3% $CO_2$.

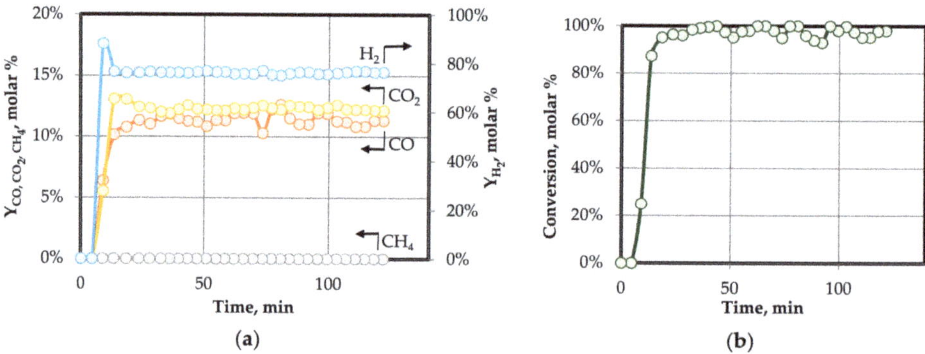


**Figure 10.** Gas composition (**a**) and gas conversion (**b**) during catalytic steam reforming of toluene at 800 °C and S/C = 4.

The results obtained show that the conversion achieved with the developed catalyst allowed us to reach comparable and even higher results to other catalysts on the basis of La or Ce (Table 7). Therefore, the Ni–Fe–Cu/LDH catalyst can be considered as an effective catalyst with a promising toluene conversion capacity.

**Table 7.** Comparison of the catalytic performances of previously reported catalysts.

| Reference | Catalyst | Toluene Conversion, % | Temperature, °C | S/C |
|---|---|---|---|---|
| [37] | Ni–Ce–Mg/olivine | 93 | 790 | 3.5 |
| [2] | Ni/Al$_2$O$_3$ | 64 | 750 | 3 |
| [38] | Ni/Al/La | 94.53 | 650 | 5.7 |
| [39] | Ni/olivine | 100 | 650–850 | 2.3 |
| [40] | La$_{0.6}$Ce$_{0.4}$NiO$_3$ | 80 | 800 | 2 |

### 3.3.1. XRD Analysis Results

The Figure 11 shows the XRD patterns after reduction and after reaction of the catalyst by comparison.

After reduction at 900 °C (Figure 11a), the diffraction lines resulting from the formation of the nickel metal phase and Ni–Fe alloy were observed, together with weaker lines of NiO and MgO. The same lines were observed in the XRD profile after reaction (Figure 11b), but longer and narrower, which meant the formation of larger particles.

In this way, the mean diameter of the Ni particles was calculated by the Scherrer Equation (7), using the peak at $2\theta = 51.2°$, which corresponded to the (200) plane of Ni fcc [31] and lattice parameter by Equation (8); all are shown in Table 8.

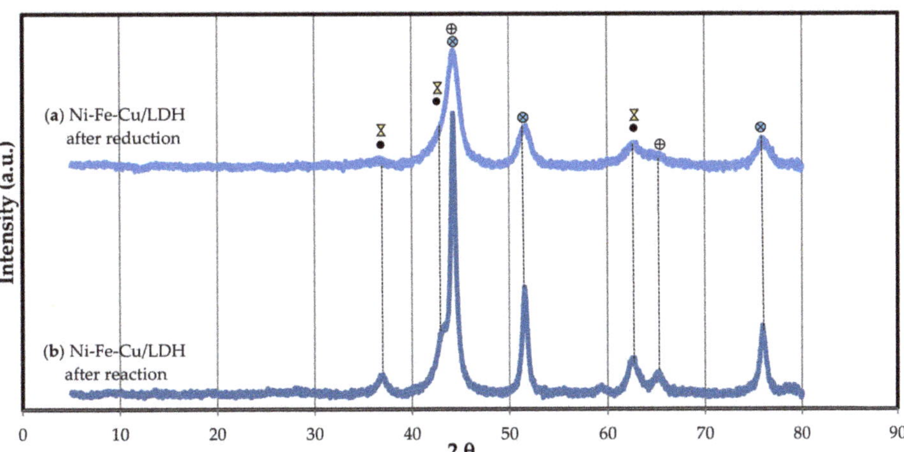


**Figure 11.** XRD profile of Ni–Fe–Cu/LDH (**a**) after reduction and (**b**) after reaction: (⊗) Ni metal, (⊕) Ni–Fe alloy, (✗) MgO-like phase, (●) NiO.

**Table 8.** Crystallite size and lattice parameters of Ni particles of Ni–Fe–Cu/LDH after reduction and after reaction.

| Samples | Crystallite Size | Lattice Parameter |
|---|---|---|
|  | $D$, nm | $a$, Å |
| Ni–Fe–Cu/LDH after reduction | 6.11 | 3.53 |
| Ni–Fe–Cu/LDH after reaction | 15.44 | 3.54 |

After the reaction, we observed that there was an increase in the size of the Ni particles, which may be attributed to the fact that as the reaction progressed, the sintering of the Ni particles took place. In addition, a slight increase in the lattice parameter took place, which was in line with the fcc structure of the metal nanoparticles, which generally expand with increasing particle size [41].

### 3.3.2. Amount and Type of Carbon Deposited on the Surface of Catalyst

The amounts of carbon at the end of different tests were analyzed by means of an elemental analyzer. The lowest values were obtained for high temperatures. Thus, for instance, for an S/C = 4 at 800 °C, the amount of carbon deposited after 2 h of reaction was only 1.4 wt %.

In order to determine the influence of the S/C ratio on the amount of carbon deposited, we analyzed the amount of carbon formed at 600 °C for 1 h for different S/C values. The results are shown in Figure 12.

From Figure 12, it can be seen that as the S/C ratio increased, the amount of carbon deposited decreased. For example, when the S/C ratio was 1, the amount of carbon formed was 49.34%, while for an S/C = 2, the amount was drastically reduced to 20.82%. The lowest amount of carbon was obtained by using high S/C ratios of 8.

These amounts of carbon deposited were even less than those detected by other authors; for instance, Josuinkas et al. [31] determined that for a Ni LDH-based catalyst, the amount of carbon deposited was 78.3–83.1 wt % at 800 °C and S/C = 1.5.

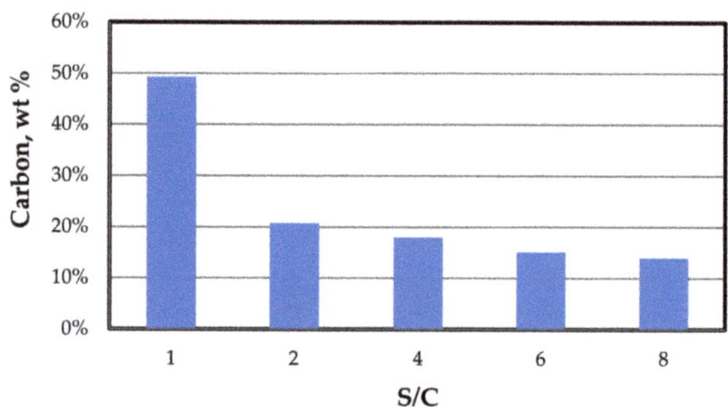

**Figure 12.** Amount of carbon deposited on the surface of the catalyst during the steam reforming reaction of toluene at 600 °C for different values of the S/C ratio.

To identify the type of carbon formed, we carried out an analysis of the TG derivative by means of a TGA analysis. TGA analysis allows us to distinguish between amorphous carbon, which is detected by a peak in the range 450–500 °C, and filamentous carbon, which decomposes in the range 600–650 °C [30]. It is well known that amorphous carbon affects the degree of conversion of the catalyst since it blocks the active centers. On the contrary, filamentous carbon affects the conversion to a lesser extent, since it had less serious effect on the deactivation than the amorphous carbon [30].

Figure 13 shows the nature of the carbon deposited on the catalyst over the steam reforming of toluene. Only one peak was appreciated in the thermograms in the range 600–650 °C, which indicates that most of the carbon formed was filamentous, which means it did not affect the activity of the catalyst.

In addition, the effects of the S/C ratio on the amount and type of carbon formed can be observed in Figure 13. Here, the type of carbon formed did not vary with the S/C ratio, being predominantly filamentous carbon in all cases. Moreover, when the S/C ratio increased, the signal of the filamentous carbon decreased, which can be explained by a lower formation of carbon when the amount of water was added.

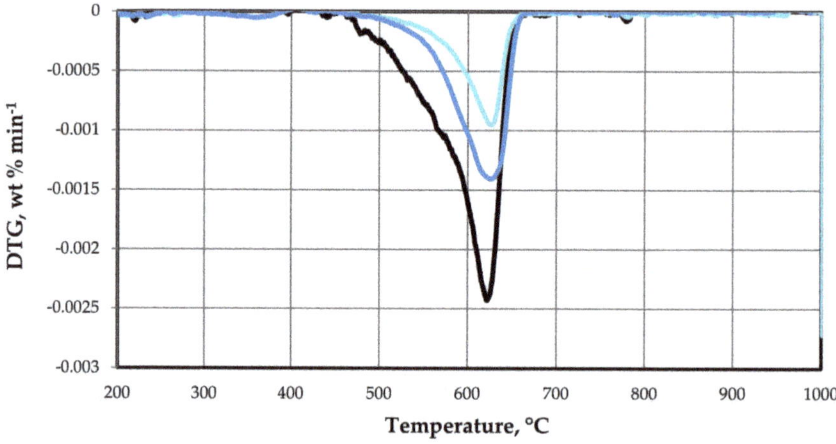

**Figure 13.** DTG signals of the carbon formed at 700 °C and S/C: 2 (—), 4 (—), and 6 (—).

### 3.3.3. SEM Analysis of the Catalyst after Reaction

Figure 14 shows the morphological changes of the reduced catalyst (Figure 14a,b) and after 2 h of reaction at 800 °C and S/C of 4 (Figure 14c,d) by scanning electron microscopy. Non-structured carbon was observed in the SEM micrographs after reaction, which confirmed the low content of the carbon analyzed, making it potentially difficult to visualize by SEM.

After reduction of the catalyst, a powder formed by particles with an average diameter of 327 nm was obtained. We observed that the particles of the catalyst were well dispersed and uniformly distributed, but after the reaction, aggregates of smaller particles were formed, resulting in an increase in particle size up to 446 nm.

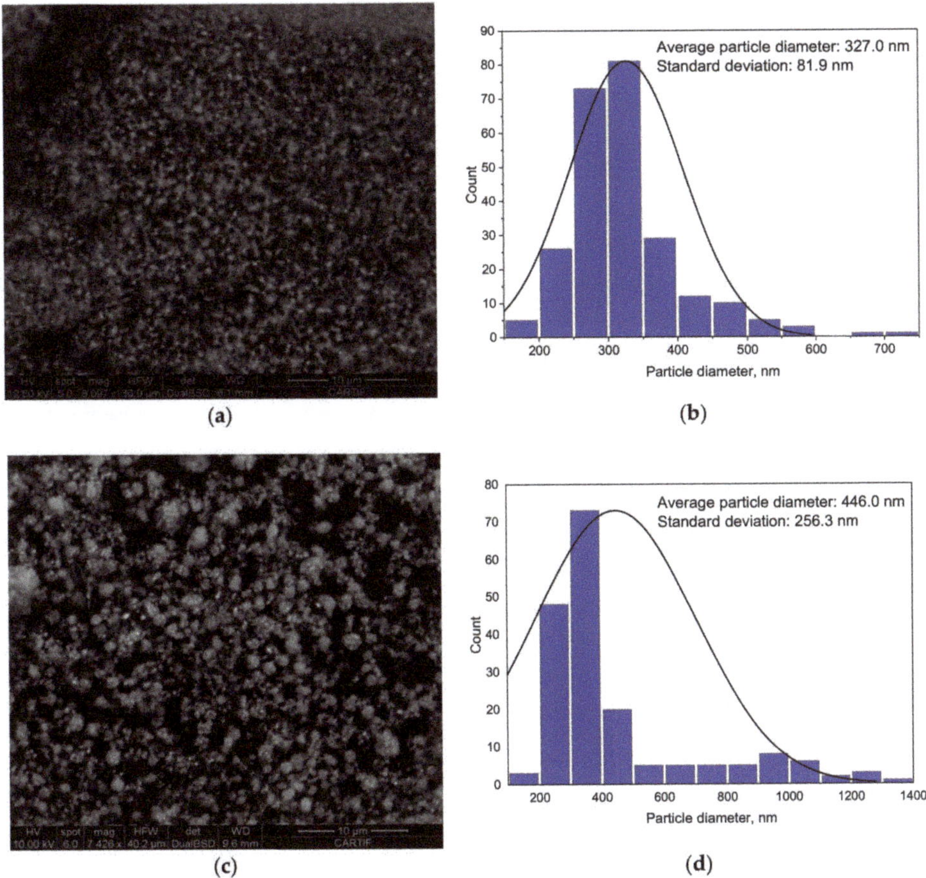


**Figure 14.** SEM micrograph and histogram of the reduced Ni–Fe–Cu/LDH at 36.9 μm (HFW) (**a**,**b**) and after the steam reforming of toluene at 40.2 μm (HFW) (**c**,**d**).

## 4. Conclusions

A study of the behavior of catalyst Ni–Fe–Cu/LDH during the steam reforming of toluene was carried out. The results suggest that for low temperatures at 600 °C, it is necessary to use S/C ratios higher than 6 to achieve gas conversions higher than 90%. On the other hand, at high temperatures above 800 °C, an S/C ratio of 2 was found to be sufficient to achieve high gas conversions. In addition, the use of low S/C ratios allowed for increased concentration of CO. S/C ratio of 4 allowed us to obtain the maximum conversion reached at 800 °C (99.87%) and low values of carbon deposited on the catalyst (1.4 wt %), along with good conversions in the range 600–700 °C above 80% and 90%, respectively. Furthermore, the type of carbon deposited on the catalyst was determined to be mainly filamentous, which, as is known, reduces conversion to a much lesser extent than amorphous carbon.

Moreover, the use of relatively cheap promoters such as Fe and Cu allowed us to obtain a competitive and more affordable catalyst than the catalysts based on expensive metals such as La and Ce. Therefore, the catalyst developed from Ni–Fe–Cu/LDH can be considered as an effective catalyst with a promising toluene conversion capacity, yielding high concentrations of $H_2$.

**Author Contributions:** Conceptualization, D.D.R., A.U.L. and G.A.G.; methodology, D.D.R. and A.U.L.; software, D.D.R. and A.U.L.; validation, D.D.R., A.U.L. and G.A.G.; formal analysis, D.D.R. and A.U.L.; investigation, D.D.R. and A.U.L.; resources, D.D.R. and A.U.L.; data curation, D.D.R. and A.U.L.; writing (original draft preparation), D.D.R.; writing (review and editing), D.D.R., A.U.L. and G.A.G.; visualization, D.D.R. and A.U.L.; supervision, G.A.G.; project administration, D.D.R. and G.A.G.; funding acquisition, D.D.R., A.U.L. and G.A.G. All authors have read and agreed to the published version of the manuscript.

**Funding:** This research received no external funding.

**Data Availability Statement:** Data is in agreement with the MDPI Research Data Policies.

**Conflicts of Interest:** The authors declare no conflict of interest.

# References

1.	Chen, J.; Tamura, M.; Nakagawa, Y.; Okumura, K.; Tomishige, K. Promoting effect of trace Pd on hydrotalcite-derived Ni/Mg/Al catalyst in oxidative steam reforming of biomass tar. *Appl. Catal. B Environ.* **2015**, *179*, 412–421. [CrossRef]
2.	Artetxe, M.; Alvarez, J.; Nahil, M.A.; Olazar, M.; Williams, P.T. Steam reforming of different biomass tar model compounds over Ni/Al$_2$O$_3$ catalysts. *Energy Convers. Manag.* **2017**, *136*, 119–126. [CrossRef]
3.	Chen, Y.H.; Schmid, M.; Kertthong, T.; Scheffknecht, G. Reforming of toluene as a tar model compound over straw char containing fly ash. *Biomass Bioenergy* **2020**, *141*, 105657. [CrossRef]
4.	Satyanarayana, K.G. *Clay Surfaces: Fundamentals and Applications*; Elsevier: Amsterdam, The Netherlands, 2004.
5.	Theiss, F.L.; Ayoko, G.A.; Frost, R.L. Synthesis of layered double hydroxides containing Mg$^{2+}$, Zn$^{2+}$, Ca$^{2+}$ and Al$^{3+}$ layer cations by co-precipitation methods—A review. *Appl. Surf. Sci.* **2016**, *383*, 200–213. [CrossRef]
6.	Mishra, G.; Dash, B.; Pandey, S. Layered double hydroxides: A brief review from fundamentals to application as evolving biomaterials. *Appl. Clay Sci.* **2018**, *153*, 172–186. [CrossRef]
7.	Duan, X.; Evans, D.G. (Eds.) *Layered Double Hydroxides*; Springer Science & Business Media: Berlin/Heidelberg, Germany, 2006; Volume 119.
8.	Liu, H.; Wierzbicki, D.; Debek, R.; Motak, M.; Grzybek, T.; da Costa, P.; Gálvez, M.E. La-promoted Ni-hydrotalcite-derived catalysts for dry reforming of methane at low temperatures. *Fuel* **2016**, *182*, 8–16. [CrossRef]
9.	Świrk, K.; Galvez, M.E.; Motak, M.; Grzybek, T.; Rønning, M.; da Costa, P. Yttrium promoted Ni-based double-layered hydroxides for dry methane reforming. *J. Co2 Util.* **2018**, *27*, 247–258. [CrossRef]
10.	Shen, Y.; Yoshikawa, K. Recent progresses in catalytic tar elimination during biomass gasification or pyrolysis—A review. *Renew. Sustain. Energy Rev.* **2013**, *21*, 371–392. [CrossRef]
11.	Mette, K.; Ressler, T.; Muhler, M. Development of Hydrotalcite-Derived Ni Catalysts for the Dry Reforming of Methane at High Temperatures. Ph.D. Thesis, Technische Universität, Berlin, Germany, 2015.
12.	Rached, A.J.; Dahdah, E.; Gennequin, C.; Tidahy, H.L.; Aboukais, A.; Abi-Aad, E.; Nsouli, B. Steam reforming of toluene for hydrogen production over NiMgALCe catalysts prepared via hydrotalcite route. In Proceedings of the 2016 7th International Renewable Energy Congress (IREC), Hammamet, Tunisia, 22–24 March 2016; IEEE: New York, NY, USA, 2016; pp. 1–6.
13.	Yu, X.; Wang, N.; Chu, W.; Liu, M. Carbon dioxide reforming of methane for syngas production over La-promoted NiMgAl catalysts derived from hydrotalcites. *Chem. Eng. J.* **2012**, *209*, 623–632. [CrossRef]
14.	Choi, S.C.; Lee, D.K.; Sohn, S.H. Morphological and Optical Properties of Cobalt Ion-Modified ZnO Nanowires. *Catalysts* **2020**, *10*, 614. [CrossRef]
15.	Koike, M.; Li, D.; Watanabe, H.; Nakagawa, Y.; Tomishige, K. Comparative study on steam reforming of model aromatic compounds of biomass tar over Ni and Ni–Fe alloy nanoparticles. *Appl. Catal. A Gen.* **2015**, *506*, 151–162. [CrossRef]
16.	Li, D.; Koike, M.; Wang, L.; Nakagawa, Y.; Xu, Y.; Tomishige, K. Regenerability of hydrotalcite-derived nickel–iron alloy nanoparticles for syngas production from biomass tar. *ChemSusChem* **2014**, *7*, 510–522. [CrossRef] [PubMed]

17. Touahra, F.; Sehailia, M.; Ketir, W.; Bachari, K.; Chebout, R.; Trari, M.; Halliche, D. Effect of the Ni/Al ratio of hydrotalcite-type catalysts on their performance in the methane dry reforming process. *Appl. Petrochem. Res.* **2016**, *6*, 1–13. [CrossRef]
18. Ferreira, R.A.R.; Ávila-Neto, C.N.; Noronha, F.B.; Hori, C.E. Study of LPG steam reform using Ni/Mg/Al hydrotalcite-type precursors. *Int. J. Hydrog. Energy* **2019**, *44*, 24471–24484. [CrossRef]
19. Lu, P.; Huang, Q.; Bourtsalas, A.C.; Chi, Y.; Yan, J. Effect of operating conditions on the carbon formation and nickel catalyst performance during cracking of tar. *Waste Biomass Valorization* **2019**, *10*, 155–165. [CrossRef]
20. Labaki, M.; Lamonier, J.F.; Siffert, S.; Aboukaïs, A. Thermal analysis and temperature-programmed reduction studies of copper–zirconium and copper–zirconium–yttrium compounds. *Thermochim. Acta* **2005**, *427*, 193–200. [CrossRef]
21. Li, D.; Wang, L.; Koike, M.; Nakagawa, Y.; Tomishige, K. Steam reforming of tar from pyrolysis of biomass over Ni/Mg/Al catalysts prepared from hydrotalcite-like precursors. *Appl. Catal. B Environ.* **2011**, *102*, 528–538. [CrossRef]
22. Cavani, F.; Trifiro, F.; Vaccari, A. Hydrotalcite-type anionic clays: Preparation, properties and applications. *Catal. Today* **1991**, *11*, 173–301. [CrossRef]
23. Casenave, S.; Martinez, H.; Guimon, C.; Auroux, A.; Hulea, V.; Cordoneanu, A.; Dumitriu, E. Acid–base properties of Mg–Ni–Al mixed oxides using LDH as precursors. *Thermochim. Acta* **2001**, *379*, 85–93. [CrossRef]
24. Silva, C.C.C.; Ribeiro, N.F.; Souza, M.M.; Aranda, D.A. Biodiesel production from soybean oil and methanol using hydrotalcites as catalyst. *Fuel Process. Technol.* **2010**, *91*, 205–210. [CrossRef]
25. Serrano-Lotina, A.M. *Obtención de Hidrógeno a Partir de Biogás Mediante Catalizadores Derivados de Hidrotalcita*; Universidad Autónoma de Madrid: Madrid, Spain, 2012.
26. Morlanés, N.; Melo, F. *Obtención de Hidrógeno Mediante Reformado Catalítico de Nafta con Vapor de Agua*; Universidad Politécnica de Valencia: Valencia, Spain, 2007.
27. Zhou, F.; Pan, N.; Chen, H.; Xu, X.; Wang, C.; Du, Y.; Li, L. Hydrogen production through steam reforming of toluene over Ce, Zr or Fe promoted Ni-Mg-Al hydrotalcite-derived catalysts at low temperature. *Energy Convers. Manag.* **2019**, *196*, 677–687. [CrossRef]
28. Bîrjega, R.; Pavel, O.D.; Costentin, G.; Che, M.; Angelescu, E. Rare-earth elements modified hydrotalcites and corresponding mesoporous mixed oxides as basic solid catalysts. *Appl. Catal. A Gen.* **2005**, *288*, 185–193. [CrossRef]
29. Tsyganok, A.I.; Inaba, M.; Tsunoda, T.; Uchida, K.; Suzuki, K.; Takehira, K.; Hayakawa, T. Rational design of Mg–Al mixed oxide-supported bimetallic catalysts for dry reforming of methane. *Appl. Catal. A Gen.* **2005**, *292*, 328–343. [CrossRef]
30. Ertl, G.; Knözinger, H.; Weitkamp, J. (Eds.) *Preparation of Solid Catalysts*; John Wiley & Sons: Hoboken, NJ, USA, 2008.
31. Josuinkas, F.M.; Quitete, C.P.; Ribeiro, N.F.; Souza, M.M. Steam reforming of model gasification tar compounds over nickel catalysts prepared from hydrotalcite precursors. *Fuel Process. Technol.* **2014**, *121*, 76–82. [CrossRef]
32. Qian, K.; Kumar, A. Catalytic reforming of toluene and naphthalene (model tar) by char supported nickel catalyst. *Fuel* **2017**, *187*, 128–136. [CrossRef]
33. RJ, B.S.; Loganathan, M.; Shantha, M.S. A review of the water gas shift reaction kinetics. *Int. J. Chem. React. Eng.* **2010**, *8*. [CrossRef]
34. Demirel, E.; Azcan, N. Thermodynamic modeling of water-gas shift reaction in supercritical water. In Proceedings of the World Congress on Engineering and Computer Science, San Francisco, CA, USA, 24–26 October 2012; Volume 2012.
35. Zhu, H.L.; Pastor-Pérez, L.; Millan, M. Catalytic Steam Reforming of Toluene: Understanding the Influence of the Main Reaction Parameters over a Reference Catalyst. *Energies* **2020**, *13*, 813. [CrossRef]
36. Ahmed, T.; Xiu, S.; Wang, L.; Shahbazi, A. Investigation of Ni/Fe/Mg zeolite-supported catalysts in steam reforming of tar using simulated-toluene as model compound. *Fuel* **2018**, *211*, 566–571. [CrossRef]
37. Zhang, R.; Wang, H.; Hou, X. Catalytic reforming of toluene as tar model compound: Effect of Ce and Ce–Mg promoter using Ni/olivine catalyst. *Chemosphere* **2014**, *97*, 40–46. [CrossRef]
38. Bona, S.; Guillén, P.; Alcalde, J.G.; García, L.; Bilbao, R. Toluene steam reforming using coprecipitated Ni/Al catalysts modified with lanthanum or cobalt. *Chem. Eng. J.* **2008**, *137*, 587–597. [CrossRef]
39. Świerczyński, D.; Libs, S.; Courson, C.; Kiennemann, A. Steam reforming of tar from a biomass gasification process over Ni/olivine catalyst using toluene as a model compound. *Appl. Catal. B Environ.* **2007**, *74*, 211–222. [CrossRef]
40. Soongprasit, K.; Aht-Ong, D.; Sricharoenchaikul, V.; Atong, D. Synthesis and catalytic activity of sol-gel derived La–Ce–Ni perovskite mixed oxide on steam reforming of toluene. *Curr. Appl. Phys.* **2012**, *12*, S80–S88. [CrossRef]
41. Nafday, D.; Sarkar, S.; Ayyub, P.; Saha-Dasgupta, T. A reduction in particle size generally causes body-centered-cubic metals to expand but face-centered-cubic metals to contract. *ACS Nano* **2018**, *12*, 7246–7252. [CrossRef] [PubMed]

*Article*

# Eucalyptus Kraft Lignin as an Additive Strongly Enhances the Mechanical Resistance of Tree-Leaf Pellets

Leonardo Clavijo [1],* , Slobodan Zlatanovic [2], Gerd Braun [3], Michael Bongards [2], Andrés Dieste [1] and Stéphan Barbe [4]

[1] Chemical Engineering Institute, Faculty of Engineering, Universidad de la República. Julio Herrera y Reissig 565, Montevideo 11300, Uruguay; andresdieste@fing.edu.uy
[2] Institute for Automation and Industrial IT, TH Köln—University of Applied Sciences, Steinmuellerallee 1, 51643 Gummersbach, Germany; slobodan.zlatanovic@th-koeln.de (S.Z.); michael.bongards@th-koeln.de (M.B.)
[3] Institute of Chemical Engineering and Plant Design, TH Köln—University of Applied Sciences, Betzdofer Straße, 2, 50679 Koeln, Germany; gerd.braun@th-koeln.de
[4] Faculty of Applied Natural Sciences, TH Köln—University of Applied Sciences, Kaiser-Wilhelm-Allee, Gebäude E39, 51373 Leverkusen, Germany; stephan.barbe@th-koeln.de
* Correspondence: lclavijo@fing.edu.uy

Received: 26 February 2020; Accepted: 16 March 2020; Published: 24 March 2020

**Abstract:** Pelleted biomass has a low, uniform moisture content and can be handled and stored cheaply and safely. Pellets can be made of industrial waste, food waste, agricultural residues, energy crops, and virgin lumber. Despite their many desirable attributes, they cannot compete with fossil fuel sources because the process of densifying the biomass and the price of the raw materials make pellet production costly. Leaves collected from street sweeping are generally discarded in landfills, but they can potentially be valorized as a biofuel if they are pelleted. However, the lignin content in leaves is not high enough to ensure the physical stability of the pellets, so they break easily during storage and transportation. In this study, the use of eucalyptus kraft lignin as an additive in tree-leaf pellet production was studied. Results showed that when 2% lignin is added the abrasion resistance can be increased to an acceptable value. Pellets with added lignin fulfilled all requirements of European standards for certification except for ash content. However, as the raw material has no cost, this method can add value or contribute to financing continued sweeping and is an example of a circular economy scenario.

**Keywords:** eucalyptus kraft lignin; tree leaf; pellet; additive; biofuel; circular economy

---

## 1. Introduction

Lignocellulosic biomass generally has a low bulk density (30–100 kg/m$^3$). Pelleting process growths the specific biomass density to more than 1000 kg/m$^3$ [1–3]. Pelleted biomass has a low, uniform moisture content and can be handled and stored cheaply and safely using well-known handling systems developed for grains [4,5]. Pellets can be produced from any one of five general categories of biomass: industrial residues and co-products, food waste, agricultural residues, energy crops, and virgin lumber [6,7]. Wood pellets are the most common type of pellet fuel and are generally made from compacted sawdust and related industrial wastes from the milling of lumber, manufacturing of wood products and furniture industry. However, as a result of limited supplies of sawdust in many countries, particularly those in central and northern Europe, agricultural products are increasingly being used as raw material for pellet production [6]. Wood pellets are becoming more popular worldwide; in the last five years, the production and exportation of pellets has increased by 73% and 93%, respectively [8]. Regardless of their many attributes, biomass pellets have a higher cost that fossil fuel sources because it is still too expensive to densify biomass. Raw materials are also a major contributor to cost [7].

The manufacturing process consists of drying the biomass and reducing the particle size to be suitable for pelleting by means of a hammer mill. The crushed biomass is compacted in the press mill to form pellets. Individual pellet density ranges from 1000–1200 kg/m$^3$, and the bulk density varies between 550–700 kg/m$^3$ depending on the pellets size. The higher heating value of wood pellets varies between 17,000 and 22,000 kJ/kg [9,10]. Pellet density and abrasion resistance are influenced by the physical and chemical properties of the raw material and the temperature and applied pressure during the pelleting process [2].

The quality of pellets is determined by a few main parameters including moisture content (MC), net heating value, abrasion resistance, particle density, ash content, and ash melting point [11]. Pellets must fulfill the requirements for these parameters for commercialization. Several institutions have created standards for pellet quality [12,13].

Mechanical resistance of pellets has been measured in terms of compressive resistance, abrasive resistance and impact resistance [14]. Compressive resistance (crushing resistance or hardness) is the maximum load a pellet can tolerate before breaking. It is related to the adhesion forces between particles in the pellet and is normally used for testing pills in pharmaceutical industry. The test provides a quick measure of the quality of pellets and can be used to improve pellet quality. However, the compressive resistance does not quantify the amount of dust that is formed during the transport, handling and storage of the pellets, and that is the main reason why this test is not normally used to adjust pelleting processing condition [14,15]. The safe and effective transportation and processing of biomass pellets is critical for bioenergy application. Low pellet mechanical resistance leads to high dust emissions, system blockages and an increased risk of fire and explosions during pellet handling, storage and transport [16].

The durability or abrasion resistance is one of the most important parameters in pellet production and is defined as the ability of densified biomass units to remain intact during handling [16,17]. High durability implies high quality pellets with a low number of fines. The most widely used laboratory methods to measure the durability of pellets, are: the tumbling box method, the Holmen durability tester, and the Ligno tester. In the tumbling box method, a sample of pellets is sieved to remove fines and then 0.5 kg of sieved pellets is placed in a tumbling can device. After tumbling for 10 min at 50 rpm, the sample is removed, sieved and weighted. Pellet durability index (PDI) is calculated as the ratio of weight after tumbling over the weight before tumbling and is expressed as a percentage [14,18,19]. The Holmen durability tester simulates pneumatic management of pellets. In this device, a sample of pellets pneumatically circulates through a square conduit of pipe, and the pellets are impacted repeatedly on hard surfaces. After procedure, pellet sample is sieved, and durability index is calculated in a similar way. The Holmen tester is severer than the tumbling can method and, therefore, yields lower PDI [19,20]. The Ligno tester device uses air to rapidly circulate 0.1 Kg of pellets around a perforated chamber during 30 s. The chamber is an inverted square pyramid with perforated sides. Forced air is the destructive force. Fines are removed continuously during the test and there is no need to screen the pellet. PDI is calculated in a similar way [19]. The tumbling box method is the most popular method for abrasion resistance determination.

Impact resistance test (drop resistance or shattering resistance) simulate the forces produced during emptying of densified products from trucks onto ground or bins. It can be used to determine the safe height of pellet production [21]. Pellets are dropped from a standardized height of 1.85-m onto a stainless-steel plate in the floor, four times. The weight of the remaining pellets, as the percentage of the initial weight is the impact resistance index (IRI) [14].

Pellet quality may be improved by using different binding agents or additives during production. Additives can form a bridge, a film, a matrix or can cause a chemical reaction to make strong inter-particle bonding. They can act in several ways, such as improving fuel quality, decreasing emissions, or increasing burning efficiency. All the current regulations for the classification of pellets limit the use of additives at 2% for wood pellets. The most used additives are lignosulphonate, starch, dolomite, corn or potato flour, and some vegetable oils. These binding agents also affect the production economics of the final product [11,17,22].

When lignin-rich biomass is compressed under high pressures and temperatures, the lignin becomes soft, because its thermosetting properties. The softened lignin acts as a glue [11,23]. If the feedstock (e.g., leaves) does not contain enough lignin, a binding agent needs to be added. Otherwise, the finished pellets have poor mechanical stability and typically break down into powder very easily. Binding agents improve production efficiency but also can increase lubricity during the grinding procedure, thus decreasing wear and tear on the machinery. Many substances can be used as additives, including molasses, starch, gluten, dry distiller, rapeseed cake, etc. Nowadays, ongoing research has sought to extract new additives from waste materials, to develop circular economy processes. Materials that are more sticky or oily are more likely to be used as binding agents. The lignin content of leaves is normally smaller than in wood. Although lignin content in leaves varies with species, it is usually less than 10–12%, and in some cases as low as 3–6%, whereas in wood, the lignin content ranges from 18–30% [24].

In some European countries, the final disposal of leaves collected in street sweeping is problematic, since burning is not allowed, leaves are generally disposed in landfills. In Germany the number of deciduous trees in the streets is particularly high, and only in Berlin there are more than 430,000 trees in the streets [25]. In this work, we study the possibility of using tree leaves as raw material for manufacturing pellets. However, due to the low lignin content of the leaves, it is necessary to add a binding agent to improve the stability of the pellets. As an additive, the use of lignosulfonates was studied by several authors with good results [5,11,12,26] however, as the availability of lignosulfonates has decreased and its price increased, the use of eucalyptus kraft lignin is proposed, given the predominance of the kraft process for the production of cellulose pulp and the use of fast-growing species such as eucalyptus, particularly in South America. In this approach, two sub products from the forest are valorized.

## 2. Materials and Methods

Wood leaves were collected from streets and gardens in the Bergisches Land (North Rhine-Westphalia, Germany), washed, dried to a moisture content of 30% (wb), and stored in plastic containers. Eucalyptus kraft lignin was obtained by acid precipitation with $CO_2$ from Eucaliptus kraft black liquor, kindly provided by the UPM-Fray Bentos pulp mill (Fray Bentos, Uruguay) [27]. Then, a washing stage with mineral acid (pH 2) was performed to remove inorganic impurities. As very high purity is not a requisite, the acid washing was performed in only one stage. Lignin was characterized in terms of moisture content (ISO 18134), ash content (ISO 18122), Klason and soluble lignin (Tappi standard T-222). Lignin glass transition temperature (Tg) was measured using a Perkin Elmer DSC 6000 device (Perkin Elmer, Waltham, MA, USA). To manufacture the pellets, a pellet press EverTec WKL120C of 3 kW was used. During pellet manufacturing, the temperature in the rolling press was monitored with an infrared thermometer.

Pellets were characterized in terms of yield, moisture content (ISO 18134), ash content (ISO 18122), durability (ISO 17831-1), calorific value (ISO 18125), ash chemical composition (ISO 16968), and ash melting point (ASTM D 1857—04). The cross section of the pellets was inspected via Scanning Electron Microscopy (SEM) after gold sputtering. To determine the calorific value, a calorimeter IKA C200 (IKA, Staufen, Germany) was used; to determine the metal content, a Perkin Elmer AAnalyst 200 Atomic Absorption Spectrometer was used; to determine the ash melting point, a LECO AF700 device (LECO, St. Joseph, MI, USA) was used.

To study the influence of the moisture content of the raw material, the leaves were crushed and conditioned with different water content values on a wet basis: 10, 14, 18, 22, and 25% (samples labeled as P-10, P-14, P-18, P-22, and P-25, respectively). To study the influence of kraft lignin as an additive, 1.0, 1.5, 2.0, and 3.0% (db) were added (samples named as L-1, L-1.5, L-2, and L-3).

## 3. Results and Discussion

Lignin characteristics are shown in Table 1. Lignin ash content was high because the washing stage at the pilot plant was performed with a minimal amount of water. The purpose of this process is to produce technical-grade lignin without further refinement in order to keep production costs as low as possible. Pure lignin content was considered for the pellet formulations. Figure 1a shows the yield of the pelleting process, which was calculated as the dry mass of the pellets obtained divided by the dry mass of the raw material used. Results shows that 22–25% is the best moisture content for pellet formation and when the moisture content is less than 18% pellet formation was poor, resulting in a low process yield. This fact is not in agreement with literature since it is widely reported that the moisture content of raw material for pellet production should be around 15% [5,6,12,14,22,28]. The explanation may be given by the friction of the rolling press, which raises the temperature to between 90–110 °C and causes water to evaporate prior to extrusion. The moisture content of the pellets after the production process is indicated in Table 2 and is independent of the moisture content of the raw material. The average moisture of the pellets in wet basis was (8.7 ± 0.2)%. This value is lower than 10%, as required for certification [13,21,22,29].

**Table 1.** Physical properties of lignin used in this work.

| Property | Value |
|---|---|
| Ash content (%) | 12.0 ± 0.9 |
| Klason lignin (%) | 73.0 ± 0.5 |
| Soluble lignin (%) | 11.3 ± 0.8 |
| Total lignin (%) | 84 ± 1 |
| Tg (°C) | 132 ± 1 |
| Net heating value (MJ/kg) | 29.1 ± 0.1 |

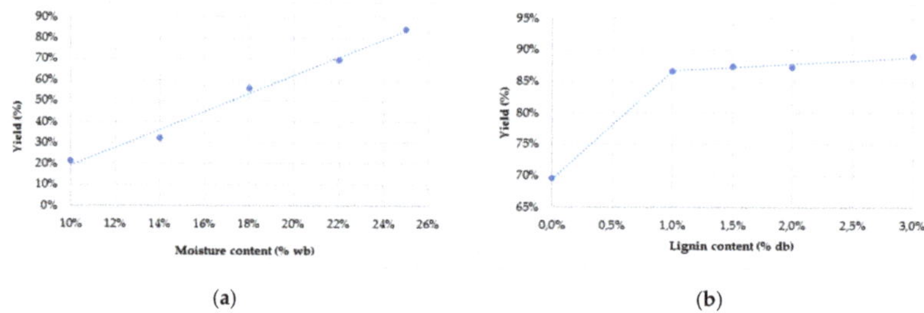

(a)　　　　(b)

**Figure 1.** (**a**) Yield of pellet production versus moisture content of raw material, (**b**) yield production versus lignin content.

**Table 2.** Moisture content of pellets versus moisture content of raw material.

| Sample | Raw Material Moisture Content (% wb) | Pellet Moisture Content (% wb) |
|---|---|---|
| P-10 | 10.0 ± 0.2 | 9.2 ± 0.1 |
| P-14 | 14.0 ± 0.3 | 8.5 ± 0.3 |
| P-18 | 18.0 ± 0.1 | 8.0 ± 0.1 |
| P-22 | 22.0 ± 0.2 | 9.1 ± 0.1 |
| P-25 | 25.0 ± 0.1 | 8.9 ± 0.1 |

For the addition of eucalyptus kraft lignin as a binding agent, raw material with a moisture content of 22% was selected. The yield increased drastically with the lignin content, as shown in

Figure 1b. This can be explained because of the bonging effect of lignin on the pellet structure. Also, lignin-added pellets showed a brighter finish.

A pellet with low durability disintegrates easily during handling, which can cause difficulties during its storage and transport, as well as health and environmental problems due to the dust generated. The durability index of the pellets varies with the moisture content of the raw material as well as with the amount of lignin added, as shown in Figure 2. The New European Pellets Standards EN 14961-1 states that the minimum durability value for a commercial pellet is 95%, which was not reached in leaf-pellet production without the inclusion of an additive. When lignin was added at 2.0% and 3.0%, the durability was acceptable for pellet commercialization and fulfilled the requirements for certification [22], but the values were lower than for wood pellets, which usually have a durability above 97.5% [13,22,29].

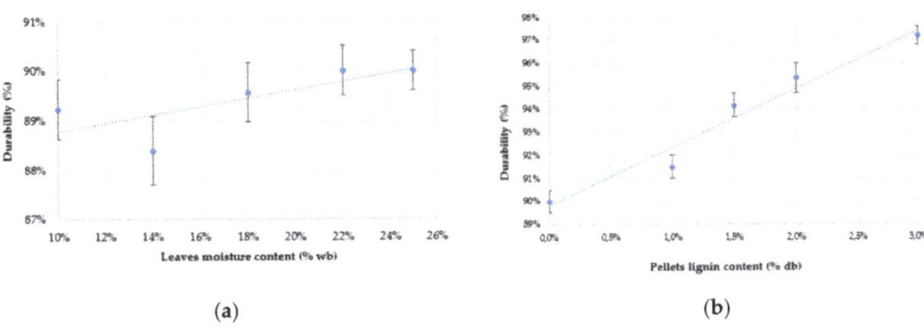



**Figure 2.** Durability of pellets. (**a**) Influence of the moisture content of leaves, (**b**) influence of the lignin content of pellets.

Figure 3 shows the SEM cross sections obtained with pellets P-22, L-1, L-2, and L-3 in which the only difference was lignin content. All pellets exhibited a homogeneous structure, and no large leaf particles could be observed. This finding confirms the quality of the applied leaf pelletization process. Unfortunately, SEM images of tree-leaf pellet cross sections could not be found for comparison in the literature.

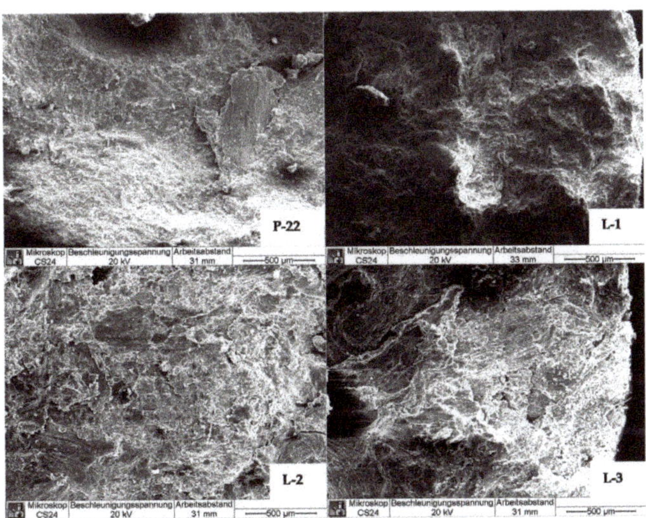



**Figure 3.** SEM images of pellets without lignin (P-22) and with different lignin contents (L-1, L-2, L-3).

The net heating value of the pellets in dry basis varies with the moisture content of the raw material, as can be seen in Figure 4a, which was quite surprising. One possible explanation is the volatilization of compound with low net heating value during the process, which is higher at low moisture contents. When lignin is used as an additive, the higher heating value increases by approximately 0.11 MJ/kg per each 1% of lignin added, as shown in Figure 4b, because of the higher net heating value of lignin. The minimum net heating value for pellet certification is 16.56 MJ/kg, which is lower than the values achieved with the leaf pellets [22,29].

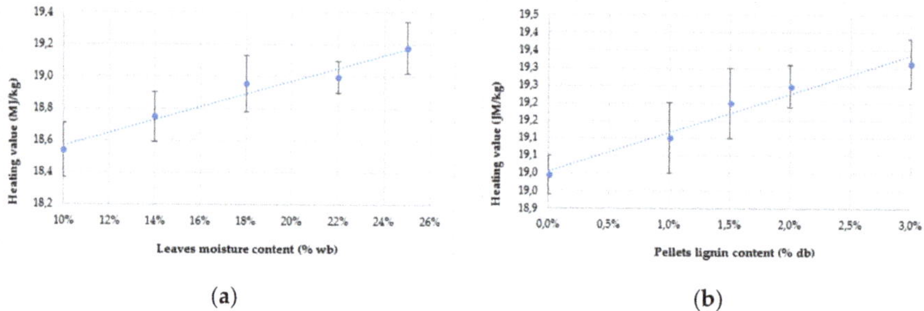

(a) (b)

**Figure 4.** Net heating value (dry basis), (**a**) influence of the moisture content of leaves, (**b**) influence of the lignin content of pellets.

In Table 3, the ash content of the raw materials and the pellets produced is listed. The ash content of leaves is much higher than in wood because of its botanic nature and its function in the plant. The amount of ashes could be problematic for both domestic and industrial uses of these pellets, as it is five times higher than the requirements for wood pellets. One option to diminish the ash content is to formulate pellets with both leaves and sawdust since sawdust has a very low ash content. In this scenario, the need for an additive has to be evaluated. In addition, end users may prefer pellets made from tree leaves if the cost is considerably lower than that of wood pellets, and this is possible since the cost of the raw material is negligible and the equipment for production is the same as for wood pellets.

**Table 3.** Ash content of leaves and pellets.

| Sample | Ash Content (% db) |
| --- | --- |
| Leaves | 10.2 ± 0.1 |
| P-10 | 11.0 ± 0.4 |
| P-14 | 10.9 ± 0.9 |
| P-18 | 10.9 ± 0.5 |
| P-22 | 11.6 ± 0.2 |
| P-25 | 12.3 ± 0.2 |
| L-1.0 | 10.7 ± 0.4 |
| L-1.5 | 11.1 ± 0.6 |
| L-2 | 11.8 ± 0.1 |
| L-3 | 12.7 ± 0.2 |

Regarding the fusibility of ash, results can be seen in Table 4. Only the initial deformation temperature (IT) is considered by quality standards. In all cases, the initial deformation temperature was above 1200 °C, except for sample P-14, in compliance with specifications for the highest quality pellets [17,22]. All pellets with lignin fulfilled this requirement. Regarding softening temperature (ST), hemispherical temperature (HT) and fluid temperature (FT) in most cases are higher than the maximum temperature that can be determined with the equipment used (1500 °C).

**Table 4.** Fusibility of pellet ash.

| Sample | IT [1] (°C) | ST [2] (°C) | HT [3] (°C) | FT [4] (°C) |
|---|---|---|---|---|
| P-10 | 1232 | >1500 | >1500 | >1500 |
| P-14 | 1119 | 1310 | 1371 | 1398 |
| P-18 | 1232 | >1500 | >1500 | >1500 |
| P-22 | 1281 | >1500 | >1500 | >1500 |
| P-25 | 1227 | 1416 | 1437 | 1440 |
| L-1 | 1304 | >1500 | >1500 | >1500 |
| L-1.5 | 1336 | >1500 | >1500 | >1500 |
| L-2 | 1314 | >1500 | >1500 | >1500 |
| L-3 | 1240 | >1500 | >1500 | >1500 |

[1]: IT: initial deformation temperature. Is the temperature at which the first rounding of the sample occurs. [2]: ST: softening temperature. Is the temperature at which the sample has fused down to a spherical lump in which the height is equal to the width at the base. [3]: HT: hemispherical temperature. Is the temperature at which the cone has fused down to a hemispherical lump at which point the height is one half the width of the base. [4]: FT: fluid temperature. Is the temperature at which the fused mass has spread out in a nearly flat layer with a maximum height of 1.6 mm.

The chemical composition of the pellets is shown in Table 5 for major (Na, K, Ca, Mg) and minor (As, Cd, Cr, Cu, Ni, Pb, Zn) elements. Although metal composition is not regulated, there is an indication of the maximum content of the minor elements in the pellets showed in the first row of Table 5 [17,22]. The results show that in all conditions, the metal content is below this indication, which ensures good environmental performance.

**Table 5.** Metal composition of pellets.

| Sample | Ca (mg/g) | Mg (mg/g) | Na (mg/g) | K (mg/g) | As (ppm) | Cd (ppm) | Cr (ppm) | Cu (ppm) | Ni (ppm) | Pb (ppm) | Zn (ppm) |
|---|---|---|---|---|---|---|---|---|---|---|---|
| Max. Content | | | | | ≤1 | ≤0.5 | ≤10 | ≤10 | ≤10 | ≤10 | ≤100 |
| P-10 | 19.0 | 3.0 | 0.5 | 7.9 | <0.2 | <0.5 | <5.0 | 8.1 | 2.0 | 6.0 | 23.9 |
| P-14 | 20.3 | 3.3 | 0.1 | 8.4 | <0.2 | <0.5 | <5.0 | 3.5 | 1.8 | 5.9 | 17.2 |
| P-18 | 17.8 | 2.8 | 0.1 | 7.4 | <0.2 | <0.5 | <5.0 | 4.3 | 1.5 | 6.0 | 15.2 |
| P-22 | 15.8 | 2.6 | 0.1 | 6.5 | <0.2 | <0.5 | <5.0 | 2.9 | 1.3 | 5.5 | 11.3 |
| P-25 | 20.0 | 3.2 | 0.3 | 8.3 | <0.2 | <0.5 | <5.0 | 2.8 | 1.5 | 7.6 | 16.8 |
| L-1 | 15.2 | 2.5 | 0.2 | 6.3 | <0.2 | <0.5 | <5.0 | 1.3 | 0.8 | 4.1 | 21.3 |
| L-1.5 | 16.7 | 2.7 | 0.8 | 6.9 | <0.2 | <0.5 | <5.0 | 2.2 | 1.1 | 5.1 | 12.5 |
| L-2 | 21.8 | 3.6 | 1.5 | 9.1 | <0.2 | <0.5 | <5.0 | 4.6 | 1.4 | 5.3 | 17.2 |
| L-3 | 19.8 | 3.3 | 1.9 | 8.2 | <0.2 | <0.5 | <5.0 | 4.1 | 1.4 | 7.0 | 17.2 |

## 4. Conclusions

The main objective of this work was to determine if tree leaves obtained in the sweeping of the streets, can be used as raw material for the production of pellets, going from being a residue to becoming a useful product. The results showed that this is possible but that it is necessary to add an additive that improves the yield of the process and the mechanical resistance of the product.

Lignin is the second most abundant polymer on the planet, and although the technology for production is developed, its uses are still preliminary. This research shows an alternative for the use of eucalyptus lignin kraft, as an additive to produce tree-leaf pellets.

This process can be useful for municipalities because they already have all the logistics of collection and stockpiling established, especially in those cities where the number of deciduous trees is high.

For both domestic and industrial use, the greatest disadvantage in the use of these pellets is given by the amount of ash obtained, a consequence of the starting raw material. For industrial use, a use for the ashes must be found, which could range from filler in cement production to silica production.

Therefore, the process is an example where two forms of biowaste -tree leaves and lignin- are utilized in a circular economy development.

**Author Contributions:** Conceptualization S.B., A.D., L.C.; methodology, L.C., S.B., G.B.; validation, L.C.; formal analysis, L.C.; investigation, L.C., S.Z., G.B.; resources, S.B., A.D.; writing—original draft preparation, L.C.; writing—review and editing, S.B., A.D.; visualization, L.C.; supervision, S.B., A.D., M.B.; project administration, S.B., M.B.; funding acquisition, S.B., M.B. All authors have read and agreed to the published version of the manuscript.

**Funding:** This research received no external funding

**Acknowledgments:** The authors thank DAAD for founding the scientific mission of Leonardo Clavijo in Germany. Additionally, the research group would like to thank the collaboration of UPM for supplying the black liquor and for logistical support, and to: metabolon Research Center for allowing the use of its facilities and for its support in this research.

**Conflicts of Interest:** The authors declare no conflict of interest.

## References

1. Lehtikangas, P. Quality properties of pelletised sawdust, logging residues and bark. *Biomass Bioenergy* **2001**, *20*, 351–360. [CrossRef]

2. Mani, S.; Tabil, L.G.; Sokhansanj, S. Evaluation of compaction equations applied to four biomass species. *Can. Biosyst. Eng./Le Genie Des Biosystemes Au Canada* **2004**, *46*, 55–61.

3. Mitchell, P.; Kiel, J.; Livingston, B.; Dupont-Roc, G. Torrefied biomass-A foresighting study into the business case for pellets from torrefied Biomass as a new solid fuel. *All Energy* **2007**, *24*, 1–27.

4. Fasina, O.O.; Sokhansanj, S. Storage and handling characteristics of alfalfa pellets. *Powder Handl. Process.* **1996**, *8*, 361–366.

5. Tumuluru, J.S.; Wright, C.T.; Hess, J.R.; Kenney, K.L. A review of biomass densification systems to develop uniform feedstock commodities for bioenergy application. *Biofuels Bioprod. Biorefin.* **2011**, *5*, 683–707. [CrossRef]

6. Nilsson, D.; Bernesson, S.; Hansson, P.A. Pellet production from agricultural raw materials–A systems study. *Biomass Bioenergy* **2011**, *35*, 679–689. [CrossRef]

7. Mani, S.; Sokhansanj, S.; Bi, S.; Turhollow, A. Economics of producing fuel pellets from biomass. *Appl. Eng. Agric.* **2006**, *22*, 421–426. [CrossRef]

8. Food and Agriculture Organization of the United Nations, FAOSTAT. Available online: http://www.fao.org/faostat/es/?#data/FO (accessed on 11 November 2019).

9. Telmo, C.; Lousada, J. Heating values of wood pellets from different species. *Biomass Bioenergy* **2011**, *35*, 2634–2639. [CrossRef]

10. Quirino, W.F.; Do Vale, A.T.; De Andrade, A.P.A.; Abreu, V.L.S.; Azevedo, A.C.D.S. Poder calorífico da madeira e de materiais ligno-celulósicos. *Revista Da Madeira* **2005**, *89*, 100–106.

11. Tarasov, D.; Shahi, C.; Leitch, M. Effect of additives on wood pellet physical and thermal characteristics: A review. *ISRN For.* **2013**, *2013*, 876939. [CrossRef]

12. García-Maraver, A.; Popov, V.; Zamorano, M. A review of European standards for pellet quality. *Renew. Energy* **2011**, *36*, 3537–3540. [CrossRef]

13. International Organization for Standardization. *Solid Biofuels–Fuel Specifications and Clases–Part 2: Graded Wood Pellets*; ISO 17225-2:2014; International Organization for Standardization: Geneva, Switzerland, 2014. Available online: https://www.iso.org/obp/ui/#iso:std:iso:17225:-2:dis:ed-2:v1:en (accessed on 25 February 2020).

14. Kaliyan, N.; Vance Morey, R. Factors affecting strength and durability of densified biomass products. *Biomass Bioenergy* **2009**, *33*, 337–359. [CrossRef]

15. Williams, O.; Taylor, S.; Lester, E.; Kingman, S.; Giddings, D.; Eastwick, C. Applicability of mechanical test for biomass pellet characterization for bioenergy applications. *Materials* **2018**, *11*, 1329. [CrossRef] [PubMed]

16. Hedlund, F.H.; Astad, J.; Nichols, J. Inherent hazards, poor reporting and limited learning in the solid biomass energy sector: A case study of a wheel loader igniting wood dust, leading to fatal explosion at wood pellet manufacturer. *Biomass Bioenergy* **2014**, *66*, 450–459. [CrossRef]

17. Obernberger, I.; Thek, G. *The Pellet Handbook: The Production and Thermal Utilization of Pellets*; Taylor & Francis: London, UK, 2010.
18. International Organization for Standardization. *Solid Biofuels–Determination of Mechanical Durability of Pellets and Briquettes–Part 1: Pellets*; ISO 17831-1:2015; International Organization for Standardization: Geneva, Switzerland, 2015. Available online: https://www.iso.org/standard/60695.html (accessed on 25 February 2020).
19. Kaliyan, N.; Vance Morey, R. Factors affecting strength and durability of densified products. In Proceedings of the American Society of Agricultural and Biological Engineers meeting, Portland, OR, USA, 9–12 July 2006. [CrossRef]
20. Thomas, M.; van del Poel, A.F.B. Physical quality of pelleted animal feed. 1-Criteria for pellet quality. *Anim. Feed. Sci. Technol.* **1996**, *61*, 89–112. [CrossRef]
21. Pietsch, W. *Agglomeration Processes–Phenomena, Technologies, Equipment*; Wiley-VCH: Weinheim, Germany, 2002.
22. European Pellet Council. *Handbook for the Certification of Wood Pellets for Heating Purposes*, 3rd ed.; ENplus: Brussels, Belgium, 2015.
23. Van Dam, J.E.G.; van den Oever, M.J.A.; Teunissen, W.; Keijsers, E.R.P.; Peralta, A.G. Process for production of high density/high performance binderless boards from whole coconut husk: Part 1: Lignin as intrinsic thermosetting binder resin. *Ind. Crop. Prod.* **2004**, *19*, 207–216. [CrossRef]
24. Petisco, C.; García-Criado, B.; Mediavilla, S.; Vázquez De Aldana, B.R.; Zabalgogeazcoa, I.; García-Ciudad, A. Near-infrared reflectance spectroscopy as a fast and non-destructive tool to predict foliar organic constituents of several woody species. *Anal. Bioanal. Chem.* **2006**, *386*, 1823–1833. [CrossRef] [PubMed]
25. The Official Website of Berlin, City Trees: Overview of the Stock Data. Available online: https://www.berlin.de/senuvk/umwelt/stadtgruen/stadtbaeume/en/daten_fakten/uebersichten/index.shtml (accessed on 20 December 2019).
26. Tumurulu, J.S.; Wright, C.T.; Kenny, K.L.; Hess, J.R. *A Review on Biomass Densification Technologies for Energy Application*; Idaho National Laboratory. U.S. Department of Energy: Idaho Falls, ID, USA, 2010.
27. Dieste, A.; Clavijo, L.; Torres, A.I.; Barbe, S.; Oyarbide, I.; Bruno, L.; Cassella, F. Lignin from *Eucalyptus* spp. kraft black liquor as biofuel. *Energy Fuels* **2016**, *30*, 10494–10498. [CrossRef]
28. Sokhansanj, S.; Fenton, J. *Cost Benefit of Biomass Supply and Pre-Processing*; BIOCAP Canada Foundation: Kingston, ON, Canada, 2006.
29. Alakangas, E. *New European Pellets Standards–EN 14961-1*; Eubionet3; VTT: Jyväskyla, Finland, 2011.

 *processes*

Article

# Outdoor Large-Scale Cultivation of the Acidophilic Microalga *Coccomyxa onubensis* in a Vertical Close Photobioreactor for Lutein Production

Juan-Luis Fuentes [1], Zaida Montero [1], María Cuaresma [1], Mari-Carmen Ruiz-Domínguez [2], Benito Mogedas [1], Inés Garbayo Nores [1,*], Manuel González del Valle [3] and Carlos Vílchez [1]

[1] Algal Biotechnology Group, Ciderta and RENSMA (Center of Research in Natural Resources, Health and Environment), University of Huelva, 21007 Huelva, Spain; juanlufc@gmail.com (J.-L.F.); zaida.montero@dqcm.uhu.es (Z.M.); maria.cuaresma@dqcm.uhu.es (M.C.); benito.mogedas@dqcm.uhu.es (B.M.); cvilchez@dqcm.uhu.es (C.V.)

[2] Laboratorio de Microencapsulación de Compuestos Bioactivos (LAMICBA) del Departamento de Ciencia de los Alimentos y Nutrición, Facultad de Ciencias de la Salud, Universidad de Antofagasta, Avda. Universidad de Antofagasta 02800. P.O. Box 170, 1240000 Antofagasta, Chile; maria.ruiz@uantof.cl

[3] BioAvan S.L., Parque Empresarial Torneo, Geología 97, 41015 Sevilla, Spain; manolo@bioavan.com

* Correspondence: garbayo@uhu.es; Tel.: +34-959219953; Fax: +34-959219942

Received: 6 February 2020; Accepted: 3 March 2020; Published: 10 March 2020

**Abstract:** The large-scale biomass production is an essential step in the biotechnological applications of microalgae. *Coccomyxa onubensis* is an acidophilic microalga isolated from the highly acidic waters of Río Tinto (province of Huelva, Spain) and has been shown to accumulate a high concentration of lutein (9.7 mg g$^{-1}$dw), a valuable antioxidant, when grown at laboratory-scale. A productivity of 0.14 g L$^{-1}$ d$^{-1}$ was obtained by growing the microalga under outdoor conditions in an 800 L tubular photobioreactor. The results show a stable biomass production for at least one month and with a lutein content of 10 mg g$^{-1}$dw, at pH values in the range 2.5–3.0 and temperature in the range 10–25 °C. Culture density, temperature, and $CO_2$ availability in highly acidic medium are rate-limiting conditions for the microalgal growth. These aspects are discussed in this paper in order to improve the outdoor culture conditions for competitive applications of *C. onubensis*.

**Keywords:** carotenoids; extremophiles; microalgal biotechnology

## 1. Introduction

The production of microalgal biomass efficiently is a key prerequisite in the production process of valuable compounds from microalgae. Large-scale production of biomass requires the processing of high volume of cultures and the use of both indoor and outdoor systems, which may differ in geometry, culture engineering principles, and variable ambient conditions. An outdoor system offers the advantage well defined cultivation infrastructures and suitable latitude-dependent environmental conditions, particularly light and temperature, which may add significant economic advantages for biomass production [1,2].

Tubular photobioreactors, open raceway systems, and flat panel systems, have so far been widely recognized as suitable for large-scale microalgal production under outdoor conditions [3,4]. The choice of the culture system is highly influenced by the characteristics of the specific microalgal species, such as its robustness, growth rate, and the value and purity requirements of the target product to be obtained. Tubular and panel photobioreactors systems should be suitable to carry out the massive production of non-robust microalgal species, where the risk of contamination by undesired microorganisms and fluctuation of specific culture parameters, such as oxygen concentration and temperature can be minimized to a certain level [5]. Corrosion of equipment due to acidity might be a problem which, however, is overcome by using reactor components made of resistant-to-acid materials. In addition, slow-growth rate microalgal species are adequate for cultivation in photobioreactors which allow for a high control of growth conditions close to optimal values [1,6,7]. On the contrary, highly robust species, such as *Chlorella vulgaris* and *Dunaliella salina*, are commercially grown in raceway open ponds with a very low level of control of temperature and microbial contamination, thus reducing operational costs without dramatic reduction in the theoretical maximum algal biomass productivity [8,9].

Lutein is a compound consumed worldwide, mainly as a food colorant. Lutein sales amount to 150 million dollars in the USA only. Currently, lutein is obtained from marigold petals, with a low lutein content of 0.03% (*w/w*) [10,11]. Actually, the only possible other sources with sufficient content to be considered for lutein production are certain strains of microalgae, several of them considered as potential lutein sources based on their lutein content which ranges from 0.5% to 1.2% dry weight [10].

*Coccomyxa onubensis* is a relatively slow-growth rate acidophilic microalga, which has been grown outdoor in a 6.0 L photobioreactor for production of lutein [12]. Among the potential applications of this microalga, an antibacterial effect against human pathogens has been reported by Navarro et al. [13], and microalgal fatty acids were suggested to be involved in the antibacterial activity. Although the mechanisms through which fatty acids may exert bactericidal activity are not fully understood, it has been reported they may promote membrane damage resulting in nutrient uptake alteration and cellular respiration inhibition [14]. In addition, this microalga could be also attractive as a nutritional supplement for food and feed industries [14]. These data indicate the potential of *C. onubensis* as source of valuable compounds and highlight the interest of producing its biomass at large-scale. In this work we investigated the growth of the microalga in an 800 L vertically stacked tubular photobioreactor, and this creates a step forward in the way for the efficient and economically feasible production of enriched *C. onubensis* biomass. To the best of our knowledge, this is the first example of biomass production of an acid-tolerant microalga in an outdoor tubular photobioreactor and offers experimental details of what should be taken into account for the successful pilot-scale production.

## 2. Materials and Methods

### 2.1. Microalga and Culture Medium

*Coccomyxa onubensis* ACCV1 was isolated from the acidic waters of the Tinto river, Huelva (Spain), phylogenetically characterized and deposited in the Experimental Phycology and Culture Collection of Algae at the University of Goettingen in Germany (SAG) with the stock number SAG 2510 [15]. According to the chemical composition of the acidic water in its natural environment, the medium used for maintaining axenic cultures of the microalga in the culture room was prepared at pH 2.5 by modifying the K9 medium described by Silverman and Lundgren [16]. The culture medium was prepared in our laboratory with the following chemical composition: 22.67 mM $K_2SO_4$, 1.34 mM KCl, 2.87 mM $K_2HPO_4$, 4.31 mM $MgCl_2$, 22.65 mM $KNO_3$, 0.09 mM $CaCl_2$, and 5 mL of Hutner's trace elements solution prepared in our laboratory as indicated in [15]. The microalgal cultures were grown in Erlenmeyer flasks (Fisher Scientific S.L., Madrid, Spain) in an algal room at 25 °C bubbled with air containing 2.5% (*v/v*) $CO_2$ and continuously illuminated by white fluorescent lamps.

For large-scale cultivation in indoor plastic bags and in an outdoor tubular photobioreactor, a culture medium based on an NPK-fertilizer solution (Agralia Fertilizantes S.L., Huesca, Spain) was used in order to reduce nutrient costs. The 1 L of culture medium (NPK-medium) contained 0.25 mM $NO_3^-$, 0.45 mM $NH_4^+$, 0.4 mM $P_2O_4$, 0.64 mM $K_2O$, and 0.4 mL of a commercial micronutrient solution (Microfer Complex, Fercampo, Málaga, Spain).

Previous experiments at laboratory scale unveiled a more efficient use of urea by the microalga, as compared to other common nitrogen sources used to grow microalgae, including nitrate, nitrite, or ammonium [15,17]. Thus, the final medium was supplemented with 8 mM urea (Panreac Química

SLU, Barcelona, Spain) which also provides carbon to the microalga, 100 µM $FeCl_3$, and pH was adjusted to 2.5.

## 2.2. Cultivation of C. onubensis in Indoor Air Fluidized-Bed Plastic Bags

The microalga was grown in two plastic bags (Plásticos Gamaza, Santander, Spain) of 400 L each, having 60 cm diameter, 2.1 m height, and 200 µm thickness, fully transparent to photosynthetically active radiation (PAR) (Figure 1A), and using the NPK-medium as described in the previous section. Two conditions of initial biomass concentration were assessed: 0.31 g $L^{-1}$ (C1, Figure 2A) and 0.57 g $L^{-1}$ (C2, Figure 2A). The bags were maintained indoors at 25 °C and continuously bubbled with $CO_2$-enriched air (2.5% *v/v*) through air diffusers placed at the bottom of the bags. The cultures were continuously illuminated with white fluorescent lamps that yielded 150 µE $m^{-2}$ $s^{-1}$ of incident light on the surface of the bag. The incident light intensity on the surface of the culture bags was measured using a photoradiometer (HD9021, Delta OHM, Padova, Italy).

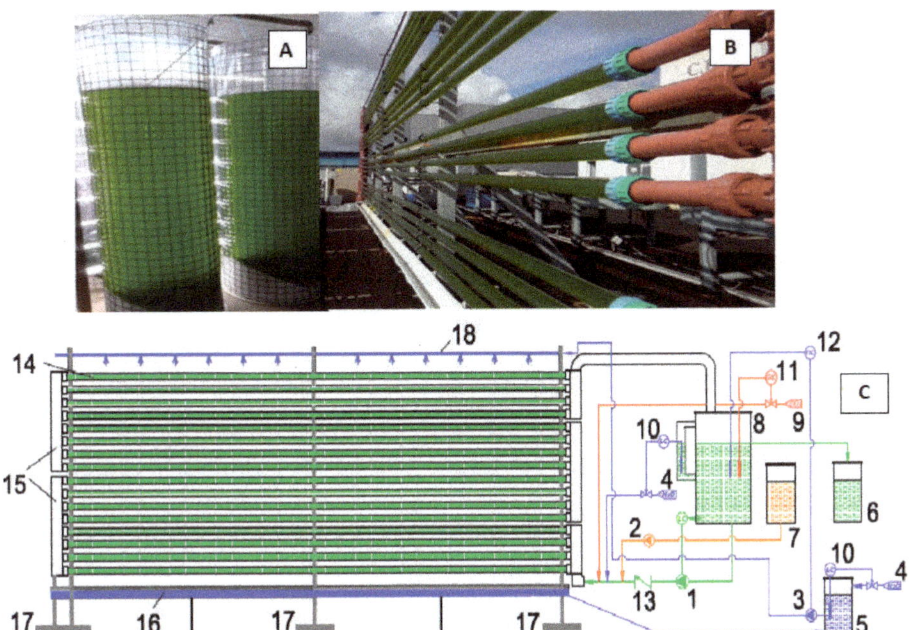

**Figure 1.** Cultivation systems for large-scale production of *Coccomyxa onubensis* biomass. Indoor 400 L cultivation bags (**A**); outdoor 800 L vertical tubular photobioreactor (**B**); configuration of the photobioreactor is detailed in (**C**): (1) culture pump; (2) nutrient pump; (3) cooling water pump; (4) water inlet; (5) cooling water tank; (6) harvested biomass tank; (7) nutrient tank; (8) degasser tank; (9) $CO_2$ inlet; (10) level controller; (11) pH controller indicator; (12) temperature controller indicator; (13) non-return valve; (14) glass pipe; (15) manifolds; (16) cooling water collector; (17) pipe support; (18) cooling pipe with sprinklers. The systems are located at CIDERTA (University of Huelva, Spain).

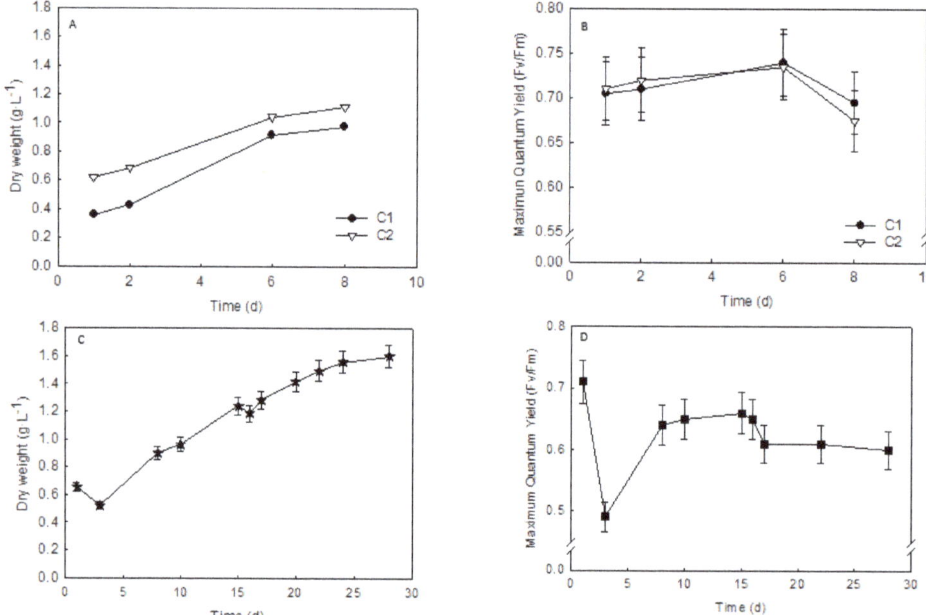

**Figure 2.** Growth of *C. onubensis* cultures in batch indoors and in an outdoor photobioreactor (PBR) system. Time-course evolution of growth in batch cultures (**A**) and photosynthetic capacity (maximum quantum yield, (**B**)). Time course of growth in an outdoor PBR (**C**). Photosynthetic capacity in photobioreactor culture (**D**). Two conditions of initial biomass concentration were assessed: 0.31 g L$^{-1}$ (C1) and 0.57 g L$^{-1}$ (C2). Details of the experimental set-up and parameter determination are described in Section 2.

### 2.3. Cultivation of C. onubensis in an Outdoor Vertical Tubular Photobioreactor

The microalga was cultured at a pilot scale in an 800 L outdoor vertical tubular photobioreactor, using the same NPK-medium previously described. The system consists of 16 horizontal glass tubes with 6 cm internal diameter vertically stacked and connected mutually by means of manifolds on both sides of the tubes (Figure 1B). The glass tubes were kindly provided by Schott (Schott Glass Iberica SL, Barcelona, Spain). The photobioreactor is equipped with temperature and pH control systems, which allowed keeping the temperature below 25 °C and pH between a range of 2.5–3 (Figure 1C). Reactor components were made of resistant-to-acid materials. This design was particularly based on the conclusions made from the work of Vaquero et al. [12] in a 6 L outdoor tubular photobioreactor (PBR).

For replication of outdoor cultivation trials, the limitation exists that only one large PBR system was available. Thus, it was not possible to replicate experiments under the same outdoor conditions. However, the trial of growth outdoors allowed to understand that the acidophilic microalga grow stably in an PBR system outdoors and also to approach the first data of productivity.

The culture broth was placed inside a 1500 L tank and pumped to the bottom manifold of the photobioreactor. The broth circulates through the four bottom tubes and reaches the large common manifold at the side extreme. Subsequently, the broth enters the four tubes above and flows in opposite direction, reaching the manifold at the extreme of the tube. The flow pattern was repeated until the broth reached the manifold placed at the highest position, finally returning to the tank where it was degassed before being again pumped back to the photobioreactor (Figure 1C).

pH was measured continuously using the pH/ATC transmitter DMM-4000/pH (Design Instrument, Barcelona, Spain) equipped with a pH sensor (SG900CD, SENSOREX, Inc., Los Angeles, CA, USA). Temperature was measured continuously using a temperature sensor (PT100). Both controllers were equipped with a data acquisition system ICP-COM type. $CO_2$ was automatically injected when the culture pH exceeded 3.5. The cooling system (see below) was set to turn on automatically when the culture temperature exceeded 24 °C.

The temperature control was carried out by dripping water over the tubular system. The water used for cooling was collected by a channel in the low part of the photobioreactor and returned to the cooling water tank. This tank was placed underground, thus reducing water losses.

### 2.4. Harvesting of Microalgal Biomass

After the cultivation period, the microalgal biomass was harvested by continuous flow centrifugation at 250 L $h^{-1}$ and 8400 rpm using an industrial centrifuge (KA–6, GEA Westfalia Separator, Oelde, Germany). The biomass was frozen at −20 °C for 24 h and subsequently lyophilized in a freeze drier (FD8512, IlShin BioBase, Ede, The Netherlands). The powder was then vacuum packed and stored at −80 °C until further use.

### 2.5. Growth Measurements and Productivity

In all cultures of this study, the algal growth evolution with time course was daily assessed by following dry weight measurements which were carried out by taking 10 mL samples of each culture. The samples were filtered through glass microfiber filters of 47 mm diameter and 0.7 μm pore size (MFV–5, Filter-Lab, Barcelona, Spain). The filters containing wet algal biomass were dried in an oven at 100 °C for 24 h [18].

### 2.6. Maximum Photosystem II Quantum Yield (Fv/Fm)

The photosynthetic efficiency was evaluated by measuring the chlorophyll fluorescence in dark-acclimated cells, considered as the maximum photosynthetic efficiency of photosystem II ($F_v/F_m$). This parameter was determined using a portable pulse amplitude modulated fluorometer (AquaPen-C AP-C 110, Photo Systems Instruments, Drasov, Czech Republic), according to the method previously described [19].

### 2.7. Pigment Analysis

Pigments were extracted with methanol and measured spectrophotometrically as described in Ruiz-Domínguez et al. [20]. Specific carotenoids—lutein and β-carotene—were separated by HPLC equipped with a diode-array detector (L-7420, TermoQuest, CA, USA) and a RP18 column (LichroCart RP18, Merck KGaA, Darmstadt, Germany), 5 μm, size 250 × 4 mm. In the mobile phase, solvent A was ethyl acetate and solvent B was acetonitrile and water (9:1, *v/v*) (mobile phase flow rate was 1 mL per minute). Carotenoids detection was at 450 nm and the carotenoids were quantified using lutein and β-carotene standards supplied by DHI-Water and Environment (Hørsholm, Denmark) [20].

### 2.8. Statistics

In this study, one outdoor experiment was carried out from 8 March 2017, to 4 April 2017. Unless otherwise indicated, the data presented in this manuscript are the means of three independent samples.

## 3. Results and Discussion

### 3.1. Batch Cultivation of C. onubensis in Plastic Bags Indoors

Cultivation of *C. onubensis* in two plastic bags of 400 L each in the indoor condition was first investigated to analyze the growth stability and the average of biomass productivity. These cultures served as a source of active biomass for inoculation to the outdoor tubular photobioreactor. Figure 2A shows the growth of *C. onubensis* cultivated indoor in batch cultures in plastic bags. Two different initial biomass concentrations were assessed by differential dilution (*v/v*) of the mother culture: 1/2 (control) and 1/1, resulting an initial biomass concentration in the cultures of 0.31 g $L^{-1}$ (C1, Figure 2A), and 0.57 g L$^{-1}$ (C2, Figure 2A), respectively. The culture with the lowest initial biomass concentration showed the fastest growth and it resulted in a higher biomass productivity. On the contrary, the culture with the highest initial biomass concentration rapidly attained maximum dry weight (Figure 2A).

A maximal lutein concentration of 9.7 mg g$^{-1}$dw was obtained, which leads to a maximal lutein productivity of 0.9 mg L$^{-1}$ d$^{-1}$ in plastic bags indoors.

The maximum biomass productivity in culture bags indoors achieved values of 0.09 g L$^{-1}$ d$^{-1}$ (Table 1) for experimental condition C1. In this condition, a maximum lutein concentration of 9.7 mg g$^{-1}$dry weight was obtained, leading to a maximum lutein productivity of 0.87 mg$^{-1}$ d$^{-1}$. The experimental condition C2 resulted in a maximum biomass productivity of 0.07 g L$^{-1}$ d$^{-1}$. Thus, condition C1 was slightly more efficient.

**Table 1.** Comparative productivity and lutein content of biomass in different acidophile and non-acidophile microalgae. "PBR" means photobioreactor.

| Microalga | Growth System | pH | Productivity (g L$^{-1}$ d$^{-1}$) | Lutein (mg g$^{-1}$) | Reference |
|---|---|---|---|---|---|
| *C. onubensis* | Indoor plastic bag 400 L | 2.5 | 0.09 | 9.7 | Present work |
| *C. onubensis* | Outdoor PBL 800 L | 2.5 | 0.14 | 10.0 | Present work |
| *C. onubensis* | Laboratory conditions 1 L | 2.5 | 1.80 | 6.0 | [18] |
| *C. onubensis* | Outdoor PBR 6 L | 2.5 | 0.40 | 4.8 | [12] |
| *Chlamydomonas acidophila* | Laboratory conditions. High light | 2.5 | 0.75 | - | [21] |
| *Chlamydomonas acidophila* | Laboratory conditions. Low light | 2.5 | 0.08 | - | [21] |
| *Galdieria sulphuraria* | Laboratory conditions | 2.0 | 0.27 | - | [22] |
| *Muriellopsis* sp. | Laboratory conditions 1 L | 7.5 | 1.60 | 6.0 | [23] |
| *Muriellopsis* sp. | Outdoor tubular | 7.5 | 13.0 [†] | 5.0 | [24] |
| *Nannocloropsis* sp. CCAP 211/78 | Outdoor PBR 6 L | 7.5 | 0.71 | - | [25] |
| *Scenedesmus almeriensis* | Laboratory conditions. High light | 7.0 | 0.95 | 5.3 | [21] |
| *Scenedesmus almeriensis* | Outdoor tubular 4000 L | 7.0 | 0.29 | 4.5 | [10] |

[†] Units: g m$^{-2}$ d$^{-1}$.

The cultures in the plastic bags are light-limited due to the large bag size, which reduces dramatically the intensity of light in the culture core. In this respect, from our own experience in bags a linear correlation exists between *C. onubensis* growth rate and light intensities below 400–500 μE m$^{-2}$ s$^{-1}$. Therefore, maintaining large amounts of *C. onubensis* biomass growing actively in the linear growth phase (dry weight between 0.6 and 1 g L$^{-1}$) even at a lower growth rate might reduce nutrient consumption and operational costs (Figure 2A).

A major constraint in cultivating microalgae at acidic pH values is the low $CO_2$ solubility at low pH. It means that the cultivation system design and operation mode must be such that the supplied $CO_2$ is homogeneously distributed throughout the cultivation system and $CO_2$ losses are minimized. These conditions will influence the biomass production yield of an acidotolerant or acidophilic microalgal species [6]. However, the results of previous cultivation assays of *C. onubensis* in small-scale tubular photobioreactors showed the long $CO_2$ retention time in the reactor, thus, indicating the convenience of using closed tubular photobioreactors for their production [12,26].

*3.2. Cultivation of C. onubensis in a Tubular Photobioreactor Outdoors*

The growth of acidophilic or acidophilic microalgae at acidic pH largely depends on the inorganic carbon availability in the form of $CO_2$ [27,28]. At low pH, $CO_2$ has a very low solubility and, consequently, $CO_2$ supply in open systems like raceway open ponds may easily result in massive losses of $CO_2$ to the atmosphere. The cultivation in closed photobioreactors (i.e., tubular systems) can avoid large losses of $CO_2$ and additionally allow for a more efficient use of $CO_2$ by the cultivated algal species [29]. In addition, the vertical configuration of a tubular photobioreactor results in lowered photoinhibition compared to the horizontal disposition [12]. Accordingly, a vertical tubular photobioreactor which allows for better control of main cultivation parameters was selected for this study.

The *C. onubensis* biomass was initially produced in batch cultures, such as in plastic bags in indoor conditions, and used as an inoculum for culturing *C. onubensis* in an 800 L tubular photobioreactor outdoors. The initial biomass concentration in the tubular photobioreactor was set at 0.65 g L$^{-1}$ and the pattern of growth is observed in Figure 2C. A biomass productivity of 0.14 g L$^{-1}$ d$^{-1}$ was achieved (Table 1). A maximal lutein concentration of 10 mg g$^{-1}$dw was obtained, which leads to a maximal lutein productivity of 1.42 mg L$^{-1}$ d$^{-1}$. The outdoor trials for *C. onubensis* production described in this manuscript were carried out in March, a month of moderate maximum and minimum temperatures of approximately 26 and 5 °C on average, respectively. The culture medium pH was adjusted continuously to 2.5, which is optimal for *C. onubensis* growth [15]. In outdoor cultures, Vaquero et al. [18] observed a noticeable effect of temperature on *C. onubensis* growth. Thus, in order to maintain the culture temperature at 25 °C in the photobioreactor, water was trickled over the glass tubes from thin perforated pipes placed at the top of the vertical tubular photobioreactor. Water was collected at the bottom of the system and recirculated (Figure 1B,C).

As shown in Figure 2C, culture adaptation to outdoor conditions resulted in a slight decrease in biomass density after the first two days, which correlates with the slight decrease of maximum quantum yield from 0.7 to 0.5 within that time period. The outdoor conditions greatly differ from indoor conditions, particularly the light intensity which is much higher in the outdoor systems. Consequently, an adaptation period is required for the cultures to perform efficiently. During the adaptation period to the outdoor high light intensity, the maximum quantum yield decreases temporarily to 0.5, which did not affect further growth performance in the reactor. Zijffers et al. [30] reported that maximum photosynthetic efficiency values below 0.6 indicate a significant loss of culture viability. Nevertheless, an average value of approximately 0.65, typically found in healthy algal cultures, was recorded after the first few days. Afterwards, the maximum quantum yield of *C. onubensis* roughly remained constant within the range 0.60–0.65 (Figure 2D).

Lutein content in the microalgal cells may vary along the growth cycle in the PBR, depending on a number of factors and the algal species [31], thus the mode of cultivation in the PBR system, the dilution rate, and/or the right moment for harvesting should be eventually optimized to get enhanced productivities. The cultivation system might influence the lutein accumulation in the cells, as light path varies largely depending on the production system, for instance the tubular photobioreactors light path is usually shorter than that in raceway open ponds [32]. In microalgae lutein may accumulate under relatively low irradiance levels as, for instance, proven in *Chlorella* [33], which suggests that the low volumetric productivity shown in this paper might still be enhanced in dense cultures.

Overall, the results suggest that *C. onubensis* can be produced massively in outdoor tubular photobioreactors, at acidic pH values, during the springtime in southern Europe, seemingly without photoinhibition if the production is carried out at suitable biomass concentrations. Non-optimized biomass and lutein productivities of 0.14 and 1.42 g L$^{-1}$ d$^{-1}$, respectively, were achieved, which can still be further improved. For instance, heating was not implemented during the night and the outdoor cultures were produced under natural light–dark cycles. The productivity should be expected to increase in cultures subjected to 24 h illumination and temperature control, thus implementing artificial light and heating overnight. Nevertheless, these experiments should be carried out only once the impact on process economy has been assessed. In addition, we hereby suggest that while performing the outdoor cultures at higher biomass concentrations in summer, major constraints, i.e., very high temperature and light irradiance, could be evaded by shading the tubular photobioreactor during midday hours and by cooling the system continuously as described in the previous section.

Since, so far, no reports are available on outdoor cultivation of acidophilic microalgal species, the comparison of the data obtained in this study with outgroup data is not possible. Nevertheless, in general, the productivity of *C. onubensis* is higher than that reported for other microalgae, including some acidophilic species, at a laboratory scale (Table 1). According to the *C. onubensis* productivity results obtained in the laboratory conditions and in low volume outdoor tubular reactors [12] (Table 1), there is still room for improvement of the productivity in high volume outdoor reactors if appropriate cultures conditions are used and prior adaptation of the microalgal species is carried out.

A further economic analysis should be done to assess the potential of outdoor production of *C. onubensis* in PBR for eventual production of lutein-enriched biomass. In this sense, a complete set of productivity data should be obtained throughout the year from the PBR operating in semicontinuous mode and with the optimal biomass concentration range.

## 4. Conclusions

The capacity of an acidic extreme environment microalga, *C. onubensis*, for massive production in outdoor photobioreactors was evaluated. The production in acidic medium minimizes the risk of biological contamination and reduces the loss of algal productivity associated with microbial proliferation. *C. onubensis* can be produced massively in outdoor tubular photobioreactors, at acidic pH values, during the springtime in southern Europe, seemingly with limited photoinhibition if the production is carried out at suitable biomass concentrations. Non-optimized, maximal biomass and lutein productivities of 0.14 g $L^{-1}$ $d^{-1}$ and 1.4 mg $L^{-1}$ $d^{-1}$, respectively, were achieved, which can still be further improved. The optimization of process engineering at extreme acidic pH, including pH-control and the use of acid-resistant materials, is a key challenge to maximize the productivity of *C. onubensis* and their derived products in photobioreactors.

**Author Contributions:** Conceptualization, J.-L.F. and C.V.; Methodology, J.-L.F., Z.M., M.-C.R.-D. and C.V.; Software, B.M.; Validation, J.-L.F. and Z.M.; Formal Analysis, M.G.d.V.; Investigation, J.-L.F., Z.M., M.-C.R.-D. and I.G.N; Resources, C.V.; Data Curation, B.M.; Writing-Original Draft Preparation, J.-L.F., C.V. and I.G.N.; Writing-Review & Editing, C.V. and I.G.N.; Supervision, M.C.; Project Administration, C.V. and J.-L.F. All authors have read and agreed to the published version of the manuscript.

**Funding:** Authors want to thank the PhD-Grant (2015/7949) from CEIMAR (Marine International Campus of Excellence, Spain) to Juan Luis Fuentes.

**Conflicts of Interest:** The authors declare no conflict of interest.

## References

1. Forján, E.; Navarro, F.; Cuaresma, M.; Vaquero, I.; Ruíz-Domínguez, M.C.; Gojkovic, Z.; Vázquez, M.; Márquez, M.; Mogedas, B.; Bermejo, E.; et al. Microalgae: Fast-growth sustainable green factories. *Crit. Rev. Environ. Sci. Technol.* **2015**, *45*, 1705–1755. [CrossRef]
2. Holdmann, C.; Schmid-Staiger, U.; Hirth, T. Outdoor microalgae cultivation at different biomass concentrations—Assessment of different daily and seasonal light scenarios by modeling. *Algal Res.* **2019**, *38*, 101405. [CrossRef]
3. Mata, T.M.; Martins, A.; Caetano, N.S. Microalgae for biodiesel production and other applications: A review. *Renew. Sustain. Energy Rev.* **2010**, *14*, 217–232. [CrossRef]
4. Norsker, N.H.; Barbosa, M.J.; Vermuë, M.H.; Wijffels, R.H. Microalgal production: A close look at the economics. *Biotechnol. Adv.* **2011**, *29*, 24–27. [CrossRef] [PubMed]
5. Acién-Fernández, F.G.; Fernández-Sevilla, J.M.; Molina-Grima, E. Photobioreactors for the production of microalgae. *Rev. Environ. Sci. Biotechnol.* **2013**, *12*, 131–151. [CrossRef]
6. Molina, E.; Fernández, J.; Acién, F.G.; Chisti, Y. Tubular photobioreactor design for algal cultures. *J. Biotechnol.* **2001**, *92*, 113–131. [CrossRef]
7. Madhubalaji, C.K.; Ajam, S.; Sijil, P.V.; Mudliar, S.; Chauhan, V.S.; Sarada, R.; Rao, A.R.; Ravishankar, G.A. Open Cultivation Systems and Closed Photobioreactors for Microalgal Cultivation and Biomass Production. In *Handbook of Algal Technologies and Phytochemicals*; CRC Press: Boca Raton, FL, USA, 2019; pp. 178–202.
8. Ben-Amotz, A. Industrial Production of Microalgal Cell-Mass and Secondary Products-Major Industrial Species: *Dunaliella*. In *Handbook of Microalgal Culture: Biotechnology and Applied Phycology*; Richmond, A., Ed.; Blackwell Publishing: Hoboken, NJ, USA, 2007; pp. 273–280. [CrossRef]
9. Liu, J.; Hu, Q. Chlorella: Industrial Production of Cell Mass and Chemicals. In *Handbook of Microalgal Culture: Applied Phycology and Biotechnology*; John Wiley & Sons: Oxford, UK, 2013; pp. 329–338. [CrossRef]

10. Fernández-Sevilla, J.M.; Acién-Fernández, F.G.; Molina-Grima, E. Biotechnological production of lutein and its applications. *Appl. Microbiol. Biotechnol.* **2010**, *86*, 27–40. [CrossRef]

11. McClure, D.D.; Nightingale, J.K.; Sachin Black, A.L.; Zhu, J.; Kavanagh, J.M. Pilot-scale production of lutein using *Chlorella vulgaris*. *Algal Res.* **2019**, *44*. [CrossRef]

12. Vaquero, I.; Ruiz-Domínguez, M.C.; Mogedas, B.; Vilchez, C.; Vega, J.M. Production of lutein-enriched biomass by growing *Coccomyxa* sp. (strain *onubensis*; Chlorophyta) under spring outdoor conditions of southwest Spain. *J. Adv. Biotechnol.* **2016**, *6*, 828–834. [CrossRef]

13. Navarro, F.; Forján, E.; Vázquez, M.; Toimil, A.; Montero, Z.; Ruiz-Domínguez, M.C.; Garbayo, I.; Castaño, M.A.; Vílchez, C.; Vega, J.M. Antimicrobial activity of the acidophilic eukaryotic microalga *Coccomyxa* sp. (strain *onubensis*). *Phycol. Res.* **2017**, *65*, 38–43. [CrossRef]

14. Navarro, F.J.; Forján, E.; Vázquez, M.; Montero, Z.; Bermejo, E.; Castaño, M.A.; Toimil, A.; Chagüaceda, E.; García-Sevillano, M.A.; Sánchez, M.; et al. Microalgae as a safe food source for animals: Nutritional characteristics of the acidophilic microalga *Coccomyxa onubensis*. *Food Nutr. Res.* **2016**, *60*. [CrossRef] [PubMed]

15. Fuentes, J.L.; Huss, V.; Montero, Z.; Torronteras, R.; Cuaresma, M.; Garbayo, I.; Vílchez, C. Phylogenetic characterization and morphological and physiological aspects of a novel acidophilic and halotolerant microalga *Coccomyxa onubensis* sp. nov. (Chlorophyta, Trebouxiophyceae). *J. Appl. Phycol.* **2016**, *28*, 3269–3279. [CrossRef]

16. Silverman, M.P.; Lundgren, D.G. Studies on the chemoautotrophic iron bacterium *Ferrobacillus ferrooxidans*. *J. Bacteriol.* **1959**, *77*, 642–647. [CrossRef]

17. Casal, C.; Cuaresma, M.; Vega, J.M.; Vílchez, C. Enhanced productivity of a lutein-enriched novel acidophile microalga grown on urea. *Mar. Drugs* **2011**, *9*, 29–42. [CrossRef] [PubMed]

18. Vaquero, I.; Vázquez, M.; Ruiz-Domínguez, M.C.; Vílchez, C. Enhanced production of a lutein-rich acidic environment microalga. *J. Appl. Microbiol.* **2014**, *116*, 839–850. [CrossRef] [PubMed]

19. Cuaresma, M.; Janssen, M.; Vílchez, C.; Wijffels, R. Horizontal or vertical photobioreactors? How to improve microalgae photosynthetic efficiency. *Bioresour. Technol.* **2011**, *102*, 5129–5137. [CrossRef]

20. Ruiz-Domínguez, M.C.; Vaquero, I.; Obregón, V.; de la Morena, B.; Vílchez, C.; Vega, J.M. Lipid accumulation and antioxidant activity in the eukaryotic acidophilic microalga *Coccomyxa* sp. (strain *onubensis*) under nutrient starvation. *J. Appl. Phycol.* **2015**, *27*, 1099–1108. [CrossRef]

21. Spinola, M.V.; Díaz-Santos, E. Microalgae Nutraceuticals: The Role of Lutein in Human Health. In *Microalgae Biotechnology for Food, Health and High Value Products*; Springer: Singapore, 2020; pp. 243–263.

22. Hirooka, S.; Miyagishima, S.Y. Cultivation of acidophilic algae *Galdieria sulfuraria* and *Pseudochlorella* sp. YKT1 in media derived from acidic hot springs. *Front. Microbiol.* **2016**, *7*, 2022. [CrossRef]

23. Del Campo, J.A.; Rodríguez, H.; Moreno, J.; Vargas, M.A.; Rivas, J.; Garcia-Guerrero, M. Lutein production by *Muriellopsis* sp. in an outdoor tubular photobioreactor. *J. Biotechnol.* **2001**, *85*, 289–295. [CrossRef]

24. Blanco, A.M.; Moreno, J.; Del Campo, J.A.; Rivas, J.; Guerrero, M.G. Outdoor cultivation of lutein-rich cells of *Muriellopsis* sp. in open ponds. *Appl. Microbiol. Biotechnol.* **2007**, *73*, 1259–1266. [CrossRef]

25. De Vree, J.H.; Bosma, R.; Janssen, M.; Barbosa, M.J.; Wijffels, R.H. Comparison of four outdoor pilot-scale photobioreactors. *Biotechnol. Biofuels* **2015**, *8*, 215. [CrossRef] [PubMed]

26. Vaquero, I.; Ruiz-Domínguez, M.C.; Márquez, M.; Vílchez, C. Cu-mediated biomass productivity enhancement and lutein enrichment of the novel microalga *Coccomyxa onubensis*. *Process. Biochem.* **2012**, *47*, 694–700. [CrossRef]

27. Gross, W. Ecophysiology of algae living in highly acidic environments. *Hydrobiologia* **2000**, *433*, 31–37. [CrossRef]

28. De Farias-Neves, F.; Hoinaski, L.; Rubi-Rörig, L.; Bianchini-Derner, R.; de Melo-Lisboa, H. Carbon biofixation and lipid composition of an acidophilic microalga cultivated on treated wastewater supplied with different $CO_2$ levels. *Environ. Technol.* **2019**, 3308–3317. [CrossRef]

29. Molina-Grima, E.; Acién-Fernández, F.G.; García-Camacho, F.; Chisti, Y. Photobioreactors: Light regime, mass transfer, and scale-up. *J. Biotechnol.* **2009**, *70*, 231–248. [CrossRef]

30. Zijffers, J.W.F.; Schippers, K.J.; Zheng, K.; Janssen, M.; Tramper, J.; Wijffels, R.H. Maximum Photosynthetic Yield of Green Microalgae in Photobioreactors. *Mar. Biotechnol.* **2010**, *12*, 708–718. [CrossRef]

31. Gerloff-Elias, A.; Spijkerman, E.; Pröschold, T. Effect of external pH on the growth, photosynthesis and photosynthetic electron transport of *Chlamydomonas acidophila* Negoro, isolated from an extremely acidic lake (pH 2.6). *Plant Cell Environ.* **2005**, *28*, 1218–1229. [CrossRef]

32. Huang, Q.; Jiang, F.; Wang, L.; Yang, C. Design of photobioreactors for mass cultivation of photosynthetic organisms. *Engineering* **2017**, *3*, 318–329. [CrossRef]

33. Gong, M.; Bassi, A. Investigation of *Chlorella vulgaris* UTEX 265 cultivation under light and low temperature stressed conditions for lutein production in flasks and the coiled tree photo-bioreactor (CTPBR). *Appl. Biochem. Biotechnol.* **2017**, *183*, 652–671. [CrossRef]

 *processes*

*Article*

# Mild Hydrothermal Pretreatment of Microalgae for the Production of Biocrude with a Low N and O Content

**Miriam Montero-Hidalgo [1], Juan J. Espada [2], Rosalía Rodríguez [2], Victoria Morales [1], Luis Fernando Bautista [1] and Gemma Vicente [2,*]**

[1] Department of Chemical and Environmental Technology, Universidad Rey Juan Carlos. C/Tulipán s/n, 28933 Móstoles, Spain; miriam.montero@urjc.es (M.M.-H.); victoria.morales@urjc.es (V.M.); fernando.bautista@urjc.es (L.F.B.)

[2] Department of Chemical, Energy and Mechanical Technology, Universidad Rey Juan Carlos. C/Tulipán s/n, 28933 Móstoles, Spain; juanjose.espada@urjc.es (J.J.E.); rosalia.rodriguez@urjc.es (R.R.)

* Correspondence: gemma.vicente@urjc.es

Received: 30 August 2019; Accepted: 12 September 2019; Published: 17 September 2019

**Abstract:** A hydrothermal pretreatment of the microalga *Nannochloropsis gaditana* at mild temperatures was studied in order to reduce the N and O content in the biocrude obtained by hydrothermal liquefaction (HTL). The work focused on the evaluation of temperature, reactor loading, and time (factors) to maximize the yield of the pretreated biomass and the heteroatom contents transferred from the microalga biomass to the aqueous phase (responses). The study followed the factorial design and response surface methodology. An equation for every response was obtained, which led to the accurate calculation of the operating conditions required to obtain a given value of these responses. Temperature and time are critical factors with a negative effect on the pretreated biomass yield but a positive one on the N and O recovery in the aqueous phase. The slurry concentration has to be low to increase heteroatom recovery and has to be high to maximize the pretreated microalga yields. Response equations were obtained for the analyzed responses, which facilitated the accurate prediction of the operating conditions required to obtain a given value of these responses.

**Keywords:** microalgae; hydrothermal liquefaction; pretreatment; low O and N biocrude

## 1. Introduction

Advanced biofuels obtained from microalgae have attracted great interest in the research field because they can be grown on non-arable land, and, therefore, they do not compete with food production. In addition, microalgae can fix $CO_2$ from the air through photosynthesis, which allows for a reduction of $CO_2$ emissions [1–3].

Among the different biomass-to-biofuel thermochemical processes, hydrothermal liquefaction (HTL) has proven to be a likely option for the production of biofuels from microalgae [4–7], as it allows for the direct processing of wet algal biomass, thus avoiding the costs related to the drying step [8,9]. Moreover, HTL is not limited to the lipid fraction of the algal biomass because carbohydrates and proteins can also be converted into the biofuel product [10,11]. Furthermore, HTL microalga processing enhances the recovery of nutrients that can be recycled for microalga growth [12,13].

A moderate temperature (250–375 °C), pressure (4–22 MPa), and reaction time within the range of 5–60 min are commonly used in HTL [9,14,15]. Under these conditions, the aqueous medium, near the critical point, promotes the degradation of macromolecules present in the algal biomass, as well as the polymerization of the resultant smaller molecules [13]. A variety of products can be obtained from HTL: liquid organic phase (biocrude), aqueous phase compounds, solid residue, and gas phases, all of whose yields and quality are strongly affected by the operating conditions and the microalga used as feedstock. Thus, the influence of temperature, pressure, reaction time and slurry concentration on biocrude yield and quality have been deeply investigated [16]. As reported elsewhere, the temperature is the most important parameter that affects HTL [17], producing biocrude yields between 40 and 50 wt.% within a temperature range of 250–375 °C, depending on the microalga used as feedstock.

It has been reported that the maximization of the biocrude yield is favored at high temperatures. However, in these conditions, the biocrude is simultaneously enriched in N (derived from microalgae chlorophyll and proteins) [17,18]. Therefore, the HTL of microalgae yields biocrudes with a larger content of O (10–20 wt.%) and N (1–8 wt.%) than the conventional crude [14]. This has a negative effect, not only on the final properties of the biocrude (high viscosity), but also on the possibilities of using it in conventional refinery operations due to catalyst poisoning [14,19]. Likewise, the potential application of biocrude as a biofuel is also limited due to the NOx emissions partially derived from N compounds [16]. In addition, the high amount of O in the biocrude reduces its heating value. In order to decrease the N and O contents of the biocrude, different strategies can be devised. One of them is a low temperature HTL (<200 °C), which has been evaluated as a previous pretreatment to increase the quality of the final biocrude and to enhance the energy efficiency of the overall process [14,20–22]. This is an advantageous option, because it allows for the hydrolysis of proteins into small molecules, which remain solubilized in the aqueous phase, diminishing the N content in the pretreated residue and consequently reducing the N content in the final biocrude. In addition, P and N compounds can be recovered from the aqueous phase and used for microalga cultivation [20]. The use of this pretreatment in combination with other processes is currently under study, but more focus has been spent on the reduction of the N content of the microalga, whereas the also necessary O content decrease has been less studied in literature. In this sense, the pyrolysis of algal biomass pretreated by HTL has been reported in a batch reactor, yielding biocrude with a low N content [23]. Another approach is the use of this low temperature pretreatment of the microalgae in combination with a final HTL process at a high temperature. This scheme has been applied to microalga processing, obtaining biocrudes with a low N content in a batch reactor [14,20,21] and also in a semi-continuous reactor [24,25]. A low N biocrude was also obtained from the yeast *Cryptococcus curvatus* through this sequential HTL in batch operation mode [26].

Though a low temperature pretreatment appears to be a promising alternative in reducing the N compounds of the final biocrude, the relationship between operating conditions, pretreated microalga yield, and properties has not been fully established. Additionally, the necessary reduction of O in the biomass has not yet been deeply studied. In this context, the use of models to determine the optimal conditions of this pretreatment to reduce both N and O concentrations in the pretreated biomass could be of great interest to achieve the desired objectives in subsequent processing. In this work, a low temperature wet pretreatment of the microalga *Nannochloropsis gaditana* was carried out to evaluate the optimal conditions and the effects of temperature, reactor loading, and reaction time in order to maximize the yield of the pretreated biomass and the N and O content transferred from biomass to the aqueous phase. The process was developed and optimized by the factorial design and response surface methodology, which is a powerful tool that has not been applied to the hydrothermal microalga pretreatment to reduce the N and O content of the final biocrude.

## 2. Materials and Methods

### 2.1. Materials

The microalga selected for this study was *N. gaditana*, which was purchased from AlgaEnergy S.A. (Madrid, Spain) and received in freeze-dried form. Dry biomass is more feasible for long-term storage and its use on the small scale. In large-scale HTL, wet microalga can be directly used as the feedstock without any dewatering and drying processes.

The elemental composition of *N. gaditana* was 48.7% C, 7.1% H, 6.8% N, 0.9% S and 21.2% O. Freshly deionized water prepared in the laboratory was used throughout the experiments. All other chemicals used in this research were obtained commercially and used as received.

## 2.2. Experimental Procedure

In a typical pretreatment essay, a solution of the desired amount of *N. gaditana* powder and 75 mL of fresh deionized water were added to the reactor. The amount of microalga depends on the assay and was between 2 and 5 g. The high pressure and high temperature batch reactor employed was E010SS 100 mL EZE-SEAL 316SS from Autoclave Engineers (Erie, PA, USA) with gas inlet and outlet connections, a thermocouple, and a cooling coil.

The reactor was sealed and purged using $N_2$. Then, it was heated to the desired temperature using an electrical heating jacket and held at that temperature for the predefined pretreatment time. At the end of the reaction, the reactor was cooled to room temperature by passing a solution of water and ethylene glycol at 5 °C through the cooling coils of the system to quench the reaction.

Before opening the reactor, pressure was relieved by a purge valve. The mixture in the reactor was transferred to a beaker, and the reactor was washed twice with 15 mL of dichloromethane (DCM), ensuring that all components were extracted. The mixture contained three phases: An aqueous phase, a solid phase (pretreated microalga), and a phase of biocrude with DCM.

The pretreated biomass was separated using filter paper with a Büchner funnel. After filtration, this solid phase and the filter paper were dried in an oven at 110 °C for 12 h before they were weighed. The weight of the solid residue was calculated by subtracting the weight of the filter paper. The aqueous and biocrude phases were separated by decantation in a separatory funnel. The top phase was the aqueous phase and the biocrude phase with DCM was at the bottom. To evaporate water and DCM from the aqueous phase and the biocrude layer, respectively, both were dried in an oven during 24 h at 100 °C for the aqueous phase and 40 °C for the biocrude phase, and then weighed. The oven used had an extractor to avoid the evaporated solvent being transferred to workspace.

The elemental analyses of the microalga and all product phases were carried out using a Flash 2000 analyzer (Thermo Fisher Scientific, Waltham, MA, USA) equipped with a thermal conductivity detector (TCD). The contents of C, N, S, and H were determined by an oxidation/reduction reactor kept at a temperature of 900 °C. The O determination was achieved through an oxygen-specific pyrolysis reactor heated at 1060 °C. Triplicate analyses were conducted for each sample, and the average values were taken.

## 3. Results and Discussion

### 3.1. Result of the Design of Experiments

The experimental design applied to the study of the hydrothermal pretreatment was a $2^3$ full factorial design. The central point experiment was carried out four times in order to determine the variability of the results and to evaluate the experimental error. According to the response surface methodology, a second order model was required because of the significant curvature effect found in the linear model. Additional experiments (star points) were included in the factorial design to produce a face-centered central composite design.

The responses selected were: The yield of the pretreated biomass phase obtained in the hydrothermal pretreatment in relation to the mass of dry matter microalga loaded (hereinafter referred to as the yield of the solid phase, $Y_{SP}$), the N recovery in the aqueous phase in relation to the N content in the microalga dry matter ($NR_{AP}$), and the O recovery in the aqueous phase referred to the O content in the dry microalga ($OR_{AP}$). The design goal was to maximize these three responses. A high solid phase yield is required in the pretreatment stage to increase, in turn, the final biocrude yield obtained in HTL. Additionally, it is necessary to transfer high amounts of N and O from the microalga to the aqueous phase (rising $NR_{AP}$ and $OR_{AP}$) to obtain a solid phase with lower N and O content that can be converted into a biocrude with a low heteroatom concentration through a conventional HTL process.

The selection of the factors was based on the operating conditions that have a significant influence on HTL reactions [27]. Therefore, the factors studied were temperature (T), reaction time (t) and the biomass:water ratio (B:W).

The selection of the levels was based on results obtained in previous studies [14,20,24]. The lower and upper temperature levels were 100 and 200 °C, respectively, since higher temperatures enter the range of HTL conditions. The levels of the reaction time were 5 and 120 min, and the levels of the biomass:water ratio were 26.7 and 66.7 mg/mL.

The experimental matrix and the results are presented in Table 1. The factorial levels on a natural scale are illustrated in columns 2, 3 and 4, whereas columns 5, 6 and 7 denote the 0 and ±1 encoded factorial levels on a dimensionless scale. Experiments were carried out randomly to minimize errors due to possible systematic tendencies in the operating conditions. Table 1 also shows the results for the three responses.

**Table 1.** Experiment matrix and experiment results.

| Reaction | T | t | B:W | $X_T$ | $X_t$ | $X_{B:W}$ | $Y_{SP}$ | $NR_{AP}$ | $OR_{AP}$ |
|---|---|---|---|---|---|---|---|---|---|
| | °C | min | mg/mL | | | | % | % | % |
| 1 | 100 | 5 | 26.7 | −1 | −1 | −1 | 35.1 | 48.2 | 43.5 |
| 2 | 200 | 5 | 26.7 | 1 | −1 | −1 | 21.2 | 63.3 | 75.1 |
| 3 | 100 | 120 | 26.7 | −1 | 1 | −1 | 19.2 | 71.0 | 68.5 |
| 4 | 200 | 120 | 26.7 | 1 | 1 | −1 | 16.3 | 64.6 | 62.8 |
| 5 | 100 | 5 | 66.7 | −1 | −1 | 1 | 56.3 | 19.8 | 31.1 |
| 6 | 200 | 5 | 66.7 | 1 | −1 | 1 | 35.8 | 46.2 | 58.8 |
| 7 | 100 | 120 | 66.7 | −1 | 1 | 1 | 41.9 | 38.5 | 42.9 |
| 8 | 200 | 120 | 66.7 | 1 | 1 | 1 | 10.2 | 68.4 | 64.5 |
| 9 | 150 | 62.5 | 46.7 | 0 | 0 | 0 | 30.5 | 39.7 | 43.0 |
| 10 | 150 | 62.5 | 46.7 | 0 | 0 | 0 | 33.4 | 43.8 | 46.7 |
| 11 | 150 | 62.5 | 46.7 | 0 | 0 | 0 | 32.2 | 34.2 | 48.3 |
| 12 | 150 | 62.5 | 46.7 | 0 | 0 | 0 | 30.0 | 40.2 | 50.0 |
| 13 | 200 | 62.5 | 46.7 | 1 | 0 | 0 | 11.5 | 67.2 | 74.0 |
| 14 | 100 | 62.5 | 46.7 | −1 | 0 | 0 | 24.7 | 60.5 | 62.1 |
| 15 | 150 | 120 | 46.7 | 0 | 1 | 0 | 37.7 | 47.4 | 59.7 |
| 16 | 150 | 5 | 46.7 | 0 | −1 | 0 | 56.7 | 27.6 | 37.7 |
| 17 | 150 | 62.5 | 66.7 | 0 | 0 | 1 | 33.8 | 39.2 | 48.2 |
| 18 | 150 | 62.5 | 26.7 | 0 | 0 | −1 | 29.0 | 52.7 | 54.1 |

Note: T = temperature, t = reaction time, and B:W = biomass:water ratio, X = coded value, $Y_{SP}$ = yield of the solid phase (%), $NR_{AP}$ = N recovery in the aqueous phase (%), and $OR_{AP}$ = O recovery in the aqueous phase (%).

Using non-linear multiple regression analysis and assuming a second-order polynomial model, mathematical models (Equations (1)–(6)) were attained from the matrix generated by the experimental pretreatment results. The statistical models (Equations (1)–(3)) were calculated from encoded levels that showed the real influence of the three operating variables on the pretreatment process, and the technological models (Equations (4)–(6)) were obtained from the real values corresponding to these operating conditions. The analysis of variance showed that the quadratic model selected to fit the experimental results was adequate because the lack-of-fit test presented $p$-values > 0.05 in all cases.

Statistical models:

$$Y_{SP} = 32.33 − 8.22\,X_T − 7.98\,X_t + 5.72\,X_{B:W} − 0.025\,X_T\,X_t − 4.425\,X_T X_{B:W} − 2.4\,X_t\,X_{B:W} + 14.108\,X^2_t − 15.036\,X^2_T − 1.697\,X^2_{B:W}; [r^2 = 0.96, r^2\ (\text{adj.}) = 0.91] \tag{1}$$

$$NR_{AP} = 42.494 + 7.17\,X_T + 8.48\,X_t − 8.77\,X_{B:W} − 2.25\,X_T\,X_t + 5.95\,X_T\,X_{B:W} + 2.1\,X_t\,X_{B:W} + 18.337\,X^2_T − 8.013\,X^2_t + 0.437\,X^2_{B:W}; [r^2 = 0.93, r^2\ (\text{adj.}) = 0.84] \tag{2}$$

$$OR_{AP} = 50.429 + 8.71\ X_T + 5.22\ X_t - 5.85\ X_{B:W} - 5.425\ X_T\ X_t + 2.925\ X_T\ X_{B:W} + 0.6\ X_t\ X_{B:W} + 14.193\ X^2_T - 5.157\ X^2_t - 2.707\ X^2_{B:W};\ = [r^2 = 0.85,\ r^2\ (adj.) = 0.68] \tag{3}$$

Industrial models:

$$Y_{SP} = -112.793 + 1.847\ T - 0.573\ t + 1.476\ B:W - 0.0000086\ T\ t - 0.004\ T\ B:W - 0.002\ t\ B:W - 0.006\ T^2 + 0.004\ t^2 - 0.004\ B:W^2;\ [r^2 = 0.96,\ r^2\ (adj.) = 0.91] \tag{4}$$

$$NR_{AP} = 229.864 - 2.286\ T + 0.483\ t - 1.547\ B:W - 0.00078\ T\ t + 0.00595\ T\ B:W + 0.0018\ t\ B:W + 0.0073\ T^2 - 0.0024\ t^2 + 0.0011\ B:W^2;\ [r^2 = 0.93,\ r^2\ (adj.) = 0.84] \tag{5}$$

$$OR_{AP} = 143.489 - 1.548\ T + 0.544\ t - 0.132\ B:W - 0.0019\ T\ t + 0.0029\ T\ B:W + 0.00052\ t\ B:W + 0.0056\ T^2 - 0.0016\ t^2 - 0.0067\ B:W^2;\ [r^2 = 0.85,\ r^2\ (adj.) = 0.68] \tag{6}$$

For each response, the second-order models could be plotted on 3D graphs (response surfaces) as a function of two of the three factors at the center point of the third one. For instance, Figure 1 shows the response surfaces for the predicted values of the solid phase yield (Figure 1a), the N recovery in the aqueous phase (Figure 1b), and the O recovery in the aqueous phase (Figure 1c) as a function of temperature and pretreatment reaction time at a biomass:water ratio of 46.7 mg/mL, which corresponds to the center point value of this factor.

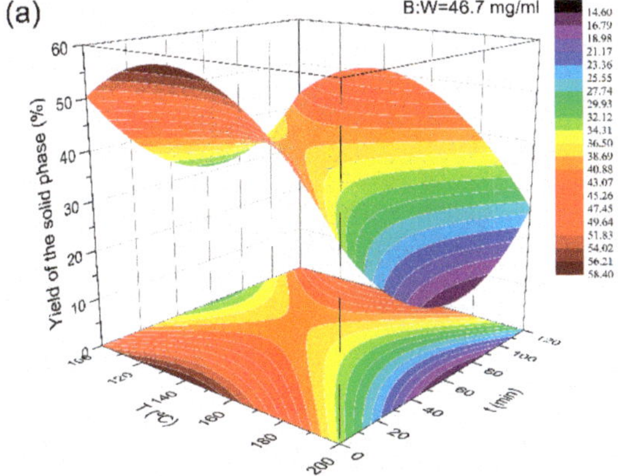


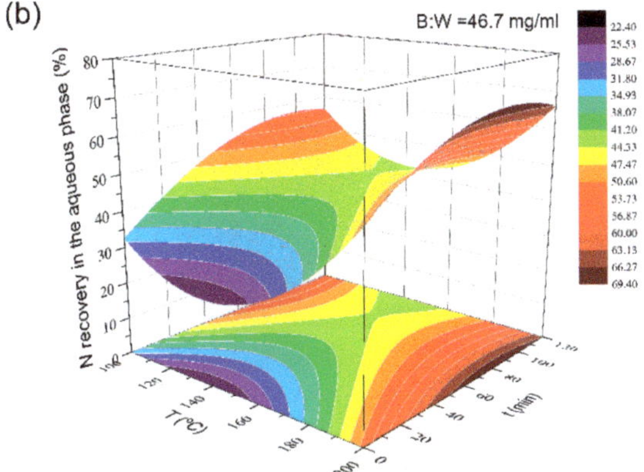


**Figure 1.** *Cont.*

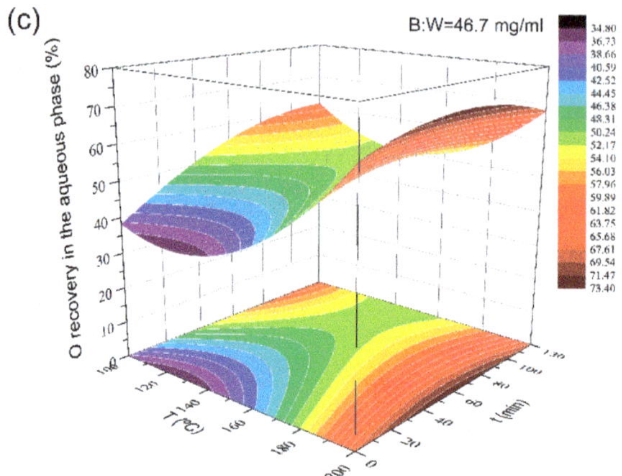

**Figure 1.** Response surfaces as a function of temperature (T) and time (t) for the predicted values of the (**a**) yield of the solid phase, (**b**) N recovery in the aqueous phase, and (**c**) O recovery in the aqueous phase (all plots correspond to a biomass:water ratio value of B:W = 46.7 mg/mL).

Figure 2 shows the relationship between experimental and predicted values for the yield of the solid phase, the N recovery in the aqueous phase, and the O recovery in the aqueous phase. For the three responses evaluated, the values calculated with the predictive non-linear models were very close to those obtained experimentally, indicating the high accuracy of the models attained. In addition, the analysis of variance (ANOVA) indicated that the second-order models were adequate to represent the experimental results for the three responses analyzed, since the $p$-values of the lack-of-fit were higher than the significance level (0.05). Thus, the values were 0.0512, 0.2050 and 0.2020 for the yield of the solid phase, the N recovery in the aqueous phase and the O recovery in the aqueous phase, respectively.

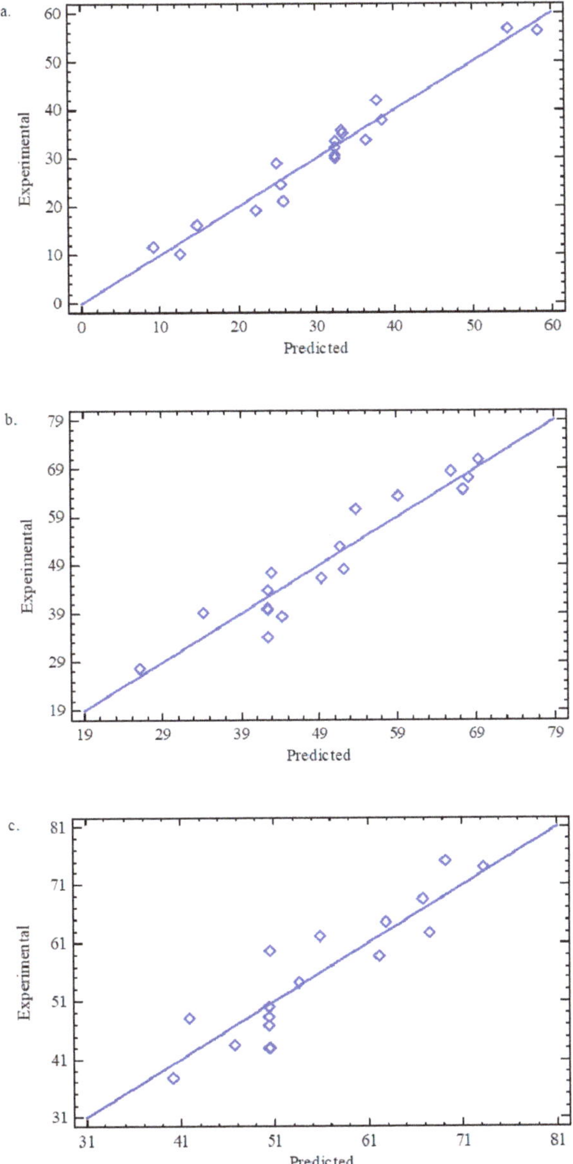

**Figure 2.** Experimental vs. predicted values for the (**a**) yield of the solid phase, (**b**) N recovery in the aqueous phase, and (**c**) O recovery in the aqueous phase.

### 3.2. Influence of Operating Conditions on the Yield of the Solid Phase

According to the statistical analysis of the experimental range evaluated, the temperature ($X_T$) was the most important factor in the yield of the solid phase obtained in the HTL pretreatment ($p = 0.0005$). The second factor in importance was pretreatment reaction time ($X_t$) ($p = 0.0005$), although the interaction between them was not significant ($p = 0.9668$). As shown in Figure 3 and Equation (1), temperature and time had a negative influence on the solid phase yield: An increase in these factors produced an overall decrease in the amount of pretreated biomass obtained during the studied hydrothermal pretreatment process. The effect of temperature and time on the yield of treated biomass was similar to the ones reported previously for HTL in mild conditions [14,23,26]. The drop of the solid phase yield with increased pretreatment temperature and time was due to the effect of the biomass cell break, which allowed for the hydrolysis of the extracted proteins and carbohydrates from the microalga into their corresponding single molecules (amino acids and sugars, respectively) in the aqueous phase. Therefore, an increase in the temperature and time promoted the hydrolysis reactions, decreasing the solid phase yield.

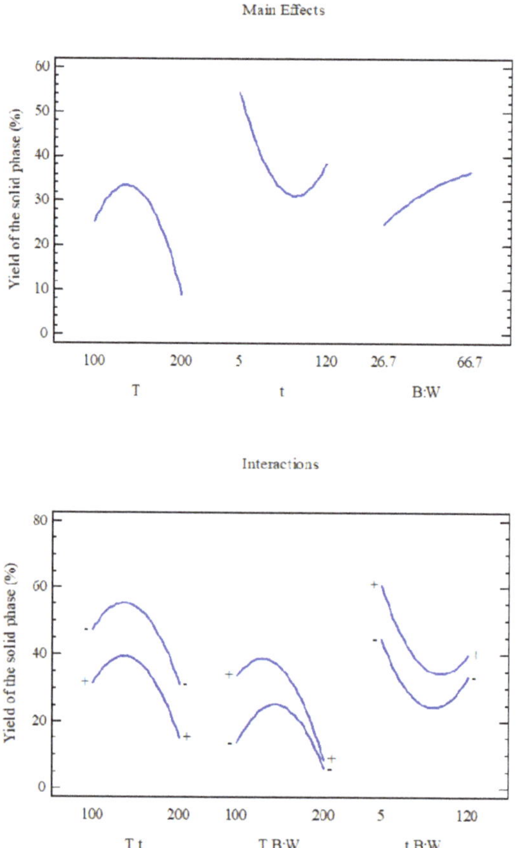

**Figure 3.** Plot of main effects and interactions for the yield of the solid phase (T = temperature, t = reaction time, and B:W = biomass:water ratio).

However, the quadratic effect of the temperature ($X^2_T$) had a significant negative influence on the yield of the solid phase (Equation (1); $p = 0.0005$). This, in turn, meant that the increase of temperature did not imply a constant decrease in this response because the curvature effect was significant at lower temperatures (Figure 3). In fact, the yield of the solid phase achieved a maximum at a temperature of approximately 140 °C. In this sense, the hydrolysis of proteins and carbohydrates did not become significant until that temperature was achieved.

In the same way, the quadratic time effect ($X^2_t$) was also significant ($p = 0.0007$), but its effect was positive (Equation (1)). Therefore, the increase of time did not again produce a constant reduction in the response analyzed, since the curvature effect was significant at long reaction times (Figure 3). Accordingly, the pretreatment at longer reaction times (>75 min) reached a minimum yield in the solid phase without further decreasing the amount of this pretreated biomass. Consequently, protein and carbohydrate hydrolysis reactions were no longer significant at these long pretreatment times. Higher yields of the solid phase were obtained at shorter reaction times, which could significantly reduce the cost and energy requirements of the mild pretreatment.

Figure 3 and Equation (1) show that the biomass:water ratio ($X_{B:W}$) had a positive effect on the amount of solid obtained from the HTL pretreatment ($p = 0.0014$). For this reason, an increase of the initial slurry concentration was somewhat beneficial to this response at the mild temperatures utilized in this study. This was due to the increase in available biomass with increasing slurry concentrations. However, the influence of the quadratic effect of this factor ($X^2_{B:W}$) was not significant ($p = 0.1723$).

Conversely, the temperature–biomass:water ratio interaction ($X_T$–$X_{B:W}$) had a small negative influence on the yield of the solid phase (Equation (1); $p = 0.0041$). At low temperatures, an increase in the biomass:water ratio significantly increased the response, but the yield of the solid phase remained nearly constant at high temperatures for any slurry biomass concentration. Thus, the slurry concentration was no longer significant at high temperatures because the effect of the hydrolysis reactions of the proteins and carbohydrates became more important at these temperatures, producing low solid yields.

The time–biomass:water ratio interaction ($X_t$–$X_{B:W}$) was also significant in the quantity of solids attained in the pretreatment ($p = 0.0226$), showing a significant negative effect (Figure 3 and Equation (1)). At lower pretreatment times, an increase in the slurry concentration led to a significant increase of the solid phase yield. However, the increase of this response with the initial slurry concentration at high temperatures was lower. This interaction can be explained as the temperature–biomass:water ratio interaction. At high temperatures, the hydrolysis reactions became more important, and therefore the influence of the biomass:water ratio was less important.

From the point of view of the solid phase yield obtained during the mild pretreatment, the optimal values were the medium temperature (130 °C), the shortest time (5 min) and the higher biomass:water ratio (67 mg/mL). At these operating conditions, the yield of solids predicted by the non-linear models (Equations (1) or (4)) was 63.5%. This predicted value was confirmed with the experimental result at the same operating conditions (63.24%).

### 3.3. Influence of Operating Conditions on the N Recovery in the Aqueous Phase

The influence of operating conditions on the N recovery in the aqueous phase is now discussed using the statistical models shown in Equation (2) as well as the main effects and interaction plots (Figure 4) and the ANOVA. The most important factor was the biomass:water ratio ($X_{B:W}$) ($p = 0.0060$). This factor had a negative influence on the N content in the aqueous phase. Therefore, an increase in the biomass:water ratio produced a remarkable decrease in the N content in the aqueous phase. This fact disadvantaged the goal of increasing N recovery because it is interesting to work at lower biomass:water ratios to reduce the cost of the biomass drying in order to obtain low concentrations of the initial biomass slurry. The influence of the quadratic effect of this factor ($X^2_{B:W}$), however, was not significant ($p = 0.8676$). The next factors in importance were pretreatment reaction time ($p = 0.0066$) and temperature ($p = 0.0106$). Overall, an increase in temperature and time improved the N recovery in the aqueous phase. That increase drove the hydrolysis of proteins into small molecules, which remained solubilized in the aqueous phase, decreasing the N content in the pretreated biomass obtained [20].

The quadratic effect of the temperature ($X^2_T$) achieved the most remarkable value ($p = 0.0047$). It had the most significant positive influence on the N recovery in the aqueous phase (Equation (2)). This means that the increase of temperature did not imply a constant increase in this response because the curvature effect was very significant at lower temperatures (Figure 4). In fact, the N recovery in the aqueous phase reached a minimum at a temperature of approximately 140 °C. This result is in agreement with the maximum yield of pretreated biomass achieved at this temperature. Temperatures higher than 140 °C increased the ionic product of the water, thus promoting the hydrolysis of proteins and enhancing both the removal of N from the microalga and the solid phase yield at the same time [20].

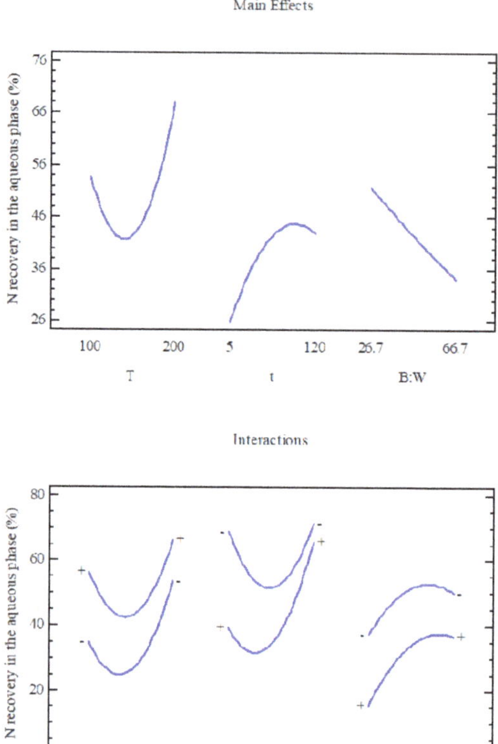

**Figure 4.** Plot of main effects and interactions for the N recovery in the aqueous phase (T = temperature, t = reaction time, and B:W = biomass:water ratio).

The temperature–biomass:water ratio interaction ($X_T$–$X_{B:W}$) in the N recovery was also significant ($p = 0.0239$) and positive (Equation (2) and Figure 4). At low temperatures, an increase in the biomass:water ratio led to a significant decrease in N recovery in the aqueous phase. However, this response achieved similarly high values at high temperatures for any slurry biomass concentration.

On the other hand, the quadratic effect of the time ($X^2_t$) had a small negative influence on the N recovery in the aqueous phase. This means that the increase of time did not imply a constant increase in this response because the curvature effect was significant at higher pretreatment times (Figure 4). In fact, the N recovery achieved a maximum at approximately 90 min, keeping constant for longer reaction times. Therefore, pretreatment times of about 90 min or higher were adequate to obtain higher values for the N recovery in the aqueous phase, but short pretreatment times were preferred from the point of view of the cost and energy requirements for this stage.

Finally, the influence of the temperature–time and the time–biomass:water ratio interactions (($X_T$–$X_t$, $X_t$–$X_{B:W}$, respectively) was not significant on the N recovery in the aqueous phase since the $p$-values were higher than 0.05 in both cases (Figure 4 and Equation (2)).

From the point of view of the N recovery in the aqueous phase during the HTL pretreatment, the optimal values were the lowest temperature (100 °C), the medium time (93.5 min) and the lowest biomass:water ratio (26.7 mg/mL). At these operating conditions, the $NR_{AP}$ predicted by the non-linear models (Equations (2) or (5)) and the corresponding experimental value were 71.1% and 73.0%, respectively.

### 3.4. Influence of Operating Conditions on the O Recovery in the Aqueous Phase

The temperature ($X_T$) was identified as the most important factor for the O recovery in the aqueous phase ($p = 0.0027$). As shown in Equation (3) and Figure 5, this operating condition had an overall positive effect on the O removal from biomass, achieving O recovery values in the aqueous phase of nearly 73% at high temperatures. Therefore, this response grew when the temperature increased. The N removal from the biomass also increased with temperature, but the O elimination was predominant. As the ionic product of water increased with temperature, high values of this operating condition favored the hydrolysis of proteins, carbohydrates and lipids catalyzed by $H^+$ and $OH^-$, enhancing the recovery of O from the microalga in the aqueous phase. The quadratic effect of this factor ($X^2_T$) had a significant ($p = 0.0044$) positive influence on this response (Equation (3)). However, its absolute value was smaller than that of its corresponding main effect. This indicates that the increase in temperature did not produce a constant rise in the O content in the aqueous phase because the curvature effect was significant at lower temperatures with a minimum at approximately 135 °C (Figure 5). Consequently, the hydrolysis of the biomass compounds did not become significant, at least at 135–140 °C.

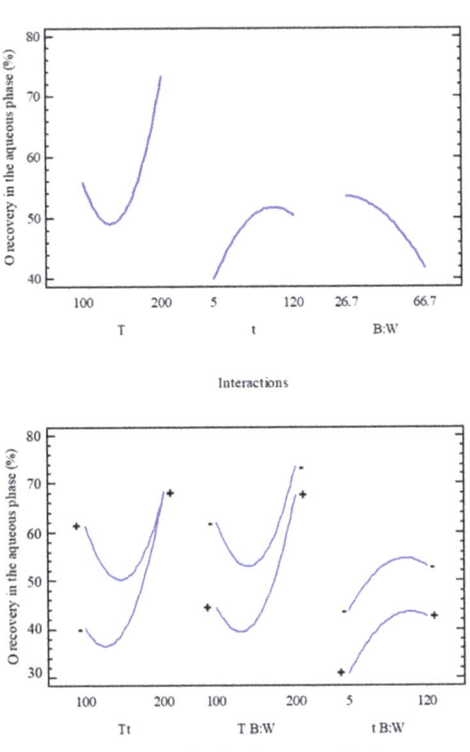


**Figure 5.** Plot of main effects and interactions for the O recovery in the aqueous phase (T = temperature, t = reaction time, and B:W = biomass:water ratio).

The biomass:water ratio ($X_{B:W}$), the time ($X_t$), and the temperature–time interaction ($X_T$-$X_t$) were significant with $p$-values of 0.0085, 0.0117 and 0.0143, respectively, but they had a lower effect on this response than temperature and its quadratic effect. The biomass:water ratio had a negative linear influence on the O recovery in the aqueous phase, as shown in Figure 5. Accordingly, the influence of the quadratic effect of this factor ($X^2_{B:W}$) was not significant ($p = 0.2326$). As far as this response is concerned, low biomass:water ratios were required to maximize the O recovery in the aqueous phase, because more $H^+$ and $OH^-$ were available to catalyze the hydrolysis of proteins, carbohydrates and lipids, which improved the removal of O from the biomass in the aqueous phase.

In addition, pretreatment time ($X_t$) had a positive effect on this heteroatom recovery in the aqueous phase, and the quadratic effect ($X^2_t$) of this factor was not significant ($p = 0.0656$). The maximum value of the O recovery was 52%, which was achieved at long pretreatment times (Figure 5). The time had a similar effect to temperature on the O removal from the biomass, but its influence, as shown, was less significant. Time also favored the hydrolysis of biomass compounds, increasing the O recovery.

On the other hand, the temperature–time interaction ($X_T$–$X_t$) had a negative influence on the O recovery in the aqueous phase. It was the only interaction that had a significant effect ($p = 0.0143$) since the temperature–biomass:water ratio interaction ($X_T$–$X_{B:W}$) and the time–biomass:water ratio interaction ($X_t$–$X_{B:W}$) did not have a significant effect on this response ($p = 0.0696$ and $p = 0.6098$, respectively). At short pretreatment times, an increase in the temperature produced a great increase in the O recovery. However, at long times, this response decreased to a minimum and later increased to 68% (Figure 4). At low temperatures, an increase of the time remarkably enhanced the O recovery. However, this response achieved its maximum value at high temperatures irrespective of the pretreatment time because the temperature influence was more significant than time.

According to these results, the optimal values to maximize the O recovery in the aqueous phase are the highest temperature (200 °C), an intermediate time (59.6 min), and a value of 35.8 mg/mL of the biomass:water ratio. At these operating conditions, the O recovery in the aqueous phase predicted by the non-linear models (Equations (3) or (6)) was 74.1%. The experimental result of O recovery at the optimal operating conditions was slightly higher (76.3%).

## 4. Conclusions

Response equations were obtained for the yield of the pretreated biomass and the recovery of N and O from the microalga in the aqueous phase, which facilitates the accurate prediction of the operating conditions required to obtain a given value of these responses. The use of these models is essential to carry out a mild hydrothermal pretreatment process of microalgae on an industrial scale to obtain an appropriate yield of pretreated biomass with a relatively low N and O contents. Temperature and time are significant factors with a negative influence on the yield of the solid phase and a positive one on N and O recovery in the aqueous phase, which means that they have a positive influence on the concentrations of these heteroatoms in the pretreated biomass. The slurry concentration is also critical, having to be low to maximize the N and O recoveries in the aqueous phase and consequently to minimize their presence in the pretreated microalga. However, the slurry concentration should be high to obtain better yields of pretreated biomass.

**Author Contributions:** M.M.-H. performed the experimental work. G.V. and L.F.B. devised the experimental work, and V.M., J.J.E. and R.R. helped in the revision of the HTL results. G.V., M.M.-H. and J.J.E. wrote the final version of the manuscript. The listed authors have contributed substantially to this work.

**Funding:** The authors of this work want to thank funding received by Comunidad de Madrid, Spain and co-financed by the FEDER "A way of making Europe" (ALGATEC-CM, P2018/BAA-4532) and BIOHIDROALGA Ministerio de Economía y Competitividad from the Spanish Government (ENE2017-83696-R).

**Conflicts of Interest:** The authors declare no conflict of interest.

## References

1. Singh, J.; Gu, S. Commercialization potential of microalgae for biofuels production. *Renew. Sustain. Energy Rev.* **2010**, *14*, 2596–2610. [CrossRef]
2. Moreno-Garcia, L.; Adjallé, K.; Barnabé, S.; Raghavan, G.S.V. Microalgae biomass production for a biorefinery system: Recent advances and the way towards sustainability. *Renew. Sustain. Energy Rev.* **2017**, *76*, 493–506. [CrossRef]

3. Rodríguez, R.; Espada, J.J.; Moreno, J.; Vicente, G.; Bautista, L.F.; Morales, V.; Sánchez-Bayo, A.; Dufour, J. Environmental analysis of Spirulina cultivation and biogas production using experimental and simulation approach. *Renew. Energy* **2018**, *129*, 724–732. [CrossRef]

4. Chiaramonti, D.; Prussi, M.; Buffi, M.; Casini, D.; Rizzo, A.M. Thermochemical Conversion of Microalgae: Challenges and Opportunities. *Energy Procedia* **2015**, *75*, 819–826. [CrossRef]

5. Raheem, A.; Wan Azlina, W.A.K.G.; Taufiq Yap, Y.H.; Danquah, M.K.; Harun, R. Thermochemical conversion of microalgal biomass for biofuel production. *Renew. Sustain. Energy Rev.* **2015**, *49*, 990–999. [CrossRef]

6. Tian, C.; Li, B.; Liu, Z.; Zhang, Y.; Lu, H. Hydrothermal liquefaction for algal biorefinery: A critical review. Renew. *Sustain. Energy Rev.* **2014**, *38*, 933–950. [CrossRef]

7. Palomino, A.; Godoy-Silva, R.D.; Raikova, S.; Chuck, C.J. The storage stability of biocrude obtained by the hydrothermal liquefaction of microalgae. *Renew. Energy* **2020**, *145*, 1720–1729. [CrossRef]

8. Jarvis, J.M.; Albrecht, K.O.; Billing, J.M.; Schmidt, A.J.; Hallen, R.T.; Schaub, T.M. Assessment of Hydrotreatment for Hydrothermal Liquefaction Biocrudes from Sewage Sludge, Microalgae, and Pine Feedstocks. *Energy Fuels* **2018**, *32*, 8483–8493. [CrossRef]

9. López Barreiro, D.; Samorì, C.; Terranella, G.; Hornung, U.; Kruse, A.; Prins, W. Assessing microalgae biorefinery routes for the production of biofuels via hydrothermal liquefaction. *Bioresour. Technol.* **2014**, *174*, 256–265. [CrossRef]

10. Li, H.; Liu, Z.; Zhang, Y.; Li, B.; Lu, H.; Duan, N.; Liu, M.; Zhu, Z.; Si, B. Conversion efficiency and oil quality of low-lipid high-protein and high-lipid low-protein microalgae via hydrothermal liquefaction. *Bioresour. Technol.* **2014**, *154*, 322–329. [CrossRef]

11. Hu, Y.; Qi, L.; Feng, S.; Bassi, A.; Xu, C.C. Comparative studies on liquefaction of low-lipid microalgae into bio-crude oil using varying reaction media. *Fuel* **2019**, *238*, 240–247. [CrossRef]

12. Patel, B.; Guo, M.; Chong, C.; Sarudin, S.H.M.; Hellgardt, K. Hydrothermal upgrading of algae paste: Inorganics and recycling potential in the aqueous phase. *Sci. Total Environ.* **2016**, *568*, 489–497. [CrossRef] [PubMed]

13. Guo, Y.; Yeh, T.; Song, W.; Xu, D.; Wang, S. A review of bio-oil production from hydrothermal liquefaction of algae. *Renew. Sustain. Energy Rev.* **2015**, *48*, 776–790. [CrossRef]

14. Jazrawi, C.; Biller, P.; He, Y.; Montoya, A.; Ross, A.B.; Maschmeyer, T.; Haynes, B.S. Two-stage hydrothermal liquefaction of a high-protein microalga. *Algal Res.* **2015**, *8*, 15–22. [CrossRef]

15. Kumar, M.; Olajire Oyedun, A.; Kumar, A. A review on the current status of various hydrothermal technologies on biomass feedstock. Renew. *Sustain. Energy Rev.* **2018**, *81*, 1742–1770. [CrossRef]

16. Galadima, A.; Muraza, O. Hydrothermal liquefaction of algae and bio-oil upgrading into liquid fuels: Role of heterogeneous catalysts. Renew. *Sustain. Energy Rev.* **2018**, *81*, 1037–1048. [CrossRef]

17. Hu, Y.; Feng, S.; Xu, C.C.; Bassi, A. Production of low-nitrogen bio-crude oils from microalgae pre-treated with pre-cooled NaOH/urea solution. *Fuel* **2017**, *206*, 300–306. [CrossRef]

18. Mathimani, T.; Baldinelli, A.; Rajendran, K.; Prabakar, D.; Matheswaran, M.; Pieter van Leeuwen, R.; Pugazhendhi, A. Review on cultivation and thermochemical conversion of microalgae to fuels and chemicals: Process evaluation and knowledge gaps. *J. Clean. Prod.* **2019**, *208*, 1053–1064. [CrossRef]

19. Arvindnarayan, S.; Sivagnana Prabhu, K.K.; Shobana, S.; Kumar, G.; Dharmaraja, J. Upgrading of micro algal derived bio-fuels in thermochemical liquefaction path and its perspectives: A review. *Int. Biodeterior. Biodegrad.* **2017**, *119*, 260–272. [CrossRef]

20. Huang, Z.; Wufuer, A.; Wang, Y.; Dai, L. Hydrothermal liquefaction of pretreated low-lipid microalgae for the production of bio-oil with low heteroatom content. *Process Biochem.* **2018**, *69*, 136–143. [CrossRef]

21. Costanzo, W.; Jena, U.; Hilten, R.; Das, K.C.; Kastner, J.R. Low temperature hydrothermal pretreatment of algae to reduce nitrogen heteroatoms and generate nutrient recycle streams. *Algal Res.* **2015**, *12*, 377–387. [CrossRef]

22. Hu, Y.; Gong, M.; Feng, S.; Xu, C.C.; Bassi, A. A review of recent developments of pre-treatment technologies and hydrothermal liquefaction of microalgae for bio-crude oil production. *Renew. Sustain. Energy Rev* **2019**, *101*, 476–492. [CrossRef]

23. Du, Z.; Mohr, M.; Ma, X.; Cheng, Y.; Lin, X.; Liu, Y.; Zhou, W.; Chen, P.; Ruan, R. Hydrothermal pretreatment of microalgae for production of pyrolytic bio-oil with a low nitrogen content. *Bioresour. Technol.* **2012**, *120*, 13–18. [CrossRef] [PubMed]

24. Prapaiwatcharapan, K.; Sunphorka, S.; Kuchonthara, P.; Kangvansaichol, K.; Hinchiranan, N. Single- and two-step hydrothermal liquefaction of microalgae in a semi-continuous reactor: Effect of the operating parameters. *Bioresour. Technol.* **2015**, *191*, 426–432. [CrossRef] [PubMed]
25. Sunphorka, S.; Prapaiwatcharapan, K.; Hinchiranan, N.; Kangvansaichol, K.; Kuchonthara, P. Biocrude oil production and nutrient recovery from algae by two-step hydrothermal liquefaction using a semi-continuous reactor. *Korean J. Chem. Eng.* **2014**, *32*, 79–87. [CrossRef]
26. Miao, C.; Chakraborty, M.; Dong, T.; Yu, X.; Chi, Z.; Chen, S. Sequential hydrothermal fractionation of yeast Cryptococcus curvatus biomass. *Bioresour. Technol.* **2014**, *164*, 106–112. [CrossRef] [PubMed]
27. Jena, U.; Das, K.C.; Kastner, J.R. Effect of operating conditions of thermochemical liquefaction on biocrude production from Spirulina platensis. *Bioresour. Technol.* **2011**, *102*, 6221–6229. [CrossRef] [PubMed]

*Article*

# Hydrothermal Liquefaction of Microalga Using Metal Oxide Catalyst

**Alejandra Sánchez-Bayo [1]**, **Rosalía Rodríguez [1]**, **Victoria Morales [2]**, **Nima Nasirian [3]**,
**Luis Fernando Bautista [2]** and **Gemma Vicente [1,*]**

[1]  Department of Chemical, Energy and Mechanical Technology, Universidad Rey Juan Carlos, Móstoles,
     28933 Madrid, Spain; alejandra.sanchezbayo@urjc.es (A.S.-B.); rosalia.rodriguez@urjc.es (R.R.)
[2]  Department of Chemical and Environmental Technology, Universidad Rey Juan Carlos, Móstoles,
     28933 Madrid, Spain; victoria.morales@urjc.es (V.M.); fernando.bautista@urjc.es (L.F.B.)
[3]  Department of Agricultural Mechanisation and Biosystems Engineering, Shoushtar Branch,
     Islamic Azad University, Daneshgah Blvd., Shoushtar 6451741117, Iran; n.nasirian@iau-shoushtar.ac.ir
*  Correspondence: gemma.vicente@urjc.es

Received: 15 November 2019; Accepted: 17 December 2019; Published: 20 December 2019

**Abstract:** The yield and composition of the biocrude obtained by hydrothermal liquefaction (HTL) of *Nannocloropsis gaditana* using heterogeneous catalysts were evaluated. The catalysts were based on metal oxides ($CaO$, $CeO_2$, $La_2O_3$, $MnO_2$, and $Al_2O_3$). The reactions were performed in a batch autoclave reactor at 320 °C for 10 min with a 1:10 (wt/wt) microalga:water ratio. These catalysts increased the yield of the liquefaction phase (from 94.14 ± 0.30 wt% for $La_2O_3$ to 99.49 ± 0.11 wt% for $MnO_2$) as compared with the thermal reaction (92.60 ± 1.20 wt%). Consequently, the biocrude yields also raised in the metal oxides catalysed HTL, showing values remarkably higher for the $CaO$ (49.73 ± 0.9 wt%) in comparison to the HTL without catalyst (42.60 ± 0.70 wt%). The N and O content of the biocrude obtained from non-catalytic HTL were 6.11 ± 0.02 wt% and 10.50 ± 0.50 wt%, respectively. In this sense, the use of the metal oxides decreased the N content of the biocrude (4.62 ± 0.15–5.45 ± 0.11 wt%), although, they kept constant or increased its O content (11.39 ± 2.06–21.68 ± 0.03 wt%). This study shows that $CaO$, $CeO_2$ and $Al_2O_3$ can be promising catalysts based on the remarkable amount of biocrude, the highest values of C, H, heating value, energy recovery, and the lowest content of N, O and S.

**Keywords:** microalgae; hydrothermal liquefaction; biocrude; metal-oxide catalyst

---

## 1. Introduction

Advanced biofuels derived from microalgae are considered a promising source of energy thanks to the well-known advantages of this type of biomass: high-production yields, neutral with respect to $CO_2$ emissions, growing in industrial facilities and wastewater can be used as nutrients in the cultivation stage [1–4]. In this context, a thermochemical process such as the hydrothermal liquefaction (HTL) is postulated as one of the most promising current routes to produce biofuels from microalgae. HTL allows for the biofuel production from wet microalgae without a drying pre-treatment [5] and it tolerates the conversion of the whole microalga biomass, also including the low-lipid microalgal biomass which normally has higher growth rates [3,6–9]. Moderate temperatures (200–380 °C) and high pressures (10–25 MPa) and microalga:water ratios within the range 5%–15% are commonly used in HTL [10,11]. HTL yields a solid residue or biochar, a gaseous phase and two liquid fractions (aqueous and organic). Although all phases are useful in a microalga biorefinery, the organic liquid phase, usually referred to as biocrude or bio-oil, is the most interesting to produce biofuels [5].

The yield and quality of the biocrude is influenced by parameters such as temperature, pressure, reaction time and biomass:water ratio [12,13]. This biocrude obtained through HTL usually has a yield between 25 and 65 wt% under thermal conditions. However, it is characterised by having a high content of N and O, since the initial microalga has a high content of these heteroatoms [14]. Consequently, the heating value is influenced in a negative way and the biocrudes obtained from HTL are not entirely in accordance with the regulations for their use as biofuel. For these reasons, current HTL research is focused mainly on improving the quality of the biocrude, reducing the content of these heteroatoms. The HTL biocrude can be upgraded with a conventional catalytic hydrotreatment [15], but novel processes involve the modification of the HTL process. Thus, some recent investigations deal to the HTL in two stages, where the N and O contained in the initial biomass is removed at low temperatures in a pre-treatment and the solid phase obtained in this first stage is transform to a low N and O content biocrude in the following HTL step [6,16,17]. Other authors have studied the use of co-solvents such as ethanol [18,19], obtaining high biocrude yields at lower temperatures. The use of appropriate catalysts in the HTL process is one of the most promising recent routes since the catalyst promotes dehydration, deoxygenation, denitrogenation and desulfurisation during the HTL, consequently decreasing the N, O and S content of the biocrude [1,20]. In particular, catalysts are capable of carrying out the hydrolysis of proteins, lipids and carbohydrates into smaller molecules, which in turn suffer decarboxylation and deamination reactions [20].

The homogeneous catalysts were the first used in HTL due to the ease of handling. Initially, catalytic tests used $Na_2CO_3$, obtaining positive effects on the biocrude yield. In addition, this catalyst allows for the reduction of the N and O present in the starting biomass, which increases the higher heating value (HHV) of the biocrude and improves its quality [1,21]. However, its use can produce secondary reactions of saponification [22], which promoted the study of acid homogeneous catalysts, such as inorganic acid or short-chain organic acids, which act as hydrogenating agents of the process, reducing the biocrude N and O contents [23,24]. However, the homogenous catalyst cannot be reused in the catalytic HTL process.

The recovery ease of the heterogeneous catalysts makes the use of these catalysts in HTL a growing line of research. In this way, there are catalysts that promote the loss of N in biocrude, such as metals supported on carbon (Pt/C, Ru/C) or alumina with nickel or platinum [10,20]. Conversely, deoxygenation is promoted mainly by alumina doped with Ni, Co or Mo [20]. In addition, new diverse materials (i.e., zeolites, carbon nanotubes, nanoparticles, etc.) have being tested recently to favour the loss of these heteroatoms in the biocrude [25,26]. In previous works, the content of N and O in the biocrude was reduced using heterogeneous catalysts such as Pd, Pt or Ru supported on C, Co, Mo, Ni, Pt, $Ni/SiO_2$ supported on $Al_2O_3$ and zeolites [27] and also metals supported in carbon nanotubes [25]. However, the possible contaminations and the activity of the materials need to be evaluated for its real reutilisation [3]. Therefore, the development of catalysts with a high hydrothermal stability that resist deactivation during HTL is essential.

In this context, the main objective of this work is to compare the biocrude obtained by the use of different metal oxide heterogeneous catalysts (CaO, $CeO_2$, $La_2O_3$, $MnO_2$ and $Al_2O_3$) in the HTL of *Nannochloropsis gaditana* to develop this process at an industrial scale. These catalysts have relatively low prices and have been highly used in the production of biodiesel thanks to their high basicity that promotes a greater catalytic activity to convert triglycerides without significant deactivation. In addition, metal oxides are non-soluble in the alcoholic media of this reaction and therefore, are recoverable [28–32]. However, there are no references about the use of oxides as catalysts in the microalga HTL process. These catalysts are advantageous for microalga HTL since their metal cations are Lewis acid sites that act as acceptors of electrons, whereas their oxygen anions are acceptors of protons or Bronze bases [33]. Besides, metal oxides are not soluble in the aqueous medium of the HTL process, so that the leaching reactions of the oxides are avoided, which would decrease their catalytic activity [30].

## 2. Materials and Methods

### 2.1. Microalga

The microalga used in this work was the *N. gaditana*, and it was supplied by AlgaEnergy S.A. (Madrid, Spain). Table 1 shows the biological composition (lipids, proteins, carbohydrates and ashes) together with the principal and minority elements composition of the microalgae, which were determined according to the methodology reported previously [34].

**Table 1.** Microalga composition (dry weight basis).

| Biochemical Composition (wt%) | | Elemental Composition (wt%) | | Metals (mg/g) | |
|---|---|---|---|---|---|
| Lipids | 35.52 ± 1.23 | H | 7.10 ± 0.18 | Na | 21.04 ± 2.01 |
| Proteins | 43.81 ± 3.50 | C | 48.7 ± 0.14 | K | 3.53 ± 0.09 |
| Carbohydrates | 15.70 ± 3.59 | N | 6.80 ± 0.04 | Mg | 0.90 ± 0.03 |
| Ashes | 4.50 ± 0.79 | S | 0.90 ± 0.04 | Fe | 0.16 ± 0.01 |
| | | O | 36.5 ± 0.20 | Ca | 0.08 ± 0.01 |
| | | P | 6.43 ± 0.02 | | |

### 2.2. Catalysts

The metal-oxide catalysts used in this work were calcium oxide (CaO), cerium (IV) oxide ($CeO_2$), manganese (IV) orxide ($MnO_2$) and lanthanum (III) oxide ($La_2O_3$) supplied by Sigma-Aldrich (St. Louis, MO, USA), and $\gamma$-alumina ($Al_2O_3$) provided by Alfa Aesar (Haverhill, MA, USA).

### 2.3. Catalytic Liquefaction

The catalytic liquefaction was performed in a 100 mL stainless-steel autoclave (EZ-SEAL®, Autoclave Engineers. Erie, PA, USA). In each experiment, 33 g of slurry with a 10 wt% of microalga and 5 wt% metal-oxide catalyst was loaded into the reactor. A control thermal experiment was carried out in the absence of any catalyst for comparison purposes. All the experiments were run at 320 °C for 10 min in an inert atmosphere (nitrogen). The stirrer was set at 500 rpm to ensure adequate mixing. At the end of the reaction, the autoclave was rapidly cooled down to room temperature to quench the reaction. Each HTL experiment was carried out three times for the statistical analysis.

When the reactor achieved room temperature, the gas phase was separated through the reactor valve that connects the reactor to the microGC Varian CP-4900 (Varian Inc., Palo Alto, CA, USA) where it was analysed. Then, 30 mL of dichloromethane (DCM) was used to collect the reaction mixture, which consists of solid residue, aqueous phase and biocrude. The solid residue and the catalyst were separated together by a vacuum filter, dried at 105 °C for 24 h and weighted. The biocrude and aqueous phase were separated by decantation in a separation funnel. Both water and DCM were evaporated by vacuum evaporation from their respective fractions (aqueous phase and biocrude, respectively) in order to determine their yield and composition.

The yields (Y) of each phases were determined on a dry basis using Equations (1)–(4) [35] and the gas phase yield was calculated by the difference.

$$Y_{Biocrude}(\%) = \frac{\text{mass of biocrude}}{\text{mass dry matter of microalgae}} \times 100 \tag{1}$$

$$Y_{Solid\ Residue}(\%) = \frac{\text{mass of solid residue}}{\text{mass dry matter of microalgae}} \times 100 \tag{2}$$

$$Y_{Water-soluble\ products}(\%) = \frac{\text{mass of water} - \text{soluble products}}{\text{mass dry matter of microalgae}} \times 100 \tag{3}$$

$$\text{Liquefaction phase}(\%) = \left(1 - \frac{\text{mass of solid residue}}{\text{mass dry matter of microalgae}}\right) \times 100 \tag{4}$$

### 2.4. Products Analysis

#### 2.4.1. Biocrude Analysis

The content of the principal elements (hydrogen, carbon, nitrogen, sulfur and oxygen) present in the biocrude oil were measured in a Vario EL III element analyser (Elementar Analysensysteme GmbH, Langenselbold, Germany) using sulphanilic acid as standard. With the content of these elements, the high heating value (HHV) of the bio-oil can be determined using the Boie equation (Equation (5)) [36].

$$\text{HHV (MJ/kg)} = 0.3516 \times C + 1.16225 \times H - 0.1109 \times O + 0.0628 \times N \tag{5}$$

The energy recovery (ER) was defined as the ratio of the total heating value of biocrude to the total heating value of microalgae, as shown by Equation (6).

$$\text{ER (\%)} = \frac{(\text{HHV of biocrude} \times \text{Mass of biocrude})}{(\text{HHV of microalga} \times \text{Mass of dry matter of microalgae})} \times 100 \tag{6}$$

The composition of biocrude was analysed by gas chromatography–mass spectrometry (GC–MS) (Bruker 450GC, Bruker Corp., Billerica, MA, USA). The samples were diluted with carbon disulphide and filtered with a 0.45 μm nylon filter. The GC-MS was coupled to a triple quadrupole mass spectrometer (Bruker 320 MS, Bruker Corp., Billerica, MA, USA) that operates in electronic impact (EI) mode and it was provided with a Rxi-5Sil MS 30 m 0.25 mm ID column (Restek, Lisses, France). Data acquisition and processing were performed by using Bruker MS Workstation software v.7. (Bruker Corp., Billerica, MA, USA).

#### 2.4.2. Analysis of Aqueous Phase

The total organic content (TOC) was determined in a Shimadzu-V equipment (Shimadzu Corp., Kioto, Japan) while the pH was measured in a pH meter Basic 30 pH meter (Crison Instruments, Barcelona, Spain). In addition, the water recovery from the aqueous phase was as evaluated by Equation (7).

$$\text{Water recovery (\%)} = \frac{\text{mass of aqueous phase}}{\text{inital mass of water}} \times 100 \tag{7}$$

#### 2.4.3. Analysis of Gas Phase

The catalytic hydrothermal liquefaction gas products were collected and analysed by using a gas chromatograph Varian CP-4900 (Varian Inc., Palo Alto, CA, USA) with a thermal conductivity detector (TCD) connected on-line to the autoclave reactor. The samples were analysed when the reaction mixture was cooled to 30 °C.

#### 2.4.4. Statistical Analysis

Overall, data were subjected to statistical analysis by one-way analysis of variance (ANOVA), and Duncan's comparison of means using SPSS statistics software version 21 (IBM Corp., Armonk, NY, USA). Duncan's comparison of means tables were also applied to evaluate the effects of catalysts on the conversion yields and the elemental compositions of biocrude obtained through HTL of microalgae. Average and standard deviation of replications were also applied where required. In addition, before the analysis, the normality of data was approved by the Kolmogorov–Smirnov test.

## 3. Results and Discussion

### 3.1. Effects of Catalysts on the Yields of Hydrothermal Liquefaction (HTL) Processes

The yields of the different fractions obtained from the catalytic HTL processes (biocrude, water-soluble products, gas phase and solid residue) are shown in Table 2. The corresponding yields for the non-catalytic HTL (thermal) process are also included in the same table for comparison purposes.

**Table 2.** Effects of catalysts on the yields of hydrothermal liquefaction (HTL) process.

| | | Biocrude (wt%) | WSP [1] (wt%) | Gas (wt%) | SR [2] (wt%) | LP [3] (wt%) |
|---|---|---|---|---|---|---|
| | Thermal | 42.60 b [4] | 31.98 c | 17.81 a | 7.60 d | 92.69 d |
| | CaO | 49.73 a | 37.43 b | 7.57 b | 5.25 c | 94.74 cd |
| | CeO$_2$ | 43.80 b | 44.12 a | 9.82 b | 2.25 ab | 97.74 ab |
| Catalysts | MnO$_2$ | 44.11 b | 37.61 b | 18.26 a | 0.50 a | 99.49 a |
| | La$_2$O$_3$ | 42.66 b | 38.98 b | 12.50 ab | 5.85 cd | 94.14 cd |
| | Al$_2$O$_3$ | 44.22 b | 34.02 c | 17.44 a | 4.30 bc | 95.69 bc |

[1] Water-soluble products; [2] Solid Residue; [3] Liquefaction Phase; [4] Different letters (a, b, c, and d) in the same column refers to statistical differences of Duncan's multiple range test at $p < 0.05$ between control and different catalysts applied in the HTL process. "a" is the desirable mean value of groups for Biocrude, WSP, Gas, SR and LP.

The yield towards the fraction of greatest interest, i.e., biocrude, were mostly between 43 and 50 wt%, which were near or somewhat higher than the equivalent obtained under thermal non-catalysed conditions (42.60 ± 0.80 wt%). According to Shi et al. [37], these catalysts prevent dehydration and promote the hydrolysis and cracking reactions. The use of CeO$_2$, MnO$_2$, La$_2$O$_3$ and Al$_2$O$_3$ as heterogeneous catalysts results in lower yields (43.80 ± 0.50 wt%, 44.11 ± 0.50 wt%, 42.66 ± 1.10 wt% and 44.22 ± 1.60 wt%, respectively) than the use of CaO (49.73 ± 0.90 wt%). If we attend at the electronegativity of these catalysts under the reaction conditions, the first four had a electronegativity between 8 and 11.5 while CaO had a value close to 5 [38]. Consequently, the highest yield of biocrude was obtained with the most basic oxide catalyst (CaO), which promoted the hydrolysis and cracking reactions. In addition, the basicity of CaO could give rise to secondary reactions of saponification, decreasing the biocrude yield [22]. However, the saponification reactions were not significant at the short reaction time used in this work (10 min) in comparison to the longer reaction times (60 min) in a previous study using palm biomass as raw material [38].

The yields of the water-soluble products for basic metal oxides catalysts (CaO, CeO$_2$, MnO$_2$, La$_2$O$_3$) ranged from 37.43 ± 0.61 wt% to 44.12 ± 0.92 wt%, being higher than the corresponding value in the thermal HTL (31.98 ± 1.20 wt%). This implies that the metallic oxides also promote an increase of this fraction yield, which was also previously observed in the HTL of rice husk using these types of catalysts [37]. In addition, the acid nature of these types of catalysts has an influence on the yield of the water-soluble products. Thus, the more acidic Al$_2$O$_3$ had a lower yield of water-soluble compounds than the more basic metal oxides.

The yields of the gas fraction obtained with CaO and CeO$_2$ were significantly lower (7.57 ± 0.50 wt% and 9.82 ± 0.01 wt%, respectively) than the thermal HTL (17.81 ± 2.27 wt%), whereas the yield of this fraction using La$_2$O$_3$ (12.50 ± 1.90 wt%), MnO$_2$ (18.27 ± 1.41 wt%) and Al$_2$O$_3$ (17.44 ± 3.21 wt%) was statistically the same as the one obtained without a catalyst. The decrease of the gas fraction yield could be related to the observed increase of the biocrude and water-soluble product yields because of the interaction between the different phases during HTL, according to the kinetic model proposed by Valdez [39].

The liquefied fraction yields obtained with metallic oxides were generally higher than this fraction yield obtained in the HTL without catalyst (the liquefaction phase values of MnO$_2$, CeO$_2$ and Al$_2$O$_3$ were significantly higher than the value of the non-catalysed HTL), which is due to the detected increase in the yields of the liquid fractions (water-soluble products and biocrude) with these catalysts. These yields are in all cases higher than 94 wt%, even reaching 99.49 ± 0.70 wt% when MnO$_2$ was used. It is noteworthy, therefore, that the high yield of the liquefied fraction indicates a conversion of the microalga near 100%. These high yields are linked to the low amounts of solid residue obtained in the catalytic processes (0.50 wt% to just under 6 wt%) in comparison with the solid yield in the thermal reaction (7.60 ± 0.70 wt%). This increase in the liquefied fraction and a low yield of the solid residue was observed previously in the HTL of rice husk and microalga *Nannochloropsis* with the same type of catalysts [37,40].

*3.2. Element Content, Higher Heating Value (HHV) and Energy Recovery (ER) of Obtained Biocrude*

Table 3 shows the elemental analysis, HHV and energy recovery (ER) of the biocrudes obtained with the metal oxide-catalysed HTL and the non-catalysed process.

Biocrudes obtained with the metal oxides CaO, $CeO_2$ and $La_2O_3$ had statistically similar N content (4.76 ± 0.02 wt%, 4.62 ± 0.15 wt% and 4.64 ± 0.01 wt%, respectively). Therefore, these catalysts are suitable to carry out denitrogenation reactions. However, the N content of the biocrude obtained with $MnO_2$ and $Al_2O_3$ were slightly higher (5.45 ± 0.11 wt% and 5.39 ± 0.21 wt%). In all cases, a significant decrease was observed in the N content with respect to the non-catalysed reaction (6.11 ± 0.02 wt%). This fact corroborates the power of metal oxides as catalysts of hydrolysis and cracking reactions breaking the macromolecules (proteins and lipids) to give compounds with N soluble in the aqueous phase fraction that decrease the content of this heteroatom in biocrude [20,38].

**Table 3.** Elemental composition, higher heating value (HHV) and energy recovery (ER) of the biocrude obtained through HTL of microalgae with different catalysts compared to non-catalysed reaction.

|  |  | N (wt%) | C (wt%) | H (wt%) | S (wt%) | O (wt%) | HHV (MJ/kg) | ER (%) |
|---|---|---|---|---|---|---|---|---|
|  | Thermal | 6.11 c | 73.84 a [1] | 9.17 c | 0.31 c | 10.54 a | 33.18 a | 58.25 a |
| Catalyst | CaO | 4.76 a | 69.98 b | 9.31 c | 0.40 d | 15.59 b | 31.47 b | 55.00 b |
|  | $CeO_2$ | 4.62 a | 73.04 a | 9.83 a | 0.16 a | 12.32 a | 33.40 a | 58.50 a |
|  | $MnO_2$ | 5.45 b | 70.17 b | 9.17 c | 0.16 a | 14.94 b | 31.49 b | 55.25 b |
|  | $La_2O_3$ | 4.63 a | 64.88 c | 8.39 d | 0.39 d | 21.68 c | 30.12 c | 53.00 c |
|  | $Al_2O_3$ | 5.39 b | 73.36 a | 9.64 b | 0.20 b | 11.39 a | 33.43 a | 58.75 a |

[1] Different letters (a, b, c, and d) in the same column refers to statistical differences of Duncan's multiple range test at $p < 0.05$ between control and different catalysts applied in the HTL process. "a" is the desirable mean value of groups for C, H, HHV, ER, N, O and S.

Regarding the O concentration in the biocrudes, the values obtained with the metal oxides ranged from 14.94 ± 0.58 wt% for $MnO_2$ to 21.68 ± 0.04 wt% for $La_2O_3$, being higher than those obtained by the thermal reaction (10.54 ± 0.50 wt%). This fact may be due to the hydrolysis of the main molecules (carbohydrates, lipids and proteins) into smaller ones that cannot undergo deoxygenation. However, the O content could be improved with these catalysts at a longer reaction time [20]. In spite of that, in comparison with thermal reaction, no considerable growth was detected in the O concentration for $CeO_2$ and $Al_2O_3$ catalysts.

The HHV values obtained from the elemental analysis were between 30.12 ± 0.01 and 33.43 ± 0.91 MJ/kg (Table 2), regardless of the type of metallic oxide used. These values are typical of biocrudes obtained from microalgae by HTL process (30–43 MJ/kg) [20]. Conversely, the ER were between 52.88% ± 0.02% to 58.75% ± 1.62%. The HHV and ER results of the $CeO_2$ and $Al_2O_3$ were similar to those of the thermal reaction and the mentioned indexes for CaO and $MnO_2$ were slightly lower. These values were approximately similar to the ones reported previously for fast HTL (400 °C, 2 min) with Pd/C, Pt/C or dimethalic alumina catalysts [15], but somewhat lower than those obtained in the HTL at 350 °C and 60 min with similar catalysts [27]. Among the catalysts applied in this study, $CeO_2$ resulted in remarkable specifications for the obtained biocrude through HTL of microalgae, *N. gaditana*. As can be clearly seen in Table 2, the highest values of C and H content, HHV and ER, in addition to the lowest content of N, O and S occurred as a consequence of the HTL process using the $CeO_2$ catalyst.

Because the biocrude properties are highly dependent on the H/C, N/C and O/C atomic ratios, these were calculated and compared with the corresponding ratios of the thermal biocrude and the initial microalgal biomass. In addition, the atomic ratios of the biocrudes were compared with a biodiesel obtained from rapeseed oil that complies European Biodiesel EN 14214 Standards, a biodiesel from *N. gaditana* oil that meets the biodiesel specifications except for the content of polyunsaturated (≥4 double bonds) methylester and a reference fossil diesel complying European Standard EN-590. The atomic ratios were represented using a Van Krevelen diagram (Figure 1a,b). The H/C, N/C and O/C ratio for the reference diesel were 1.84, 0.005 and 0.008, respectively. In the HTL process, deoxygenation and denitrogenation mechanisms are produced during the reaction [12]. Therefore, O/C and N/C ratios decrease with respect to the initial biomass around 80% and 40%, respectively. The H/C ratios shown in Figure 1 were similar for the five biocrudes regardless of the catalysts used, showing values around 1.50–1.65. However, the ratio of the thermal biocrude obtained was 1.25 because of its higher percentage of C in the composition. On the other hand, the N/C ratios were all between 0.05 and 0.06, slightly higher than that for the reference diesel oil (0.005) and somewhat lower than the thermal biocrude (0.084). Despite the fact that the N content was reduced with respect to the thermal reaction, a total recovery of C in the biocrude was not achieved and therefore, the N/C ratio obtained was somewhat higher than the limits reported in the literature [20]. The values of O/C were between 0.10 and 0.20, which were higher values than the ones obtained previously for this heterogeneous catalysed process (0.004–0.053) [15], which is associated to the fact that the O and C were not completely recovered in the biocrude phase. However, the O/C ratios obtained with oxygenated biofuels, such as biodiesels from rapeseed and *N. gaditana* oils, were close to 0.10 and similar to the O/C ratios of the biocrudes obtained using CeO$_2$ (0.135) and Al$_2$O$_3$ (0.113). Longer reaction times are required to decrease the O concentrations of the biocrude obtained by HTL with heterogeneous catalysts. In spite of higher values of the H/C, N/C and O/C, we can also remark on the fact that, apart from higher NO$_x$ emission due to the higher N content and lower heating value, using oxygenated biofuels such as biodiesel and bioethanol has significantly reduced the formation of air pollutants like CO, soot and unburned hydrocarbons and increased the combustion efficiency. In addition, developing engine technologies and fuel additives, besides optimizing biocrude production and upgrading systems, will provide a sustainable source of biofuel for commercial purposes [41–44].

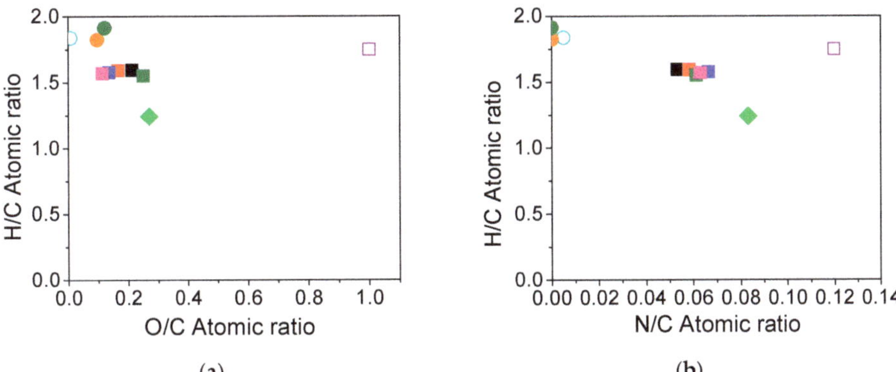

(a)        (b)

**Figure 1.** Van Krevelen diagram for biocrude obtained with metal oxide catalyst (■) CaO, (■) CeO$_2$, (■) MnO$_2$, (■) La$_2$O$_3$ and (■) Al$_2$O$_3$. For comparison purposes, thermal conditions (♦), reference diesel (○), biodiesel from rapeseed oil (●), biodiesel from *N. gaditana* oil (●) and microalga *N. gaditana* (□) are shown. (a) O/C vs H/C and (b) N/C vs H/C).

The composition of the biocrudes obtained by GC–MS, grouped by families, is shown in Table 4. A high content of acids was obtained in the biocrudes using CaO, CeO$_2$, La$_2$O$_3$ and Al$_2$O$_3$, achieving values of 28.08 wt%, 31.18 wt%, 24.99 wt% and 17.82 wt%, respectively. These catalysts promote hydrolysis reactions to a greater extent, increasing the amount of organic acids and alcohols. The latter achieved values of 9.46–18.48 wt%. Conversely, the content of organic acids in the biocrude obtained in the HTL with $MnO_2$ was remarkably higher (60.96 wt%). In this sense, this catalyst promotes the deamination reactions of aminoacids to organic acids, in addition to the hydrolysis of lipids to these organic acids. The high content of oxygenates (mainly acids, alcohols and amides) was around 65–70 wt% of the composition, which causes a low biocrude stability, a low HHV and high corrosive power [45]. However, these values were lower than those of the thermal reaction were, where the oxygenated compounds were higher than 85 wt%. Attending to the N compounds (amines, amides and nitriles), they supposed between a 20.24 wt% and a 36.85 wt%. These compounds mainly come from the hydrolysis of the proteins [12]. The content of amines was higher than that of the thermal reaction (9.74 wt%), while the amides were lower than the one obtained in the control thermal reaction (31.26 wt%). Finally, the amount of hydrocarbons in the biocrudes obtained with these metal oxides were higher (12.71–15.36 wt%) than the corresponding value obtained in the thermal control HTL (6.02 wt%), that is associated again with the decrease of heteroatoms in biocrude.

**Table 4.** Biocrudes composition determinate by gas chromatography–mass spectrometry (GC–MS).

| | | Composition (%) | | | | | | | | | |
|---|---|---|---|---|---|---|---|---|---|---|---|
| | | **Acid** | **Alcohol** | **Aldehyde** | **Amine** | **Amide** | **Ketone** | **HC** | **AHC** | **Ether** | **Nitrile** |
| | Thermal | 43.25 | 6.49 | 0.46 | 9.74 | 31.26 | 1.99 | 3.71 | 2.31 | 0.55 | 0.23 |
| Catalyst | CaO | 28.09 | 18.48 | 0.00 | 16.51 | 10.10 | 7.31 | 11.26 | 4.10 | n.d | 3.11 |
| | $CeO_2$ | 31.18 | 9.46 | 0.00 | 18.84 | 16.30 | 9.44 | 12.51 | 0.33 | n.d | 1.71 |
| | $MnO_2$ | 60.96 | 0.00 | 0.00 | 10.71 | 6.08 | 9.41 | 10.68 | 2.03 | n.d | 0.12 |
| | $La_2O_3$ | 24.99 | 17.32 | 0.00 | 15.12 | 9.36 | 16.53 | 13.26 | 0.00 | n.d | 3.13 |
| | $Al_2O_3$ | 17.82 | 17.27 | 0.00 | 12.02 | 34.78 | 8.31 | 7.10 | 1.81 | n.d | 0.14 |

HC: Hydrocarbon; AHC: Aromatic hydrocarbon; n.d: non-detectable.

The boiling temperature of the biocrudes obtained by metal oxide catalysis was determined and compared with a diesel that complies with the EN-590 standard (Table 5). The results obtained for the boiling point showed that the obtained biocrudes boiled completely at temperatures close to 529.20–551.80 °C and were similar to the one achieved in the thermal reaction (546.00 °C). However, these boiling points of the biocrude were higher than in the reference diesel (476.10 °C). Despite the decrease in the heteroatom content of the biocrudes obtained with the catalysts, the content of N and O remained high compared to the diesel that follows the regulations. This implies that the biocrude containing these electronegative heteroatoms is capable of forming more intense dipole–dipole interactions, which would lead to a higher apparent boiling point in the mixture. This fact was previously observed in the metal oxide-catalysed HTL of Malaysian oil palm biomass [38] due to the presence of oxygenated compounds and aromatic compounds.

**Table 5.** Boiling point distribution of biocrudes obtained from HTL of microalgae using metal oxides catalysts compared with reference diesel.

| | | Boiling Point (°C) | | | | |
|---|---|---|---|---|---|---|
| | **Distillation (%)** | **20** | **40** | **60** | **80** | **100** |
| | Reference diesel | 223.40 | 290.10 | 332.90 | 379.70 | 476.10 |
| | Thermal | 417.80 | 422.50 | 438.90 | 476.30 | 545.60 |
| Catalyst | CaO | 330.90 | 395.20 | 439.30 | 486.00 | 539.50 |
| | $CeO_2$ | 348.70 | 393.50 | 420.10 | 467.10 | 529.20 |
| | $MnO_2$ | 319.80 | 388.50 | 418.10 | 467.90 | 539.20 |
| | $La_2O_3$ | 301.90 | 383.00 | 418.30 | 480.60 | 546.20 |
| | $Al_2O_3$ | 347.70 | 395.20 | 436.00 | 488.00 | 551.80 |

### 3.3. Analysis of Aqueous Phase

Although no statistically significant differences have been observed by one-way ANOVA, the recovery of the aqueous phases obtained in the reactions of HTL catalysed by metal oxides was slightly higher (92.11 ± 1.10–95.48 ± 2.07 wt%) than the recovery of this fraction in the thermal process (90.13 ± 1.11 wt%) (Table 6). The pH of this phase (8.20–10.80) is marked by the basicity of the catalyst. The values of the different fractions according to the used catalyst have coherence in function of the basicity of the catalysts used, the lowest value being for the $Al_2O_3$ and the highest for the oxide of greater basic character, the CaO [38].

**Table 6.** Characteristics of the aqueous phase.

|  |  | **Water Recovery (wt%)** | **pH** | **TOC (mg/L)** |
|---|---|---|---|---|
|  | Thermal | 90 ± 1 | 8.40 | 1283 ± 4 |
|  | CaO | 93 ± 1 | 10.80 | 745 ± 1 |
|  | $CeO_2$ | 94 ± 2 | 8.53 | 698 ± 2 |
| Catalysts | $MnO_2$ | 92 ± 1 | 8.70 | 694 ± 1 |
|  | $La_2O_3$ | 94 ± 1 | 8.90 | 681 ± 2 |
|  | $Al_2O_3$ | 95 ± 2 | 8.20 | 741 ± 5 |

TOC: Total Organic Carbon.

The total organic carbon (TOC) values (681–745 ppm) indicated a high formation of new organic compounds soluble in aqueous medium. These values are within the ranges found in the starting biomass at temperatures between 300 and 350 °C (300–1146 ppm) [23]. However, the value obtained in the thermal reaction was higher (1283 ppm). Therefore, the carbon content was lower in the aqueous fraction obtained through metal oxide-catalysed HTL in comparison to the thermal reaction, which implies the existence of other elements such as N or O. As previously mentioned, the use of metal oxides as catalysts reduced the presence of this N in the biocrude, increasing the amount of this heteroatom in the aqueous phase.

### 3.4. Analysis of Gas Phase

The gaseous phases obtained by metal oxide heterogeneous catalysis had a composition similar to that obtained in the thermal reaction. These fractions were composed mainly of $CO_2$, by more than 95% in all cases. The gas fraction also contained small amounts of saturated and monounsatured linear light hydrocarbons ($C_1$–$C_4$) (<1%), and less than 1% of $H_2$ and CO. Therefore, the results showed that, despite the different catalysts used in the HTL reactions, the composition was always similar to the previous results in the literature [27,38]. The high and low contents of $CO_2$ and light hydrocarbons respectively, were related to the working temperature, which was below the critical point of water. In addition, the low concentration of CO can be due to the fact that the elimination of hydrogen occurred mainly by decarboxylation instead of decarbonylation [46]. Taking into account the $CO_2$-rich gas fraction obtained in all cases, this fraction can be integrated into a microalgal biorefinery, with the growth of the microalga stage being recirculated, thus providing the carbon source necessary for its development.

## 4. Conclusions

Our results indicated that metal oxide catalysts can improve the results of the HTL process of microalgae in comparison with the non-catalysed reaction. The catalytic HTL catalysts increased the yields of biocrude (42.60 ± 0.80–49.73 ± 0.90 wt%) compaed to the thermal reaction (42.60 ± 0.02 wt%). The biocrude yield was remarkably high with the CaO catalyst (49.73 ± 0.90 wt%). In addition, the biocrude yields increase for $CeO_2$ (43.80 ± 0.50 wt%), $Al_2O_3$ (44.22 ± 1.60 wt%) and $MnO_2$ (44.11 ± 0.60 wt%) were also high. Biocrudes showed a decrease in the N content with all the metal oxides compared to the one obtained without the catalyst (6.11 ± 0.0.02 wt%). The N content achieved with $CeO_2$, $La_2O_3$ and CaO was low (4.63 ± 0.15 wt%, 4.64 ± 0.01 wt% and 4.76 ± 0.02 wt%, respectively). However, despite being reduced in comparison to microalgae (36.50 ± 0.20 wt%), there is still a high content of O in all the biocrudes (11.39 ± 2.06–21.68 ± 0.03 wt%). Thus, apart from the positive effects of catalysts utilizing on HTL process, $CeO_2$, along with $Al_2O_3$ and CaO, can be considered for further investigations to achieve the greatest yield of produced biocrude with optimised specifications.

**Author Contributions:** A.S.-B. performed the experimental work (catalytic experiments and sample characterisation). L.F.B. and G.V. devised the experimental work. R.R. and V.M. helped in the revision of the HTL results. N.N. performed the statistical analysis. A.S.-B., G.V. and N.N. wrote the final version of the manuscript. All authors have read and agreed to the published version of the manuscript.

**Funding:** The authors gratefully acknowledge the financial support from Comunidad de Madrid provided through project ALGATEC-CM (P2018/BAA-4532), co-financed by the European Social Fund and the European Regional Development Fund, and the BIOHIDROALGA project (ENE2017-83696-R), financed by Ministerio de Economía y Competitividad of Spain.

**Conflicts of Interest:** The authors declare no conflict of interest.

## References


1. Biller, P.; Riley, R.; Ross, A.B. Catalytic hydrothermal processing of microalgae: Decomposition and upgrading of lipids. *Bioresour. Technol.* **2011**, *102*, 4841–4848. [CrossRef] [PubMed]
2. Huang, Y.; Chen, Y.; Xie, J.; Liu, H.; Yin, X.; Wu, C. Bio-oil production from hydrothermal liquefaction of high-protein high-ash microalgae including wild Cyanobacteria sp. and cultivated Bacillariophyta sp. *Fuel* **2016**, *183*, 9–19. [CrossRef]
3. López Barreiro, D.; Prins, W.; Ronsse, F.; Brilman, W. Hydrothermal liquefaction (HTL) of microalgae for biofuel production: State of the art review and future prospects. *Biomass Bioenergy* **2013**, *53*, 113–127. [CrossRef]
4. López Barreiro, D.; Zamalloa, C.; Boon, N.; Vyverman, W.; Ronsse, F.; Brilman, W.; Prins, W. Influence of strain-specific parameters on hydrothermal liquefaction of microalgae. *Bioresour. Technol.* **2013**, *146*, 463–471. [CrossRef]
5. Tian, C.; Li, B.; Liu, Z.; Zhang, Y.; Lu, H. Hydrothermal liquefaction for algal biorefinery: A critical review. *Renew. Sustain. Energy Rev.* **2014**, *38*, 933–950. [CrossRef]
6. Jazrawi, C.; Biller, P.; He, Y.; Montoya, A.; Ross, A.B.; Maschmeyer, T.; Haynes, B.S. Two-stage hydrothermal liquefaction of a high-protein microalga. *Algal Res.* **2015**, *8*, 15–22. [CrossRef]
7. Jena, U.; Das, K.C.C.; Kastner, J.R.R. Effect of operating conditions of thermochemical liquefaction on biocrude production from Spirulina platensis. *Bioresour. Technol.* **2011**, *102*, 6221–6229. [CrossRef]
8. Tang, X.; Zhang, C.; Li, Z.; Yang, X. Element and chemical compounds transfer in bio-crude from hydrothermal liquefaction of microalgae. *Bioresour. Technol.* **2015**, *202*, 8–14. [CrossRef]
9. Toor, S.S.; Rosendahl, L.; Rudolf, A. Hydrothermal liquefaction of biomass: A review of subcritical water technologies. *Energy* **2011**, *36*, 2328–2342. [CrossRef]
10. Galadima, A.; Muraza, O. Hydrothermal liquefaction of algae and bio-oil upgrading into liquid fuels: Role of heterogeneous catalysts. *Renew. Sustain. Energy Rev.* **2018**, *81*, 1037–1048. [CrossRef]
11. Jarvis, J.M.; Billing, J.M.; Hallen, R.T.; Schmidt, A.J.; Schaub, T.M. Hydrothermal Liquefaction Biocrude Compositions Compared to Petroleum Crude and Shale Oil. *Energy Fuels* **2017**, *31*, 2896–2906. [CrossRef]
12. Gollakota, A.R.K.; Kishore, N.; Gu, S. A review on hydrothermal liquefaction of biomass. *Renew. Sustain. Energy Rev.* **2018**, *81*, 1378–1392. [CrossRef]
13. Mathimani, T.; Mallick, N. A review on the hydrothermal processing of microalgal biomass to bio-oil - Knowledge gaps and recent advances. *J. Clean. Prod.* **2019**, *217*, 69–84. [CrossRef]
14. Chen, W.-H.; Lin, B.-J.; Huang, M.-Y.; Chang, J.-S. Thermochemical conversion of microalgal biomass into biofuels: A review. *Bioresour. Technol.* **2014**, *184*, 314–327. [CrossRef]
15. Patel, B.; Arcelus-Arrillaga, P.; Izadpanah, A.; Hellgardt, K. Catalytic Hydrotreatment of algal biocrude from fast Hydrothermal Liquefaction. *Renew. Energy* **2017**, *101*, 1094–1101. [CrossRef]
16. Montero-Hidalgo, M.; Espada, J.J.; Rodríguez, R.; Morales, V.; Bautista, L.F.; Vicente, G. Mild Hydrothermal Pretreatment of Microalgae for the Production of Biocrude with a Low N and O Content. *Processes* **2019**, *7*, 630. [CrossRef]

17. Prapaiwatcharapan, K.; Sunphorka, S.; Kuchonthara, P.; Kangvansaichol, K.; Hinchiranan, N. Single- and two-step hydrothermal liquefaction of microalgae in a semi-continuous reactor: Effect of the operating parameters. *Bioresour. Technol.* **2015**, *191*, 426–432. [CrossRef]

18. Chen, Y.; Wu, Y.; Zhang, P.; Hua, D.; Yang, M.; Li, C.; Chen, Z.; Liu, J. Direct liquefaction of Dunaliella tertiolecta for bio-oil in sub/supercritical ethanol–water. *Bioresour. Technol.* **2012**, *124*, 190–198. [CrossRef]

19. Yang, L.; Li, Y.; Savage, P.E. Near- and supercritical ethanol treatment of biocrude from hydrothermal liquefaction of microalgae. *Bioresour. Technol.* **2016**, *211*, 779–782. [CrossRef]

20. Xu, D.; Lin, G.; Guo, S.; Wang, S.; Guo, Y.; Jing, Z. Catalytic hydrothermal liquefaction of algae and upgrading of biocrude: A critical review. *Renew. Sustain. Energy Rev.* **2018**, *97*, 103–118. [CrossRef]

21. Jena, U.; Das, K.C.; Kastner, J.R. Comparison of the effects of Na2CO3, Ca3(PO4)2, and NiO catalysts on the thermochemical liquefaction of microalga Spirulina platensis. *Appl. Energy* **2012**, *98*, 368–375. [CrossRef]

22. Yu, G.; Zhang, Y.; Guo, B.; Funk, T.; Schideman, L. Nutrient Flows and Quality of Bio-crude Oil Produced via Catalytic Hydrothermal Liquefaction of Low-Lipid Microalgae. *Bioenergy Res.* **2014**, *7*, 1317–1328. [CrossRef]

23. Ross, A.B.; Biller, P.; Kubacki, M.L.; Li, H.; Lea-Langton, A.; Jones, J.M. Hydrothermal processing of microalgae using alkali and organic acids. *Fuel* **2010**, *89*, 2234–2243. [CrossRef]

24. Yang, W.; Li, X.; Liu, S.; Feng, L. Direct hydrothermal liquefaction of undried macroalgae Enteromorpha prolifera using acid catalysts. *Energy Convers. Manag.* **2014**, *87*, 938–945. [CrossRef]

25. Chen, Y.; Mu, R.; Yang, M.; Fang, L.; Wu, Y.; Wu, K.; Liu, Y.; Gong, J. Catalytic hydrothermal liquefaction for bio-oil production over CNTs supported metal catalysts. *Chem. Eng. Sci.* **2017**, *161*, 299–307. [CrossRef]

26. Robin, T.; Jones, J.M.; Ross, A.B. Catalytic hydrothermal processing of lipids using metal doped zeolites. *Biomass Bioenergy* **2017**, *98*, 26–36. [CrossRef]

27. Duan, P.; Savage, P.E. Hydrothermal Liquefaction of a Microalga with Heterogeneous Catalysts. *Ind. Eng. Chem. Res.* **2011**, *50*, 52–61. [CrossRef]

28. Chang, F.; Zhou, Q.; Pan, H.; Liu, X.-F.; Zhang, H.; Xue, W.; Yang, S. Solid Mixed-Metal-Oxide Catalysts for Biodiesel Production: A Review. *Energy Technol.* **2014**, *2*, 865–873. [CrossRef]

29. Gryglewicz, S. Rapeseed oil methyl esters preparation using heterogeneous catalysts. *Bioresour. Technol.* **1999**, *70*, 249–253. [CrossRef]

30. Granados, M.L.; Poves, M.D.Z.; Alonso, D.M.; Mariscal, R.; Galisteo, F.C.; Moreno-Tost, R.; Santamaría, J.; Fierro, J.L.G. Biodiesel from sunflower oil by using activated calcium oxide. *Appl. Catal. B Environ.* **2007**, *73*, 317–326. [CrossRef]

31. Kouzu, M.; Hidaka, J.S. Transesterification of vegetable oil into biodiesel catalyzed by CaO: A review. *Fuel* **2012**, *93*, 1–12. [CrossRef]

32. Dossin, T.F.; Reyniers, M.-F.; Marin, G.B. Kinetics of heterogeneously MgO-catalyzed transesterification. *Appl. Catal. B Environ.* **2006**, *62*, 35–45. [CrossRef]

33. Refaat, A.A. Biodiesel production using solid metal oxide catalysts. *Int. J. Environ. Sci. Technol.* **2011**, *8*, 203–221. [CrossRef]

34. Mendoza, A.; Vicente, G.; Bautista, L.F.; Morales, V. Opportunities for Nannochloropsis gaditana biomass through the isolation of its components and biodiesel production. *Green Process. Synth.* **2015**, *4*, 97–102. [CrossRef]

35. Li, H.; Liu, Z.; Zhang, Y.; Li, B.; Lu, H.; Duan, N.; Liu, M.; Zhu, Z.; Si, B. Conversion efficiency and oil quality of low-lipid high-protein and high-lipid low-protein microalgae via hydrothermal liquefaction. *Bioresour. Technol.* **2014**, *154*, 322–329. [CrossRef]

36. López Barreiro, D.; Samorì, C.; Terranella, G.; Hornung, U.; Kruse, A.; Prins, W. Assessing microalgae biorefinery routes for the production of biofuels via hydrothermal liquefaction. *Bioresour. Technol.* **2014**, *174*, 256–265. [CrossRef]

37. Shi, W.; Li, S.; Jin, H.; Zhao, Y.; Yu, W. The hydrothermal liquefaction of rice husk to bio-crude using metallic oxide catalysts. *Energy Sources Part A Recover. Util. Environ. Eff.* **2013**, *35*, 2149–2155. [CrossRef]

38. Yim, S.C.; Quitain, A.T.; Yusup, S.; Sasaki, M.; Uemura, Y.; Kida, T. Metal oxide-catalyzed hydrothermal liquefaction of Malaysian oil palm biomass to bio-oil under supercritical condition. *J. Supercrit. Fluids* **2016**, *120*, 384–394. [CrossRef]

39. Valdez, P.J.; Tocco, V.J.; Savage, P.E. A general kinetic model for the hydrothermal liquefaction of microalgae. *Bioresour. Technol.* **2014**, *163*, 123–127. [CrossRef]

40. Wang, W.; Xu, Y.; Wang, X.; Zhang, B.; Tian, W.; Zhang, J. Hydrothermal liquefaction of microalgae over transition metal supported TiO2 catalyst. *Bioresour. Technol.* **2018**, *250*, 474–480. [CrossRef]
41. Pedersen, T.H.; Jensen, C.U.; Sandström, L.; Rosendahl, L.A. Full characterization of compounds obtained from fractional distillation and upgrading of a HTL biocrude. *Appl. Energy* **2017**, *202*, 408–419. [CrossRef]
42. Karmakar, R.; Kundu, K.; Rajor, A. Fuel properties and emission characteristics of biodiesel produced from unused algae grown in India. *Pet. Sci.* **2018**, *15*, 385–395. [CrossRef]
43. Ahmed, I. *Oxygenated Diesel: Emissions and Performance Characteristics of Ethanol-Diesel Blends in CI Engines*; SAE Technical Paper; SAE International: Warrendale, PA, USA, 2001.
44. Kurtz, E.M.; Kuhel, D.; Anderson, J.E.; Mueller, S.A. A Comparison of Combustion and Emissions of Diesel Fuels and Oxygenated Fuels in a Modern DI Diesel Engine. *SAE Int. J. Fuels Lubr.* **2012**, *5*, 1199–1215. [CrossRef]
45. Duan, P.; Wang, B.; Xu, Y. Catalytic hydrothermal upgrading of crude bio-oils produced from different thermo-chemical conversion routes of microalgae. *Bioresour. Technol.* **2015**, *186*, 58–66. [CrossRef] [PubMed]
46. Garcia Alba, L.; Torri, C.; Samorì, C.; Van Der Spek, J.; Fabbri, D.; Kersten, S.R.A.; Brilman, D.W.F. Hydrothermal treatment (HTT) of microalgae: Evaluation of the process as conversion method in an algae biorefinery concept. *Energy Fuels* **2012**, *26*, 642–657. [CrossRef]

 *processes*

*Article*

# Scale-Up Cultivation of *Phaeodactylum tricornutum* to Produce Biocrude by Hydrothermal Liquefaction

Irene Megía-Hervás [1,2], Alejandra Sánchez-Bayo [1], Luis Fernando Bautista [3], Victoria Morales [3], Federico G. Witt-Sousa [2,†], María Segura-Fornieles [2] and Gemma Vicente [1,*]

[1]  Department of Chemical, Energy and Mechanical Technology, ESCET, Universidad Rey Juan Carlos, Móstoles, 28933 Madrid, Spain; irene.megia@urjc.es (I.M.-H.); alejandra.sanchezbayo@urjc.es (A.S.-B.)
[2]  AlgaEnergy S.A. Parque Empresarial La Moraleja, Avda. Europa, 19. Alcobendas, 28108 Madrid, Spain; fws@algaenergy.es (F.G.W.-S.); msf@algaenergy.es (M.S.-F.)
[3]  Department of Chemical and Environmental Technology, ESCET, Universidad Rey Juan Carlos, Móstoles, 28933 Madrid, Spain; fernando.bautista@urjc.es (L.F.B.); victoria.morales@urjc.es (V.M.)
*  Correspondence: gemma.vicente@urjc.es
†  (F.G.W.-S.) In memoriam.

Received: 26 July 2020; Accepted: 26 August 2020; Published: 1 September 2020

**Abstract:** *Phaeodactylum tricornutum* is an interesting source of biomass to produce biocrude by hydrothermal liquefaction (HTL). Its biochemical composition, along with its biomass productivity, can be modulated according to this specific application by varying the photoperiod, the addition of $CO_2$ or the variation of the initial nitrate concentration. The lab-scale culture allowed the production of a *P. tricornutum* biomass with high biomass and lipid productivities using a 18:6 h light:dark photoperiod and a specific $CO_2$ injection. An initial concentration of nitrates (11.8 mM) in the culture was also essential for the growth of this species at the lab scale. The biomass generated in the scale-up photoreactor had acceptable biomass and lipid productivities, although the values were higher in the biomass cultivated at the lab scale because of the difficulty for the light to reach all cells, making the cells unable to develop and hindering their growth. The biocrudes from a 90-L cultivated microalga (B-90L) showed lower yields than the ones obtained from the biomass cultivated at the lab scale (B-1L) because of the lower lipid and high ash contents in this biomass. However, the culture scaling-up did not affect significantly the heteroatom concentrations in the biocrudes. A larger-scale culture is recommended to produce a biocrude to be used as biofuel after a post-hydrotreatment stage.

**Keywords:** culture; scale-up; microalgae; hydrothermal liquefaction; biocrude; *Phaeodactylum tricornutum*

## 1. Introduction

Nowadays, there is an increase in energy demand, which requires the replacement of a high percentage of fossil fuels with green energy supplies and technological development [1]. Thus, many investigations are focused on the search for new ways of producing energy and fuels, the vast majority from renewable sources. In this context, the European Directive on the promotion of the use of energy from renewable sources establishes a binding renewable energy target of at least 32% for the European Union by 2030 and the share of renewable energy within the final consumption of energy in the transport sector being at least 14% by the same year [2].

Microalgae are an interesting alternative source for biofuel production due to their rapid growth rate and their ability to accumulate lipids or carbohydrates [3–7]. Moreover, microalgae have additional advantages: (1) microalga cultivation does not compete with agricultural lands, (2) its use would avoid the conflict with food production, (3) microalgae are capable of growing in various aqueous media, including sewage or seawater, and (4) they are neutral with respect to $CO_2$ emissions [8–11]. Algae are included as feedstocks for the production of advanced biofuels in the European Directive for the promotion of renewable energy, which comprise a contribution of these advanced biofuels of at least 3.5% as a share of the final consumption of energy in the transport sector in 2030 [2].

The biochemical composition of the microalgae can be adapted for specific purposes through the manipulation of culture conditions [12,13]. Several authors have pointed out that the change in the abiotic factors such as the nutrient composition, temperature, salinity, pH, photoperiod and intensity and quality of light may affect the biochemical composition of the microalgae [9,13,14]. Photoperiod, $CO_2$ injection and the amount of nitrogen available are some of the most important factors on the microalga culture. They have influence on the growth rate and biomass composition, increasing the lipid accumulation or changing the fatty acid profile by nitrogen deficiency in the culture media [15,16]. These factors, in turn, affect the microalgal-derived biofuel production [9,17,18]. The scale-up viability is another key factor on the culture of the microalgae, since large-scale photoreactors are required for the subsequent biofuel production. In this sense, the increase in the bioreactor capacity usually produces several changes in biomass production and composition [19]. For instance, the biomass yield decreases unpredictably when the light-dark interchange does not remain constant and the dark zone increases, which occurs when the reactor is scaled up, except when only the bioreactor length increases [20].

Among the different types of microalgae, diatoms are the dominant primary producers in the ocean. These marine species are easily adaptable environmentally and are considered responsible for 25% of global primary productivity of organic compounds [21]. There are many marine diatoms, although one of the most studied is *Phaeodactylum tricornutum* because of its high biomass productivity (235 mg/(L·d)) [22]. Therefore, it can be potentially grown at high quantities in large-scale bioreactors [12]. The typical biochemical profile of this microalgae is 30–70% proteins [23], 10–30% carbohydrates [24] and lipid contents between 20% and 30% [15], being an important source of eicosapentaenoic acid (EPA) and carotenoids [25]. This microalga has a high potential to produce advanced biofuels [25,26].

One of the main problems affecting the biofuels production from microalgae is reducing the economic cost, as 50% of the total required energy for the process is consumed in biomass drying. Hydrothermal liquefaction (HTL) is a process able of converting wet biomass into a biofuel. The biomass conversion technique is carried out in a water medium at temperatures of 200–370 °C and pressures of 10–25 MPa [27]. These conditions are established in order to decompose the biomass and hydrolyze the macromolecules (lipids, proteins and carbohydrates) into smaller organic compounds, which are capable in turn of producing hydrocarbons after decarboxylation and denitrogenation reactions [28]. This process is considered the most promising technique for the conversion of wet microalgal biomass, since the biocrude has a calorific value between 30 and 50 MJ/kg, within the range of that of a petroleum crude but, also, because of the exploitation of the additional phases (aqueous, gas and solid residue) obtained from HTL [29,30]. The aqueous phase has water-soluble products such as alcohols, aldehydes, ketones, carboxylic acids and nutrients. Therefore, this phase can be used in the biogas production or be recirculated to the cultivation phase [31]. The gaseous phase contains a high percentage of $CO_2$, so it can be recirculated to the culture as a supplementary supply of inorganic carbon. Finally, the solid residue usually contains some nutrients so it can be used as fertilizer. Additionally, it can be used as biochar thanks to its high content of carbon [29].

In this research, the culture of the promising marine species *P. tricornutum* was studied for further HTL-processing to produce a high-quality and yield biocrude. Different operating conditions in the culture were studied in a lab-scale bioreactor, such as the photoperiod, injection of $CO_2$ in the media and initial nitrate concentration. In addition, the scale-up of the cultivation stage was performed at the best operating conditions. The effect of scaling up the *P. tricornutum* culture to produce a biocrude by HTL was carried out. This microalga has been little-used for HTL, and no references can be found for the specific objectives of this study [32,33].

## 2. Materials and Methods

### 2.1. Microalga and Chemicals

*Phaeodactylum tricornutum* (CCAP 1055/1) inoculum was supplied by AlgaEnergy S.A. (Alcobendas, Spain). The diatom was grown in Mann and Myers culture medium [34].

The characterization of the biomass was carried out using the following reagents: sulfuric acid, phenol, chloroform, methanol, sodium hydroxide, copper sulphate, sodium carbonate and EDTA sodium salt supplied by Scharlab (Barcelona, Spain) and tryton-X, phenylmethylsulfonyl fluoride, Folin reactant and sodium dodecyl sulfate supplied by Sigma Aldrich (St. Louis, MO, USA). Dichloromethane supplied by Scharlab (Barcelona, Spain) was used in the HTL essays.

### 2.2. Lab-Scale Cultivation Experiments

All the cultivation experiments were carried out in 1-L glass bottles fitted with a bubbling aeration system. In all cases, the initial biomass concentration after inoculation was 0.2 g/L, and the initial sodium nitrate concentration in the medium was 11.8 mM. The illumination system consisted in fluorescent lights yielding 200 $\mu E/(m^2 s)$ at the nearest point on the external surface of the bottles, which was enough for the right growth of the species [35]. The tests were performed in triplicate at a constant temperature of 23 °C, and the cultures were run for 14 days. For the study of the effect of photoperiod, two different light:dark values were used, i.e., 12:12 and 18:6 h. The effect of the additional supply of $CO_2$ was also studied for the 18:6-h photoperiod. For this purpose, 1% $CO_2$-enriched air was injected continuously (1 L/min) into the cultures of the bubbling aeration system within the tolerance range [36,37]. The load of nitrate in the media was also assessed in presence of an additional supply of $CO_2$ (1%, 1 L/min) using the following initial nitrate concentrations: 11.8 mM (reference), the minimum value of the optimal range for this species (11.8–17.7 mM) [38], 5.9 mM (50% reduction) and 0 mM (absence of nitrate).

### 2.3. Scale-Up Cultivation Trials

A comparative study of *P. tricornutum* production was carried out by cultivating the microalga during 14 days in a 90-L bubble column photobioreactor to assess the influence of process scaling. The scale-up was carried out under the optimal operating conditions established in the lab-scale cultures (18:6-h lig ht:dark photoperiod using a continuous $CO_2$ flow and an initial nitrate concentration of 11.8 mM). The initial biomass concentration after inoculation was 0.2 g/L, as in the previous lab-scale experiments.

### 2.4. Culture Analysis and Biomass Characterization

Cell growth was measured daily by spectrophotometric absorbance at 625 nm using a spectrophotometer Lange DR 5000 (Hach Lange Spain S.L.U. L'Hospitalet de Llobregat, Barcelona, Spain). Nitrate consumption was monitored every day. For this purpose, a culture sample of 1 mL was centrifuged at 12,000 rpm for 5 min, and then, the absorbance at 220 and 270 nm of the supernatant was measured.

The specific growth rate ($\mu$), defined as the logarithmic increase in cell density per unit of time, was calculated by Equation (1).

$$\mu = \frac{\ln(N_t/N_0)}{\Delta t} \tag{1}$$

where $N_t$ is the biomass concentration at the end of the exponential phase, $N_0$ is the biomass concentration at the beginning of the exponential phase and $\Delta t$ is the time interval ($t$–$t_0$).

For biochemical analysis, dry biomass cells (0.5 g) were lysed with 5 mL of lysis buffer (distilled water containing 1.1 mM EDTA disodium salt, 0.2 mM phenylmethylsulfonyl fluoride and 0.5% Triton-X100). Proteins were measured by the Lowry method [39], adding 100 $\mu$L of sodium dodecyl sulphate solution and 1 mL of Lowry reactive to the lysis supernatant. After 10 min in darkness, Folin reactive was added and kept for 30 min. Protein concentration was calculated, reading at 750 nm and using bovine serum albumin as a standard calibration protein. Carbohydrates were measured by the Du Bois method [40] using 0.2 mL of the lysed sample, 50 µL of phenol (90%) and 5 mL of sulfuric acid (98%). After 30 min at room temperature, the carbohydrate content was measured by spectrophotometric analysis at 485 nm using glucose as the calibration standard. Lipids were extracted using a chloroform:methanol (4/5 *v/v*) solvent mixture, and the lipid content was calculated gravimetrically [41]. The ash content was determined by gravimetric analysis after calcination at 600 °C for 4 h with a ramp of 50 °C per minute.

The elemental analysis of the biomass was performed in a Flash 2000 (Thermo Fisher Scientific, Waltham, MA, USA) fitted with a thermal conductivity detector (TCD). The C, H, N and S contents were determined by an oxidation/reduction reactor at 900 °C, while the O content was independently determined through a specific pyrolysis reactor at 1060 °C.

The moisture content was studied by drying 0.5-g samples in an oven at 90 °C during 48 h until constant weight.

The productivities of the biomass ($P_B$) and lipids ($P_L$) contained in the biomass were evaluated from Equations (2) and (3), respectively [42].

$$P_B = \frac{N_t - N_0}{\Delta t} \tag{2}$$

where $N_t$ is the biomass concentration at the end of the exponential phase, $N_0$ is the initial biomass concentration (g/L) and $\Delta t$ is the time interval ($t_t$–$t_i$).

$$P_L = P_B \times \frac{\text{mass of lipids}}{\text{mass of dry microalgae}} \tag{3}$$

### 2.5. Hydrothermal Liquefaction Process

The HTL reactions were performed in a 100-mL stainless steel autoclave (EZ-SEAL®, Autoclave Engineers, Erie, PA, USA) using a *P. tricornutum* biomass from the cultivation using 1-L and 90-L photobioreactors. The reactions were carried out at a temperature of 320 °C with 30 g of wet biomass, whose dry weight was 7.74%, at three reaction times (0, 10 and 30 min) in order to evaluate the influence of the reaction time on the quality and quantity of biocrude produced. In this work, time 0 was considered when the reactor reached the temperature setpoint.

At the end of the reaction, the autoclave was rapidly cooled down to room temperature to quench the reaction. Then, the gas phase was allowed to flow through a valve that connected the reactor to the gas chromatograph. Dichloromethane was used to collect the reaction mixture that consisted of solid residue, an aqueous phase and biocrude. The solid residue was separated by vacuum filtration, dried and weighted. The biocrude and aqueous phase were separated by decantation in a separation funnel, and finally, water and dichloromethane were evaporated from their respective phases in order to measure the dry weight of the aqueous phase and biocrude, respectively.

The yields of biocrude ($Y_B$), water-soluble products ($Y_{AP}$) and solid residue ($Y_{SR}$) were determined on a dry basis using Equations (4) to (6) and the overall liquefied phase through Equation (7) [43]. The yield of the gas phase was estimated by difference to the initial dry biomass used.

$$Y_B(\%) = \frac{\text{mass of biocrude}}{\text{mass dry microalgae}} \times 100 \tag{4}$$

$$Y_{AP}(\%) = \frac{\text{mass of aqueous phase}}{\text{mass dry microalgae}} \times 100 \tag{5}$$

$$Y_{SR}(\%) = \frac{\text{mass of solid residue}}{\text{mass dry microalgae}} \times 100 \tag{6}$$

$$\text{Liquefied phase (\%)} = \left(1 - \frac{\text{mass of solid residue}}{\text{mass dry microalgae}}\right) \times 100 \tag{7}$$

### 2.6. Analysis of HTL Products

The biocrude was analyzed by elemental analysis following the same protocol as that described above for biomass (Section 2.4). The high heating value (HHV) for the biocrude and the microalgal biomass was determined by using Equation (8) [44].

$$\text{HHV (MJ/kg)} = 0.3414 \times C + 1.4445 \times H - \frac{N - O - 1}{8} + 0.093 \times S \tag{8}$$

Finally, the energy recovery (ER) was calculated with Equation (9).

$$\text{ER (\%)} = \frac{(\text{HHV of biocrude} \times \text{mass biocrude})}{(\text{HHV of microalgae} \times \text{mass of dry microalgae})} \times 100 \tag{9}$$

The total organic content (TOC) of the aqueous phase was determined in Shimadzu-V equipment (Shimadzu Corp, Kyoto, Japan), and the pH was measured with a Basic 30 pH meter (Crison Instruments, Barcelona, Spain).

The gas products were analyzed using a gas chromatograph Varian CP-4900 MicroGC (Varian Inc., Palo Alto, CA, USA) fitted with a thermal conductivity detector (TCD) connected online to the autoclave reactor.

### 2.7. Statistical Analysis

All the experiments were performed in triplicate in order to determine the variability of the results and to assess the experimental errors. In this way, the arithmetical averages and the standard deviations were calculated for all the results.

In addition, statistical analysis was performed by one-way ANOVA using Statgraphics Centurion XVII software (Statpoint Technologies Inc., Warrenton, VA, USA) to determine differences in the biochemical and elemental compositions, yields and productivities between different photoperiods, $CO_2$ injections, nitrate concentrations and cultivation scales. Previously, the variances were checked for homogeneity by the Levene's test, and the Student-Newman-Keuls (SNK) test was used to discriminate among different treatments after a significant F-test. The Student-Newman-Keuls test was used to discriminate among different treatments. Except where another value was explicitly indicated, the confidence level was set at 95% ($p$-value $< 0.05$).

## 3. Results and Discussion

### 3.1. Lab-Scale Culture

#### 3.1.1. Effect of Photoperiod

The growth of the microalgae under photoperiods 12:12 and 18:6 h of light:dark in the 1-L bioreactor is shown in Figure 1. Maximum biomass concentrations of 1.56 and 1.73 g/L were obtained for the 12:12 and 18:6 h light:dark photoperiods, respectively. Both biomass concentrations were higher than those obtained by Morais et al. [45] with the same microalga, who obtained a maximum concentration of 1.3 g/L under 12:12 and 24:0-h light:dark photoperiods at a lower light intensity of 74 $\mu$E/(m$^2$s). However, the volume of the photobioreactor was somewhat higher (2 L) to the one studied in this work. The aeration homogenizes the culture more efficiently in the smaller photobioreactors, which means that all the cells remain for less time in dark areas.

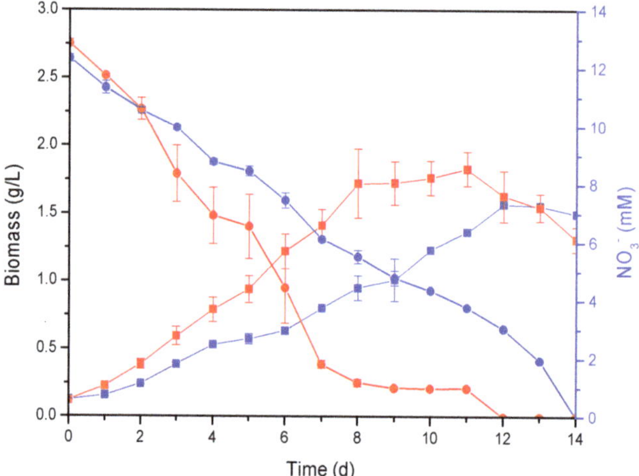

**Figure 1.** Growth curves (squares) and nitrate uptakes (circles) for cultures of *Phaeodactylum tricornutum* at different light:dark photoperiods: 12:12 h (blue symbols) and 18:6 h (red symbols).

The growth rate was 0.24 and 0.33 d$^{-1}$ (*p*-value = 0.01) for the 12:12 and 18:6 h light:dark photoperiods, respectively. Therefore, the growth rate raised with the light hours, which has been previously reported. Thus, a similar increase with the light cycle was also observed for *Chlorella vulgaris* [9,46] and *Nannochloropsis* [47]. In addition, a growth rate of 0.34 d$^{-1}$ for *Nannochloropsis* at the 18:6 photoperiod was previously reported [47]. The most notable difference was observed in the day the stationary phase was reached: 8 and 12 days for the 18:6 and 12:12 h photoperiods, respectively. In both photoperiods, the total nitrate consumption was reached after 14 days of cultivation; however, the number of light hours accelerated the consumption of nutrients.

The lipids obtained increased (significantly at a confidence level 0 94%, *p*-value = 0.055) from 24% ± 1% to 30% ± 3% (Table 1) as the light exposure increased from 12 to 18 h, showing a similar trend than that reported for other species in previous studies [48]. Thus, lipids were mainly accumulated after light exposures above 12 h. In addition to the light time, the amount of lipids depends also on other factors like the $CO_2$ supply or concentration of nutrients in the culture [49]. The amount of lipids in the biomass obtained in the culture is frequently analyzed in the literature. However, a complete study of the biomass composition and productivity, which has been scarcely reported previously, is recommended to fully understand their influence in the biocrude production by HTL. Consequently, the concentration of proteins, carbohydrates and ash in the *P. tricornutum* biomass were measured and presented also in Table 1, together with the corresponding biomass and lipid productivities. Unlike what was observed for *Scenedesmus obliquus* and *Chlorella* [50,51], the protein content did not change significantly (*p*-value = 0.127) when the photoperiod increased the light times. By contrast, the content of carbohydrates decreased (*p*-value = 0.019) as the light hours increased. George et al. [52] obtained similar results in cultures of *Ankistrodesmus falcatus*. Biller and Ross [53] concluded that there is a trend in the biocrude yield from HTL where lipid contents are the main positive influential factor, followed by protein and carbohydrate contents in this order of influence. Thus, a high content of lipids and proteins favors the production of hydrocarbons after hydrolysis and denitrogenation reactions and hydrolysis and decarboxylation, respectively [28].

**Table 1.** Effect of the photoperiod on biomass composition and productivities. $P_B$: biomass and $P_L$: lipid productivities.

|  | Light:Dark Photoperiod (h) | |
| --- | --- | --- |
|  | **12:12** | **18:6** |
| Lipids (%) | 23.96 ± 1.09 | 29.55 ± 2.73 |
| Proteins (%) | 49.10 ± 0.79 | 54.45 ± 3.76 |
| Carbohydrates (%) | 6.66 ± 0.80 | 3.73 ± 1.02 |
| Ash (%) | 20.27 ± 0.56 | 12.27 ± 0.04 |
| $P_B$ (mg/(L·d)) | 123.87 ± 0.06 | 200.02 ± 15.18 |
| $P_L$ (mg/(L·d)) | 25.06 ± 0.01 | 52.67 ± 4.00 |

The ash content decreased from 20.27% ± 0.56% to 12.27% ± 0.04% ($p$-value = 0.002) with the increase in light exposure. The ash portion varies according to the culture conditions [54], and in this case, the observed reduction of ash was probably due to the corresponding increase in the photosynthetic efficiency, which, in turn, means a higher biomass production with the light cycle. A high ash content may inhibit the transformation of the microalgae in HTL and has a negative effect on the biocrude properties [55].

Interestingly, the biomass productivity increased at higher light photoperiods, which was observed previously for the microalga *S. obliquus* [50]. The biomass productivity achieved 123.87 ± 0.06 and 200.02 ± 15.18 mg/(L·d) ($p$-value = 0.013) at the 12:12 and 18:6-h light:dark photoperiods, respectively. As the lipid content and biomass productivity raised with the light, the lipid productivity also increased with the hours of light (25.06 ± 0.01 to 52.67 ± 4.00 mg/(L·d) ($p$-value < 0.007) for the 12:12 and 18:6-h photoperiods, respectively). Therefore, the results showed a greater efficiency in the production of both biomass and lipids when the culture was exposed to 18 h of light.

One of the objectives of the liquefaction process is to obtain a biofuel with a low content of heteroatoms, mainly O and N. Table 2 summarizes the biomass elemental analysis at the different photoperiods. The O and N contents of the biomass obtained at both photoperiods did not show significant differences, which was related with the biochemical composition of the biomass obtained. The contents of these heteroatoms were lower than the reported values of other microalgae, such as *Arthrospira platensis* [56], *C. vulgaris* [53] and *Dunaliella tertiolecta* [33].

**Table 2.** Effect of the photoperiod on the elemental composition of the biomass.

| Light:Dark (h) | C (%) | N (%) | H (%) | S (%) | O (%) |
| --- | --- | --- | --- | --- | --- |
| 12:12 | 53.65 ± 0.80 | 9.29 ± 0.10 | 7.65 ± 0.10 | 0.12 ± 0.01 | 29.29 ± 0.60 |
| 18:6 | 52.74 ± 0.20 | 9.18 ± 0.10 | 7.50 ± 0.10 | 0.90 ± 0.00 | 29.68 ± 0.30 |

Therefore, biomass grown under a larger number of light hours seems a priori more adequate to HTL due to its high biomass and lipid productivities, high protein and lipid contents and lower ash content. Thus, the 18:6-h photoperiod was chosen to produce the *P. tricornutum* biomass for HTL.

### 3.1.2. Influence of $CO_2$ Injection

The growth curve (Figure 2) shows the results obtained in the essays with and without an additional $CO_2$ injection in the *P. tricornutum* culture. It can be observed that the exponential phase was extended for two more days when a continuous flow of additional $CO_2$ was injected. This caused a remarkable increase in the total biomass production, reaching a concentration above 2.61 g/L with a continuous injection of $CO_2$ compared to 1.73 g/L obtained in the absence of the supplemental $CO_2$. A similar notable increase was observed previously for the other diatom [17]. Total nitrate depletion was reached at the end of both experiments (Figure 2), as observed in the study of the effect of the photoperiod.

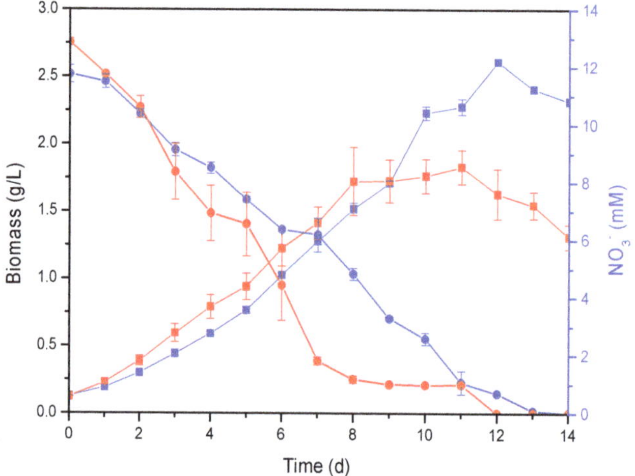

**Figure 2.** Growth curves (squares) and nitrate uptake (circles) for cultures of *P. tricornutum* under an 18:6-h photoperiod with continuous $CO_2$ injection (blue symbols) and without $CO_2$ injection (red symbols).

Table 3 shows the biochemical composition and biomass and lipid productivities of *P. tricornutum* microalgae grown with and without supplemental $CO_2$ injections. The results show that the contents of lipids and proteins did not change significantly (*p*-values: 0.171 and 0.121, respectively) when a flow of pure $CO_2$ was bubbled in the culture. This could be due to the fact that the metabolic flow involving the biosynthesis of these types of biomolecules is not altered if the carbon source is maintained above a minimum threshold. Other authors, however, reported an increase in lipid and protein accumulations when $CO_2$ was injected in the culture medium up to a maximum concentration of 10% [17]. Carbohydrates increased with the injection of $CO_2$ into the culture medium, from 3.73% ± 1.02% to 12.54% ± 0.68% (*p*-value = 0.0005). This is in accordance with a previous study for the microalga *S. obliquus* [57] where the availability of $CO_2$ in the culture medium favored the production of carbohydrates during the dark phase.

**Table 3.** Effect of $CO_2$ injection on biomass composition and productivities.

|  | Without $CO_2$ Injection | With $CO_2$ Injection |
|---|---|---|
| Lipids (%) | 29.55 ± 2.73 | 33.15 ± 2.56 |
| Proteins (%) | 54.45 ± 3.76 | 49.16 ± 1.59 |
| Carbohydrates (%) | 3.73 ± 1.02 | 12.54 ± 0.68 |
| Ash (%) | 12.27 ± 0.04 | 5.15 ± 1.46 |
| $P_B$ (mg/(L·d)) | 200.02 ± 15.18 | 210.54 ± 6.12 |
| $P_L$ (mg/(L·d)) | 52.67 ± 4.00 | 69.80 ± 1.68 |

The ash concentration decreased with the $CO_2$ supplement from 12.27% ± 0.04% to 5.15% ± 1.46% (*p*-value = 0.014). The acid nature of $CO_2$ in the solution controls the pH of the medium at lower values with the supplementation of carbon dioxide, avoiding the precipitation of insoluble salts of the medium during the growth [58].

However, the productivity of the biomass remained without significant changes ($p$-value = 0.353), regardless of whether the additional $CO_2$ was injected or not. Although an inhibition by an excess of $CO_2$ to the culture medium was previously reported [59], the results of the biomass productivity did not indicate an inhibition effect by the presence of the extra $CO_2$. These similar values of biomass productivities may be due to the fact that the microalgae grew faster and took two more days to reach the stationary phase in the presence of extra amounts of $CO_2$.

The $CO_2$ supplemented to the microalgae system led to a significant rise in lipid productivity, from $52.67 \pm 4.00$ to $69.80 \pm 1.68$ mg/(L·d) ($p$-value = 0.010), despite the similar biomass productivities obtained. A similar increase with $CO_2$ was observed for *C. vulgaris* [60].

The elemental composition of the microalgal biomass (Table 4) slightly changed when the $CO_2$ availability increased. The results only showed a small increase in the O content ($p$-value = 0.033) and a small decrease in the N content ($p$-value = 0.041). N is mainly contained in proteins, which remained approximately constant with the additional $CO_2$ injection. Therefore, the above observed results were likely due to the remarkable increase in carbohydrates, rich in O, in the biomass cultured with extra $CO_2$. Therefore, there is no significant modifications regardless of the presence of a specific $CO_2$ injection in the culture medium, because the carbon content from air allows a nonlimiting growth of cells [61]. Similar results have been reported for *C. vulgaris* [59].

**Table 4.** Effect of $CO_2$ injections on the elemental composition of the biomass.

|  | C (%) | N (%) | H (%) | S (%) | O (%) |
|---|---|---|---|---|---|
| Without $CO_2$ injection | $52.74 \pm 0.20$ | $9.18 \pm 0.10$ | $7.50 \pm 0.10$ | $0.90 \pm 0.00$ | $29.68 \pm 0.30$ |
| With $CO_2$ injection | $53.13 \pm 0.06$ | $8.52 \pm 0.02$ | $7.33 \pm 0.01$ | $0.98 \pm 0.01$ | $30.10 \pm 0.05$ |

Therefore, an extra $CO_2$ injection was selected to obtain the *P. tricornutum* biomass for the subsequent larger-scale culture and biocrude production by HTL because of the high lipid productivity, together with a lower ash content achieved in the experiments with a direct supplement of $CO_2$.

### 3.1.3. Effect of Initial Nitrate Concentration

Based on the previous cultures with 18:6 h of a light:dark photoperiod and $CO_2$ injection, the initial nitrate content was reduced with respect to the original culture medium (11.8 mM) to stress the microalga and evaluate its growth, composition and productivity. The results show that the decrease in nitrate negatively affected the biomass concentration (Figure 3), since the cell growth was inhibited by the lack of a readily available N source. Similar results were observed earlier [62,63]. Culture mediums with nitrate reductions of 50% implied a 40% decrease in biomass growth (from 2.61 g/L to 1.56 g/L). In the culture media without nitrates, the biomass growth decreased to 0.86 g/L. Our results indicate that nitrates can be considered as essential nutrients for *P. tricornutum* growth [62], and therefore, there is a need to have an initial nitrate concentration of at least 11.8 mM in the media.

The original medium (nitrate concentration of 11.8 mM) reached the stationary phase on day 10, whereas the total nitrates uptake was reached on day 13. On the other hand, the culture with a 50% of nitrates reduction reached the stationary phase on day 9 because of the nitrogen limitation. The concentration of nitrates was exhausted on day 8, and therefore, cells could not reproduce. In the medium without them, the stationary phase was reached on day 6 because of the inhibitory effect of the lack of a nitrogen source.

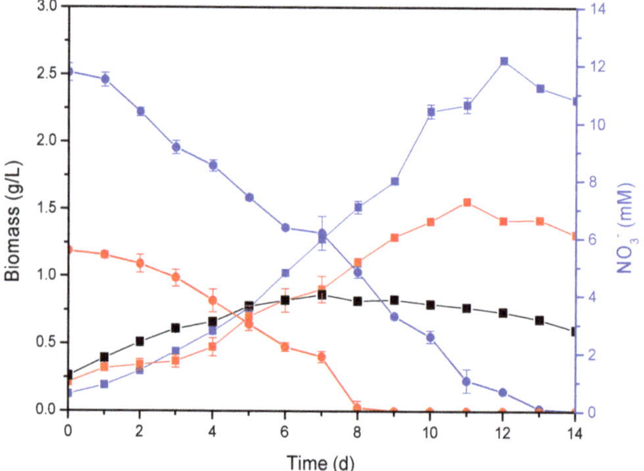

**Figure 3.** Growth curve (squares) and nitrate uptake (circles) for cultures of *P. tricornutum* under a 18:6-h photoperiod with continuous $CO_2$ injection starting with different initial nitrate concentrations: 11.8 mM (blue symbols), 5.9 mM (red symbols) and 0 mM (black symbols).

Table 5 shows the effect of the initial nitrate concentration in the culture in the biomass composition. The final lipid content was not significantly affected by the reduction of the initial concentration of the nitrogen source. Although microalgae are subjected to stress in the absence of nitrates, favoring the accumulation of lipids in the cells [64,65], the harvesting and characterization of the biomass was carried out after 14 days of cultivation. At that time, nitrate was depleted from the media in all the experiments, so that all cultures were similarly stressed in terms of N availability. The concentration of proteins decreased from 49.16% ± 1.59% (nitrate concentration of 11.8 mM) to 33.47% ± 2.38% (lack of nitrate). The presence of nitrate in the medium is the main source of N assimilation for the microalgae and, therefore, essential for protein formation [62].

**Table 5.** Effect of the initial nitrate concentration on the biomass composition and productivities.

| | Initial ($NO_3^-$) (mM) | | |
| --- | --- | --- | --- |
| | **11.8** | **5.9** | **0** |
| Lipids (%) | 33.15 ± 2.56 a | 34.89 ± 2.25 a | 34.91 ± 2.01 a |
| Proteins (%) | 49.16 ± 1.59 a | 43.81 ± 2.68 b | 33.47 ± 2.38 c |
| Carbohydrates (%) | 12.54 ± 0.68 a | 11.96 ± 0.01 a | 22.65 ± 2.12 b |
| Ash (%) | 5.15 ± 1.46 a | 9.34 ± 0.78 b | 8.97 ± 0.62 b |
| $P_B$ (mg/(L·d)) | 210.54 ± 5.08 a | 119.75 ± 0.24 b | 84.21 ± 2.18 c |
| $P_L$ (mg/(L·d)) | 69.80 ± 1.68 a | 41.78 ± 0.08 b | 29.40 ± 0.76 c |

Values show average ± standard deviation, and letters show significant differences between different initial nitrate concentrations for each component and productivity ($p$-value < 0.05, Student-Newman-Keuls (SNK) test).

The amount of carbohydrates was significantly higher (22.65% ± 2.12%) for the cultivation performed in the absence of an initial supply of nitrate than that obtained when the initial concentrations of nitrate were 11.8 and 9.6 mM (12.54% ± 0.68% and 11.96% ± 0.01%, respectively). Nutrient reductions such as nitrate drive the microalgae to accumulate energy-rich reserve compounds—essentially, lipids and carbohydrates [64].

The ash concentration increased with the reduction of nitrate in the culture from 5.15% ± 1.46% (11.8 mM) to 9.3%5 ± 0.78% (5.9 mM) and 8.97% ± 0.62% (absence of an initial supply of nitrate). Therefore, the reduction of the N source negatively affected the biomass composition, increasing the ash content, which is not adequate for the following HTL stage. The lower ash content obtained with the higher initial nitrate concentration (11.8 mM) was due to the better photosynthetic efficiency and, therefore, higher biomass production at this nitrate concentration.

Regarding the biomass and lipid productivity, Table 5 shows about a 2.5-fold increase in these values at the higher initial nitrate concentration. Although the composition of the biomass is affected at different levels, the main effect of the availability of large amounts of N was exerted on the growth of microalgae, boosting the values of the specific growth rate and, hence, the productivity associated to the biochemical components of the cells.

From the results obtained in the elemental analysis of the biomass (Table 6), the main remarkable effect of the decrease in the amount of N in the growth medium was a significant reduction in the N content of the biomass when the culture was not supplemented with an initial nitrate concentration. The rest of the elements underwent changes that, although statistically significant in some cases, did not turn out to be very noteworthy.

**Table 6.** Effect of the initial nitrate concentration on the elemental composition of the biomass.

| $(NO_3^-)$ (mM) | C (%) | N (%) | H (%) | S (%) | O (%) |
|---|---|---|---|---|---|
| 11.8 | 53.13 ± 0.06 a | 8.52± 0.02 a | 7.33 ± 0.01 a | 0.98 ± 0.01 a | 30.10 ± 0.05 a |
| 5.9 | 53.58 ± 0.20 b | 9.27 ± 0.03 b | 7.38 ± 0.19 a | 1.53 ± 0.17 b | 28.23 ± 0.10 b |
| 0 | 59.07 ± 0.15 c | 2.77 ± 0.01 c | 8.81 ± 0.03 b | 0.42 ± 0.03 c | 28.93 ± 0.15 c |

Values show average ± standard deviation, and letters show significant differences between different initial nitrate concentrations for each element ($p$-value < 0.05, SNK test).

According to the drastic reduction in biomass and lipid productivities and the increase of ash content in the absence of nitrates, the importance of the nitrate presence was confirmed for the adequate *P. tricornutum* growth. In this sense, an initial concentration of 11.8 mM of nitrate in the growth medium was chosen to continue the study.

## 3.2. Culture Scaling

Scaling tests (Figure 4) show that the microalga did not require an adaptation phase, regardless of the reactor used, and began to grow rapidly in the culture medium. The stationary phase was reached four days later in the 90-L volume reactor compared to in the 1-L one. In addition, the specific growth rate was 0.29 and 0.17 d$^{-1}$ in the bioreactors of 1 L and 90 L, respectively, which are suitable values for this type and size of bubble column reactors indoors [66]. The biomass produced in the 90-L bioreactor is close to other productions obtained outdoors in similar column bioreactors for this species (0.96 g/L) [25], as well as for other pilot plant reactors, such as circular ponds or tubular photobioreactors [21], indicating a good biomass production. It must be noted that total consumption of nitrate was not reached for the 90-L culture after 14 days.

The value of the biomass productivity obviously decreased with the bioreactor volume because of the observed greater shielding of the microalga, which prevents the light from reaching the microalga cells at higher volumes (Table 7). Thus, the biomass productivities were 210.54 ± 5.08 and 56.01 ± 4.45 mg/L·d for the 1-L and 90-L bioreactors, respectively ($p$-value < 0.0001). In addition, a light stress reduction at the higher volume, produced by the increase in dark areas, affected the lipid productivity, decreasing from a value of 69.8 ± 1.68 for the 1-L reactor to 9.85 ± 0.78 mg/(L·d) for the 90-L ($p$-value < 0.0001). The high biomass and lipid productivities obtained at the 1-L bioreactor can be achieved using optimal culture conditions, which only have a remarkable positive effect at this small-scale cultivation. In this sense, the productivities obtained on a larger scale, although lower, results are adequate for the production of this microalgae and its subsequent use in the production of biofuels.

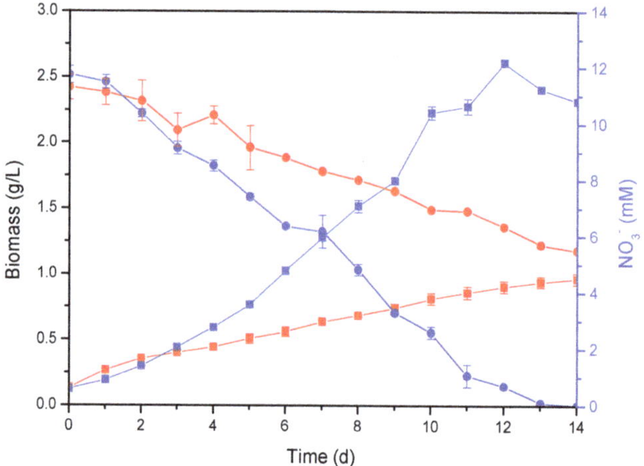

**Figure 4.** Growth curves (squares) and nitrogen uptake (circles) for cultures of *P. tricornutum* under an 18:6-h photoperiod with continuous $CO_2$ injections in photobioreactors at different scales: 1-L bottle (blue symbols) and 90-L column bioreactor (red symbols).

**Table 7.** Biomass analysis for the 1-L bottle culture and the 90-L scale.

| | $V_{reactor}$ (L) | |
|---|---|---|
| | **1** | **90** |
| Lipids (%) | $33.15 \pm 2.56$ | $17.59 \pm 0.03$ |
| Proteins (%) | $49.16 \pm 1.59$ | $58.49 \pm 0.18$ |
| Carbohydrates (%) | $12.54 \pm 0.68$ | $9.80 \pm 0.05$ |
| Ash (%) | $5.15 \pm 1.46$ | $14.12 \pm 0.17$ |
| $P_B$ (mg/(L·d)) | $210.54 \pm 5.08$ | $56.01 \pm 4.45$ |
| $P_L$ (mg/(L·d)) | $69.80 \pm 1.68$ | $9.85 \pm 0.78$ |

Finally, the elemental analysis of the biomass produced in both reactors hardly showed significant differences (Table 8). A slight but significant decrease of C and H ($p$-value $< 0.0001$) was observed in the biomass grown at a photoreactor of 90 L, related to a higher concentration of ashes and a lower concentration of lipids and carbohydrates. In addition, the concentration of N was slightly lower ($p$-value $= 0.0001$) in the biomass cultivated in the bioreactor of 90 L ($7.50\% \pm 0.02\%$) in comparison to the biomass obtained in the 1-L reactor ($8.52\% \pm 0.02\%$) despite the higher concentration of proteins found in the former biomass. Consequently, a little higher ($p$-value $= 0.0004$) concentration of O was obtained in the biomass grown in a 90-L culture ($37.29\% \pm 0.22\%$) than that in the corresponding biomass cultivated in the 1-L bioreactor ($30.10\% \pm 0.05\%$), which is probably connected with the observed increase in proteins as the size of the photobioreactor increased. In this sense, the elemental distribution in the biomass obtained did not change substantially, despite the differences of the biomass analysis obtained (Table 7). Although a small size photobioreactor seems to be more adequate to obtain a biomass for biofuel production, according to the above results of the biomass analysis, the development of the HTL process on an industrial scale requires high amounts of biomass and, therefore, the use of large-scale photoreactors.

**Table 8.** Elemental analysis for the biomass cultivated on different scales.

| $V_{bioreactor}$ (L) | C (%) | N (%) | H (%) | S (%) | O (%) |
|---|---|---|---|---|---|
| 1 | $53.13 \pm 0.06$ | $8.52 \pm 0.02$ | $7.33 \pm 0.01$ | $0.98 \pm 0.01$ | $30.10 \pm 0.05$ |
| 90 | $47.14 \pm 0.22$ | $7.50 \pm 0.02$ | $6.94 \pm 0.03$ | $1.12 \pm 0.00$ | $37.29 \pm 0.22$ |

### 3.3. Hydrothermal Liquefaction Process

The yield of the different fractions obtained from the HTL process at 320 °C (biocrude, aqueous phase, gas phase and solid residue) for the three reaction times and both culture scales evaluated are shown in Figure 5. The biomass grown in the 1-L photobioreactor (B-1L) exhibited a higher biocrude yield compared to the biomass produced at 90 L (B-90L) in the HTL process at all reaction times ($p$-value ≤ 0.05), which was mainly due to the fact that the former had a higher lipid content (Table 7) that contributed to the increase in the yield of this fraction [53]. The yield of the biocrude fraction obtained from the B-1L at 10 min was similar (36.64% ± 4.93%) to those obtained by Christensen et al. (38.8% ± 1.3% at 350 °C and 15 min) for commercial *P. tricornutum* at harsher operating conditions [32], while those of B-90L were lower (25.61–30.03%).

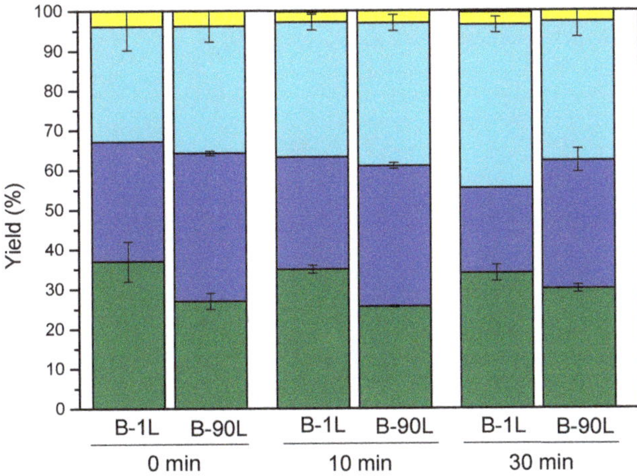


**Figure 5.** Yields of the different phases after a hydrothermal liquefaction (HTL) reaction at 320 °C with the *P. tricornutum* biomass: biocrude (green), aqueous phase (dark blue), gas phase (light blue) and solid residue (yellow). Time 0 min was considered when the reactor reached the setpoint temperature. B-1L and B-90L: *P. tricornutum* biomass cultivated in 1-L and 90-L bioreactors, respectively.

The biocrude yields produced using the B-1L were significantly unaffected at different reaction times (33.99% ± 1.67%–36.64% ± 4.93%, $p$-value > 0.05). Christensen et al. [32] obtained similar yields at 300 and 325 °C, which are closer to the value of 320 °C used in this work. Similarly, the biocrude yield from B-90L did not change with time (25.61% ± 0.27%–30.03% ± 0.95%). Therefore, the reaction time did not have a significant effect on the biocrude yields at the high temperature used in this work (320 °C).

The yield of the aqueous fraction decreased with the reaction time using B-1L but remained approximately constant over time with B-90L. However, the yields of the aqueous fractions were higher ($p$-values < 0.05) for the B-90L because of the lower biocrude yields obtained in this case. More water-insoluble molecules were produced at longer reaction times, which may be due to decarboxylation, deamination, dehydration, oligomerization and condensation reactions, thus producing a change in the product distribution and a decrease in the yield of the aqueous phase [67].

The yield of the gas phase was nearly constant (30.51% ± 0.40% to 35.96% ± 1.98%) for B-90L. A similar behavior were reported previously for HTL carried out at 350 °C and similar reaction times with *Nannochloropsis* [53,68]. However, there was a significant increase in the gas phase with time in the 1-L culture (28.50% ± 5.54% to 41.03% ± 1.77%) related to the higher content of C and, to a lesser extent, of carbohydrates in this biomass, which produces higher gas yields. The production of gaseous compounds is related to the yields of the biocrude and aqueous phases in the reaction mechanism proposed by Sheehan and Savage for *Nannochloropsis* [30].

The solid residues represented the smallest fractions of the products obtained in the HTL process, their yields varying from 2.59% to 4.74%, in agreement with bibliographic results, where solid fraction yields below 10% are usual for the HTL of microalgae [29]. The low yields achieved for the solid residue are the cause for the high transformation efficiency of HTL, showing liquefaction yields (sum of biocrude and aqueous and gas phases) over 95% in all cases [29].

*3.4. Analysis of Biocrude*

Table 9 shows the elemental composition, the higher heating value (HHV) and the energy recovery (ER) in the biocrude obtained by HTL using B-1L and B-90L at different reaction times.

**Table 9.** Elemental analysis (wt%, dry basis), higher heating value (HHV) and energy recovery of the biocrude phase after the hydrothermal liquefaction (HTL) process at 320 °C for different reaction times. B-1L and B-90L: *P. tricornutum* biomass cultivated in 1-L and 90-L bioreactors, respectively. ER: energy recovery.

| Time (min) | Biomass | C (%) | H (%) | N (%) | S (%) | O (%) | HHV (MJ/kg) | ER (%) |
|---|---|---|---|---|---|---|---|---|
| 0 | B-1L | 74.46 ± 1.41 | 9.73 ± 0.47 | 5.60 ± 0.97 | 0.42 ± 0.20 | 9.79 ± 0.20 | 38.27 ± 1.52 | 51.46 ± 1.86 |
| | B-90L | 74.92 ± 0.86 | 9.52 ± 0.09 | 5.43 ± 0.10 | 0.35 ± 0.04 | 9.78 ± 0.10 | 39.36 ± 0.46 | 38.61 ± 0.61 |
| 10 | B-1L | 76.54 ± 0.59 | 9.76 ± 0.07 | 5.52 ± 0.28 | 0.64 ± 0.01 | 7.54 ± 0.01 | 39.47 ± 0.34 | 50.61 ± 0.60 |
| | B-90L | 75.06 ± 0.73 | 9.37 ± 0.09 | 5.39 ± 0.04 | 0.43 ± 0.04 | 9.74 ± 0.12 | 39.71 ± 0.40 | 36.77 ± 0.52 |
| 30 | B-1L | 77.52 ± 0.03 | 9.62 ± 0.12 | 5.43 ± 0.23 | 0.57 ± 0.10 | 6.86 ± 0.31 | 39.37 ± 0.26 | 49.11 ± 0.53 |
| | B-90L | 75.29 ± 0.74 | 9.65 ± 0.07 | 5.42 ± 0.03 | 0.44 ± 0.11 | 9.20 ± 0.06 | 38.767 ± 0.37 | 41.98 ± 0.50 |

A significant decrease in the O amount was observed in all the biocrude phases with respect to the starting biomass. The O content varied with time from 9.79% ± 0.20% to 6.86% ± 0.31% and from 9.78% ± 0.10% to 9.20% ± 0.06% in the biocrudes obtained in the HTL of the B-1L and B-90L, respectively, whereas this heteroatom content was 30.10% ± 0.05% and 37.29% ± 0.22% in the starting biomass obtained in the same bioreactors. The O concentration was moderately higher for the biocrudes from B-90L, particularly at longer HTL times. These results indicated the presence of decarboxylation reactions during the HTL process that intensified while increasing the reaction time [69]. Furthermore, these values were lower than the O contents in the biocrude obtained in the HTL with other species, such as *Tetraselmis* (12.3%), *Scenedesmus almeriensis* (9.6%), *Chlorella* (30.38%), *Nannochloropsis gaditana* (14.49%) [27] or *Scenedesmus* (10.5%) [70].

The N content in the biocrudes were very similar in all cases and lower (5.42% ± 0.03% to 5.39% ± 0.04%) than the corresponding N amount in the two initial biomasses (8.52% ± 0.01% and 7.50% ± 0.02% for the biomasses obtained in the reactors with 1 and 90 L, respectively) because of the denitrogenation reactions during HTL [27]. These values were comparable to those obtained for the same species at 325 °C (5.62%) [32].

The observed decrease in N and O amounts in the biocrudes is typical of the HTL process of microalgae, which causes the contents of C (74.46% ± 1.41%–77.52% ± 0.03%) and H (9.37% ± 0.09%–9.76% ± 0.07%) to increase with respect to the corresponding C and H amounts in the raw biomass. The concentrations of C and H changed within the range 53.13% ± 0.06%–47.14% ± 0.22% and 7.33% ± 0.01%–6.94% ± 0.03% for the 1 and 90-L cultivated microalgae, respectively. As noted above, the composition of the initial biomass hardly interfered with the C and H contents in the biocrudes, since these values were very similar. Interestingly, the amounts of C and H of the biocrudes described herein were considerably higher than those recently reported for the same microalga [33], where the biocrudes with slightly lower N percentages and appreciably higher concentrations of O were obtained. However, the elemental composition indicates that the biocrudes cannot be directly used as transport fuel, and, consequently, a subsequent hydrotreating stage would be required to reduce the contents of N and O and, therefore, improve the chemical composition of the biocrudes to fulfil the standard regulations concerning the presence of these heteroatoms in the commercial fuel.

The calculated HHV values of the biocrudes for each of the HTL reactions of both biomasses were similar (38.27 ± 1.52–39.71 ± 0.50 MJ/kg), within the range obtained from other microalgae (30–43 MJ/kg) [28] and close to petroleum crude oil (43 MJ/kg) [71]. In addition, the values of HHV were significantly higher with respect to the corresponding values in the initial microalgal biomass (27.25–26.81 MJ/kg), due to the decrease in the O content and the increase of C and H contents in the biocrudes. The calorific value obtained were higher than that found by López-Barreiro et al. (30.3 and 35.9 MJ/kg) for biocrude from HTL produced at similar temperatures with *P. tricornutum* cultured in bubble columns of 25 L [33]. The presence of nutrients from the culture medium in HTL increased the biocrude yields and the contents of C and O in comparison to the HTL of commercial microalga, being responsible for the increase in calorific value [32]. Furthermore, the HHV achieved in the present work were similar to those obtained by Christensen et al. [32] for temperatures around 400 °C.

The energy recovered in the biocrude obtained from the B-90L were lower (38.61–41.98%) than the corresponding values for the B-1L (49.11–51.47%) because of the previously observed lower biocrude yields in the HTL with the B-90L.

The biofuel properties were significantly affected by their H/C, N/C and O/C ratios, as it is well known. The above ratios for the biocrudes, along with those of a petroleum diesel, a biodiesel and a biocrude obtained from the microalgae *N. gaditana* [72], were plotted in Van Krevelen diagrams (Figure 6) for comparison purposes. The results showed that the O/C and N/C ratios decreased in our biocrudes with respect to the initial biomass, which were due to the denitrogenation and decarboxylation reactions taking place during the HTL process [27].

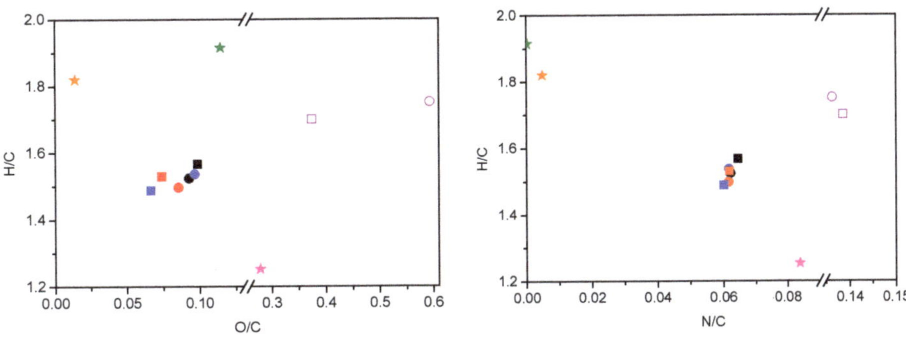

**Figure 6.** Van Krevelen diagrams for biocrude from HTL at 320 °C using B-90L (circles) at different times: 0 (●), 10 (●) and 30 min (●) and B-1L (squares) at different times: 0 (■), 10 (■) and 30 min (■). For (**a,b**), values for B-90L (○) and b-1L (□), reference diesel (★), biodiesel from *N. gaditana* oil (★) and biocrude from *N. gaditana* obtained through HTL at 320 °C and 10 min (★) [72]. B-1L and B-90L: *P. tricornutum* biomass cultivated in 1-L and 90-L bioreactors, respectively.

The O/C ratios of the biocrudes decreased by 82.3% and 85.6% with respect to the initial biomass cultivated in the 1 and 90-L photobioreactors, respectively. The O/C ratios obtained (0.066–0.099) (Figure 6a) were, in all cases, within the usual range found in the literature (0.0–0.3) [73], being lower than those obtained for biodiesel (0.11) and *N. gaditana*-derived biocrude (0.28). A decrease in O/C ratio from 0.099 to 0.066 was observed with time in the HTL of the B-1L. Besides, the lowest O/C ratio (0.066) and, therefore, the closest to the reference diesel (0.014) was achieved in the biocrude obtained in the HTL at 30 min using the B-1L.

In the same way, the N/C ratios of the biocrudes (Figure 6b) were very similar to each other (~0.06), regardless of the biomass used. Therefore, the scaling-up did not seem to affect the fuel properties of the biocrudes significantly. A large decrease of 56.6% and 54.8% in the N/C ratios was observed with respect to the raw biomasses obtained in the 1 and 90-L cultivations, respectively. The N/C of the biocrudes decreased slightly with time, being lower than those found for the biocrude produced with the microalga *N. gaditana* (0.084) but far from the N/C values of the *N. gaditana* biodiesel (0.0004) and the conventional diesel (0.0048) [72]. Furthermore, all the values obtained were between the limits found in the literature (0.056–0.1) [73].

Based on the O/C and N/C ratios of the biocrudes, the biomass of *P. tricornutum* is a promising feedstock for HTL compared to the microalga *N. gaditana*. However, the H/C ratio decreased from 1.70 for the initial microalga to a range between 1.48–1.56 for the biocrudes when using B-1L. A similar decrease occurred with B-90L from 1.75 for the biomass to values around 1.5 for the biocrudes. These decreases were due to the high yields obtained from the aqueous phases. The H/C ratios of the *P. tricornutum*-derived biocrudes were higher than the H/C ratio for a biocrude from *N. gaditana* (1.25) [72]. Therefore, the H/C values of the biocrude from *P. tricornutum* were closer to those of the reference diesel (1.84). All these values were within the range 1.37–1.62 for biocrudes from microalgae with high protein and lipid contents [7].

*3.5. Analysis of the Aqueous Phase*

The aqueous phase is one of the fractions of the HTL process that are part of the liquefied phase [74]. The elemental analysis of these fractions (data not shown) led to a C content between 7.93% ± 0.15% and 7.99% ± 0.06% using the B-1L, whereas the C content was slightly higher (8.43% ± 0.16% to 9.48% ± 0.66%) with the biomass cultivated in the larger bioreactor. The relatively low amounts of C in the aqueous layer were due to the breakdown of macromolecules into smaller ones that are soluble in water. However the heteroatom (N and O) amounts were relatively high, mainly because of the hydrolysis of carbohydrates and proteins and the subsequent decarboxylation and deamination, which produced N and O compounds soluble in water media [28]. The N content in the aqueous phase decreased with time using both microalgae, due to the fact that the deamination reactions are promoted at longer times [27], but the values were lower in the case of the B-1L (11.57% ± 0.25% to 8.13% ± 0.04%) in comparison to those obtained with the B-90L (13.48% ± 0.66% to 12.56% ± 0.91%). The opposite trend with time was observed in the O content in the aqueous phase due to the increased decarboxylation of organic molecules [27]. Thus, the O content increased with the HTL time in both cases, showing lower values (29.8% ± 0.7% to 33.6% ± 0.5%) when the B-1L was used as compared with B-90L (37.5% ± 0.4% to 30.7% ± 0.4%).

The pH values obtained in the aqueous phase were, in all cases, around 8 (7.95–8.77), consistent with the slightly alkaline values found in the literature [29,75], due to the formation of soluble basic compounds in this phase. A slight increase in pH was observed over time for both biomasses, being somewhat higher in the aqueous layer obtained in the HTL of the B-90L (8.33–8.77) than those obtained in the same layer from the B-1L (7.95–8.35). This may be due to the higher amount of N in the aqueous layer in the former, since nitrates were not completely consumed by the microalga cultivated in the 90-L bioreactor. In this sense, the basic composition of the aqueous phase was mainly due to the high portions of $NH_3$ and N compounds [76].

Another key factor in this aqueous phase is the total organic carbon. These values indicated a high formation of new organic compounds soluble in water. The TOC values obtained in this work (880.4–956.2 ppm) were within the bibliographic range (300–1146 ppm) [77].

*3.6. Analysis of the Gaseous Fraction*

Gaseous fractions are also included in the liquefied phase. The gaseous phases obtained by HTL from the different cultivated biomasses had a similar composition. These fractions were mainly composed of $CO_2$ (>80 mol%) in all conditions, as usually reported for the microalgal HTL [74]. The $CO_2$ content of the gaseous fraction was higher than 98 mol% for the HTL reactions using the biomass grown in a 90-L culture. Here, the gas fraction also contained small amounts of linear saturated and monounsaturated light hydrocarbons (C1–C4) (<1 mol%), $H_2$ (<2 mol%) and CO (<1 mol%). In the HTL using B-1L, the $CO_2$ content was lower (79.82–97.19 mol%). In this case, a methane concentration within the range 8–17 mol% was obtained, which may be due to their high content of lipids. Christensen et al. reported a similar methane content when they performed the HTL at higher temperatures with a biomass with lower concentration of lipids [32]. Additionally, the gas fraction contained small amounts of linear saturated and monounsaturated light hydrocarbons (C2–C4) (<1 mol%), $H_2$ (<2 mol%) and CO (<1 mol%). The large concentration of $CO_2$ of the gas fraction from HTL makes this stream suitable for recirculation towards the cultivation stage in photobioreactors, providing the inorganic carbon source necessary for its development [72].

*3.7. Analysis of the Solid Fraction*

Finally, the smaller phase was the solid residue. This phase consisted mainly of ashes, carbon-rich compounds and minority elements present in the biomass (metals and phosphorous). Therefore, this fraction has been used to obtain biochar from the thermochemical process or in use as a fertilizer [27].

## 4. Conclusions

The composition and productivity of the microalga *P. tricornutum* is affected by the variation in daylight hours, the supply of $CO_2$ into the culture and the availability of nutrients such as nitrate in the culture media. Consequently, it is possible to direct the microalgal metabolism to increase the biomass yield and accumulate specific compounds. For instance, microalgae with high biomass and lipid productivities are adequate for the HTL process to obtain a high yield and quality biocrude. The lab-scale culture of *P. tricornutum* produced a higher biomass and lipid productivities with the daylight hours and a supplemental $CO_2$ injection. In addition, the lab study confirmed that the initial amount of nitrates was essential for efficient growth, achieving high biomass and lipid productivities. The biomass generated during the scaled-up culture had a lower biomass and lipid productivities than the corresponding biomass obtained at the lab scale, despite having carried out the experiments at the same operating conditions. In any case, the values obtained are suitable for the following HTL stage and similar to other works. Using high-volume photobioreactors, a greater shielding of the microalga is frequent and was observed in this work, which prevents the light from reaching the microalga cells, reducing the growth of the microalgae and the lipid accumulation. The biocrudes obtained by HTL using B-90L exhibited somewhat lower yields than those obtained from the biomass cultivated at the lab scale (B-1L), because the former biomass presented lower lipid and higher ash contents. However, the heteroatom contents of the biocrudes were similar and lower than those in the corresponding starting microalga in both cases. Nevertheless, the reduction in the heteroatom content was not enough, and a biocrude post-treatment is required in order to reach the regulated values as direct fuel in transport. In summary, both biocrudes had similar characteristics, although the biomass generated during scale-up had a lower biomass and lipid productivities. In this sense, the use of large-scale cultivated wet biomasses for *P. tricornutum* is recommended to make the overall process more economical and to produce enough amounts of biocrude to be use as a biofuel after the post-hydrotreatment stage.

**Author Contributions:** Conceptualization, F.G.W.-S., G.V. and L.F.B.; methodology, I.M.-H. and A.S.-B.; validation, G.V., V.M., and M.S.-F.; formal analysis, G.V. and V.M.; investigation, I.M.H.; resources, I.M.H. and A.S.-B.; data curation, G.V., L.F.B. and V.M.; writing—original draft preparation, I.M.H. and A.S.-B.; writing—review and editing, G.V., L.F.B. and M.S.-F.; visualization, G.V., V.M. and M.S.-F.; supervision, G.V., V.M. and M.S.-F. and funding acquisition, G.V. All authors have read and agreed to the published version of the manuscript.

**Funding:** The authors acknowledge the support of project IND2017/IND financed by Comunidad de Madrid and the company AlgaEnergy for collaborating in this Industrial Doctorate project. AlgaEnergy acknowledges AENA for the concession of its land to locate the Technological Platform for Experimentation with Microalgae (PTEM).

**Conflicts of Interest:** The authors declare no conflict of interest.

## References


1. Akhtar, J.; Amin, N.A.S. A review on process conditions for optimum bio-oil yield in hydrothermal liquefaction of biomass. *Renew. Sustain. Energy Rev.* **2011**, *15*, 1615–1624. [CrossRef]
2. EU. Directive (EU) 2018/2001 of the European Parliament and of the Council on the promotion of the use of energy from renewable sources. *Off. J. Eur. Union* **2018**, *L 328*, 82–209.
3. Gambelli, D.; Alberti, F.; Solfanelli, F.; Vairo, D.; Zanoli, R. Third generation algae biofuels in Italy by 2030: A scenario analysis using Bayesian networks. *Energy Policy* **2017**, *103*, 165–178. [CrossRef]
4. Amaro, H.M.; Guedes, A.C.; Malcata, F.X. Advances and perspectives in using microalgae to produce biodiesel. *Appl. Energy* **2011**, *88*, 3402–3410. [CrossRef]
5. Gonzalez-Fernandez, C.; Sialve, B.; Molinuevo-Salces, B. Anaerobic digestion of microalgal biomass: Challenges, opportunities and research needs. *Bioresour. Technol.* **2015**, *198*, 896–906. [CrossRef]
6. Elliott, D.C.; Biller, P.; Ross, A.B.; Schmidt, A.J.; Jones, S.B. Hydrothermal liquefaction of biomass: Developments from batch to continuous process. *Bioresour. Technol.* **2015**, *178*, 147–156. [CrossRef]
7. Huang, Y.; Chen, Y.; Xie, J.; Liu, H.; Yin, X.; Wu, C. Bio-oil production from hydrothermal liquefaction of high-protein high-ash microalgae including wild *Cyanobacteria* sp. and cultivated *Bacillariophyta* sp. *Fuel* **2016**, *183*, 9–19. [CrossRef]
8. Safi, C.; Zebib, B.; Merah, O.; Pontalier, P.Y.; Vaca-Garcia, C. Morphology, composition, production, processing and applications of *Chlorella vulgaris*: A review. *Renew. Sustain. Energy Rev.* **2014**, *35*, 265–278. [CrossRef]
9. Krzemińska, I.; Pawlik-Skowrońska, B.; Trzcińska, M.; Tys, J. Influence of photoperiods on the growth rate and biomass productivity of green microalgae. *Bioprocess Biosyst. Eng.* **2014**, *37*, 735–741. [CrossRef]
10. Ummalyma, S.B.; Sukumaran, R.K.; Pandey, A. Evaluation of Freshwater Microalgal Isolates for Growth and Oil Production in Seawater Medium. *Waste Biomass Valoriz.* **2020**, *11*, 223–230. [CrossRef]
11. Sánchez-Bayo, A.; Morales, V.; Rodríguez, R.; Vincente, G.; Bautista, L.F. Effect of Cultivation Variables on Biomass Composition and Growth of *Microalgae* and *Cyanobacteria*. *Molecules* **2020**, *25*, 2834. [CrossRef]
12. Jorquera, O.; Kiperstok, A.; Sales, E.A.; Embiruçu, M.; Ghirardi, M.L. Comparative energy life-cycle analyses of microalgal biomass production in open ponds and photobioreactors. *Bioresour. Technol.* **2010**, *101*, 1406–1413. [CrossRef]
13. Juneja, A.; Ceballos, R.M.; Murthy, G.S. Effects of environmental factors and nutrient availability on the biochemical composition of algae for biofuels production: A review. *Energies* **2013**, *6*, 4607–4638. [CrossRef]
14. Harun, R.; Singh, M.; Forde, G.M.; Danquah, M.K. Bioprocess engineering of microalgae to produce a variety of consumer products. *Renew. Sustain. Energy Rev.* **2010**, *14*, 1037–1047. [CrossRef]
15. Haro, P.; Sáez, K.; Gómez, P.I. Physiological plasticity of a Chilean strain of the diatom *Phaeodactylum tricornutum*: The effect of culture conditions on the quantity and quality of lipid production. *J. Appl. Phycol.* **2017**, *29*, 2771–2782. [CrossRef]
16. Hu, Q. Environmental Effects on Cell Composition. In *Handbook of Microalgal Culture*; Blackwell Publishing Ltd.: Oxford, UK, 2003; pp. 83–94.
17. Wang, X.W.; Liang, J.R.; Luo, C.S.; Chen, C.P.; Gao, Y.H. Biomass, total lipid production, and fatty acid composition of the marine diatom *Chaetoceros muelleri* in response to different $CO_2$ levels. *Bioresour. Technol.* **2014**, *161*, 124–130. [CrossRef]
18. Bertozzini, E.; Galluzzi, L.; Ricci, F.; Penna, A.; Magnani, M. Neutral Lipid Content and Biomass Production in *Skeletonema marinoi* (*Bacillariophyceae*) Culture in Response to Nitrate Limitation. *Appl. Biochem. Biotechnol.* **2013**, *170*, 1624–1636. [CrossRef]
19. Tabernero, A.; Martín del Valle, E.M.; Galán, M.A. Evaluating the industrial potential of biodiesel from a microalgae heterotrophic culture: Scale-up and economics. *Biochem. Eng. J.* **2012**, *63*, 104–115. [CrossRef]
20. Molina-Grima, E. Acién Fernández, F.G.; García Camacho, F.; Chisti, Y. Photobioreactors: Light regime, mass transfer, and scaleup. *Prog. Ind. Microbiol.* **1999**, *35*, 231–247.
21. Silva Benavides, A.M.; Torzillo, G.; Kopecký, J.; Masojídek, J. Productivity and biochemical composition of *Phaeodactylum tricornutum* (*Bacillariophyceae*) cultures grown outdoors in tubular photobioreactors and open ponds. *Biomass Bioenergy* **2013**, *54*, 115–122. [CrossRef]

22. Quelhas, P.M.; Trovão, M.; Silva, J.T.; Machado, A.; Santos, T.; Pereira, H.; Varela, J.; Simões, M.; Silva, J.L. Industrial production of *Phaeodactylum tricornutum* for $CO_2$ mitigation: Biomass productivity and photosynthetic efficiency using photobioreactors of different volumes. *J. Appl. Phycol.* **2019**, *31*, 2187–2196. [CrossRef]

23. Branco-Vieira, M.; San Martin, S.; Agurto, C.; Freitas, M.A.V.; Mata, T.M.; Martins, A.A.; Caetano, N. Biochemical characterization of *Phaeodactylum tricornutum* for microalgae-based biorefinery. *Energy Procedia* **2018**, *153*, 466–470. [CrossRef]

24. Sánchez Mirón, A.; Cerón García, M.C.; Contreras Gómez, A.; García Camacho, F.; Molina Grima, E.; Chisti, Y. Shear stress tolerance and biochemical characterization of *Phaeodactylum tricornutum* in quasi steady-state continuous culture in outdoor photobioreactors. *Biochem. Eng. J.* **2003**, *16*, 287–297. [CrossRef]

25. Branco-Vieira, M.; San Martin, S.; Agurto, C.; Dos Santos, M.A.; Freitas, M.A.V.; Mata, T.M.; Martins, A.A.; Caetano, N.S. Potential of *Phaeodactylum tricornutum* for biodiesel production under natural conditions in Chile. *Energies* **2018**, *11*, 54. [CrossRef]

26. Branco-Vieira, M.; San Martin, S.; Agurto, C.; Freitas, M.A.V.; Martins, A.A.; Mata, T.M.; Caetano, N.S. Biotechnological potential of *Phaeodactylum tricornutum* for biorefinery processes. *Fuel* **2020**, *268*, 117357. [CrossRef]

27. Gollakota, A.R.K.; Kishore, N.; Gu, S. A review on hydrothermal liquefaction of biomass. *Renew. Sustain. Energy Rev.* **2018**, *81*, 1378–1392. [CrossRef]

28. Xu, D.; Lin, G.; Guo, S.; Wang, S.; Guo, Y.; Jing, Z. Catalytic hydrothermal liquefaction of algae and upgrading of biocrude: A critical review. *Renew. Sustain. Energy Rev.* **2018**, *97*, 103–118. [CrossRef]

29. López Barreiro, D.; Prins, W.; Ronsse, F.; Brilman, W. Hydrothermal liquefaction (HTL) of microalgae for biofuel production: State of the art review and future prospects. *Biomass Bioenergy* **2013**, *53*, 113–127. [CrossRef]

30. Sheehan, J.D.; Savage, P.E. Modeling the effects of microalga biochemical content on the kinetics and biocrude yields from hydrothermal liquefaction. *Bioresour. Technol.* **2017**, *239*, 144–150. [CrossRef]

31. Gu, X.; Martinez-Fernandez, J.S.; Pang, N.; Fu, X.; Chen, S. Recent development of hydrothermal liquefaction for algal biorefinery. *Renew. Sustain. Energy Rev.* **2020**, *121*, 109707. [CrossRef]

32. Christensen, P.S.; Peng, G.; Vogel, F.; Iversen, B.B. Hydrothermal liquefaction of the microalgae *Phaeodactylum tricornutum*: Impact of reaction conditions on product and elemental distribution. *Energy Fuels* **2014**, *28*, 5792–5803. [CrossRef]

33. López Barreiro, D.; Zamalloa, C.; Boon, N.; Vyverman, W.; Ronsse, F.; Brilman, W.; Prins, W. Influence of strain-specific parameters on hydrothermal liquefaction of microalgae. *Bioresour. Technol.* **2013**, *146*, 463–471. [CrossRef] [PubMed]

34. Mann, J.E.; Myers, J. On pigments, growth, and photosynthesis of *Phaeodactylum tricornutum* 2. *J. Phycol.* **1968**, *4*, 349–355. [CrossRef] [PubMed]

35. Cui, Y.; Thomas-Hall, S.R.; Schenk, P.M. *Phaeodactylum tricornutum* microalgae as a rich source of omega-3 oil: Progress in lipid induction techniques towards industry adoption. *Food Chem.* **2019**, *297*, 124937. [CrossRef]

36. Liu, J.; Hu, Q. Chlorella: Industrial Production of Cell Mass and Chemicals. In *Handbook of Microalgal Culture: Applied Phycology and Biotechnology*, 2nd ed.; John Wiley & Sons, Ltd.: Oxford, UK, 2013; pp. 327–338, ISBN 9781118567166.

37. Shimamura, R.; Watanabe, S.; Sakakura, Y.; Shiho, M.; Kaya, K.; Watanabe, M.M. Development of *Botryococcus* Seed Culture System for Future Mass Culture. *Procedia Environ. Sci.* **2012**, *15*, 80–89. [CrossRef]

38. Gao, B.; Chen, A.; Zhang, W.; Li, A.; Zhang, C. Co-production of lipids, eicosapentaenoic acid, fucoxanthin, and chrysolaminarin by *Phaeodactylum tricornutum* cultured in a flat-plate photobioreactor under varying nitrogen conditions. *J. Ocean Univ. China* **2017**, *16*, 916–924. [CrossRef]

39. Lowry, O.H.; Rosebrough, N.J.; Lewis Farr, A.; Randall, R.J. Protein measurement with the Folin phenol reagent. *Anal. Biochem.* **1951**, *217*, 220–230.

40. DuBois, M.; Gilles, K.A.; Hamilton, J.K.; Rebers, P.A.; Smith, F. Colorimetric method for determination of sugars and related substances. *Anal. Chem.* **1956**, *28*, 350–356. [CrossRef]

41. Bligh, E.G.; Dyer, W.J. A rapid method of total lipid extraction and purification. *Can. J. Biochem. Physiol.* **1959**, *37*, 911–917. [CrossRef]

42. Matos, Â.P.; Feller, R.; Moecke, E.H.S.; Sant'Anna, E.S. Biomass, lipid productivities and fatty acids composition of marine *Nannochloropsis gaditana* cultured in desalination concentrate. *Bioresour. Technol.* **2015**, *197*, 48–55. [CrossRef]

43. Li, H.; Liu, Z.; Zhang, Y.; Li, B.; Lu, H.; Duan, N.; Liu, M.; Zhu, Z.; Si, B. Conversion efficiency and oil quality of low-lipid high-protein and high-lipid low-protein microalgae via hydrothermal liquefaction. *Bioresour. Technol.* **2014**, *154*, 322–329. [CrossRef] [PubMed]

44. Channiwala, S.A.; Parikh, P.P. A unified correlation for estimating HHV of solid, liquid and gaseous fuels. *Fuel* **2002**, *81*, 1051–1063. [CrossRef]

45. Morais, K.C.C.; Vargas, J.V.C.; Mariano, A.B.; Ordonez, J.C.; Kava, V. Sustainable energy via biodiesel production from autotrophic and mixotrophic growth of the microalga *Phaeodactylum tricornutum* in compact photobioreactors. In Proceedings of the 2016 IEEE Conf. Technol. Sustain. SusTech, Phoenix, AZ, USA, 9–11 October 2016; pp. 257–264.

46. Khoeyi, Z.A.; Seyfabadi, J.; Ramezanpour, Z. Effect of light intensity and photoperiod on biomass and fatty acid composition of the microalgae, *Chlorella vulgaris. Aquac. Int.* **2012**, *20*, 41–49. [CrossRef]

47. Wahidin, S.; Idris, A.; Shaleh, S.R.M. The influence of light intensity and photoperiod on the growth and lipid content of microalgae *Nannochloropsis* sp. *Bioresour. Technol.* **2013**, *129*, 7–11. [CrossRef]

48. Sun, X.M.; Ren, L.J.; Zhao, Q.Y.; Ji, X.J.; Huang, H. Microalgae for the production of lipid and carotenoids: A review with focus on stress regulation and adaptation. *Biotechnol. Biofuels* **2018**, *11*, 1–16. [CrossRef]

49. Simionato, D.; Sforza, E.; Corteggiani Carpinelli, E.; Bertucco, A.; Giacometti, G.M.; Morosinotto, T. Acclimation of *Nannochloropsis gaditana* to different illumination regimes: Effects on lipids accumulation. *Bioresour. Technol.* **2011**, *102*, 6026–6032. [CrossRef]

50. Vendruscolo, R.G.; Fagundes, M.B.; Maroneze, M.M.; do Nascimento, T.C.; de Menezes, C.R.; Barin, J.S.; Zepka, L.Q.; Jacob-Lopes, E.; Wagner, R. Scenedesmus obliquus metabolomics: Effect of photoperiods and cell growth phases. *Bioprocess Biosyst. Eng.* **2019**, *42*, 727–739. [CrossRef]

51. Seyfabadi, J.; Ramezanpour, Z.; Khoeyi, Z.A. Protein, fatty acid, and pigment content of *Chlorella vulgaris* under different light regimes. *J. Appl. Phycol.* **2011**, *23*, 721–726. [CrossRef]

52. George, B.; Pancha, I.; Desai, C.; Chokshi, K.; Paliwal, C.; Ghosh, T.; Mishra, S. Effects of different media composition, light intensity and photoperiod on morphology and physiology of freshwater microalgae *Ankistrodesmus falcatus*—A potential strain for bio-fuel production. *Bioresour. Technol.* **2014**, *171*, 367–374. [CrossRef]

53. Biller, P.; Ross, A.B. Potential yields and properties of oil from the hydrothermal liquefaction of microalgae with different biochemical content. *Bioresour. Technol.* **2011**, *102*, 215–225. [CrossRef]

54. Sukarni; Sudjito; Hamidi, N.; Yanuhar, U.; Wardana, I.N.G. Potential and properties of marine microalgae *Nannochloropsis oculata* as biomass fuel feedstock. *Int. J. Energy Environ. Eng.* **2014**, *5*, 279–290. [CrossRef]

55. Liu, H.; Yingquan, C.; Haiping, Y.; Gentili, F.; Söderlind, U.; Wang, X.; Zhang, W.; Chen, H. Conversion of high-ash microalgae through hydrothermal liquefaction. *Sustain. Energy Fuels* **2020**, *4*, 2782–2791. [CrossRef]

56. Chernova, N.; Kiseleva, S.; Vlaskin, M.; Grigorenko, A.; Rafikova, Y. Hydrothermal liquefaction of microalgae for biofuel production: The recycling of nutrients from an aqueous solution after HTL. *IOP Conf. Ser. Mater. Sci. Eng.* **2019**, *564*, 12112. [CrossRef]

57. Ji, M.K.; Yun, H.S.; Hwang, J.H.; Salama, E.S.; Jeon, B.H.; Choi, J. Effect of flue gas $CO_2$ on the growth, carbohydrate and fatty acid composition of a green microalga *Scenedesmus obliquus* for biofuel production. *Environ. Technol.* **2017**, *38*, 2085–2092. [CrossRef] [PubMed]

58. Moheimani, N.R.; Borowitzka, M.A.; Isdepsky, A.; Sing, S. Standard Methods for Measuring Growth of Algae and Their Composition. In *Algae for Biofuels and Energy*; Springer: Dordrecht, The Netherlands, 2013; pp. 265–284, ISBN 9789400754799.

59. Mohsenpour, S.F.; Willoughby, N. Effect of $CO_2$ aeration on cultivation of microalgae in luminescent photobioreactors. *Biomass Bioenergy* **2016**, *85*, 168–177. [CrossRef]

60. Widjaja, A. Lipid Production from Microalgae as a Promising Candidate for Biodiesel Production. *Makara Technol. Ser.* **2010**, *13*, 47–51. [CrossRef]

61. Lourenço, S.; Barbarino, E.; Filho, J.; Schinke, K.; Aidar, E. Effects of Different Nitrogen Sources on the Growth and Biochemical Profile of 10 Marine Microalgae in Batch Culture: An Evaluation for Aquaculture. *Phycologia* **2002**, *41*, 158–168. [CrossRef]

62. Osborne, B.; Geider, R. Effect of nitrate-nitrogen limitation on photosynthesis of the diatom *Phaeodactylum tricornutum* Bohlin (*Bacillariophyceae*). *Plant. Cell Environ.* **2006**, *9*, 617–625. [CrossRef]

63. San Pedro, A.; González-López, C.V.; Acién, F.G.; Molina-Grima, E. Marine microalgae selection and culture conditions optimization for biodiesel production. *Bioresour. Technol.* **2013**, *134*, 353–361. [CrossRef]

64. Ho, S.H.; Chen, C.Y.; Chang, J.S. Effect of light intensity and nitrogen starvation on $CO_2$ fixation and lipid/carbohydrate production of an indigenous microalga *Scenedesmus obliquus* CNW-N. *Bioresour. Technol.* **2012**, *113*, 244–252. [CrossRef]

65. Ho, S.H.; Huang, S.W.; Chen, C.Y.; Hasunuma, T.; Kondo, A.; Chang, J.S. Characterization and optimization of carbohydrate production from an indigenous microalga *Chlorella vulgaris* FSP-E. *Bioresour. Technol.* **2013**, *135*, 157–165. [CrossRef]

66. Fernández Sevilla, J.M.; Cerón García, M.C.; Sánchez Mirón, A.; Belarbi, E.H.; García Camacho, F.; Molina Grima, E. Pilot-plant-scale outdoor mixotrophic cultures of *Phaeodactylum tricornutum* using glycerol in vertical bubble column and airlift photobioreactors: Studies in fed-batch mode. *Biotechnol. Prog.* **2004**, *20*, 728–736. [CrossRef] [PubMed]

67. Xu, D.; Savage, P.E. Effect of reaction time and algae loading on water-soluble and insoluble biocrude fractions from hydrothermal liquefaction of algae. *Algal Res.* **2015**, *12*, 60–67. [CrossRef]

68. Valdez, P.J.; Savage, P.E. A reaction network for the hydrothermal liquefaction of *Nannochloropsis* sp. *Algal* **2013**, *2*, 416–425. [CrossRef]

69. Toor, S.S.; Rosendahl, L.; Rudolf, A. Hydrothermal liquefaction of biomass: A review of subcritical water technologies. *Energy* **2011**, *36*, 2328–2342. [CrossRef]

70. Vardon, D.R.; Sharma, B.K.; Blazina, G.V.; Rajagopalan, K.; Strathmann, T.J. Thermochemical conversion of raw and defatted algal biomass via hydrothermal liquefaction and slow pyrolysis. *Bioresour. Technol.* **2012**, *109*, 178–187. [CrossRef]

71. Brown, T.M.; Duan, P.; Savage, P.E. Hydrothermal liquefaction and gasification of *Nannochloropsis* sp. *Energy Fuels* **2010**, *24*, 3639–3646. [CrossRef]

72. Sánchez-Bayo, A.; Rodríguez, R.; Morales, V.; Nasirian, N.; Bautista, L.F.; Vicente, G. Hydrothermal liquefaction of microalga using metal oxide catalyst. *Processes* **2020**, *8*, 15. [CrossRef]

73. Tian, C.; Li, B.; Liu, Z.; Zhang, Y.; Lu, H. Hydrothermal liquefaction for algal biorefinery: A critical review. *Renew. Sustain. Energy Rev.* **2014**, *38*, 933–950. [CrossRef]

74. Filipe, R.; Hu, Y.; Shui, H.; Charles, C. Biomass and Bioenergy Hydrothermal liquefaction of biomass to fuels and value-added chemicals: Products applications and challenges to develop large-scale operations. *Biomass Bioenergy* **2020**, *135*, 105510.

75. Kumar, M.; Olajire Oyedun, A.; Kumar, A. A review on the current status of various hydrothermal technologies on biomass feedstock. *Renew. Sustain. Energy Rev.* **2018**, *81*, 1742–1770. [CrossRef]

76. Gai, C.; Zhang, Y.; Chen, W.T.; Zhou, Y.; Schideman, L.; Zhang, P.; Tommaso, G.; Kuo, C.T.; Dong, Y. Characterization of aqueous phase from the hydrothermal liquefaction of *Chlorella pyrenoidosa*. *Bioresour. Technol.* **2015**, *184*, 328–335. [CrossRef] [PubMed]

77. Ross, A.B.; Biller, P.; Kubacki, M.L.; Li, H.; Lea-Langton, A.; Jones, J.M. Hydrothermal processing of microalgae using alkali and organic acids. *Fuel* **2010**, *89*, 2234–2243. [CrossRef]

 *processes*

*Article*

# Numerical Investigation of Fluid Flow and In-Cylinder Air Flow Characteristics for Higher Viscosity Fuel Applications

Mohd Fadzli Hamid [1], Mohamad Yusof Idroas [1], Shukriwani Sa'ad [2], Teoh Yew Heng [1], Sharzali Che Mat [3], Zainal Alimuddin Zainal Alauddin [1], Khairul Akmal Shamsuddin [4], Raa Khimi Shuib [5] and Muhammad Khalil Abdullah [5,*]

1. School of Mechanical Engineering, Universiti Sains Malaysia, Engineering Campus, Seri Ampangan, Nibong Tebal 14300, Malaysia; Mohd_Fadzli@outlook.com (M.F.H.); meyusof@usm.my (M.Y.I.); yewhengteoh@usm.my (T.Y.H.); mezainal@usm.my (Z.A.Z.A.)
2. Faculty of Computer, Media and Technology Management, TATI University College, Kemaman 24000, Malaysia; shukriwani@tatiuc.edu.my
3. Faculty of Mechanical Engineering, Universiti Teknologi MARACawangan Pulau Pinang, Permatang Pauh 13500, Malaysia; sharzali.chemat@uitm.edu.my
4. Mechanical Section, Universiti Kuala Lumpur, Malaysian Spanish Institute, Kulim 09000, Malaysia; khairulakmal@unikl.edu.my
5. School of Material and Mineral Resources Engineering, Universiti Sains Malaysia, Engineering Campus, Seri Ampangan, Nibong Tebal 14300, Malaysia; raakhimi@usm.my
* Correspondence: mkhalil@usm.my

Received: 6 January 2020; Accepted: 13 February 2020; Published: 8 April 2020

**Abstract:** Generally, the compression ignition (CI) engine that runs with emulsified biofuel (EB) or higher viscosity fuel experiences inferior performance and a higher emission compared to petro diesel engines. The modification is necessary to standard engine level in order to realize its application. This paper proposes a guide vane design (GVD), which needs to be installed in the intake manifold, is incorporated with shallow depth re-entrance combustion chamber (SCC) pistons. This will organize and develop proper in-cylinder airflow to promote better diffusion, evaporation and combustion processes. The model of GVD and SCC piston was designed using SolidWorks 2017; while ANSYS Fluent version 15 was utilized to run a 3D analysis of the cold flow IC engine. In this research, seven designs of GVD with the number of vanes varied from two to eight vanes (V2–V8) are used. The four-vane model (V4) has shown an excellent turbulent flow as well as swirl, tumble and cross tumble ratios in the fuel-injected region compared to other designs. This is indispensable to break up heavier fuel molecules of EB to mix with the air that will eventually improve engine performance.

**Keywords:** piston bowl; alternative fuel; vanes; emulsified biofuel; biofuel

---

## 1. Introduction

A diesel engine is one of the most indispensable power generation systems and is mainly used in industrial, public transportation, power generation, heavy-duty machinery, and agricultural applications due to their higher fuel-conversion efficiency, power output, torque and reliability compared to gasoline engines [1]. Furthermore, the emissions such as carbon monoxide (CO), hydrocarbon (HC) and carbon dioxide ($CO_2$) from a diesel engine is much lesser compared to the gasoline engine emissions. However, diesel engine remains to be an important source of pollution as their usage leads to release of nitrogen oxides (NOx), black smoke, particulate matter (PM) and sulfur oxides (SOx) that are detrimental to both environment and human health [2]. In fact, the emissions from diesel engines have been classified as carcinogenic by the International Agency for Research on Cancer (IARC). It is based on sufficient evidence that high exposure to diesel emissions can be a risk of lung cancer [3], soot emissions can cause cardiovascular diseases [4], while NOx emissions can cause ground level ozone [5], smog [6] and acid rain [7].

On the other hand, fluctuating petroleum prices, fossil fuel depletion, energy demand escalations and stringent emission regulations have intensified the look-out of the scientific community for alternative renewable fuels in place of the existing fossil fuel. Therefore, non-conventional types of fuel, made from biological resources such as biofuel and biodiesel, have been researched. These studied aimed to tackle the problems that arise due to the comparable properties with that of fossil fuels. However, this alternative fuel not only has high viscosity and boiling point, but also low volatility and calorific values. Biofuel, particularly refined palm oil (RPO) is readily made, safe to be handled and stored and is renewable [8]. However, direct use of this oil degrades the engine performance if operated for a prolonged period due to their high viscosity and low volatility, which causes filters, fuel lines and injectors to be clogged. In addition, engine problems such as piston ring sticking, carbon deposit build-up and lubricating oil thickening were also observed [8], therefore it would be necessary to overhaul, repair and replace some parts of the diesel engine.

The prominent factors that control and govern the combustion process depend on the air motion within the engine, charge temperature, compression ratio, spray structure, burning rate, piston bowl geometry, injection strategies, auto ignition fuels, and fuel molecular structure [9]. There is a significant influence of piston bowl geometry with respect to the combustion and the amount of emission as it strongly affects the mixing of air and fuel prior to the start of injection (SOI). Jing Li et al. [10] investigated the effect of piston bowl geometry on combustion and emissions using a high viscous fuel (biodiesel). The studies inferred that at low engine speed, the shallow depth re-entrance combustion chamber (SCC) piston exhibited better engine performance. Using computational fluid dynamics (CFD) analysis, Hamid et al. [11] discovered that the SCC piston had an ability to generate high swirl, tumble and cross tumble ratios. They also observed that the turbulence kinetic energy was increased, had a well-organized flow and a better air fuel mixture, especially for high viscous fuel applications. Consequently, the combustion efficiency had been improved and reduced incomplete combustion. In addition, the SCC piston is a preferable design since it can be run with high viscosity alternative fuel in the diesel engine.

The inherent long carbon chains become a limitation to biofuels as their nature results in high viscosity and density. The in-cylinder airflow rate was low and produced an adverse effect, as previously mentioned, because of the physiochemical nature of the alternative fuel, that produced an undesired injection profile which degraded cone angle and increased the length of penetration spraying due to high viscosity. Several measures have been taken to lower the viscosity of the alternative fuel for producing an injection profile closed to the diesel, e.g., blending alternative fuel with fossil fuel [12], preheating the alternative fuel [13], emulsification of alternative fuel [14], nano-fluid additives [15], mixing with low and high viscosity biofuel [16] and adjusting injection timing [17–19]. In general, these techniques partially minimized those aforementioned issues, but are still lacking certain requisites compared to the engines using petrol and diesel fuels.

Theoretically, when the in-cylinder airflow rate increased, the swirl, tumble, cross tumble ratios and kinetic energy of turbulence also increases and accelerates the in-cylinder evaporation and diffusion. These effects generally will enhance engine performance such as higher engine power and lower brake specific fuel consumption (BSFC). To enhance and improve the in-cylinder airflow rate and its characteristics, several strategies can be implemented, such as redesigning the airflow intake manifold, modifying the piston bowl and guide vanes to guide the inlet airflow [20].

Therefore, this research will investigate numerically the effect of the numbers of GVD incorporated with the SCC pistons to enhance and organize the in-cylinder airflow rate and its characteristics. Based on the previous literature, the geometry of the guide vanes consists of four main parameters: vane number, angle, height and length. Nevertheless, this research is limited to the numbers of GVD regardless of other parameters as including several parameters tend to make the analysis complex and increases the computational time consumption. Therefore, the remaining parameters were kept constant according to the previous researchers [21,22]. The design model of the SCC piston is modification suited on the YANMAR L70 engine specifications. The details of the designs is described in the following sections.

## 2. Methodology

### 2.1. Computer Simulation

There are four main steps in order to investigate the airflow rate and its characteristics, namely to draw the SCC piston and GVD, mesh the parts, define their boundary condition and analyze the cold flow IC engine. SolidWorks 2017 and ANSYS-FLUENT v15 software were utilized to prepare the model and analyze the in-cylinder airflow in the transient engine cycle without combustion. The complete details of the computer simulation setup is described in the following section.

### 2.2. Guide Vane Design (GVD)

GVD is designed specifically to enhance the air velocity due to swirling flow generated. The geometry of GVD consists of vanes number (N), height (Hv), length (*l*) and angle (θ) which are illustrated in Figure 1 and the specification of GVD is shown in Table 1. GVD dimensions play an important parameter in generating optimized in-cylinder air flow characteristics. They will guide the intake airflow into the combustion chamber, generate turbulence phenomena and sustain swirl momentum until the end of the expansion stroke. The increasing swirl flow will produce high convective heat transfer coefficient inside the combustion chamber [23]. Nevertheless, if the number of vanes increases, it tends to obstruct the airflow and affect volumetric efficiency [24]. By considering this, our research has a limit to eight guide vanes (V8) starting with a base (without vane). The vane twist angle was fixed at 35° angle.

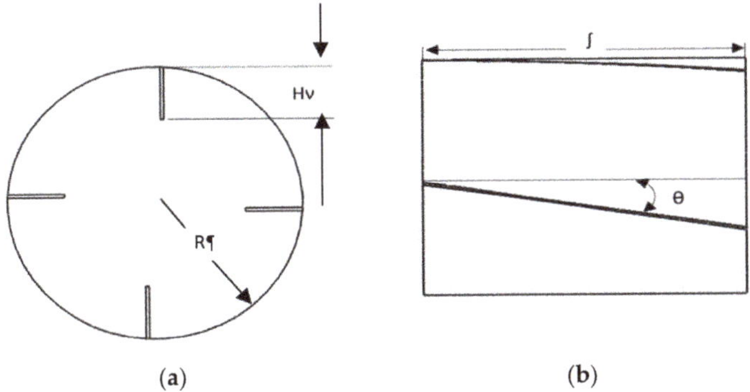


**Figure 1.** GVD design. (**a**) Front view; (**b**) Side view**.

**Table 1.** Specification of guide vane design (GVD).

| No | Parameter | Value |
|---|---|---|
| 1 | Number of Vanes (N) | Base, 2,3,4,5,6,7,8 |
| 2 | Vane Length (*l*) | 30 mm |
| 3 | Width of vane | 0.5 mm |
| 4 | Vane Height (*Hv*) | 0.6 R |
| 5 | Vane twist angle (θ) | 35° |
| 6 | Angle of incidence | 90° |

### 2.3. Shallow Depth Re-Entrance Combustion Chamber (SCC)

To improve the mixing of air and fuel, most of the important modifications were performed on the engine design. The nature of formation of mixture in the engine cylinder is predominantly dependent on the shape of the combustion chamber and the piston bowl design. Running the emulsified biofuel with high viscosity will deteriorate the injection profile. Hwang et al. [25] carried out studies on the injection profile using waste cooking oil biodiesel. They observed that the penetration length was longer, and the cone angle was shorter compared to petrol and diesel fuels. To mitigate these issues, many researchers [26] suggested that the piston bowl design needs to be modified for smooth running with high viscosity fuel. They discovered that the SCC piston bowl design (as shown in Figure 2) is recommended since it can organize the airflow well; swirl ratio ($Rs$), tumble ratio ($R_T$), cross tumble ratio ($R_{CT}$) and break the penetration length of injection to facilitate enhanced mixing with the surrounding air. Therefore, the effect of the GVD and SCC piston combination will be focused on this research.

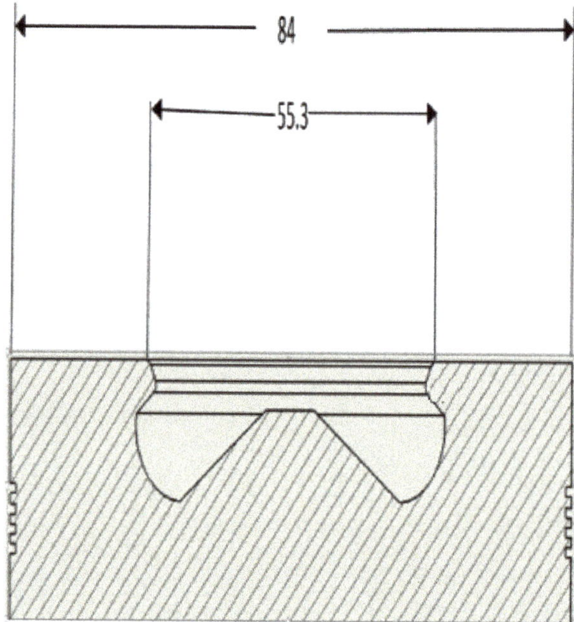


**Figure 2.** Schematic diagram for shallow depth re-entrance combustion chamber (SCC) piston bowl geometry design (all in mm) [11]. Reproduced with permission from Hamid et al., (Renewable Energy); published by (Elsevier), (2018).

### 2.4. Engine Model

The geometry of the engine model was adapted from the experimental Yanmar model type L70AE-DTM CI generator with a four-stroke direct injection, vertical cylinder, one cylinder, one intake valve and one exhaust valve. The technical specification of the engine is given in Table 2.

**Table 2.** Technical specifications of Yanmar L70AE CI engine generator.

| Engine Parameters | Details |
| --- | --- |
| Engine Model | Yanmar L70AE |
| Bore | 78 mm |
| Stroke | 62 mm |
| Compression ratio | 19.1 |
| Number of cylinder | 1 |
| Engine weight | 36 kg |
| Type of injection | Direct inject |
| Fuel injection pressure | 19.6 Mpa |
| Displacement | 0.296 L |
| High idle speed | 3600 rpm |
| Max. rated power | 4.9 kW @ 3600 rpm |
| Injection timing | $14° \pm 1°$ BTDC |
| Intake | Naturally aspirated |
| Cooling | Forced air |
| Lubrication | Forced lubrication with trochoid pump |
| Direct of rotation | Counter clockwise |
| Starting system | Electric start/Recoil start |
| IVO, IVC | 155°, 59° |
| EVO, EVC | −59°, 155° |

## 3. Simulation Setting

The engine geometry of Yanmar L70AE-DTM was modelled using SolidWorks 2017. The GVD, intake runner, exhaust runner, cylinder, intake valve and the exhaust valve have been modelled separately and assembled together as illustrated in Figure 3. The assembled model of the engine was exported to the CFD software, namely ANSYS-FLUENT v15. The software was used to construct a solver, comprising of mathematical computations that simulates and analyses cold flow. To compute the parameters representing fluid flow, both valves were set as solid domains as in reality while the others were set as a fluid flow domain. The moving boundaries were the main challenges in order to simulate a 3D IC engine such as piston bowl, valve and cylinder. The moving grid and remapping mesh were the common strategies used by many researchers [27,28]. The mesh generation on this research was based on assembly level meshing technique. The CFD simulation setting was based according to the cold flow IC engine published by ANSYS Inc. [29].

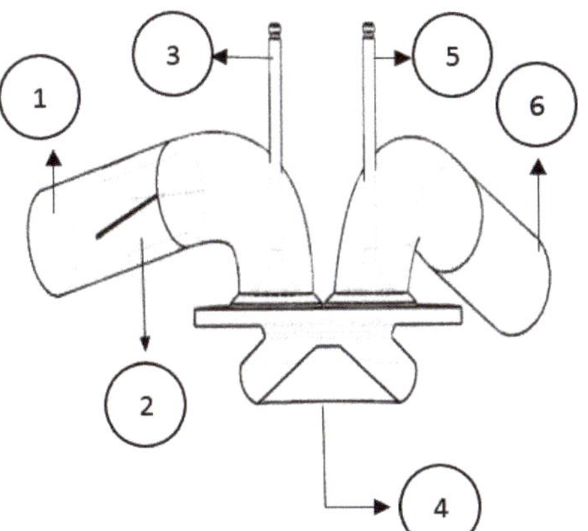



| No | Name |
| --- | --- |
| 1. | Intake runner |
| 2. | GVD |
| 3. | Intake valve |
| 4. | SCC piston |
| 5. | Exhaust valve |
| 6. | Exhaust runner |

**Figure 3.** Schematic diagram of modelling engine Yanmar L70AE-DTM configuration.

The equations governing the fluid flow which formed the basis for simulation are the conservation of mass, momentum and energy (energy equation) [30,31]. The conservation of mass is derived based on the control volume and the corresponding differential Equation [32]. It is written as:

$$\frac{\partial p}{\partial t} + \nabla(\rho U) = 0 \tag{1}$$

$\rho$ is the fluid density and $U$ is three-dimensional flow velocities in the $x$, $y$ and $z$ directions.

The conservation of linear momentum is derived based on the Newton's second law where the surface forces are the control volumes and forces are the body of the control volume. It can be written as:

$$\frac{\partial(\rho U)}{\partial t} + \nabla(\rho U \times U) = \nabla p - \nabla \tau + S_M \tag{2}$$

$p$ is the fluid pressure, $\tau$ is the strain rate and $S_M$ is a momentum source. This equation is also known as the Navier–Stoke equation [33].

The rate of energy change inside the fluid element is also known as the energy equation and it is given by:

$$\frac{\partial(\rho h_{tot})}{\partial t} - \frac{\partial p}{\partial t} - \nabla(\rho U h_{tot}) = \nabla(\lambda \nabla T) + \nabla(U \cdot \tau) + U \cdot S_M \tag{3}$$

$h_{tot}$ and $\lambda$ are the total enthalpy and thermal conductivity, respectively.

Shear Stress Transport (SST) is a two-equation eddy-viscosity model that was used in this numerical study. This model is a combination of the $k$-$\omega$ and $k$-$\varepsilon$ turbulence models. It is a low Reynolds number model. It resolution has similar requirements to the $k$-$\omega$ model and the low Reynolds number $k$-$\varepsilon$ turbulence model, but its formulation abolishes some weakness displayed by pure $k$-$\omega$ and $k$-$\varepsilon$ turbulence models. While the $k$-$\omega$ model pertains to the inner boundary layer, the $k$-$\varepsilon$ model plays a role in the outer region. The combinational model overcame the limitation of shear stress until 5% turbulence intensity in the adverse gradient region, wherein it is sufficient to consider fully developed flow turbulence. The conditions of temperature and pressure were set at 300 K and 1 atm respectively. The detailed information on the setting used for the models and their limitations can be referred to ANSYS FLUENT v15–Solver Theory Guide [26].

On the basis of the physical boundary conditions of the engine, the simulation was carried out in two distinct phases of analysis; intake analysis and intake port analysis. Intake analysis is related to the intake runner, while intake port analysis corresponds to the clearance volume prior to the downward movement of the piston and intake runner in the $y$-direction and drawing of air into the cylinder. The components that are not applicable in this analysis were suppressed since there was no contribution to the results and to reduce the computation time during calculations.

The results of the intake analysis were transferred to the compression and expansion analysis for further simulations. During the compression and expansion analysis, the intake and exhaust valves remained closed while air was compressed during the upward movement of the piston towards the top dead center (TDC) and expanded during the downward movement towards the bottom dead center (BDC). The only domain applicable during this analysis was the cylinder volume, where the volume changed due to the motion of the piston being progressed up and down via the moving mesh. To retain the stability of the simulation progress, the time step must be small enough to simulate the moving mesh.

## 4. Results and Discussion

The in-cylinder airflow rate and its characteristics are well-recognized such that they can enhance the evaporation, diffusion and combustion process. As a result, it can be utilized for emulsified biofuel to improve engine performance and reduce engine emissions. Results and discussion will focus on the events taking place within the fuel injection period or ignition delay, the time difference between start of injection (SOI) and start of combustion (SOC). Due to the default setting of Yanmar L70AE-DTM (manufacturer setting), the fuel injection is at 14° before TDC, therefore the results will be covered at a crank angle (CA) from 346° SOI until 352° SOC.

### 4.1. Numerical Validations

The purpose of the experimental setup is to validate the numerical simulation, which was carried out in the test rig of single a cylinder of Yanmar engine as shown in the Figure 4. During the experiment setup, a high sensitivity water-cooled precision type sensor Kistler 7061B, magnetic pickup shaft encoder and TDC position optical sensor were used to measure the in-cylinder pressure and crank angle data. A sensor of type 7061B was screwed directly into the standard M14 hole. The position of the sensor was mounted near the valve for better accuracy of the values. K-type thermocouple was used to measure the boundary temperatures, at the intake and exhaust boundaries, and the thermocouple was positioned as close as possible to the cylinder head. The SCC piston was used to validate the experimental data.

**Figure 4.** The engine Yanmar L70 setup test rig.

Figure 5 shows the in-cylinder pressure against the crank angle (θ) diagram from the simulation and experiment results without combustion at a rotational speed of 2000 rpm. The data measurement for the intake temperature and pressure are 302 K and 1.02 bar, respectively. The graph shows the variation of pressure between 0° to 540°. Based on the figure, a reasonable agreement between experimental and numerical results with a slight difference of about 7% peak pressure is witnessed. This minor deviation might be due to the gas seepage from the cylinder into the crankcase.

### 4.2. Grid Independence Test (GIT)

Grid independence test (GIT) can be described as optimum estimation on the numerical accuracy of the computed results that rely on a number of elements. The computational domain for the numerical calculation covers the valves and intake port, cylinder head and piston-bowl. It is necessary to conduct the test as it will affect the computation time and cost. The cell size typically ranges from coarse to ultra and the number of elements was set in between 100–400 k, where about half of the cells used to generate the mesh at the cylinder head and piston-bowl for the sake of grid sensitivity and reasonable computation time. The hexahedral mesh has been adopted in this mesh generation because of better accuracy and stability compared to the tetrahedral cells. Table 3 shows the summary of GIT and it was found that case 3 had shown an appropriate meshing grid due to less nominal deviation. Case 3 shows the optimum mesh number, if increase the elements of mesh number, the pressure shows similar pressure value with case 3. Therefore, case 3 was chosen for further analysis. Figure 6 depicts the dynamic mesh of SCC grid piston type during intake, compression and exhaust at different crank angles.

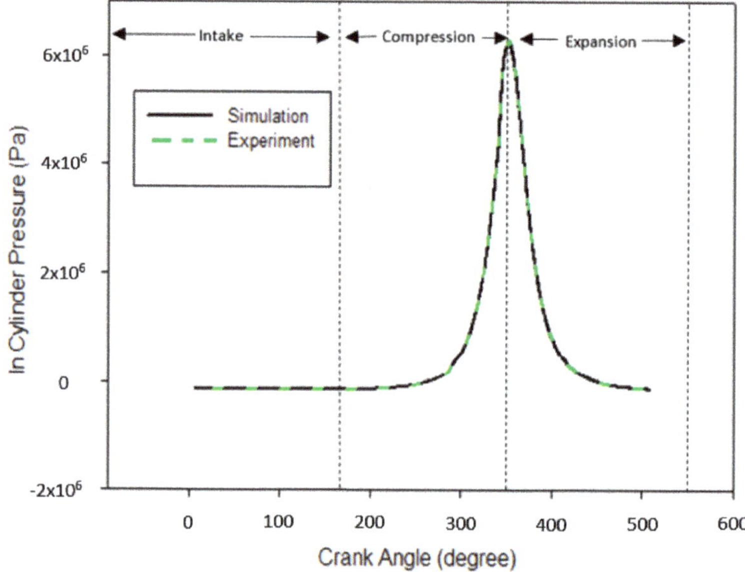



**Figure 5.** In-cylinder pressure against crank angle (degree).

**Table 3.** Summary of grid independence test (GIT).

| Case | 1 | 2 | 3 | 4 | 5 |
| --- | --- | --- | --- | --- | --- |
| Elements average cylinder | 102,343 | 254,223 | 302,309 | 371,424 | 447,573 |
| Pressure | $2.082 \times 10^6$ | $3.099 \times 10^6$ | $3.153 \times 10^6$ | $3.153 \times 10^6$ | $3.153 \times 10^6$ |

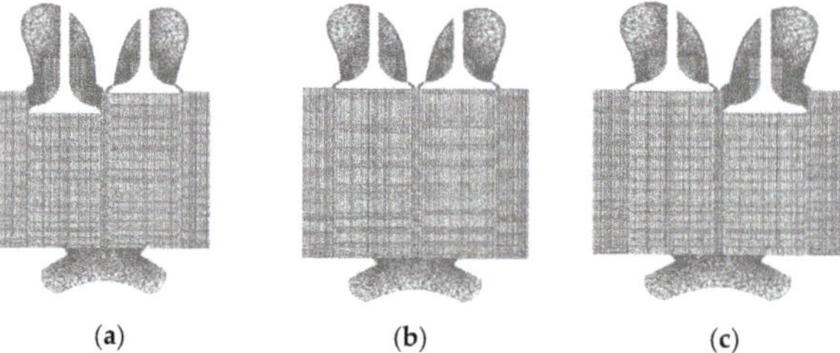



**Figure 6.** Computational domain during (**a**) intake CA at 45° (**b**) compression CA at 300° (**c**) exhaust CA at 630°.

### 4.3. Turbulence Kinetic Energy (TKE)

The simulation results of TKE are presented in Figure 7. TKE, a measure of turbulence, is defined as the mean kinetic energy per unit mass associated with eddies in a turbulent flow. As can be seen from the graph, before the piston reached TDC, TKE declined linearly along with the piston movement. This is due to the reduction of the cylinder volume during the motion of the piston towards TDC. This finding is well-agreed by Payri et al. [34]. Prasad et al. [35] studied by varying the design of piston bowl using the AVL Fire CFD software and found that the TKE trend shows a consistent decline due to the limitation of the in-cylinder airflow movement. Another trend that was noticed is TKE and the vane numbers had no linear relationship as the nature of relationship can be observed to be mixed and inconsistent. It means that the increase of the vane numbers does not necessarily improve TKE. However, this trend was already discovered by Miles [36] during his studies on the influence of in-cylinder airflow on using a baffle-type swirl generator to choke the intake manifold. From the graph, all of GVD's models improved the magnitude of TKE compared to the base model. The V4 model had the highest TKE as compared to the others of vane numbers. The difference between the base model and V4 were approximately 21%. The second highest TKE was the V2 model as it had an improvement of approximately 16% compared to the base engine. This might be due to the airflow obstruction; too many numbers of vanes unable to properly guide the airflow and possibly to limiting airflow efficiency.

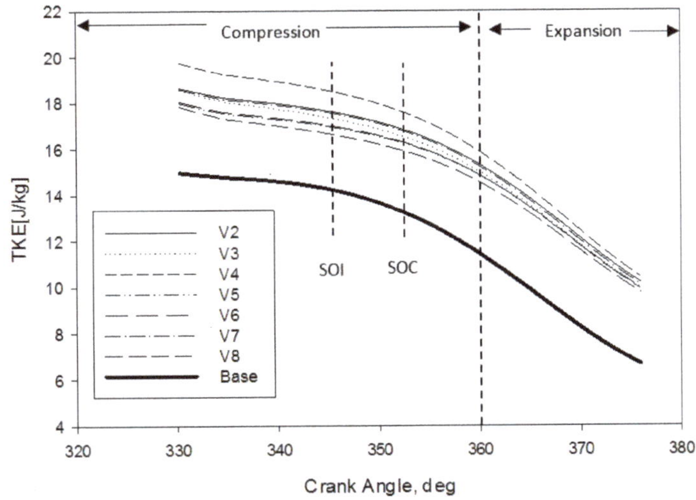


**Figure 7.** TKE against crank angle ($\theta$).

### 4.4. In-Cylinder Swirl, Tumble and Cross Tumble Ratio

One of the key factors that determine the large scale mixing of air-fuel during the intake and compression stroke is the in-cylinder airflow motion in the combustion chamber, especially in the velocity streamline. The three main components that affect the in-cylinder airflow motion are $R_S$, $R_T$, $R_{CT}$ are calculated from the crank angle engine stroke to the above mentioned designs. Figure 8 shows the orientation diagram for the definition of $R_S$, $R_T$ and $R_{CT}$; their directions will be discussed accordingly.

Figure 9 shows the $R_S$ of in-cylinder for different numbers of GVD against the crank angle before TDC. $R_S$ is defined as a rotation airflow around the swirl axis relative to the flow (around the cylinder axis) [37] and used to promote rapid combustion. High-magnitude in-cylinder $R_S$ encourages better air fuel mixing, breaks up more fuel molecules and improves engine performance. As can be seen from the graph, the $R_S$ increases due to higher airflow acceleration and is proportional to the crank angle; thus, the angular momentum is conserved at the time of compression before approaching TDC. According to the graph, during the expansion process, the declining trend is due to the reversal flow exiting from the piston and wall friction. However, the main focus was on SOI at 346° and SOC at 352°. The results imply that the utilization of GVD had improved the swirl flow generally and 4 vanes (V4) had shown about 35% swirl flow improvement compared to the original baseline in-cylinder swirl flow. Kim et al. [38] illustrated that from their photographic results, the flame size without the swirl control valve (SCV) was smaller than with SCV at 1.6° crank angle after SOI, due to strong swirl flow. Therefore, it's confirmed that enhancing the swirl flow will benefit engine operation and can be manipulated for high viscous fuel, e.g., emulsified biofuel.

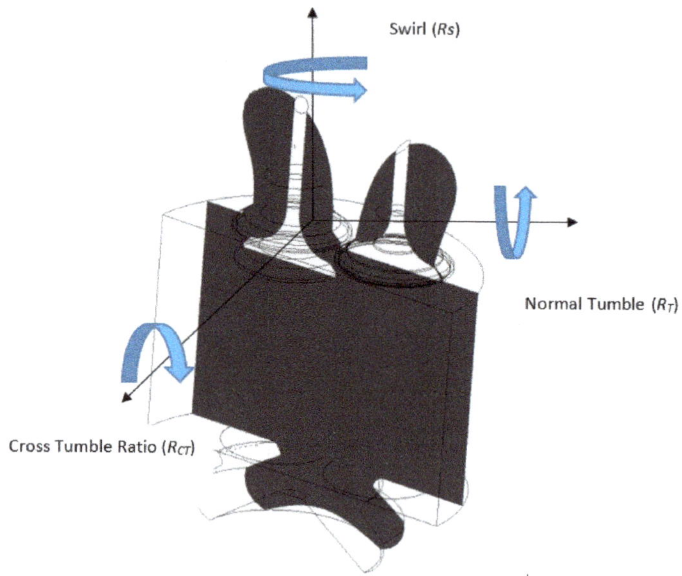



**Figure 8.** Orientation diagram for swirl, tumble and cross tumble components.

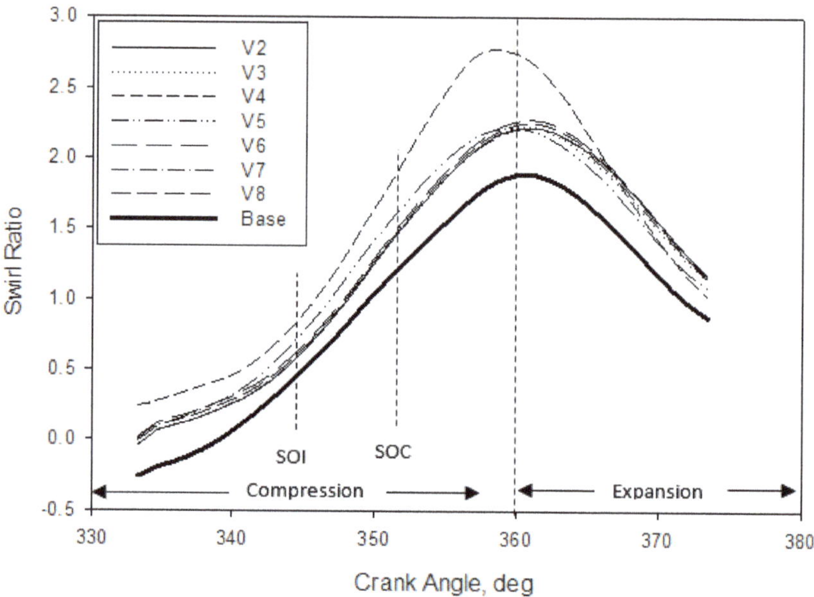



**Figure 9.** Swirl ratio against crank angle (θ).

Figure 10 shows the $R_{CT}$ for different numbers of GVD against crank angle before TDC. $R_{CT}$ is defined as a rotational ratio of airflow on the cross tumble axis relative to the other axes [39]. As in $R_S$, the negative or positive value of $R_{CT}$ is neglected as it is arbitrary, and depends on the magnitude obtained. It can be seen that the installation of GVD has increased the magnitude ratio of $R_{CT}$. Nevertheless, $R_{CT}$ has a close correlation between $R_S$ and $R_T$. Khalighi et al. [40] noted that in order to maximize $R_{CT}$ and $R_S$ and $R_T$ need to be maximized. Rabault et al. [41] reported that enhancing $R_{CT}$ assisted the premixed of air fuel mixture to become much better, indirectly achieve good combustion. As previously mentioned, emulsified biofuel increases the penetration length and shortens the cone angle. Therefore, with the correct number of vanes, $R_{CT}$ can significantly break up the length of penetration; which becomes wider during injection into the combustion chamber. Based on the figure, it can be clearly seen that the design of V4 is an appropriate GVD for producing homogeneous mixture during compression thus, the lateral flow of air is improved within the cylinder.

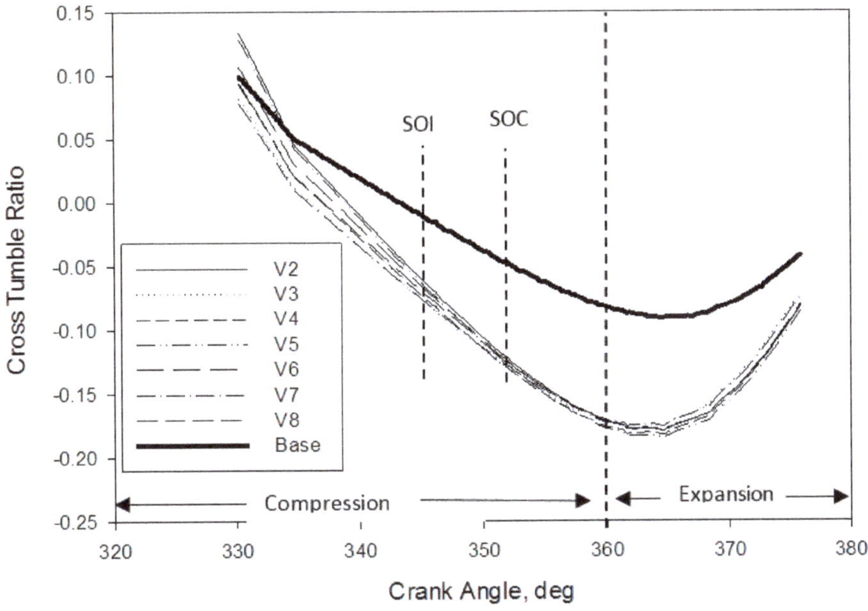


**Figure 10.** Cross tumble ratio against crank angle (θ).

Figure 11 shows the $R_T$ for different numbers of GVD against crank angle before TDC. $R_T$ is defined as a ratio of rotational airflow around the tumble axis (orthogonal to the cylinder axis) [42]. The function of $R_T$ is generally to aid the flow of the molecular fuel to the wider area of the combustion chamber. $R_T$ also helps to maintain a uniform distribution of flow along the piston bowl. As can be seen from the graph, the high-magnitude of tumbles is clearly visible on all GVD models compared to the base model. Again, V4 and V3 of GVD show greater $R_T$ occurrence in the area of SOI and SOC. Therefore, V4 and V3 are able to generate higher $R_T$, which can enhance the mixing process and improve engine operation which is fuelled with high viscous fuel. In addition, it also implies that incorporating GVD and SCC piston facilitates a strong lateral flow of air within the cylinder. The role of lateral flow function is to assist in spreading the atomization molecule fuel throughout the piston bowl, thereby offering sufficient time to combust and avoid deposition of residual carbon deposit. This has shown an agreement with the findings of Payri et al. [34], wherein it was inferred that the piston geometry had little influence on $R_T$ during the compression stroke. However, the piston bowl design and stronger in-cylinder lateral airflow had a significant effect; especially in the mean velocity field and turbulent zone near TDC, approaching at the early stage of the expansion stroke.

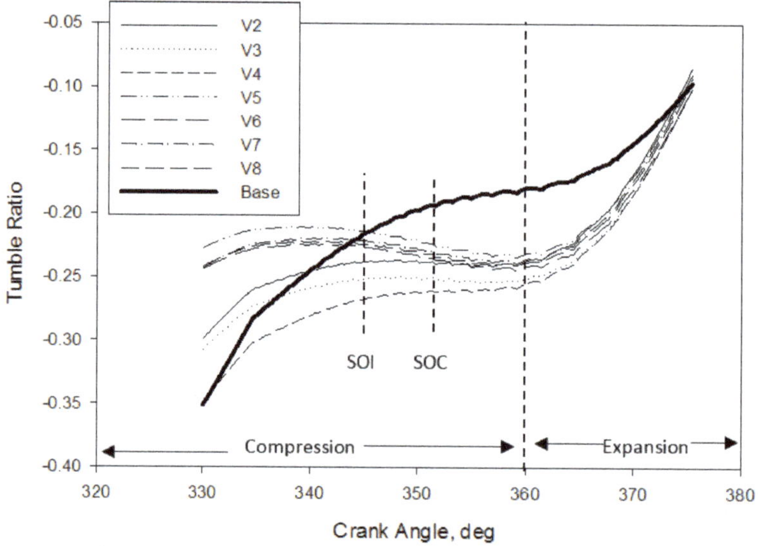

**Figure 11.** Tumble ratio against crank angle.

### 4.5. In-Cylinder Airflow Characteristics during Intake Stroke

The instantaneous streamline of the intake stroke at a crank angle of 10° and 90° after TDC is presented in Figure 12. For intake stroke at a crank angle of 10°, the airflow pattern was initially induced through a piston bowl. The V4 design, which had guided the airflow via vanes, shows the preliminary turbulent flow in the intake manifold before being induced into the combustion chamber. It also demonstrates the ability of the fluid to develop rotational motion in the cylinder and benefit in assisting the atomization of heavy molecules, i.e., emulsified biofuel. The intake stroke at crank angles of 90° shows the $R_S$ phenomena due to the influence of tumble flow. However, the engine without vanes (base) has shown low-velocity flow compared to the engine using vanes. These results were supported by Heywood [43], who suggested the consideration of the vortex flow during intake process, in order to enhance the turbulence intensity at the compression stage.

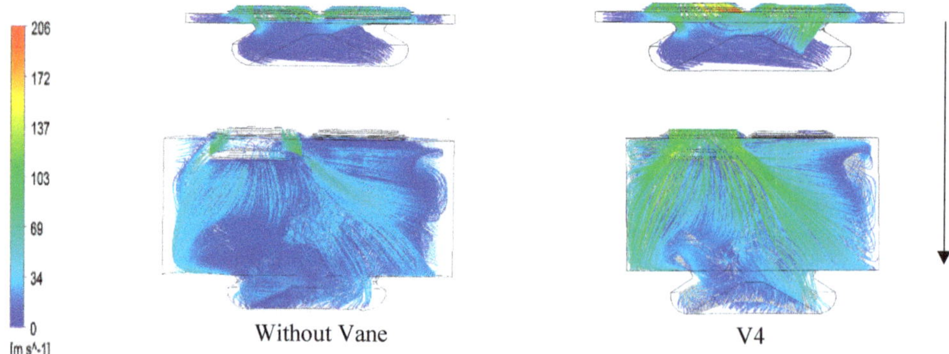


**Figure 12.** The computed streamline intake stroke crank angle at 10° (top) and 90° (bottom) showing the swirl and tumble airflow structure.

### 4.6. In-Cylinder Airflow Characteristics during Compression Stroke

Figure 13 shows the instantaneous streamline of the compression stroke at crank angles of 346° and 310°. At the time of compression stroke, when the volume tends to change as a result of compression, the density of air, temperature and pressure witnessed an increase. From Figure 11, a notable effect on the amplification of turbulent flow and the acceleration of air can be seen. The velocity of air was generally higher during cranking angle at a position of 310° for both the designs. However, the air velocity decreased gradually when it reached TDC. The SCC piston with V4 design showed uniform velocity in-cylinder flow on both sides at a crank angle of 346° (at SOI stage). It implied that the engine using vanes in their operation can produce a strong velocity of air, high turbulent flow and be able to transport heavy molecules of fuel, i.e., emulsified biofuel with the homogenous mixture. With the abilities mentioned above, the flame speed and the reliability of combustion for a very low air fuel ratio will be promoted.

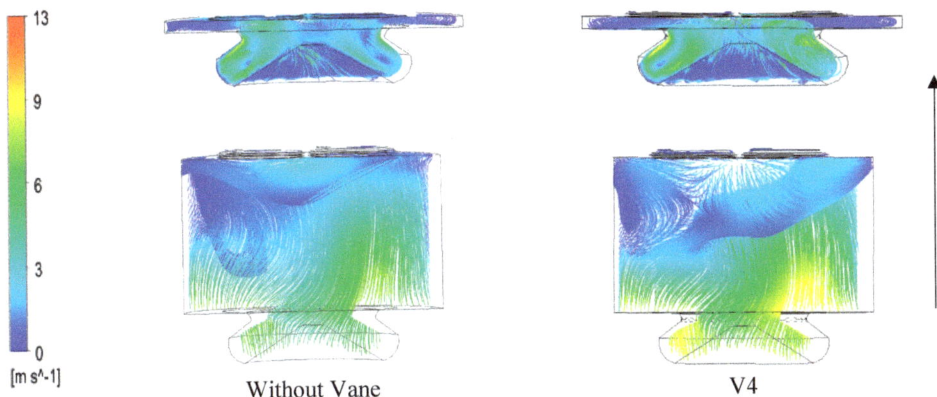


**Figure 13.** The computed streamline compression stroke crank angle at 346° (top) and 310° (bottom).

### 4.7. In-Cylinder Pressure during Compression Stroke

The improvement of combustion efficiency is rely on the in-cylinder pressure inside the engine. Higher in-cylinder pressure value will benefited to the fuel penetration during spraying and aids to expend the cone angle which is necessary for application of higher viscous fuel. Figure 14 shows the variation of in-cylinder pressure without vane and with GVD V4. From the figure can be seen that, GVD model V4 produces an extra in-cylinder pressure compared to the without GVD. The area that built up pressure in the V4 model is at the near injection port area and obviously produced more pressure in the piston-bowl with the organized in-cylinder air flow throughout along the piston-bowl area. With this result, the higher in-cylinder pressure will definitely give more resistance to the injected fuel in term of friction to the air flow and consequently reduce the penetration length during injection.

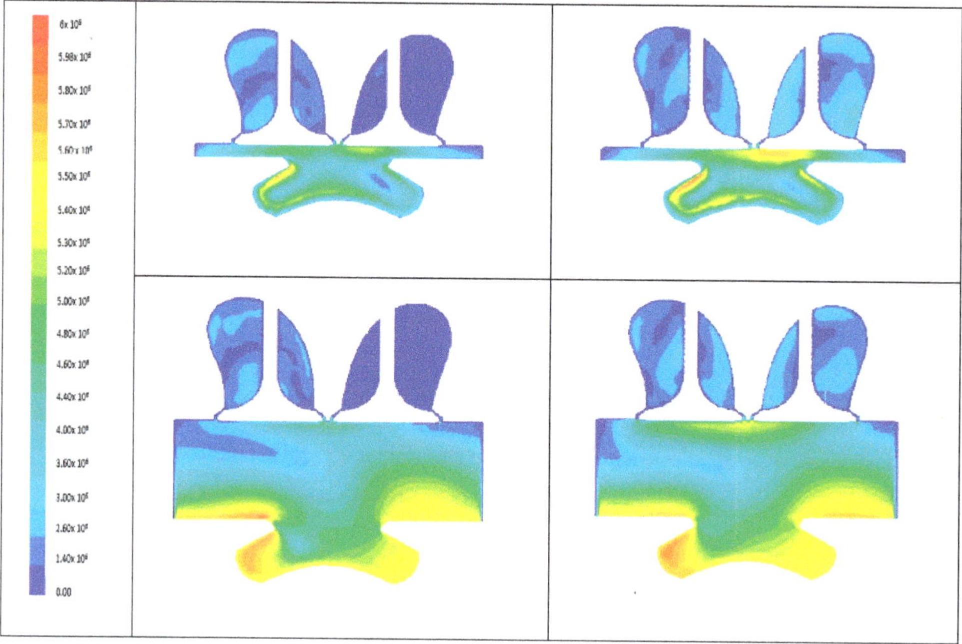

**Figure 14.** The in-cylinder pressure during compression stroke crank angle at 346° (top) and 310° (bottom).

## 5. Conclusions

Emulsified biofuel possesses a high potential to replace fossil fuel and to be used in the diesel engine. The similar properties of emulsified biofuel and fossil fuel enable this fuel to be used in the engine with minor modifications. However, emulsified biofuel is more viscous, less volatile and has heavy molecules that restrict to maximum evaporation of spraying. This phenomenon eventually deposits the carbon on the cylinder wall and piston head, piston ring sticking and resulted in incomplete combustion. Therefore, introducing GVD can improve and generate more swirl and tumble airflow to enhance the turbulent flow for extra-evaporating emulsified biofuel. The in-cylinder air flow characteristics of $(R_S)$, $(R_T)$, TKE and $R_{CT}$ were investigate and compared between the 7 models of GVD and base model of various vane numbers. Hence, based on the simulation result, the four numbers of vanes (V4) in GVD have shown the best performance in terms of $TKE$, $R_S$, $R_T$ and $R_{CT}$. The streamline compression stroke shows that V4 allows the air flow throughout piston-bowl to be more recognized, which enhanced the in-cylinder air flow velocity that expected could break up the length penetration during spraying process. For future work, the study will be extended and investigated using the strategy of combination of GVD with SCC piston and their effect of injection profile, structure of spraying and combustion characteristics using higher viscosity fuel such as emulsified biofuel application in the diesel engine.

**Author Contributions:** Conceptualization, M.F.H. and M.Y.I.; Methodology, M.F.H., M.K.A. M.Y.I. and S.S.; Experimentation, S.C.M., M.F.H. and T.Y.H.; Data analysis, M.F.H., K.A.S., and R.K.S.; Resources, M.Y.I. and M.F.H.; Writing—original draft preparation, M.F.H and M.K.A.; Writing, review and editing : Z.A.Z.A. All authors have read and agreed to the published version of the manuscript.

**Funding:** The authors would like to thank to Universiti Sains Malaysia (USM) on Research University Grant Scheme under Grant No. 1001/PBAHAN/8014006 and USM Fellowship RU (1001/CIPS/AUPE001) for supporting the research journey.

**Conflicts of Interest:** The authors declare no conflict of interest.

## Nomenclature

| English Symbols | Description | Units |
| --- | --- | --- |
| $U$ | Three dimensional flow velocities x,y and z directions | m/s |
| $P$ | Pressure | Pa |
| $T$ | Temperature | K |
| $t$ | Time | s |
| $N$ | Engine speed | rpm |
| $T$ | Torque | N·m |
| $m$ | Mass | kg |
| $L$ | Litre | l |
| $h_{tot}$ | Total enthalpy | J/kg |
| $v$ | Velocity | m/s |
| $v_{max}$ | Maximum velocity | m/s |
| $S_M$ | Momentum source | N/m$^3$ |
| $\vec{u}$ | Three-dimensional flow | mm/s |
| $R$ | Radius | mm |
| $l$ | Length | mm |
| $x,y,z$ | Cartesian coordinates | mm |
| **Greek Symbols** | **Description** | **Units** |
| $\rho$ | Fluid density | kg/m$^3$ |
| $\omega$ | Angular acceleration | rad/s |
| $\lambda$ | Thermal conductivity | W/m·K |
| $\theta$ | Crank angle degree | - |
| $\tau$ | Strain rate | 1/s |
| $\nabla$ | Gradient operator | - |
| **Abbreviations** | **Description** | |
| IVO | Intake valve open | |
| IVC | Intake valve close | |
| EVO | Exhaust valve open | |
| EVC | Exhaust valve close | |
| GVD | Guide Vane Design | |
| SCC | Shallow depth re-entrance combustion chamber | |
| IARC | International Agency for Research on Cancer | |
| RPO | Refine Palm Oil | |
| $S_M$ | Momentum Source | |
| $R_T$ | Tumble Ratio | |
| $R_S$ | Swirl Ratio | |
| $R_{CT}$ | Cross Tumble Ratio | |
| TKE | Turbulence Kinetic Energy | |

## References


1. Reitz, R.D.; Duraisamy, G. Review of high efficiency and clean reactivity controlled compression ignition (RCCI) combustion in internal combustion engines. *Prog. Energy Combust. Sci.* **2015**. [CrossRef]
2. Al-attab, K.; Wahas, A.; Almoqry, N.; Alqubati, S. Biodiesel production from waste cooking oil in Yemen: A techno-economic investigation. *Biofuels* **2017**, *8*, 17–27. [CrossRef]
3. Benbrahim-Tallaa, L.; Baan, R.A.; Grosse, Y.; Lauby-Secretan, B.; El Ghissassi, F.; Bouvard, V.; Guha, N.; Loomis, D.; Straif, K.; International Agency for Research on Cancer Monograph Working Group. Carcinogenicity of diesel-engine and gasoline-engine exhausts and some Nitroarènes. *Pollut. Atmos.* **2012**, *13*, 663–664. [CrossRef]
4. IARC. Diesel and gasoline engine exhausts and some nitroarenes. In *IARC Monographs Evaluation Carcinogenic Risks to Humans*; IARC: Lyon, France, 2013.
5. IARC. Diesel and gasoline engine exhausts. In *IARC Monographs Evaluation Carcinogenic Risks to Humans*; IARC: Lyon, France, 2012.

6. Rai, R.; Glass, D.C.; Heyworth, J.S.; Saunders, C.; Fritschi, L. Occupational exposures to engine exhausts and other PAHs and breast cancer risk: A population-based case-control study. *Am. J. Ind. Med.* **2016**, *59*, 437–444. [CrossRef]

7. Mauderly, J.L.; Seilkop, S.K. The National Environmental Respiratory Center (NERC) experiment in multi-pollutant air quality health research: II. Comparison of responses to diesel and gasoline engine exhausts, hardwood smoke and simulated downwind coal emissions. *Inhal. Toxicol.* **2014**, *26*, 668–690. [CrossRef]

8. Mat Yasin, M.H.; Mamat, R.; Najafi, G.; Ali, O.M.; Yusop, A.F. Potentials of palm oil as new feedstock oil for a global alternative fuel: A review. *Renew. Sustain. Energy Rev.* **2017**, *79*, 1034–1049. [CrossRef]

9. Agarwal, D.; Agarwal, A.K. Performance and emissions characteristics of Jatropha oil (preheated and blends) in a direct injection compression ignition engine. *Appl. Therm. Eng.* **2007**, *27*, 2314–2323. [CrossRef]

10. Li, J.; Yang, W.M.; An, H.; Maghbouli, A.; Chou, S.K. Effects of piston bowl geometry on combustion and emission characteristics of biodiesel fueled diesel engines. *Fuel* **2014**, *120*, 66–73. [CrossRef]

11. Hamid, M.F.; Idroas, M.Y.; Sa'ad, S.; Bahri, A.J.S.; Sharzali, C.M.; Zainal, Z.A. Numerical investigation of in-cylinder air flow characteristic improvement for Emulsified biofuel (EB) application. *Renew. Energy* **2018**, *127*, 84–93. [CrossRef]

12. Ali, O.M.; Mamat, R.; Abdullah, N.R.; Abdullah, A.A. Analysis of blended fuel properties and engine performance with palm biodiesel-diesel blended fuel. *Renew. Energy* **2015**, *86*, 59–67. [CrossRef]

13. Ren, Y.; Li, X.G. Numerical study on the combustion and emission characteristics in a direct-injection diesel engine with preheated biodiesel fuel. *J. Automob. Eng.* **2011**, *225*, 531–543. [CrossRef]

14. Cheng, C.H.; Cheung, C.S.; Chan, T.L.; Lee, S.C.; Yao, C.D.; Tsang, K.S. Comparison of emissions of a direct injection diesel engine operating on biodiesel with emulsified and fumigated methanol. *Fuel* **2008**, *87*, 1870–1879. [CrossRef]

15. Khond, V.W.; Kriplani, V.M. Effect of nanofluid additives on performances and emissions of emulsified diesel and biodiesel fueled stationary CI engine: A comprehensive review. *Renew. Sustain. Energy Rev.* **2016**, *59*, 1338–1348. [CrossRef]

16. Mofijur, M.; Rasul, M.G.; Hyde, J.; Azad, A.K.; Mamat, R.; Bhuiya, M.M.K. Role of biofuel and their binary (diesel-biodiesel) and ternary (ethanol-biodiesel-diesel) blends on internal combustion engines emission reduction. *Renew. Sustain. Energy Rev.* **2016**. [CrossRef]

17. Agarwal, A.K.; Dhar, A.; Gupta, J.G.; Kim, W., II; Choi, K.; Lee, C.S.; Park, S. Effect of fuel injection pressure and injection timing of Karanja biodiesel blends on fuel spray, engine performance, emissions and combustion characteristics. *Energy Convers. Manag.* **2015**, *91*, 302–314. [CrossRef]

18. Taib, N.M.; Mansor, M.R.A.; Mahmood, W.M.F.W. Modification of a direct injection diesel engine in improving the ignitability and emissions of diesel-ethanol-palm oil methyl ester blends. *Energies* **2019**, *12*, 2644. [CrossRef]

19. Kim, H.Y.; Ge, J.C.; Choi, N.J. Effects of fuel injection pressure on combustion and emission characteristics under low speed conditions in a diesel engine fueled with palm oil biodiesel. *Energies* **2019**, *12*, 3264. [CrossRef]

20. Abed, K.A.; Gad, M.S.; El Morsi, A.K.; Sayed, M.M.; Elyazeed, S.A. Effect of biodiesel fuels on diesel engine emissions. *Egypt. J. Pet.* **2019**, *28*, 183–188. [CrossRef]

21. Bari, S.; Saad, I. Performance and emissions of a Compression Ignition (CI) engine run with biodiesel using guide vanes at varied vane angles. *Fuel* **2015**, *143*, 217–228. [CrossRef]

22. Bari, S.; Saad, I. CFD modelling of the effect of guide vane swirl and tumble device to generate better in-cylinder air flow in a CI engine fuelled by biodiesel. *Comput. Fluids* **2013**, *84*, 262–269. [CrossRef]

23. Bae, C.; Kim, J. Alternative fuels for internal combustion engines. *Proc. Combust. Inst.* **2017**. [CrossRef]

24. Mallick, M.; Kumar, A.; Tamboli, N.; Kulkarni, A. Study on drag coefficient for the flow past a cylinder. *Physics* **2014**, *5*, 301–306.

25. Hwang, J.; Qi, D.; Jung, Y.; Bae, C. Effect of injection parameters on the combustion and emission characteristics in a common-rail direct injection diesel engine fueled with waste cooking oil biodiesel. *Renew. Energy* **2014**, *63*, 9–17. [CrossRef]

26. Yadav, S.P.R.; Saravanan, C.G. Engine characterization study of hydrocarbon fuel derived through recycling of waste transformer oil. *J. Energy Inst.* **2015**, *88*, 386–397. [CrossRef]

27. Voutchkov, I.; Keane, A.; Shahpar, S.; Bates, R. (Re-) Meshing using interpolative mapping and control point optimization. *J. Comput. Des. Eng.* **2018**, *5*, 305–318. [CrossRef]

28. Blanchard, G.; Loubère, R. High order accurate conservative remapping scheme on polygonal meshes using a posteriori MOOD limiting. *Comput. Fluids* **2016**, *136*, 83–103. [CrossRef]

29. ANSYS Inc. *ANSYS Fluent Theory Guide*; Release 18.2; ANSYS Inc.: Ganonsburg, PA, USA, 2013.

30. Rahman, M.M.; Mohammed, M.K.; Bakar, R.A. Effect of air fuel ratio on engine performance of single cylinder port injection hydrogen fueled engine: A numerical study. In Proceedings of the International MultiConference of Engineers and Computer Scientists, Hong Kong, China, 18–20 March 2009.

31. Rajak, U.; Nashine, P.; Singh, T.S.; Verma, T.N. Numerical investigation of performance, combustion and emission characteristics of various biofuels. *Energy Convers. Manag.* **2018**, *156*, 235–252. [CrossRef]

32. Baumann, M.; di Mare, F.; Janicka, J. On the validation of large eddy simulation applied to internal combustion engine flows part II: Numerical analysis. *Flow Turbul. Combust.* **2014**, *92*, 299–317. [CrossRef]

33. Stock, H.W.; Haase, W. Navier-Stokes airfoil computations with e sup N transition prediction including transitional flow regions. *AIAA J.* **2012**, *38*, 2059–2066. [CrossRef]

34. Payri, F.; Benajes, J.; Margot, X.; Gil, A. CFD modeling of the in-cylinder flow in direct-injection Diesel engines. *Comput. Fluids* **2004**, *33*, 995–1021. [CrossRef]

35. Prasad, B.V.V.S.U.; Sharma, C.S.; Anand, T.N.C.; Ravikrishna, R.V. High swirl-inducing piston bowls in small diesel engines for emission reduction. *Appl. Energy* **2011**, *88*, 2355–2367. [CrossRef]

36. Miles, P.; Choi, D.; Megerle, M.; Ewert, B.R.; Reitz, R.; Lai, M.C.; Sick, V. The influence of swirl ratio on turbulent flow structure in a motored HSDI diesel engine—A combined experimental and numerical study. *SAE Tech. Pap.* **2010**. [CrossRef]

37. Hamid, M.F.; Idroas, M.Y.; Basha, M.H.; Saad, S.; Mat, S.C.; Khalil, M.; Zainal, Z.A. Numerical study on dissimilar guide vane design with SCC piston for air and emulsified biofuel mixing improvement. *Chemistry* **2016**. [CrossRef]

38. Kim, K.; Chung, J.; Lee, K.; Lee, K. Investigation of the swirl effect on diffusion flame in a direct-injection (DI) diesel engine using image processing technology. *Energy Fuels* **2008**, *22*, 3687–3694. [CrossRef]

39. Buhl, S.; Gleiss, F.; Kohler, M.; Hartmann, F.; Messig, D.; Brucker, C.; Hasse, C. A combined numerical and experimental study of the 3D tumble structure and piston boundary layer development during the intake stroke of a gasoline engine. *Flow Turbul. Combust.* **2017**, *98*, 579–600. [CrossRef]

40. Khalighi, B. Study of the intake tumble motion by flow visualization and particle tracking velocimetry. *Exp. Fluids* **1991**, *10*, 230–236. [CrossRef]

41. Rabault, J.; Vernet, J.A.; Lindgren, B.; Alfredsson, P.H. A study using PIV of the intake flow in a diesel engine cylinder. *Int. J. Heat Fluid Flow* **2016**. [CrossRef]

42. Wang, T.; Liu, D.; Tan, B.; Wang, G.; Peng, Z. An investigation into in-cylinder tumble flow characteristics with variable valve lift in a gasoline engine. *Flow Turbul. Combust.* **2015**, *94*, 285–304. [CrossRef]

43. Heywood, J.B. *Internal Combustion Engine Fundementals*; McGraw-Hill Education: New York, NY, USA, 1998.

*Article*

# Numerical Simulation of a Wall-Flow Particulate Filter Made of Biomorphic Silicon Carbide Able to Fit Different Fuel/Biofuel Inputs

**M. Pilar Orihuela** [1,*], **Onoufrios Haralampous** [2], **Ricardo Chacartegui** [1], **Miguel Torres García** [1] **and Julián Martínez-Fernández** [3]

1   Departamento de Ingeniería Energética, Universidad de Sevilla, Avenida de los Descubrimientos s/n, 41092 Sevilla, Spain; ricardoch@us.es (R.C.); migueltorres@us.es (M.T.G.)
2   University of Thessaly, Geopolis, 41500 Larissa, Greece; onoufrios@uth.gr
3   Departamento de Física de la Materia Condensada, Universidad de Sevilla, Avenida Reina Mercedes s/n, 41012 Sevilla, Spain; martinez@us.es
*   Correspondence: orihuelap@us.es; Tel.: +34-954485970

Received: 12 November 2019; Accepted: 4 December 2019; Published: 11 December 2019

**Abstract:** To meet the increasingly strict emission limits imposed by regulations, internal combustion engines for transport applications require the urgent development of novel emission abatement systems. The introduction of biodiesel or other biofuels in the engine operation is considered to reduce greenhouse gas emissions. However, these alternative fuels can affect the performance of the post-combustion systems due to the variability they introduce in the exhaust particle distribution and their particular physical properties. Bioceramic materials made from vegetal waste are characterized by having an orthotropic hierarchical microstructure, which can be tailored in some way to optimize the filtration mechanisms as a function of the particle distribution of the combustion gases. Consequently, they can be good candidates to cope with the variability that new biofuel blends introduce in the engine operation. The objective of this work is to predict the filtration performance of a wall-flow particulate filter (DPF) made of biomorphic silicon carbide (bioSiC) with a systematic procedure that allows to eventually fit different fuel inputs. For this purpose; a well-validated DPF model available as commercial software has been chosen and adapted to the specific microstructural features of bioSiC. Fitting the specific filtration and permeability parameters of this biomaterial into the model; the filtration efficiency and pressure drop of the filter are predicted with sufficient accuracy during the loading test. The results obtained through this study show the potential of this novel DPF substrate; the material/microstructural design of which can be adapted through the selection of an optimum precursor.

**Keywords:** internal combustion engine; biodiesel; particulate matter emissions; biomorphic silicon carbide; vegetal waste; diesel particulate filter

---

## 1. Introduction

Emission levels for automotive engines are submitted to increasingly stringent limits. From September 2015, all new European diesel cars must be compliant with the Euro 6 emissions standards, which set a particle number emission limit of $6 \times 10^{11}$ particles $km^{-1}$ and a limit of 4.5 mg $km^{-1}$ for the mass of particulate [1]. In the United States, the phase-in period of Tier 3 (2017–2025) applies at a federal level, and requires automakers in the US to certify an increasing percentage of their fleet are complying with the new emissions standard (3 mg/mi of particulate matter) [2]. US standards are led by the California low emission vehicle (LEV) legislation, which sets stricter emission limits to cope with its exceptional smog problems [3]. In the near future, new standards will likely be developed for even stricter emissions. In 2017, a new worldwide harmonized light-vehicle test procedure (WLTP) came into force, and it is compulsory for all new car registrations from September 2018 [4]. Also, for non-road mobile machinery, the Stage V of the EU Regulation 2016/1628 will be effective from 1 January 2019, reducing the particles mass limit for all the engines above 19 kW, and introducing a new limit for particle number emissions [5].

Different strategies to control emissions are developed by different manufacturers. New designs on combustion chamber, injection and supercharging systems, and control are introduced in new engines to reduce particle emissions, but they are not enough to ensure the compliance with current regulation thresholds [6]. Switching to cleaner alternative fuels can be also an effective way to reduce pollutant emissions in internal combustion engines (ICEs). The introduction of biofuels in the automotive sector provide an all-inclusive solution to the dependence on fossil fuels, and the associated environmental impact [7]. Although there are multiple biofuel formulations and many different studies on the emissions derived from their application to ICEs [8–10], there is a general agreement in their contribution to reduce gas emissions (CH, CO) and particulate matter (PM) [11–13] and overall life cycle carbon dioxide ($CO_2$) [14]. Some research suggests, nonetheless, that increasing the blending rate of biofuel in an engine may increase particulate emissions depending on the engine design [15], on the biofuel properties [16], on the engine operating conditions [17], and even on the measurement method [18]. Despite the progress in the engine technology and the development of new biofuels, complying with the current emissions standards requires highly effective aftertreatment systems in the abatement of soot particles [6,19]. Nowadays, the most popular aftertreatment system for the abatement of particulate emissions in ICEs is the wall-flow diesel particulate filter (DPF). The requirements for the proper performance of a DPF are, mainly, high filtration efficiency, low pressure drop [20], and high capacity to resist the regeneration processes [21]. With an appropriate design, and a suitable substrate, the wall-flow DPF is able to comply with the current PM legislation [20]. The challenge for a DPF is the correct balance between filtration efficiency and pressure drop for an adequate engine performance. These parameters are generally closely related, an increase in filtration efficiency brings an increase in pressure drop, and vice versa. Also, the quick saturation of the filter, and the thermal stress produced by the regeneration cycles leads, in many cases, to the cracking and collapse of the DPF structure [22], increasing engine backpressure and penalizing the engine operation.

The introduction of biodiesel or other alternative fuels in the engine operation brings a new variable to the post-combustion systems. Biofuels alter the particle size distribution in the exhaust gases [23,24], and this affects in turn the performance of the DPF [25]. The combustion of biodiesel reduces the primary particle diameter [26] and shifts the distribution curve towards smaller particles [24,27]. The behavior of a wall-flow DPF is given by its geometry (diameter, length, wall thickness, cell density) at the macroscopic scale [28], and by the properties of the material used as substrate (permeability, porosity, pore size, tortuosity) at the microscopic scale [29]. For a given geometry, the microstructural properties of the substrate define the filtration efficiency and the pressure drop of the filter. The soot size has a significant influence on the initial deep-bed loading process [30]. The search for new materials that improve DPFs permeability, thermal properties, and filtration performance is a recurrent research topic today. The alterations that new biofuel blends introduce in vehicular PM emissions should be also taken into account when designing or developing new substrates for DPFs.

Recent studies have presented biomorphic silicon carbide (bioSiC) as a viable candidate for use as a substrate in hot gas filtration applications [31,32] and, specifically, as a substrate in DPFs for automotive diesel engines [33]. BioSiC is a porous bioceramic material characterized by preserving the hierarchical biological microstructure of the wood precursor from which it was made [34] so it is considered a bio waste or biomass. In this sense, the microstructure of this material can be tailored, to some extent, by the choice of the precursor to fit any application [35]. As opposed to traditional ceramic granular media, bioSiC can be good candidate to optimize the pressure drop/efficiency balance of DPFs while coping with the variability that new biofuel blends introduce in the engine operation. BioSiC can be manufactured from a wide variety of plant species or precursors, including vegetal waste or biomass. Each precursor leads to a different microstructure. If the microstructural characterization of a precursor is known, numerical simulation techniques can be applied to predict the performance of the global filtration system and to identify the potential for its use at real scale with different fuel inputs.

Mathematical and physical models used for internal combustion engine particulates filter performance prediction, evolve from the pioneering work of Bissett and Shadman [36,37], published in 1985. The original approach proposed by Bissett was one-dimensional model with two channels, based on the basic principles of fluid-mechanics, and the application of an energy balance in the solid wall and gas phase. A transient filtration model was implemented later by Konstandopoulos et al. [38], who used exhaust conditions, including particle size distribution and wall microstructure properties, to calculate filtration efficiency in discretized wall slabs. These models have been further refined by Tandon et al. [29] who gave emphasis on the efficiency evolution during transition to cake filtration, and Bollerhof et al. [39], who studied filtration in inhomogeneous wall structures. The latter model and several variations are available in the commercial software package, named Axisuite [40], which is employed in this work. A comprehensive review of DPF modeling is given in references [41,42]. In this work, a numerical model of wall-flow DPFs has been adapted to the experimental microstructural features of the bioSiC material, in order to establish a starting point in the generalized analysis of different precursors, used for bioSiC generation. The objective is to predict the filtration performance of a bioSiC wall-flow DPF with a systematic procedure that allows to eventually improve the system performance and to fit different fuel/biofuel inputs through the identification of optimum precursors. In this study, the experimental microstructural features of bioSiC made from medium density fiberboard (MDF) were used. The validation of the model was made in a small prototype of bioSiC wall-flow DPF. It was designed, manufactured, and tested under controlled conditions, with the aid of a soot generator [33]. The resulting experimental measurements of filtration efficiency and pressure drop were then used to calibrate the real scale model, with the adjusted microstructural parameters to simulate the performance under NEDC driving cycle conditions. The same procedure might be used in the future to fit the microstructure of any other non-granular substrate.

This paper is structured in four sections. In Section 1, the motivations and objectives of the work are presented, and the general background of the model is introduced. Section 2 summarizes the main aspects and equations of the numerical model. It will describe the calibration-validation process and how the results of a small prototype were extrapolated to a full-size system. In Section 3, the main results are reported and a prediction about the performance of the filter, compared to that of other commercial systems, is presented. Finally, Section 4 summarizes the main results and conclusions.

## 2. Materials and Methods

The numerical model used for this work is applicable to any wall-flow DPF with constant cross-section and straight channels. A large part of the governing equations in the model act at the macroscopic scale and do not depend on the material the filter is made of, so these could be applicable to any DPF with the same geometry regardless of the composition of its substrate. There is a group of equations that model the walls of the filter and the passage of the gas through them, and here is where the microstructural characteristics of the substrate may have an effect. This group of equations depend on features of the porous material such as porosity, pore size, and permeability. Thus, to apply this latter group of equations, a previous knowledge of the biomorphic substrate is needed.

### 2.1. Model Description

The numerical model used in this study is built in two levels. The first level is the 'single channel problem'. In the 'single channel problem', the spatial discretization is comprised of one inlet channel and four quarters of outlet channels. Figure 1 presents the frontal area of the control volume, which is marked with a dashed red line. The background image is a SEM micrograph of the MDF bioSiC filter used for the calibration and validation of the model, as explained in Section 2.2. At this level, all the equations for a single channel are solved. The second level is the 'multi-channel problem', which extrapolates the results of the single channel to the whole system, including the additional phenomena associated with the physics of the complete monolith.

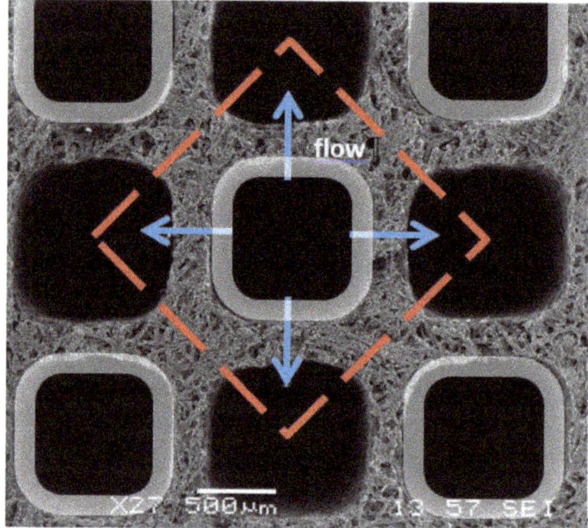


**Figure 1.** Front area of the control volume in the 'single channel problem' represented over a SEM image of a real MDF-bioSiC filter.

The evolution of pressure and velocity in each single channel is described mainly through two interrelated groups of equations. On the one hand, along the free path of the inlet or the outlet channels, the behavior of the flux is governed by the mass and momentum conservation equations of fluid dynamics. On the other hand, across the porous mediums, such as the walls of the filter or the soot layer, the behavior of the flux is governed by Darcy's law and the inertial effects (Forchheimer). Table 1 summarizes these equations. The full model is described in detail in [43].

**Table 1.** Summary of the governing equations of the flux in the single channel control volume.

| Control Volume | Equation |
| --- | --- |
| Along the free path of the channels | Continuity equation:<br>$\frac{\partial}{\partial z}(\rho_i v_i) = (-1)^i \frac{N}{d} \rho_w v_w$<br>Momentum conservation equation:<br>$\frac{\partial p_i}{\partial z} + \frac{\partial}{\partial z}\left(\rho_i v_i^2\right) = -\frac{\alpha_1 \mu v_i}{d^2}$ |
| Through porous media | Across the soot layer:<br>$\Delta p_{soot} = \frac{\mathcal{R}}{M_g}\frac{T}{\bar{p}}\frac{\mu d \rho_w v_w}{2 k_p(\bar{p})} ln\left(\frac{d}{d-2w_p}\right)$<br>Across the substrate walls:<br>$\Delta p_{wall} = \frac{\mu}{k_w}\frac{w_w}{} v_w + \frac{C_E}{\sqrt{k_w}}\rho_w v_w^2$<br>Total pressure drop between the inlet and the outlet channel:<br>$p_1 - p_2 = \Delta p_{soot} + \Delta p_{wall}$ |

In these equations, the subscript $i$ is the identifier of the channel: 1 for inlet channels, and 2 for outlet channels. $N$ is the number of walls of the channel, 4 in this case, and $d$ is their width. The pressure drop through the substrate wall is set considering that the flow velocity $v_w$ is constant. The result is the direct application of Darcy's law with the Forchheimer's extension. On the contrary, the pressure drop through the soot layer is calculated, considering that the gas velocity varies along the soot layer, due to changes in gas density and flow area. Taking into account the geometrical definitions shown in Figure 2, the pressure drop through the soot layer is calculated by expressing $v$ and $\rho$ as a function of the coordinate $w$, and integrating along the thickness of the soot layer [43]. The result depends on the permeability of the particulate deposit $k_p$, and on the soot layer thickness $w_p$, the value of which is recalculated in every successive step with the accumulated amount.

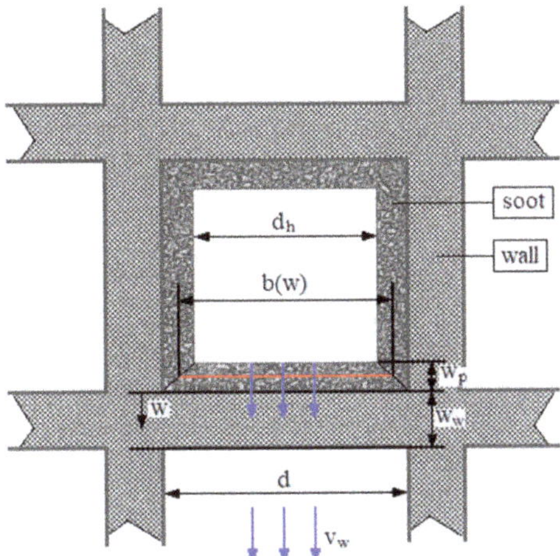


**Figure 2.** Scheme of the channel with geometrical definitions used in the model [43].

Equations for the porous medium summarized in Table 1 depend on permeability values that must be calibrated based on experimental results. The pressure drop across the soot layer depends on the permeability of the particulate deposit $k_p$, while the pressure drop across the substrate walls depends on the permeability of the substrate wall $k_w$. The permeability of the particulate deposit may be expressed as a function of the local temperature and pressure, using the correlation of Pulkrabek [44]

$$k_p = k_{p,0}\left(1 + C_4 \frac{p_0}{\bar{p}} \mu \sqrt{\frac{T}{M_g}}\right) \tag{1}$$

The local pressure $p$ depends initially on the position $w$, but introducing this variable in the analytical approach would extremely complicate the solution. Instead, a mean value of the pressure $\bar{p}$ is considered, calculated as the average between the inlet and the outlet pressure. $C_4$ and $k_{p,0}$ are characteristic parameters of the porous media and must be estimated based on experimental data.

Similarly, the permeability of the wall may be expressed as

$$k_w = \frac{1}{\frac{1}{k_{w,0}} + C_1 \rho_p + C_2 \rho_p^2} \cdot \left(1 + C_4 \frac{p_0}{\bar{p}} \mu \sqrt{\frac{T}{M_g}}\right) \tag{2}$$

The permeability of the clean wall $k_{w,0}$ is the permeability of the clean bioSiC for each particular precursor under consideration. For calibration and validation purposes, this study has used a real MDF-bioSiC DPF as a model, the main features of which are summarized in Section 2.2 including its permeability in the clean stage. $C_1$ and $C_2$ are also characteristic parameters of the porous substrate but they govern the behaviour of the medium as it becomes loaded with particles. The permeability of the loaded wall is a function of the amount of soot trapped in the wall, $\rho_p$ is the instantaneous concentration of soot in the wall. Therefore, the values of $C_1$ and $C_2$ must also be estimated, based on experimental data. In the next section, the calibration procedure is explained for both the soot layer permeability and the wall permeability. With regard to the filtration model, the filtration efficiency of a porous medium is the result of the behavior of its unit collectors at the micro-scale. In this work, the model was specifically prepared to be applied to a bioSiC filter made from MDF. Thus, to model the filtration efficiency, the fiber microstructure of bioSiC made from MDF was taken into account. The filtration efficiency of a clean fiber unit collector by diffusion and interception exposed to aggregate particles is summarized in Table 2 [43,45].

**Table 2.** Single collector filtration efficiency by diffusion and interception for fiber unit collectors.

| Filtration Mechanism | Equation |
|---|---|
| Diffusion | Diffusional efficiency [46]: $$\eta_D = \eta_{clean,D} \, 1.6 \left(\frac{\varepsilon_{pore,0}}{Ku}\right)^{1/3} Pe^n Pe \, C_d$$ Peclet number: $$Pe = \frac{u_w d_{fib,0}}{D_{part}}$$ Parameter in the diffusional efficiency [47]: $$C_d = 1 + 0.388 \, Kn_{fib}\left(\frac{\varepsilon_{pore,0} Pe}{Ku}\right)^{1/3}$$ |
| Direct interception | Direct interception efficiency [46]: $$\eta_R = \eta_{clean,R} \, 0.6 \left(\frac{\varepsilon_{pore,0}}{Ku}\right)^{1/3} \frac{R^2}{(1+R)} C_r$$ Interception parameter: $$R = \frac{2R_c}{d_{fib,0}}$$ Parameter in the direct interception efficiency [47]: $$C_r = 1 + \frac{1.996 \, Kn_{fib}}{R}$$ |

## 2.2. Experimental Validation and Model Calibration for a bioSiC DPF

In order to determine the value of the empirical parameters included in the model, calibrate it, and validate it, the results of a previous experimental study were used. In that study, two small prototypes of bioSiC wall-flow filters were manufactured from MDF, and tested under controlled conditions in a laboratory test rig, with a soot laden gas stream. The description of the samples, the test rig, and the experimental procedure, can be found in [33]. Here, only the parameters relevant to the model are reported.

The prototypes of bioSiC wall-flow filters, with a square cross section of 9.2 × 9.2 mm and a length of 31 mm, had a cell density of 57.59 cells cm$^{-2}$ (371.6 cpsi). A soot generator (PALAS GFG 1000) was used to create a gas stream laden with a particulate distribution similar to that of an ICE fueled either with fossil or biodiesel. In setting a pressure of 1.2 bar, and a spark frequency of 200 sparks s$^{-1}$ at the soot generator, an argon flow rate of 5 L min$^{-1}$ is obtained with a soot production of 4 mg h$^{-1}$. The experimental study performed by Rodríguez-Fernández et al. with one fossil fuel, two paraffinic biofuels: an hydrotreated vegetable oil (HVO) and a gas-to-liquid biofuel (GTL) and one biodiesel [48] was taken as a reference for the identification of a standard particle size distribution. Figure 3 reproduces the particle size distributions obtained in this study; Table 3 summarizes the composition and properties of the corresponding fuels.

**Table 3.** Main properties and composition of the fuels tested by Rodríguez-Fernández et al. [48].

| Properties and Composition | Diesel | HVO Biofuel | GTL Biofuel | Biodiesel |
|---|---|---|---|---|
| Standard | UNE EN 590 | UNE EN 15940 | UNE EN 15940 | UNE EN 14214 |
| Lower heating value (MJ/kg) | 43.04 | 43.96 | 44.03 | 37.26 |
| Sulfur (mg/kg) | <10 | <10 | <10 | <10 |
| Water (mg/kg) | 60 | 19.2 | 20 | 102 |
| C (% w/w) | 85.74 | 84.68 | 84.82 | 76.45 |
| H (% w/w) | 14.26 | 14.53 | 15.18 | 12.36 |
| O (% w/w) | 0 | 0 | 0 | 11.19 |
| Density at 15 °C (kg/m$^3$) | 811 | 779.6 | 774 | 874.3 |
| Viscosity at 40 °C (mm$^2$/s) | 2.02 | 2.99 | 2.34 | 4.5 |

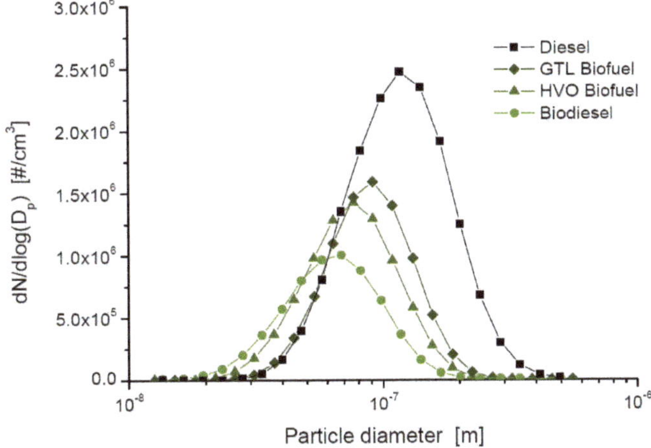


**Figure 3.** Particle size distributions obtained by J. Rodríguez-Fernández, M. Lapuerta, and J. Sánchez-Valdepeñas from testing different biofuels in an internal combustion engine [48].

As mentioned before, the numerical model used in this study was solved with the aid of the commercial software Axisuite [40]. The settings for the gas inlet and the DPF in the theoretical model (Table 4) were established taking into account the real conditions under which the tests were carried out. The soot particle size distribution, characterized by the mean particle diameter and the standard deviation, was directly observed from the experimental particle size distribution curve of the soot generator, as shown in Figure 4. The soot aggregate structure of the aerosol produced by the soot generator, can be characterized by the following parameters [33]: fractal dimension $D_f = 2.1$ [49], soot primary particle radius $r_0 = 6 \times 10^{-9}$ m [50], and soot primary particle density $\rho_0 = 1700$ kg/m$^3$ [51].

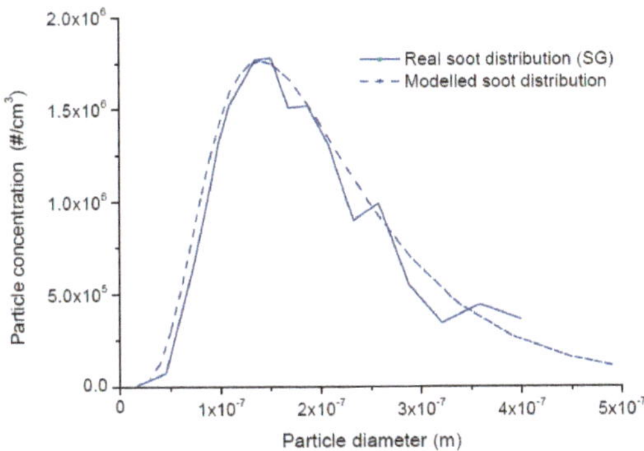


**Figure 4.** Real particle size distribution of the soot generator and modeled curve.

For the resolution of the problem, a one-dimensional discretisation of the DPF domain was used. As a result, a representative pair of channels, inlet and outlet, is used, instead of the complete square cross-section. The substrate of the DPF was modeled according to the microstructural characteristics of biomorphic silicon carbide made from MDF. For the porosity and the mean pore diameter, the values reported by Gomez-Martin et al. [31], were considered. For the initial permeability of the clean substrate, which may be dependent on the relative orientation of the wood fibers, with respect to the direction of the gas flow, the value $4 \cdot 10^{-14}$ m$^2$ was taken, calculated in a previous experimental study [33]. For the remaining physical or chemical parameters of the substrate, standard values for silicon carbide, with 44% porosity, and 9 μm mean pore size, were selected [43].

**Table 4.** Parameter settings in the theoretical model of the small prototypes.

| Inlet Settings | Value |
|---|---|
| Composition of the gas stream | Pure Argon |
| Volume flow rate | 5 L min$^{-1}$ (0.1365 × 10$^{-3}$ kg s$^{-1}$) |
| Temperature | Room temperature: ~30 °C |
| Pressure | - |
| Soot mass flow | 0.004 g h$^{-1}$ |
| Soot particle size distribution | Log-Normal (μ = 140 × 10$^{-9}$ m; σ = 1.7) |
| Soot aggregate structure | Fractal dimension: 2.1 [49]<br>Soot primary particle radius: 6.6 × 10$^{-9}$ m<br>Soot primary particle density: 1700 kg m$^{-3}$ |

| DPF Settings | Value |
|---|---|
| Substrate length | 0.032 m |
| Plug length | 0.001 m |
| Substrate equivalent diameter | 0.00997 m |
| Cell density | 370 cpsi |
| Wall thickness | 0.00038 m (14.96 mil) |

| Substrate Properties Specific for bioSiC | Value |
|---|---|
| Initial permeability (clean) | 4 × 10$^{-14}$ m$^2$ |
| Substrate pore diameter | 15.7 × 10$^{-6}$ m |
| Pore volume fraction | 0.49 |

To calibrate the empirical parameters of the model, mentioned previously, the following sequence was followed:

- The parameters related to the filtration efficiency are calibrated. The initial diffusion and interception collection efficiencies ($\eta_{clean,D}$ and $\eta_{clean,R}$), as well as the diffusion mechanism exponent of the Peclet number ($nPe$), are adjusted by matching up the initial filtration efficiency of the real filter, with the initial filtration efficiency of the modeled filter, for the different ranges of particle diameters. The filtration of small particles is mainly governed by the diffusion mechanism, while the filtration of larger particles is mainly governed by the interception mechanism. The exponent of the Peclet number affects the gradient of the efficiency curve on the right-hand side of the curve (from 10 to 100 nm). Then, the growth rate of the efficiency curve is adjusted by changing the gradient parameter in wall filtration efficiency ($\eta_{load}$). Calibrating these parameters, the overlap between the theoretical evolution curve of the filtration efficiency and the real curve is achieved.
- The parameters related to the pressure drop are calibrated. If the permeability of the clean substrate is correct, the initial pressure drop resulting from the simulation should match with the real drop. From then on, there are two consecutive filtration stages. The first stage, the wall filtration stage, allows the calibration of the permeability of the loaded substrate ($k_w$). Its value is adjusted by matching up the pressure drop of the real filter with the pressure drop of the modeled filter in the first growing part of the curves up to the transition phase. The second stage, the cake filtration stage, allows calibrating the soot permeability ($k_p$). The soot permeability directly affects the gradient of the pressure drop during cake filtration, so the estimated value can be obtained by matching the experimental curve's slope. Calibrating these parameters, the overlap between the theoretical evolution curve of the pressure drop and the experimental is achieved.

After calibrating the empirical parameters in the numerical model of the small 30-mm-long bioSiC prototypes, the model is validated, and an extrapolation of the model to a real-size DPF can be made.

*2.3. Extrapolation to a Real-Size Automotive DPF*

The extrapolation to a real-size automotive DPF affected: (i) the external dimensions of the DPF; (ii) the gas inlet conditions, corresponding in this case to the exhaust pipe of a real light-duty diesel engine (gas composition, volume flow rate, soot concentration and properties, particle size distribution, etc.); and, (iii) the thermal conditions (initial and boundary conditions of the DPF).

A standard $\varnothing 5.66'' \times$ L6$''$ ($\varnothing 0.144 \times$ L0.152 m) long monolith was considered as target DPF, representative of a commercial DPF. The same cell density (370 cpsi) and wall thickness (0.38 mm) of the pilot-scale prototype were maintained. Table 5 shows the DPF settings in this case.

**Table 5.** Parameter settings in the theoretical model of a real-scale bioSiC DPF filter.

| Inlet Settings—Simulating the Exhaust of a Real Engine (Light Duty) Fueled with Either Diesel or Biodiesel | |
| --- | --- |
| Composition of the gas stream | 78% $N_2$, 10% $O_2$, 5% $CO_2$, 6% $H_2O$, 0.1% CO |
| Volume flow rate | 175 $m^3 h^{-1}$ (0.035 kg $s^{-1}$) [52] |
| Temperature | 473 K (200 °C) |
| Pressure | - |
| Soot mass flow | 4.15 g $h^{-1}$ |
| Soot particle size distribution | Log-Normal ($\mu = 80 \times 10^{-9}$ m; $\sigma = 1.41$) |
| Soot aggregate structure | Fractal dimension: 2.07 [53]<br>Soot primary particle radius: $12 \times 10^{-9}$ m [54]<br>Soot primary particle density: 2000 kg $m^{-3}$ |
| **DPF Settings** | **Value** |
| Substrate length | 0.152 m (6$''$) |
| Plug length | 0.005 m |
| Substrate equivalent diameter | 0.144 m (5.66$''$) |
| Cell density | 370 cpsi |
| Wall thickness | 0.00038 m (14.96 mil) |
| **Substrate Properties Specific for bioSiC** | **Value** |
| Initial permeability (clean) | $4 \times 10^{-14}$ $m^2$ |
| Substrate pore diameter | $15.7 \times 10^{-6}$ m |
| Pore volume fraction | 0.49 |

Simulating the real exhaust conditions of a light-duty diesel engine (passenger car) with the following assumptions implied:

- A standard concentration of species (78% $N_2$, 10% $O_2$, 5% $CO_2$, 6% $H_2O$, 0.1% CO) was introduced
- The volume flow rate was 175 $m^3$ $h^{-1}$ (0.035 kg $s^{-1}$.) accordingly to the size of the new DPF. The estimation of the mass flow rate was derived from the engine map of a four-cylinder 1997 cc diesel engine (PSA DW10ATED) operating at 1800 rpm and 100 Nm [52].
- The initial particle size distribution and concentration was set to the standard values of a commercial automotive, light-duty diesel engine. The parameters used to characterize the distribution curve were taken from a 90 CV four-cylinder 1248 cc diesel engine (General Motors Z13DTH), belonging to the Department of Applied Science and Technology (DISAT) of the Politecnico di Torino. Its particulate (PM) emissions were measured with a scanning mobility particle sizer (SMPS), showing a log-normal distribution with mean particle diameter equal to 80 nm and standard deviation equal to 1.41. From this distribution, multiplying by the effective density of the particles in each size range, a soot production of 4.15 g $h^{-1}$ was calculated. The effective density was calculated with an empirical correlation [55]: $\rho(d) = 7543.9 \, d^{-0.56}$.
- The soot aggregate structure was defined through the following parameters: the primary particle radius was set to 12 nm according to [54]; the fractal dimension was set to 2.07 [53]; and, for the primary particle density, the default value 2000 kg $m^{-3}$ was left.

Table 5 also summarizes all these parameters used to simulate the real exhaust gases of a diesel engine.

There are three fixed parameters throughout the whole study. These are the three microstructural characteristics of the bioSiC substrate: permeability, porosity, and mean pore size. Also, the calibrated parameters shown in the Results section for the filtration and pressure drop performance of the filter were left unaltered, except for the soot permeability, as the permeability of real diesel soot may differ from that of the graphite soot produced artificially in a soot generator. In the extrapolation process to simulate a real-scale bioSiC DPF, the last step was to adapt the gas inlet temperature to simulate the real operating conditions of the DPF in the engine. Although the regeneration process was not studied in this work, the thermal equations and the reaction scheme were activated in the model, so the temperature of the gas raises the temperature of the substrate, slightly affecting the early stages of soot oxidation. The temperature of the inlet gas was set to 200 °C.

## 3. Results

Equations presented in Tables 1 and 2 summarize the numerical model used to simulate the performance of a bioSiC wall-flow filter. The validation of the model is achieved through the calibration of the empirical parameters discussed in Section 2.2 based on experimental tests. Table 6 shows all the resulting values.

**Table 6.** Calibrated values of the empirical parameters of the numerical model.

| Filtration Efficiency-Based Parameters | Pressure Drop-Based Parameters |
|:---:|:---:|
| $\eta_{clean,D} = 0.72$ | $k_w(\rho_P) = 0.7 \times 10^{-14} \text{ m}^2 \ (2.2\text{g/L})$ |
| $\eta_{clean,R} = 11$ | $k_p(\overline{p}) = 0.61 \times 10^{-15} \text{ m}^2$ |
| $nPe = -0.57$ | |
| $\eta_{load} = 0.28$ | |

In adjusting these parameters, an overlap between the calculated (model) and the experimental curves of filtration efficiency and pressure drop is sought. Figure 5 shows how these curves look after the calibration process.

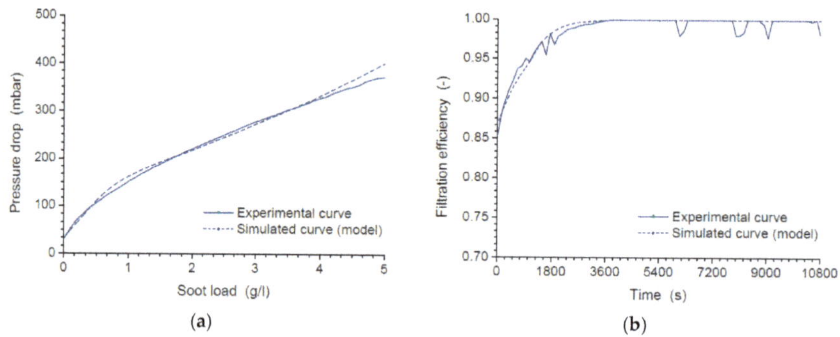


**Figure 5.** Comparison between the modeled and the experimental curves: (**a**) evolution of the pressure drop of the small wall-flow bioSiC DPF prototype with the soot load; (**b**) evolution of the filtration efficiency vs. time of the wall-flow bioSiC DPF prototype.

The real evolution of the pressure drop in the filter follows a very slow transition, from the deep-bed filtration stage to the soot-cake filtration stage. Biomorphic silicon carbide has a high soot storage capacity, so that the permeability of the loaded wall almost reaches (diminishing) the permeability of the soot, in the last loading stages, before the cake starts forming. In the simulation, this transition is a little sharper, as can be seen in Figure 5a. The filtration efficiency is high from the beginning, at the clean stage, but the growing trend is slower. The gradient parameter in wall filtration efficiency ($\eta_{load}$) is small, compared to a typical SiC substrate. The model shows a small gradient change at the transition from the deep bed filtration to the soot cake filtration. The late transition time point is a result of the high soot storage capacity of the substrate. In any case, after one hour of soot load, the filtration efficiency reaches a value close to 100% (Figure 5b). All simulations were performed in Axitrap [43].

The maximum deviation between the experimental pressure drop curve and the modeled curve is 8.7% and occurs in the transition point from deep-bed filtration to cake filtration (Figure 5a). For the rest of the points, the deviation remains below this limit, although a diverging trend can be noticed in the last part of the curve (soot load > 4 mg). In Figure 5b, the maximum error between the experimental filtration efficiency and that modeled, in averaged trends, is 4%. Higher local deviations (up to 9%) are observed at some points, due to oscillations in the experimental curves. The authors attribute these oscillations in the measured concentration of particles to the sudden release of particle agglomerations from the filter substrate due to the dragging effect of the gas flow.

*Filtration Efficiency and Pressure Drop of a Full-Size bioSiC DPF*

Next, the results of simulating the model for the full-size DPF are presented. Figure 6 shows the evolution of the pressure drop with the soot load, and the evolution of the filtration efficiency with time, both for the Ø5.66″ × L6.0″ (Ø0.144 × L0.152 m) bioSiC DPF with 370 cpsi and 0.38 mm wall thickness.

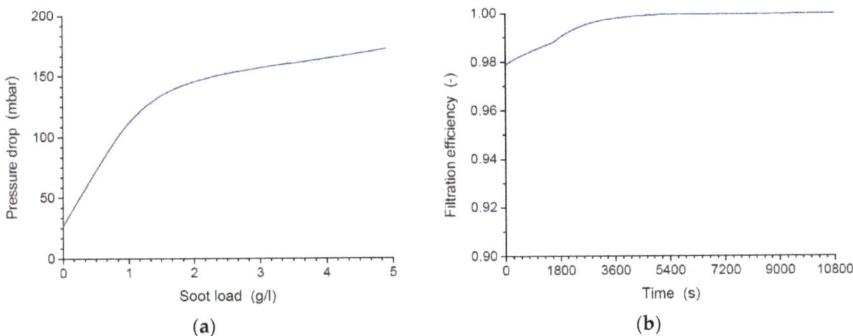

**Figure 6.** Expected performance of a full-size wall-flow bioSiC DPF (0.035 kg s$^{-1}$, 473 K): pressure drop (**a**), and filtration efficiency (**b**).

When the full-size filter is simulated, the resulting filtration efficiency is higher than that of the original prototype. This may be explained based on the different particle size distributions at the inlet and the dependence of the filtration efficiency on the particle diameter [56]. From 50 nm, the filtration efficiency of an MDF-bioSiC DPF increases with the particle size. It may vary between 0.55 and 0.85 for particles smaller than 100 nm, and between 0.7 and 0.95, for particles larger than 100 nm [33]. The soot generator used for the laboratory tests had a mean particle diameter of 144 nm, while a standard diesel engine has a mean particle diameter of 80 nm. Hence, when exposed to the particle distribution used in the laboratory tests, the bioSiC DPF has lower filtration efficiency than in the case of using the particle distribution of an engine.

## 4. Discussion

Results presented in Figure 6 cannot be directly compared to other published studies, due to differences in the boundary conditions and filter design. For example, gas flow rate, temperature, cell density, and wall thickness can significantly affect the results. In order to evaluate the performance of the bioSiC DPF compared to other options available in the market, a comparative study was carried out, adapting the bioSiC DPF geometrical features and test conditions to those reported in the literature.

### 4.1. Comparison with Commercial DPFs

In order to assess the performance of the bioSiC DPF as a particulate filter, and to compare it with other reported specimens, a set of reference cases from the literature were selected. They are summarized in Tables 7 and 8. Those values not given in the reference case were left unaltered as in Table 5.

**Table 7.** DPF reference cases simulation parameters (part one).

| Reference | | Mizutani, 2007 [57] | | | Dabhoiwala, 2008/09 [58,59] | | Tandon, 2010 [29] | | | Tsuneyoshi, 2011 [60] | |
|---|---|---|---|---|---|---|---|---|---|---|---|
| Case Identification | | Stationary engine test | Initial $\Delta$P test | Soot load $\Delta$P test | Heavy Duty, 20% load | Heavy Duty, 60% load | 275 cpsi $\varepsilon = 51\%$ d50 ~21.8 $\mu$m | 200 cpsi $\varepsilon = 49\%$ d50 ~13.4 $\mu$m | 300 cpsi $\varepsilon = 51\%$ d50 ~14 $\mu$m | Engine test | $\Delta$P test |
| **Inlet settings** | | | | | | | | | | | |
| Composition of the gas stream | | Exhaust gas | Air | Exhaust gas | Exhaust gas | | Exhaust gas | | | Exhaust gas | Air |
| Volume/mass flow rate (m$^3$/h–kg/s) | | 2000 rpm 50 Nm | 540–0.174 | 85.8–0.028 | 1469–0.245 | 2455–0.327 | | 26.49–0.012 | | 1400 rpm 190 Nm | 240–0.077 |
| Temperature (K–°C) | | | 298–25 | 473–200 | 563–290 | 707–434 | | 298–25 | | 623–350 | 298–25 |
| Soot mass flow (g/h) | | 0.6 | | | 15.46 | 18.25 | | 1.5 [61] | | | |
| Soot part size distribution | Log-Normal Mean ($\mu$) (m) | 70 | – | Gas burner | 144 | | | Gas burner 80 | | 80 | – |
| | Std. deviation ($\sigma$) (-) | – | | Soot gen. | 1.82 | | | 1.8 | | | |
| Soot aggregate structure | Fractal dimension Primary part radius (m) Primary part density (kg/m$^3$) | | | | | | | 20/35 | | | |
| **DPF settings** | | | | | | | | | | | |
| Substrate material | | SiC | | | Cordierite | | | | | SiC | |
| Substrate length (m–inch) | | 0.152–6 | | | 0.3048–12 | | | 0.152–6 | | 0.153–6 | |
| Plug length (m) | | | | | | | | | | | |
| Substrate diameter (m–inch) | | 0.144–5.66 | | | 0.2667–10.5 | | | 0.144–5.66 | | 0.144–5.66 | |
| Cell density (cpsi–cpsc) | | 300–46.5 | | | 200–31 | | 275–42.6 | 200–31 | 300–46.5 | 300–46.5 | |
| Wall thickness (mm–mil) | | 0.3048–12 | | | 0.3048–12 | | 0.3048–12 | 0.3048–12 | 0.33–13 | 0.25–10 | |

**Table 8.** DPF reference cases simulation parameters (part two).

| Reference | | Haralampous, 2004 [62] | | | Wolff, 2010 [63] | | | | Bollerhoff, 2012 [39] | |
|---|---|---|---|---|---|---|---|---|---|---|
| Case Identification | | Transient $\Delta$P (200 °C) | Transient $\Delta$P (400 °C) | Lab test | Engine test XP-SiC | Engine test SiC | Initial $\Delta$P test XP-SiC | Initial $\Delta$P test SiC | | |
| Inlet settings | | | | | | | | | | |
| Composition of the gas stream | | Exhaust gases | | Exhaust gas | Exhaust gas | Exhaust gas | Air | Air | Exhaust gas | |
| Volume/mass flow rate (m$^3$/h–kg/s) | | 245–0.05 approx. | 393–0.08 approx. | 362–0.0694 | 205–0.0417 | 205–0.0417 | 600–0.268 | 600–0.268 | 200–0.037 | |
| Temperature (K–°C) | | 473–200 | 673–400 | 513–240 | 473–200 | 473–200 | 298–25 | 298–25 | 523–250 | |
| Soot mass flow (g/h) | | 0 (filtered gases) | | 9 | | | | | 2.4 | |
| Soot part size distribution | Log-Normal Mean ($\mu$) (m) Std. deviation ($\sigma$) (-) | - | | Soot gen | | | | | 60 1.78 | |
| Soot aggregate structure | Fractal dimension Primary part radius (m) Primary part density (kg/m$^3$) | - | | | | | | | | |
| DPF settings Substrate material | | SiC | | XP-SiC | XP-SiC | SiC | XP-SiC | SiC | Cordierite | Improved SiC |
| Substrate length (m–inch) | | 0.152–6 | | | | 0.177–7 | | | 0.152–6 | |
| Plug length (m) | | 0.005 | | | | | | | | |
| Substrate diameter (m–inch) | | 0.144–5.66 | | | | 0.144–5.66 | | | 0.144–5.66 | |
| Cell density (cpsi–cpsc) | | 200–31 | | | | 300–46.5 | | | 200–31 | |
| Wall thickness (mm–mil) | | 0.38–15 | | 0.33–13 | 0.33–13 | 0.3048–12 | 0.33–13 | 0.3048–12 | 0.3048–12 | 0.3556–14 |

For each reference case, the specific characteristics of the DPF and the test conditions were maintained but leaving the microstructural features of a biomorphic substrate. When some data were not provided in the reference, the value for the specific variable was estimated or approximated, based on similar cases. The resulting filtration efficiency and pressure drop are then compared with those presented in each reference for each DPF example.

The comparative study has focused both on the initial clean-state performance of the DPF, as well as on the evolution during loading. The performance at clean state depends only on the characteristics of the substrate, while the evolution also depends on the soot and its interaction with the microstructure of the porous medium. Table 9 shows the results of the comparison study in terms of pressure drop at the clean stage. It summarizes the initial pressure drop of a number of commercial DPFs found in the literature, and, for each case, the calculated pressure drop that a bioSiC DPF would have if it had the same geometrical features, and were tested under the same conditions.

**Table 9.** Initial pressure drop of a bioSiC DPF compared to a similar commercial DPF.

| Bibliographic Source | Case from Table 7 | ΔP of the Reference DPF (kPa) | ΔP of a Similar bioSiC DPF (kPa) | Deviation |
|---|---|---|---|---|
| Mizutani, 2007 [57] | Soot load ΔP test | 1.5 | 1.5 | 0% |
| Mizutani, 2007 [57] | Initial ΔP test | 5.3 | 5.8 | +9.4% |
| Dabhoiwala, 2008 [58] | Heavy Duty, 20% load | 4.3 | 3.4 | −20.9% |
| Dabhoiwala, 2008 [58] | Heavy Duty, 60% load | 8.0 | 6.0 | −25.0% |
| Tsuneyoshi, 2011 [60] | ΔP test | 1.9 | 2.0 | +5.3% |
| Wolff, 2010 [63] | Lab test | 2.5 | 4.2 | +68.0% |
| Wolff, 2010 [63] | Initial ΔP test SiC | 4.9 | 9.0 | +83.7% |
| Wolff, 2010 [63] | Initial ΔP test XP-SiC | 4.5 | 9.2 | +104.4% |

The permeability of bioSiC made from MDF depends on the flow direction. When the gas flows in the compression direction of the panel (perpendicular to the fibres), the Darcian permeability is around $1\cdot10^{-12}$ $m^2$ [31]. When the gas flows perpendicular to the compression direction (in the same plane of the fibers), the Darcian permeability is around $4 \times 10^{-14}$ $m^2$ [33]. This latter permeability better characterizes the substrate of the real DPF prototype, and is used in this work. This permeability is relatively low compared to other commercial substrates for DPFs. As a result, the pressure drop of a bioSiC filter is, for the reference cases, in general, higher than the pressure drop of the majority of the other filters used for comparison. Only cordierite filters may have similar or higher pressure drops.

A similar comparison can be made with the initial filtration efficiency of the filters. Table 10 summarizes the initial efficiency of a number of commercial DPFs found in the literature, and, for each case, the calculated efficiency that a bioSiC DPF would have if it had the same geometrical features and were tested under the same conditions. The filtration efficiency of the bioSiC DPF is, in all the cases, higher than the values obtained for the reference cases, with values always above 90%. They are considerably high, taking into account that this represents only the clean stage, and that it will increase as the DPF gets loaded with particles.

**Table 10.** Initial filtration efficiency of a bioSiC DPF compared to a number of commercial DPFs.

| Bibliographic Source | Case from Table 7 | Initial Efficiency of the Reference DPF (%) | Initial Efficiency of a Similar bioSiC DPF (%) | Deviation |
|---|---|---|---|---|
| Tandon, 2010 [29] | 275 cpsi | 38 | 98.5 | +159% |
| Tandon, 2010 [29] | 200 cpsi | 50 | 98.2 | +96% |
| Tandon, 2010 [29] | 300 cpsi | 77 | 98.9 | +28% |
| Mizutani, 2007 [57] | Stationary engine test | 86 | 96.8 | +13% |
| Wolff, 2010 [63] | Engine test SiC | 53 | 95.8 | +81% |
| Wolff, 2010 [63] | Engine test XP-SiC | 60 | 96.5 | +61% |
| Wolff, 2010 [63] | Lab test | 90 | 94.0 | +4% |
| Bollerhoff, 2012 [39] | Cordierite | 60 | 96.6 | +61% |
| Bollerhoff, 2012 [39] | Improved SiC | 45 | 97.8 | +117% |
| Tsuneyoshi, 2011 [60] | Engine test | 55 | 94.7 | +72% |
| Dabhoiwala, 2009 [59] | Heavy Duty, 20% load | 52 | 94.6 | +82% |

## 4.2. Behavior under a New European Driving Cycle (NEDC)

In order to evaluate the performance under transient conditions and assess its potential to fulfil current regulation, the bioSiC wall-flow DPF was simulated under a standard driving cycle. A pre-set NEDC (New European Driving Cycle) 2.2 L diesel Euro 4 scenario was used, which includes data acquired from a Honda Accord, 2.2 i-CTDi, Euro 4 CI engine, tested on a roller bench. The NEDC is established by the current European regulation for the homologation of passenger cars [64] and comprises four identical ECE segments (urban cycles), followed by one EUDC segment (extra urban cycle). The total distance of the combined cycle is 11,023 m, the total test time is 1180 s, and the average speed is 33.6 km h$^{-1}$. The NEDC standard has been recently replaced by the WLTP at experimental level, but it is still necessary for computer model simulations [65].

In this scenario, soot emissions are automatically set as a log-normal distribution with $\mu$ = 60·10-9 m and $\sigma$ = 1.8. The soot inlet concentration was set at 9.4·10$^{-6}$ kg soot/kg exhaust, according to [66], where an amount of soot of 16 mg km$^{-1}$ of PM in a 2.2 L Euro 4 diesel vehicle without DPF, for the NEDC cycle, is reported. The geometrical definition of the DPF was left as in Table 5: Ø5.66″ × L6″ (Ø0.144 × L0.152 m) with 370 cpsi and 0.38 mm wall thickness. The substrate was characterized as biomorphic silicon carbide in microstructural terms: porosity, pore diameter, and permeability. The NEDC cycle was simulated with a time step of 1 s. The total particulate emissions produced during the cycle were calculated summing up the PM concentration along the whole period. The simulation of the MDF bioSiC wall-flow DPF under the NEDC cycle yields the graphs shown in Figure 7.

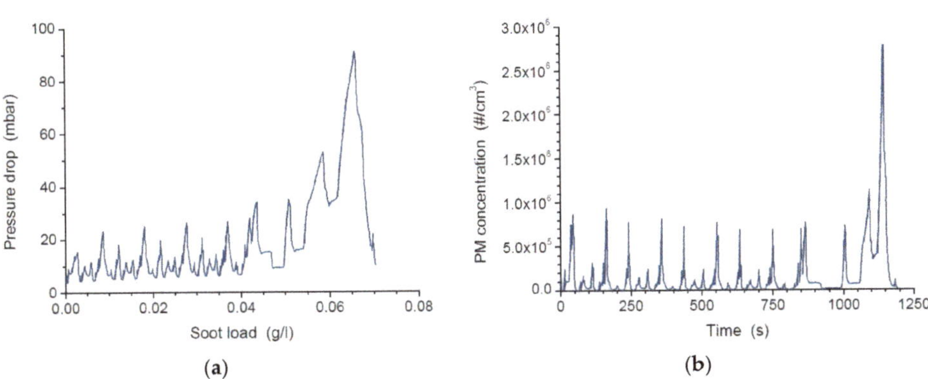

(a)

(b)

**Figure 7.** Simulated pressure drop (**a**) and particulate emissions (**b**) during a NEDC cycle with an MDF bioSiC wall-flow DPF.

In this case, the PM concentration at the outlet has been presented instead of the efficiency. Focusing the attention on the first four ECE segments (urban), a gradual increase in pressure drop and a gradual decrease in the particles release, can be observed. While the pressure drop peak in the first cycle is around 23.5 mbar, the peaks in the fourth cycle are around 27 mbar. The maximum pressure drop occurs in the extra-urban driving segment, due to the more severe operating conditions in this segment. This segment accounts for the majority of accumulated soot, but also for the majority of released particles. This is not only because the particles production in this segment is greater, but also because the filtration efficiency at this point has grown, due to the soot accumulated within the substrate in previous stages.

Summing up all the released PM emissions in the NEDC cycle, $8.75 \cdot 10^8$ particles $km^{-1}$ and $0.129$ mg $km^{-1}$ are obtained. These values are significantly below the thresholds imposed by current regulation: $6 \cdot 10^{11}$ particles $km^{-1}$ for the number of particles, and $4.5$ mg $km^{-1}$ for the mass. Although Euro 4 engines are no longer in production in Europe, there are car markets (e.g., India) where this technology can prove its efficiency regarding the pollutant emissions reduction. These results have been obtained with a non-optimized precursor/geometry integration. In this work, a bioSiC made from MDF has been used. It is characterized not only by having very good filtration efficiency, but also by having lower permeability than other tested substrates. There is potential to optimize it to maintain high filtration efficiency, reducing pressure drop, by choosing different precursors.

## 5. Conclusions

Internal combustion engines for transport applications require the urgent development of novel emission abatement systems. In this work, the potential of bioSiC wall-flow DPFs as aftertreatment systems in automotive engines is studied. These filters take advantage of the advanced microstructural features of the substrate, which can be tailored through the selection of the initial precursors, to reduce the pressure drop, and to optimize the filtration mechanisms as a function of the particle distribution of the combustion gases. By using bioceramic substrates with tailored microstructure, the trade-off between filtration efficiency and pressure drop may be overcome, and a better adaptation to new biofuel particulate emissions may be achieved. Biomorphic ceramics offer this chance of choosing from among a wide range of different microstructures: different combinations of porosity and pore size distribution with an orthotropic behavior. Different plant precursors or wood wastes can be used for generating the ceramic material, resulting in different microstructures and performance.

This work addresses, for the first time, the adaptation of a generic numerical model of wall-flow DPFs to specific microstructural parameters of the biomorphic ceramic substrate, with the aim of establishing a starting point in the generalized analysis of bioceramic precursors as automotive diesel particulate filters. A numerical model has been adapted and used to simulate a real-scale wall-flow particulate filter, made of biomorphic silicon carbide, and to predict its performance under different scenarios. The model relies on specific microstructural features of MDF-bioSiC.

It has been validated upon experimental data with a small-scale real prototype and a gas stream, laden with a distribution of laboratory generated particles with high accuracy.

Once calibrated and validated, the model was extrapolated to a real-scale DPF, in which the operating conditions of a real engine exhaust were simulated. An exhaustive analysis was then carried out to compare the predicted/modeled performance of the MDF-bioSiC DPF, and a number of DPF reference cases reported in the literature. The bioSiC DPF with a MDF substrate shows improvements in the filtration efficiencies in all cases. MDF-bioSiC DPFs show significantly high filtration efficiency, always above the efficiency of any other reported DPF, with the same geometry, and under the same testing conditions. The efficiency is around 95% at the clean stage in all the cases, and close to 100%, after a short period of time.

The model was simulated under the New European Driving Cycle (NEDC) test to evaluate the performance under transient and real driving conditions. The simulation of the bioSiC DPF under the NEDC shows that particulate emissions can be achieved, with this aftertreatment system, that are well below the regulation limits. In particular, the simulation predicts the release of 0.13 mg of soot per km, which is only 3% of the maximum mass of particles allowed by the Euro 6 standard. This shows the high interest of advancing in the development of these systems.

The pressure drop introduced by the MDF-bioSiC DPF is typically high. It is usually higher than the pressure drop introduced by other reference commercial DPFs, albeit not in all cases. This is due to the low permeability of the chosen precursor. In future works, optimization of system design/substrate integration with the engine will be developed, to improve pressure drop performance maintaining very high performance.

**Author Contributions:** Conceptualization, R.C.; Methodology, M.P.O. and O.H.; Software, O.H.; Validation, M.P.O.; Formal analysis, M.P.O.; Investigation, R.C., M.P.O., and O.H.; Resources, O.H.; Data curation, M.P.O.; Writing—original draft preparation, M.P.O.; Writing—review and editing, O.H., R.C., and M.T.G.; Visualization, M.T.G.; Supervision, R.C., O.H., and M.T.G.; Project administration, J.M.-F. and R.C.; Funding acquisition, J.M.-F. and R.C.

**Funding:** This research was funded by the Spanish Ministry of Economy and Competitiveness (MINECO), grant numbers MAT2013-41233-R, BES-2014-069023, and EEBB-I-17-12338.

**Acknowledgments:** This work was supported by the Spanish Ministry of Economy and Competitiveness (MINECO) (grant no. MAT2013-41233-R). M. Pilar Orihuela is very grateful to the MINECO for her pre-doctoral contract (grant no. BES-2014-069023), and for the mobility scholarship which allowed her to undertake the research stay at the Technological Educational Institute of Thessaly (grant no. EEBB-I-17-12338). The authors want to acknowledge the members of Exothermia SA for their technical support and for providing a free academic license for Axisuite. The experimental campaign was carried out with the support of Fino's research group, at the Department of Applied Science and Technology at the Politecnico di Torino.



## Nomenclature

| | |
|---|---|
| $C_1$ | Parameter for wall permeability correction (m kg$^{-1}$) |
| $C_2$ | Parameter for wall permeability correction (m$^4$ kg$^{-2}$) |
| $C_4$ | Slip correction factor (m s (kg mole K)$^{-0.5}$) |
| $C_E$ | Ergun coefficient (-) |
| $C_d$ | Parameter in the diffusional efficiency (-) |
| $C_r$ | Parameter in the direct interception efficiency (-) |
| $d$ | Channel width (m) |
| $d_{fib}$ | Fiber diameter (m) |
| $D_f$ | Fractal dimension (-) |
| $D_{part}$ | Particle diffusion coefficient (m$^2$ s$^{-1}$) |
| $k$ | Permeability (m$^{-2}$) |
| $Kn$ | Knudsen number (-) |
| $Ku$ | Kubawara hydrodynamic factor (-) |
| $M_g$ | Molecular weight (kg mole$^{-1}$) |
| $N$ | Number of walls (-) |
| $nPe$ | Exponent of the Peclet number for the diffusion mechanism (-) |
| $p$ | Pressure (Pa) |
| $Pe$ | Peclet number (-) |
| $r_0$ | Soot primary particle radius (m) |
| $R$ | Interception parameter |
| $\mathfrak{R}$ | Gas constant (J mole$^{-1}$ K$^{-1}$) |
| $R_c$ | Maximum radius of aggregate cluster (m) |
| $T$ | Temperature (K) |
| $v$ | Velocity (m s$^{-1}$) |
| $w$ | Thickness (m) |
| $z$ | Axial dimension |

| $\alpha_1$ | Constant in the channel pressure drop correlation (-) |
|---|---|
| $\varepsilon_{pore}$ | Porosity (-) |
| $\mu$ | Dynamic viscosity (Pa s) |
| $\eta$ | Filtration efficiency (-) |
| $\rho$ | Density (kg m$^{-3}$) |
| $\rho_0$ | Soot primary particle density (kg m$^{-3}$) |

**Subscripts**

| 0 | Initial (clean) state |
|---|---|
| $D$ | Diffusion |
| $i$ | Identifier of the channel: 1 for inlet channels, and 2 for outlet channels |
| $fib$ | Unit fiber |
| $p$ | Particles deposit |
| $R$ | Interception |
| $soot$ | Soot layer |
| $w$ | Wall |

## References


1. Barroso, J.M. Regulation No 459/2012. *Off. J. Eur. Comm.* **2012**, *142*, 16–24.
2. United States: Cars and Light-Duty Trucks: Tier 3. Available online: www.dieselnet.com/standards/us/ld_t3 (accessed on 26 June 2019).
3. Hooftman, N.; Messagie, M.; Van Mierlo, J.; Coosemans, T. A review of the European passenger car regulations—Real driving emissions vs local air quality. *Renew. Sustain. Energy Rev.* **2018**, *86*, 1–21. [CrossRef]
4. European Automobile Manufacturers' Association. WLTP Facts. Available online: http://wltpfacts.eu (accessed on September 15, 2019).
5. Regulation (EU) 2016/1628 of the European Parliament and the council, of 14 September 2016. *Off. J. Eur. Union* **2016**, *252*, 53–117.
6. Reşitoğlu, İ.A.; Altinişik, K.; Keskin, A. The pollutant emissions from diesel-engine vehicles and exhaust aftertreatment systems. *Clean Technol. Environ. Policy* **2015**, *17*, 15–27. [CrossRef]
7. The European Parliament and the Council Directive 2003/30/EC of the European Parliament and of the Council of 8 May 2003 on the promotion of the use of biofuels or other renewable fuels for transport. *Off. J. Eur. Union* **2003**, *L 123*, 42–46.
8. Wu, F.; Wang, J.; Chen, W.; Shuai, S. A study on emission performance of a diesel engine fueled with five typical methyl ester biodiesels. *Atmos. Environ.* **2009**, *43*, 1481–1485. [CrossRef]
9. Tayari, S.; Abedi, R.; Rahi, A. Comparative assessment of engine performance and emissions fueled with three different biodiesel generations. *Renew. Energy* **2020**, *147*, 1058–1069. [CrossRef]
10. Verma, P.; Stevanovic, S.; Zare, A.; Dwivedi, G.; Van, T.C.; Davidson, M.; Rainey, T.; Brown, R.J.; Ristovski, Z.D. An overview of the influence of biodiesel, alcohols, and various oxygenated additives on the particulate matter emissions from diesel engines. *Energies* **2019**, *12*, 1987. [CrossRef]
11. Ogunkunle, O.; Ahmed, N.A. A review of global current scenario of biodiesel adoption and combustion in vehicular diesel engines. *Energy Rep.* **2019**, *5*, 1560–1580. [CrossRef]
12. Mofijur, M.; Rasul, M.G.; Hyde, J.; Azad, A.K.; Mamat, R.; Bhuiya, M.M.K. Role of biofuel and their binary (diesel-biodiesel) and ternary (ethanol-biodiesel-diesel) blends on internal combustion engines emission reduction. *Renew. Sustain. Energy Rev.* **2016**, *53*, 265–278. [CrossRef]
13. E, J.; Pham, M.; Zhao, D.; Deng, Y.; Le, D.H.; Zuo, W.; Zhu, H.; Liu, T.; Peng, Q.; Zhang, Z. Effect of different technologies on combustion and emissions of the diesel engine fueled with biodiesel: A review. *Renew. Sustain. Energy Rev.* **2017**, *80*, 620–647. [CrossRef]
14. Karavalakis, G.; Bakeas, E.; Fontaras, G.; Stournas, S. Effect of biodiesel origin on regulated and particle-bound PAH (polycyclic aromatic hydrocarbon) emissions from a Euro 4 passenger car. *Energy* **2011**, *36*, 5328–5337. [CrossRef]
15. Durbin, T.D.; Collins, J.R.; Norbeck, J.M.; Smith, M.R. Effects of Biodiesel, Biodiesel Blends, and a Synthetic Diesel on Emissions from Light Heavy-Duty Diesel Vehicles. *Environ. Sci. Technol.* **2000**, *34*, 349–355. [CrossRef]

16. Kontses, A.; Dimaratos, A.; Keramidas, C.; Williams, R.; Hamje, H.; Ntziachristos, L.; Samaras, Z. Effects of fuel properties on particulate emissions of diesel cars equipped with diesel particulate filters. *Fuel* **2019**, *255*, 115879. [CrossRef]

17. Fontaras, G.; Karavalakis, G.; Kousoulidou, M.; Tzamkiozis, T.; Ntziachristos, L.; Bakeas, E.; Stournas, S.; Samaras, Z. Effects of biodiesel on passenger car fuel consumption, regulated and non-regulated pollutant emissions over legislated and real-world driving cycles. *Fuel* **2009**, *88*, 1608–1617. [CrossRef]

18. Szabados, G.; Bereczky, Á.; Ajtai, T.; Bozóki, Z. Evaluation analysis of particulate relevant emission of a diesel engine running on fossil diesel and different biofuels. *Energy* **2018**, *161*, 1139–1153. [CrossRef]

19. Buono, D.; Senatore, A.; Prati, M.V. Particulate filter behaviour of a Diesel engine fueled with biodiesel. *Appl. Therm. Eng.* **2012**, *49*, 147–153. [CrossRef]

20. Guan, B.; Zhan, R.; Lin, H.; Huang, Z. Review of the state-of-the-art of exhaust particulate filter technology in internal combustion engines. *J. Environ. Manag.* **2015**, *154*, 225–258. [CrossRef]

21. Lupše, J.; Campolo, M.; Soldati, A. Modelling soot deposition and monolith regeneration for optimal design of automotive DPFs. *Chem. Eng. Sci.* **2016**, *151*, 36–50. [CrossRef]

22. Martirosyan, K.S.; Chen, K.; Luss, D. Behavior features of soot combustion in diesel particulate filter. *Chem. Eng. Sci.* **2010**, *65*, 42–46. [CrossRef]

23. Guo, Y.; Stevanovic, S.; Verma, P.; Jafari, M.; Jabbour, N.; Brown, R.; Cravigan, L.; Alroe, J.; Godday, C.; Brown, R.; et al. An experimental study of the role of biodiesel on the performance of diesel particulate filters. *Fuel* **2019**, *247*, 67–76. [CrossRef]

24. Ajtai, T.; Pintér, M.; Utry, N.; Kiss-Albert, G.; Gulyás, G.; Pusztai, P.; Puskás, R.; Bereczky, A.; Szabados, G.; Szabó, G.; et al. Characterisation of diesel particulate emission from engines using commercial diesel and biofuels. *Atmos. Environ.* **2016**, *134*, 109–120. [CrossRef]

25. Bermúdez, V.; Serrano, J.R.; Piqueras, P.; Sanchis, E.J. On the impact of particulate matter distribution on pressure drop of wall-flow particulate filters. *Appl. Sci.* **2017**, *7*, 234. [CrossRef]

26. Savic, N.; Rahman, M.M.; Miljevic, B.; Saathoff, H.; Naumann, K.H.; Leisner, T.; Riches, J.; Gupta, B.; Motta, N.; Ristovski, Z.D. Influence of biodiesel fuel composition on the morphology and microstructure of particles emitted from diesel engines. *Carbon* **2016**, *104*, 179–189. [CrossRef]

27. Nyström, R.; Sadiktsis, I.; Ahmed, T.M.; Westerholm, R.; Koegler, J.H.; Blomberg, A.; Sandström, T.; Boman, C. Physical and chemical properties of RME biodiesel exhaust particles without engine modifications. *Fuel* **2016**, *186*, 261–269. [CrossRef]

28. Deng, Y.; Zheng, W.; Jiaqiang, E.; Zhang, B.; Zhao, X.; Zuo, Q. Influence of geometric characteristics of a diesel particulate filter on its behavior in equilibrium state. *Appl. Therm. Eng.* **2017**, *123*, 61–73. [CrossRef]

29. Tandon, P.; Heibel, A.; Whitmore, J.; Kekre, N.; Chithapragada, K. Measurement and prediction of filtration efficiency evolution of soot loaded diesel particulate filters. *Chem. Eng. Sci.* **2010**, *65*, 4751–4760. [CrossRef]

30. Chiavola, O.; Chiatti, G.; Sirhan, N. Impact of Particulate Size During Deep Loading on DPF Management. *Appl. Sci.* **2019**, *9*, 3075. [CrossRef]

31. Gómez-Martín, A.; Orihuela, M.P.; Becerra-Villanueva, J.A.; Martínez-Fernández, J.; Ramírez-Rico, J. Permeability and mechanical integrity of porous biomorphic SiC ceramics for application as hot-gas filters. *Mater. Des.* **2016**, *107*, 450–460. [CrossRef]

32. Orihuela, M.P.; Gómez-Martín, A.; Becerra-Villanueva, J.A.; Chacartegui, R.; Ramírez-Rico, J. Performance of biomorphic silicon carbide as particulate filter in diesel boilers. *J. Environ. Manag.* **2017**, *203*, 907–919. [CrossRef]

33. Orihuela, M.P.; Gómez-Martín, A.; Miceli, P.; Becerra-Villanueva, J.A.; Chacartegui, R.; Fino, D.; Becerra, J.A.; Chacartegui, R.; Fino, D. Experimental measurement of the filtration efficiency and pressure drop of wall-flow Diesel Particulate Filters (DPF) made of biomorphic Silicon Carbide using laboratory generated particles. *Appl. Therm. Eng.* **2018**, *131*, 41–53. [CrossRef]

34. Singh, M.; Martínez-Fernández, J.; Ramírez de Arellano-López, A. Environmentally conscious ceramics (ecoceramics) from natural wood precursors. *Curr. Opin. Solid State Mater. Sci.* **2003**, *7*, 247–254. [CrossRef]

35. Ramírez de Arellano-López, A.; Martínez-Fernández, J.; González, P.; Domínguez, C.; Fernández-Quero, V.; Singh, M. Biomorphic SiC: A New Engineering Ceramic Material. *Int. J. Appl. Ceram. Technol.* **2005**, *1*, 56–67. [CrossRef]

36. Bissett, E.J. Mathematical model of the thermal regeneration of a wall-flow monolith diesel particulate filter. *Chem. Eng. Sci.* **1984**, *39*, 1233–1244. [CrossRef]

37. Bisset, E.J.; Shadman, F. Thermal regeneration of diesel-particulate monolithic filters. *AIChE J.* **1985**, *31*, 753–758. [CrossRef]

38. Konstandopoulos, A.G.; Johnson, J.H. Wall-flow diesel particulate filters—Their pressure drop and collection efficiency. *SAE Trans.* **1989**, *98*, 625–647.

39. Bollerhoff, T.; Markomanolakis, I.; Koltsakis, G. Filtration and regeneration modeling for particulate filters with inhomogeneous wall structure. *Catal. Today* **2012**, *188*, 24–31. [CrossRef]

40. Exothermia SA. Available online: www.exothermia.com (accessed on 1 January 2017).

41. Konstandopoulos, A.G.; Kostoglou, M.; Kladopoulou, E.; Vlachos, N. Advances in the science and technology of diesel particulate filter simulation. *Adv. Chem. Eng.* **2007**, *33*, 213–294.

42. Koltsakis, G.; Haralampous, O.; Depcik, C.; Ragone, J.C. Catalyzed diesel particulate filter modeling. *Rev. Chem. Eng.* **2013**, *29*, 1–61. [CrossRef]

43. *Axisuite®Axitrap: Catalyzed Diesel Particulate Filter Simulation*; Exothermia SA; 4.02.4.; Exothermia SA: Thessaloniki, Greece, 2015.

44. Pulkrabek, W.W.; Ibele, W.E. The effect of temperature on the permeability a porous material. *Int. J. Heat Mass Transf.* **1987**, *30*, 1103–1109. [CrossRef]

45. Steffens, J.; Coury, J.R. Collection efficiency of fiber filters operating on the removal of nano-sized aerosol particles: I—Homogeneous fibers. *Sep. Purif. Technol.* **2007**, *58*, 99–105. [CrossRef]

46. Lee, K.W.; Liu, B.Y.H. Theoretical Study of Aerosol Filtration by Fibrous Filters. *Aerosol Sci. Technol.* **1982**, *1*, 147–161. [CrossRef]

47. Liu, B.Y.H.; Rubow, K.L. Efficiency, pressure drop and figure of merit of high efficiency fibrous and membrane filter media. In Proceedings of the Fifth World Filtration Congress, Nice, France, 5–8 June 1990.

48. Rodríguez-Fernández, J.; Lapuerta, M.; Sánchez-Valdepeñas, J. Regeneration of diesel particulate filters: Effect of renewable fuels. *Renew. Energy* **2017**, *104*, 30–39. [CrossRef]

49. Schneider, J.; Weimer, S.; Drewnick, F.; Borrmann, S.; Helas, G.; Gwaze, P.; Schmid, O.; Andreae, M.O.; Kirchner, U. Mass spectrometric analysis and aerodynamic properties of various types of combustion-related aerosol particles. *Int. J. Mass Spectrom.* **2006**, *258*, 37–49. [CrossRef]

50. Wentzel, M.; Gorzawski, H.; Naumann, K.H.; Saathoff, H.; Weinbruch, S. Transmission electron microscopical and aerosol dynamical characterization of soot aerosols. *J. Aerosol Sci.* **2003**, *34*, 1347–1370. [CrossRef]

51. Gysel, M.; Laborde, M.; Mensah, A.A.; Corbin, J.C.; Keller, A.; Kim, J.; Petzold, A.; Sierau, B.; Sierau, B. Technical note: The single particle soot photometer fails to reliably detect PALAS soot nanoparticles. *Atmos. Meas. Tech.* **2012**, *5*, 3099–3107. [CrossRef]

52. Stratakis, G.A.; Psarianos, D.L.; Stamatelos, A.M. Experimental investigation of the pressure drop in porous ceramic diesel particulate filters. *Proc. Inst. Mech. Eng. D J. Automob. Eng.* **2002**, *216*, 773–784. [CrossRef]

53. Ma, Y.; Zhu, M.; Zhang, D. Effect of a homogeneous combustion catalyst on the characteristics of diesel soot emitted from a compression ignition engine. *Appl. Energy* **2014**, *113*, 751–757. [CrossRef]

54. Lapuerta, M.; Martos, F.J.; Herreros, J.M. Effect of engine operating conditions on the size of primary particles composing diesel soot agglomerates. *Aerosol Sci.* **2007**, *38*, 455–466. [CrossRef]

55. Haralampous, O.; Payne, S. Experimental testing and mathematical modelling of diesel particle collection in flow-through monoliths. *Int. J. Engine Res.* **2016**, *17*, 1045–1061. [CrossRef]

56. Yang, J.; Stewart, M.; Maupin, G.; Herling, D.; Zelenyuk, A. Single wall diesel particulate filter (DPF) filtration efficiency studies using laboratory generated particles. *Chem. Eng. Sci.* **2009**, *64*, 1625–1634. [CrossRef]

57. Mizutani, T.; Kaneda, A.; Ichikawa, S.; Miyairi, Y.; Ohara, E.; Takahashi, A.; Yuuki, K.; Matsuda, H.; Kurachi, H.; Toyoshima, T.; et al. Filtration Behavior of Diesel Particulate Filters (2). *SAE Tech. Pap.* **2007**. [CrossRef]

58. Dabhoiwala, R.H.; Johnson, J.H.; Naber, J.D.; Bagley, S.T. A Methodology to Estimate the Mass of Particulate Matter Retained in a Catalyzed Particulate Filter as Applied to Active Regeneration and On-Board Diagnostics to Detect Filter Failures. *SAE Tech. Pap. Ser.* **2008**, *2008-01–0764*, 1–23.

59. Dabhoiwala, R.H.; Johnson, J.H.; Naber, J.D. Experimental Study Comparing Particle Size and Mass Concentration Data for a Cracked and Un-Cracked Diesel Particulate Filter. *SAE Tech. Pap. Ser.* **2009**, *2009-01–0629*, 1–12.

60. Tsuneyoshi, K.; Takagi, O.; Yamamoto, K. Effects of Washcoat on Initial PM Filtration Efficiency and Pressure Drop in SiC DPF. *SAE Tech. Pap. Ser.* **2011**, *2011-01–08*, 1–10.

61. www.testo.com Soot generator testo REXS–Reproducible EXhaust Simulator-Data sheet 1–2. Available online: https://static-int.testo.com/media/4d/22/d72019911546/testo-REXS-Datasheet-1981-9514.pdf (accessed on 10 December 2019).

62. Haralampous, O.A.; Kandylas, I.; Koltsakis, G.C.; Samaras, Z. Diesel particulate filter pressure drop Part 1: Modelling and experimental validation. *Int. J. Engine Res.* **2004**, *5*, 149–162. [CrossRef]

63. Wolff, T.; Friedrich, H.; Johannesen, L.; Hajizera, S. *A New Approach to Design High Porosity Silicon Carbide Substrates*; SAE Tech. Pap. 2010-01–0539; SAE International: Warrendale, PA, USA, 2010; pp. 1–11.

64. Gil-Robles, J.M. Directive 98/69/EC. *Off. J. Eur. Comm.* **1998**, *350*, 1–56.

65. Tucki, K.; Mruk, R.; Orynycz, O.; Wasiak, A.; Botwińska, K.; Gola, A. Simulation of the operation of a spark ignition engine fueled with various biofuels and its contribution to technology management. *Sustainability* **2019**, *11*, 2799. [CrossRef]

66. Williams, R.; Hamje, H.; Zemroch, P.J.; Clark, R.; Samaras, Z.; Dimaratos, A.; Jansen, L.; Fittavolini, C. Effect of fuel properties on emissions from Euro 4 and Euro 5 diesel passenger cars. *Transp. Res. Procedia* **2016**, *14*, 3149–3158. [CrossRef]

*Communication*

# Numerical Analysis of High-Pressure Direct Injection Dual-Fuel Diesel-Liquefied Natural Gas (LNG) Engines

**Alberto Boretti** 

Department of Mechanical Engineering, College of Engineering, Prince Mohammad Bin Fahd University, Al Khobar 31952, Saudi Arabia; a.a.boretti@gmail.com

Received: 19 January 2020; Accepted: 20 February 2020; Published: 25 February 2020

**Abstract:** Dual fuel engines using diesel and fuels that are gaseous at normal conditions are receiving increasing attention. They permit to achieve the same (or better) than diesel power density and efficiency, steady-state, and substantially similar transient performances. They also permit to deliver better than diesel engine-out emissions for $CO_2$, as well as particulate matter, unburned hydrocarbons, and nitrous oxides. The adoption of injection in the liquid phase permits to further improve the power density as well as the fuel conversion efficiency. Here, a model is developed to study a high-pressure, 1600 bar, liquid phase injector for liquefied natural gas (LNG) in a high compression ratio, high boost engine. The engine features two direct injectors per cylinder, one for the diesel and one for the LNG. The engine also uses mechanically assisted turbocharging (super-turbocharging) to improve the steady-state and transient performances of the engine, decoupling the power supply at the turbine from the power demand at the compressor. Results of steady-state simulations show the ability of the engine to deliver top fuel conversion efficiency, above 48%, and high efficiencies, above 40% over the most part of the engine load and speed range. The novelty of this work is the opportunity to use very high pressure (1600 bar) LNG injection in a dual fuel diesel-LNG engine. It is shown that this high pressure permits to increase the flow rate per unit area; thus, permitting smaller and lighter injectors, of faster actuation, for enhanced injector-shaping capabilities. Without fully exploring the many opportunities to shape the heat release rate curve, simulations suggest two-point improvements in fuel conversion efficiency by increasing the injection pressure.

**Keywords:** compression ignition; direct injection; cryogenic gas; diesel engines; dual fuel engines; natural gas; greenhouse gas emissions; particulate matter

---

## 1. Introduction

Renewable energy, that is practically ony wind and solar, cannot cover the world's total primary energy supply by 2050. The power needed by wind and solar, and the power and energy needed from the storage, are impossible to be achieved. Hence, there is a need to valorize fossil fuel resources, natural gas, oil, coal, as well as use uranium to better convert fossil fuel energy, and further reduce the environmental impact of their use. Not many are working on engines anymore; despite this fact, the world's growing total primary energy supply is still covered—by more than 90%—by combustion fuels, oil, coal, natural gas, biomass, and waste [1]. Since 1990, the contribution by combustion fuels has not changed at all; their use is increasing, with a constant share of the growing total [1]. Solar and wind, presently at 2% of the total primary energy supply, will not be able to satisfy the total primary energy supply by 2050 in a growing world. In addition to the limited opportunities to grow their installed capacity, there are even more limited opportunities to grow the energy storage needed to compensate wind and solar intermittency and unpredictability [2,3]. Hence, it still makes a lot of sense to improve the fuel conversion efficiency of internal combustion engines (ICEs), and improve their environmental friendliness, especially with alternative fuels. Under this reasonable perspective, there are opportunities to make more efficient ICEs, either with traditional fuel, or even better, with low carbon fuels (e.g., natural gas (NG)), or with zero-carbon fuels (e.g., hydrogen). As we still need internal combustion engines, there is the opportunity to design more fuel-efficient engines, especially for low carbon fuels. This could drastically improve $CO_2$ emissions, while also improve the use of natural resources. By replacing a conventional diesel engine with a higher efficiency, diesel-liquefied natural gas (LNG) engine, there is the advantage of the C/H ratio of the fuel, the advantage of reduced depletion of natural resources, the advantage of reduced emissions of pollutants, starting from PM, the diversification of fuel sources, the security of the fuel sources, see references from [4–22], just to mention a few. The starting point for this work is given in [4], which describes the advantages of diesel-LNG engines. Here, the focus is on the better shaping of the heat release rate curve, thanks to the use of very high injection pressures of 1600 bar.

The development of high-power density high-efficiency ICEs, fueled with fuels that are gas at normal conditions, is becoming increasingly relevant. Some of the most promising designs are the dual-fuel, diesel injection ignition, direct injection (DI), and the compression ignition (CI). Further improvement of these engines calls for better-dedicated injectors, see references from [23–27]. These injectors must work at high pressures, with cryogenic fluids, delivering substantial amounts of fuel flow energy within brief time frames, with high speeds of actuation. The aim of this work is to investigate the benefit that a higher pressure direct injector may offer in an ICE environment characterized by a high boost and a high compression ratio, typical of the latest racing diesel ICEs. A high compression ratio translates into high thermal efficiency, while a high boost translates into high power density. High boost and high compression ratio are two requirements of today's diesel engines—for passenger cars, but especially for racing car applications. Diesel engines for passenger cars, such as the 3.0l V6 turbo direct injection (TDI) engine developed by Audi for the VW Phaeton and Touareg, have a compression ratio of 17, coupled to a high boost turbocharger to deliver a maximum torque of 500 Nm, which translates into a brake mean effective pressure (BMEP) of 20.94 bar.

High-pressure DI injectors were proposed in the past to work with gaseous fuels up to 200 bar, delivering up to 23 g/s of methane with effective minimum passage areas of about 0.8 mm$^2$. The adoption of injection pressures of 1600 bar, coupled to the cryogenic delivery of the liquefied natural gas (LNG) at 113 K, allows much larger flow rates per unit of effective flow area; thus, allowing much larger fuel energy flows with smaller injectors, that are lighter and much faster actuating. Simulations are presented for a dual fuel diesel-LNG engine, featuring two injectors per cylinder, one for the diesel, and one for the LNG. The LNG injector is one of these new generation injectors. While the coupling of a pilot/pre-injection of diesel with a main injection of LNG is straightforward; mixed modes of combustion are also possible, injecting a part of the LNG before, and a part of the LNG after the diesel injection ignition occurs, see again references from [23–27]. The LNG injected, prior or contemporary to the diesel, then burns premixed, and the LNG injected after the diesel combustion starts, then it burns diffusion controlled. The engine features a high boost and high compression ratio. It has a super turbocharger, where the turbocharger shaft is connected to the crankshaft by gears and a continuously variable transmission (CVT), to produce the required boost in any operating condition, either steady or transient. The super turbocharger also allows the recovery of the extra energy at the turbine, either steady or transient, references [28–31]. In a hybrid powertrain, the super-turbocharger may also be replaced by an F1 style [32] electrically assisted turbocharger where the extra energy to the turbocharger, or the extra energy from the turbocharger, is drawn from or delivered to the traction battery via a motor-generator unit. In this case, the apparent steady-state efficiency of the engine is larger, especially at low speeds and high loads, as the extra power at the compressor is delivered by the electric motor. At high speed and loads, the apparent steady-state efficiency is otherwise marginally smaller as the extra power at the turbine is delivered to the electric generator.

One example of promising gas injectors that did not evolve to widespread products was the Hoerbiger injector. This direct injection injector was developed as part of the BMW effort towards improved hydrogen engines during the first decade of this century, see references [33–35]. The high-pressure direct injector GV1 had an equivalent flow area of 0.8 mm$^2$, maximum inlet pressure 200 bar, nom. inlet pressure 150 bar. It was intended for operation with maximum cylinder pressure 150 bar. It was also operated at 300 bar of pressure. The steady-state flow rate with compressed natural gas (CNG), i.e., gaseous methane, was 23 g/s. The operating temperature was −40 °C to 120 °C. Internal leakage was <0.2%, and injection accuracy ±2%. The minimum injection duration was 1.0 ms and response time ~0.5 ms. There was also a version of the direct injector with active closing (double acting). The equivalent flow area was 0.7 mm$^2$. The maximum inlet pressure was 200 bar and the nom. inlet pressure 150 bar. The maximum cylinder pressure was 180 bar. The steady-state flow rate (CNG) was 20 g/s. The operating temperature was −40 °C to 120 °C, the internal leakage <0.1%, the injection accuracy ±2%, the minimum injection duration 0.5 ms and the response time ~0.1 ms. These injectors, which had solenoid designs, were very far from delivering the same speed of actuation and energy flow rates of highly atomized diesel fuel permitted by the latest 2850 bar piezo diesel injectors, extremely helpful in shaping the heat release rate curve by shaping the fuel injection profile to match performance and emission criteria within constraints of maximum pressure and rate of pressure build-up dp/dθ ($p$ pressure, θ crank angle).

It is also very well known, from the experience by Westport, references [36–42], to name a few references, with thousands of heavy-duty trucks, diesel engines converted to diesel-LNG by using their patented dual-fuel diesel-LNG injector; when replacing the main injection of the diesel with a main injection of the LNG, there is no penalty in the fuel conversion efficiency, nor in the BMEP output. The Westport results are supported by engine performance simulations, engine laboratory experiments, and the experience of heavy-duty truck drivers.

The innovation proposed in this paper vs. prior studies, such as [4], is the opportunity to use very high pressure (1600 bar) for the injection of liquefied natural gas (LNG) in a dual fuel diesel-LNG engine. As demonstrated for the diesel, injection pressures matter. In the diesel, injection pressures have dramatically increased during the 25 years between 1990 and 2015, from 1000 to almost 3000 bar. This increment allowed the delivery of much better power densities and fuel conversion efficiencies, while also reducing the pollutant formation within the cylinder, within constraints of peak pressure and rate of pressure build-up, through the shaping of the heat release curve. For diesel engines designed for the early 1990s, injection pressures were between 1000 and 1200 bar versus the pre-1990s level of 650 to 700 bar. In 2013, injectors using pressures up to 2850 bar were already considered. The increment of the injection pressure has many benefits. The higher pressure allows the increase of the flow rate per unit area; thus, permitting smaller and lighter injectors, of faster actuation, for enhanced injector-shaping capabilities. Therefore, the paper is not just about a higher injection pressure, it is about the enhanced capabilities of shaping the heat release curve thanks to higher injection pressure, in the case of LNG, as it has been done for the diesel.

The sample simulations are based on a diesel racing engine studied in 2013 and 2014. The numerical model for the diesel-LNG engine is obtained by only adding a second direct injector for the LNG, to the numerical model for the diesel only. The proposed concept is applicable to engines for passenger cars, light-duty trucks, racing cars, but mostly heavy-duty trucks. The limit of peak pressure obviously changes according to the specific application, the same for the details of the injection. Injection shaping is generally performed by targeting specific fuel conversion efficiency within the specific constraints of peak pressure, the gradient of peak pressure, peak temperature, pollutants, and others.

## 2. Materials and Methods

A numerical method is used to describe the operation of the engine featuring the proposed high-pressure direct injector specific to LNG. This section describes the model. Then, the following section proposes the results. Engine performance simulations are performed using state-of-the-art engine performance simulations. Details of the engine system are provided below, as well as the injector modeling, and the modeling of the super-turbocharger, which are not conventional engine system components. Engine performance simulations are performed by using state of the art computer-aided engineering (CAE) computer codes modeling the steady-state operation of super-turbocharged diesel and methane engines. Engine performance simulations were performed by using very well-known commercial CAE software tools. As these codes have now been around for more than 40 years, with well-known skills and limitations, their description is unnecessary. Pollutant emissions, as well as transients, are disregarded in the present study.

The fuel flow rates per unit cross-sectional area, for natural gas, at normal temperatures and different pressures, are considered first. Usually, compressed natural gas (CNG) is pressurized to 300 bar, but as Westport considers higher pressures of 600 bars for the injection of cryogenic fuel (LNG), there is a column detailing the operation with CNG at 600 bar. Then, the injection of natural gas at cryogenic temperatures (LNG) is considered, with injection pressures up to 1600 bar.

Gas and liquid injector flow equations are different. For gas, while the flow rate is independent of backpressure if the flow is choked, it obviously depends on the upstream pressure ($p_0$) and temperature ($T_0$) and the specific gas properties. The gas constant (R) and specific heat ratio ($\gamma$) of methane impact on the choked flow conditions. The flow equation for a choked nozzle is the following [43]:

$$\dot{m} = \frac{A \cdot p_0}{\sqrt{T_0}} \cdot \sqrt{\frac{\gamma}{R}} \cdot \left(\frac{\gamma + 1}{2}\right)^{\frac{-(\gamma+1)}{2 \cdot (\gamma-1)}} \tag{1}$$

where A is the effective minimal area. For methane, the gas constant R is 518.28 J/kg K and the specific heat ratio $\gamma$ is 1.31. For 8 holes, 0.4 mm diameter holes' injector, having a discharge coefficient of 0.8, the geometric minimal area is $1.01 \cdot 10^{-6}$ m$^2$ and the effective minimal area A is $8 \cdot 10^{-7}$ m$^2$. Table 1 presents the mass flow rates for methane CH$_4$ gas injected through an injector with a choked flow area of $8 \cdot 10^{-7}$ m$^2$ at various pressures, and ambient temperature 300 K. As the latest injectors by Westport for LNG are being developed for 600 bar [44], this option is also included in Table 1 for CNG. The table provides a steady fuel flow rate ($\dot{m}$) and the fuel energy flow rate ($\dot{m} \cdot$LHV), with LHV the lower heating value. The lower heating value LHV is the amount of heat released by combustion of a kg of fuel and returning the temperature of the combustion products to 150 °C. Opposite to the higher heating value, the latent heat of vaporization of water in the reaction products is not recovered. As the combustion of the fuel injected after the diesel injection ignition has occurred is limited by the diffusion of the quickly vaporized fuel, and the mixing with air, the fuel energy flow rate is an important parameter in shaping the heat release rate curve. In addition, the actuation speed is relevant. Clearly, moving from 200 to 300 and then 600 bar, there is a drastic increment of $\dot{m}$ and $\dot{m} \cdot$LHV.

**Table 1.** Injection of methane gas. By increasing the injection pressure of the compressed natural gas (CNG) the fuel energy flow rate increases, but not enough.

| A | m$^2$ | $8 \times 10^{-7}$ | $8 \times 10^{-7}$ | $8 \times 10^{-7}$ |
|---|---|---|---|---|
| $p_0$ | Pa | $2 \times 10^{+7}$ | $3 \times 10^{+7}$ | $6 \times 10^{+7}$ |
| $T_0$ | K | 300.0 | 300.0 | 300.0 |
| $\gamma$ | | 1.31 | 1.31 | 1.31 |
| R | J/kg K | 518.28 | 518.28 | 518.28 |
| $\dot{m}$ | g/s | 26.70 | 40.04 | 80.09 |
| LHV | kJ/g | 50 | 50 | 50 |
| $\dot{m} \cdot$LHV | kJ/s | 1335 | 2002 | 4004 |

To further increase the fuel flow rate, as well as the fuel energy flow rate, making the injector smaller and much quicker to operate, there is the need to move to cryogenic injection temperatures, 113 K for methane. At this temperature, methane is liquid. Having high pressure and lower temperature allows the increase of flow rate per unit effective flow area; thus, making injectors smaller and faster actuating, but delivering high flow with high atomization. Injection shaping is mandatory for building high-efficiency, high power density, and dual fuel diesel injection ignition DI CI engines. Properties of methane are given in Table 2. Data is from [45]. The Joule–Thomson effect represents the temperature change of a real gas or liquid when forced adiabatically through a valve [46]. The inversion temperature on methane expansion at a low temperature is a relevant phenomenon presently neglected in engine models that should be otherwise accounted for. The isobaric properties of methane are available with a minimum temperature limit. This is the highest of the following values 90.69 K, and the temperature at which a density of 451.48 kg/m$^3$ is reached. Additionally, there is a maximum value of the temperature of 625 K. Concerning the isothermal properties, the acceptable range of temperatures is 90.7 to 625.0 K. The pressure is limited to the value at which a density of 451.48 kg/m$^3$ is reached [45], is not covering the opportunity of having pressure 1600 bar and temperature 113 K. The minimum temperature available with 1600 bar is 127 K. The maximum pressure available with 113 K is 900 bar. The properties of LNG in the range of pressures and temperatures of interest, down from 1600 bar and up from 113 K, are obtained by approximating the isothermal and isobaric curves of [45], extrapolated for the missing values with polynomials and interpolating between them. Every parameter is given as a function of pressure and temperature.

**Table 2.** Properties of cryogenic methane. Data from [45]. These data permit to set up the polynomial fluid model in the simulations.

| | | | |
|---|---|---|---|
| Temperature (K) | 113 | Temperature (K) | 127 |
| Pressure (bar) | 900 | Pressure (bar) | 1600 |
| Density (kg/m$^3$) | 473.6 | Density (kg/m$^3$) | 487.94 |
| Volume (m$^3$/kg) | 0.002112 | Volume (m$^3$/kg) | 0.002049 |
| Internal Energy (kJ/kg) | −40.279 | Internal Energy (kJ/kg) | −17.178 |
| Enthalpy (kJ/kg) | 149.76 | Enthalpy (kJ/kg) | 310.73 |
| Entropy (J/g·K) | −0.44156 | Entropy (J/g·K) | −0.31288 |
| $C_v$ (J/g·K) | 2.2275 | $C_v$ (J/g·K) | 2.2706 |
| $C_p$ (J/g·K) | 3.1545 | $C_p$ (J/g·K) | 3.0968 |
| Sound Speed (m/s) | 1862.7 | Sound Speed (m/s) | 2057.8 |
| Joule–Thomson (K/bar) | −0.053094 | Joule–Thomson (K/bar) | −0.054012 |
| Viscosity (Pa·s) | 0.00024 | Viscosity (Pa·s) | 0.000252 |
| Thermal Conductivity (W/m·K) | 0.25827 | Thermal Conductivity (W/m·K) | 0.29263 |
| Phase | liquid | Phase | liquid |

For incompressible flow [43], the mass flow rate may be simply approximated as:

$$\dot{m} = A \cdot \rho \cdot \sqrt{\frac{2 \cdot (p_0 - p_c)}{\rho}} \tag{2}$$

where $\rho$ is the density and $p_c$ is the downstream chamber pressure. Again, A is the effective minimal area, i.e., the geometric area multiplied by a discharge coefficient. By taking $p_0 = 600 \cdot 10^5$ Pa and $p_c = 150 \cdot 10^5$ Pa, we have for methane a steady fuel flow rate $\dot{m}$ of 162.81 g/s, and a similarly increased fuel energy flow rate $\dot{m} \cdot$ LHV of 8141 kJ/s. By increasing the injection pressure to 1600 bar, mass and energy flow rates further increase, see Table 3. This permits to drastically reduce the geometric flow areas, for reduced weight, and faster actuation of the injector in a piezo or solenoid design. Another relevant factor is the improved atomization of LNG with higher pressure, because of the smaller diameters of the nozzles, and the larger injection pressure. The above approach provides values for the mass flow rate to be used as an indication of the potentials of the cryogenic high-pressure strategy. Actual injection parameters may be slightly different [47,48], because of the compressibility of the liquid fluid and the vaporization across the injector. Modeling of diesel injectors is usually performed by considering a compressible liquid fuel that may change phase to a vaporized fuel, [49–52]. With methane liquid, compressibility and change of phase are extremely relevant for the way the fluid leaves the injector and then vaporizes. The latent heat of vaporization (HVAP) is then 0.5 kJ/g with methane. Other properties of methane are finally given in Table 4.

**Table 3.** Injection of methane liquid. By increasing the injection pressure of the liquefied natural gas (LNG) to 1600 bar and reducing the injection temperature to 113 K the fuel energy flow rate increases dramatically.

| | | | | | | |
|---|---|---|---|---|---|---|
| $A$ | $m^2$ | $8 \times 10^{-7}$ | $8 \times 10^{-7}$ | $8 \times 10^{-7}$ | $8 \times 10^{-7}$ | $3.56 \times 10^{-7}$ |
| $p_0$ | Pa | $3 \times 10^{+7}$ | $6 \times 10^{+7}$ | $1.6 \times 10^{+8}$ | $1.6 \times 10^{+8}$ | $1.6 \times 10^{+8}$ |
| $T_0$ | K | 113 | 113 | 127 | 113 | 113 |
| $p_c$ | Pa | $1.5 \times 10^{+7}$ | $1.5 \times 10^{+7}$ | $1.5 \times 10^{+7}$ | $1.5 \times 10^{+7}$ | $1.5 \times 10^{+7}$ |
| $\rho$ | kg/m$^3$ | 443.53 | 460.2 | 487.94 | 500 | 500 |
| $\dot{m}$ | g/s | 92 | 163 | 301 | 305 | 135 |
| $LHV$ | kJ/g | 50 | 50 | 50 | 50 | 50 |
| $\dot{m}{\cdot}LHV$ | kJ/s | 4614 | 8141 | 15,047 | 15,232 | 6770 |
| $HVAP$ | kJ/g | 0.4806 | 0.4806 | 0.4806 | 0.4806 | 0.4806 |
| $\dot{m}{\cdot}LHV^*$ | kJ/s | 4570 | 8062 | 14,902 | 15,085 | 6705 |

**Table 4.** Other properties of methane. Data from [45,53]. Opposite to the diesel, methane needs ignition. Methane is much easier to vaporize, mix with air, and burn than the diesel once combustion is started.

| | | | |
|---|---|---|---|
| Critical temperature ($T_c$) (K) | 190.564 | Flashpoint (K) | 85 |
| Critical pressure ($P_c$) (bar) | 45.992 | Flammabilityin air (%) | 5.3–15 |
| Critical density ($D_c$) (kg/m$^3$) | 162.66 | Auto-ignition temp.(K) | 813 |
| Normal boiling point (K) | 111.667 | Octane number | 125 |
| Molecular Weigh | 16.04 | Heat (latent) of vaporization (kJ/kg) | 511 |
| Energy Content (higher heating value) (MJ/kg) | 55.5 | Energy Content (lower heating value) | 50 |
| Air-to-fuel ratio (by mass) | 17.2:1 | Stoichiometric percentage fuel (by mass) | 5.8% |

The fuel pump power, needed per unit of fuel flow rate, is roughly $\Delta p/(\rho{\cdot}\eta_p)$, where $\Delta p$ is the pressure rise, $\rho$ is the density, and $\eta_p$ the pump efficiency, typically 0.85. This has to be compared with the power produced by the engine per unit fuel flow rate, that is $LHV{\cdot}\eta_e$, where $\eta_e$ is the engine efficiency, that is approaching 50%. If $\Delta p = 1.6{\cdot}10^8$ Pa and $\rho = 500$ kg/m$^3$, since $LHV = 50{\cdot}10^6$ J/kg, the ratio of the pump power to engine power is a considerable, but still accepTable 1.51%. This is less than the power requirement of a 2850 bar diesel injector, where $\Delta p = 2.85{\cdot}10^8$ Pa $\rho = 846$ kg/m$^3$ and $LHV = 42.6{\cdot}10^6$ J/kg, translating in a ratio of the pump power to engine power of 1.86% (the efficiency of the diesel is also slightly less, further increasing this ratio).

To be precise, as the engine is dual fuel, the total fuel pump power to engine power is the sum of the fuel pump power to engine power of the LNG pump, plus the fuel pump power to engine power of the diesel pump, where, however, both fuels contribute to the total fuel energy. Thus, if $\alpha$ is the fuel energy replacement by LNG, typically 0.95, then the ratio of LNG fuel energy to diesel fuel energy is $1/(1-\alpha)$, or about 20, and the ratio of LNG mass flow rate to diesel mass flow rate is $LHV_{diesel}/LHV_{LNG}/(1-\alpha)$, or about 17.44. Hence, the actual ratio of the pump power to engine power is a linear interpolation of the values for LNG only and diesel only weighted by the relative mass fraction, thus an intermediate number much closer to 1.51% than 1.86%, at about 1.53% in the case of $\alpha = 0.95$.

The engine considered is a 3.7-L TDI V6. The main parameters are proposed in Table 5. The V6 has a V-angle of 120 degrees.

**Table 5.** Main Engine Geometrical Parameters.

| | | | |
|---|---|---|---|
| Bore [mm] | 93.0 | Compression Ratio | 17.50 |
| Stroke [mm] | 90.7 | Bore/Stroke | 1.025 |
| Connecting Rod Length [mm] | 180.0 | intake valve closure [CA] | −123 |
| Piston Pin Offset [mm] | 0.00 | exhaust valve opening [CA] | 147 |
| Displacement/Cylinder [liter] | 0.616 | intake valve opening [CA] | 329 |
| Total Displacement [liter] | 3.697 | exhaust valve closure [CA] | 407 |
| Number of Cylinders | 6 | Layout | V120 |
| Number of Intake Valves | 2 × 6 | Number of Exhaust Valves | 2 × 6 |
| Intake Valve Ref. Diameter [mm] | 50 | Exhaust Valve Ref. Diameter [mm] | 45 |
| Intake Valve Max. Lift [mm] | 12.6 | Exhaust Valve Max. Lift [mm] | 12.5 |

This engine model was developed as part of a study for the engine of a Fédération Internationale de l'Automobile (FIA) world endurance championship, Le Mans Prototype class 1 hybrid car powered by a diesel engine. The target performances were those expected from the Audi LMP1-H [54]. The research and development activities were performed in 2013 and 2014. The modeling studies were performed by using Gamma Technologies GT-Power as well as Ricardo Wave. GT-Power and Wave have many similarities, with minimal differences in the physics being represented. The major difference for the specific application is the opportunity for GT-Power to connect the turbocharger shaft to the crankshaft, which is not permitted in Wave. Wave is, however, much simpler and less demanding in terms of computer requirements. The problem was solved in Wave through post-processing, correcting the engine outputs. The models were used to simulate super-turbocharging, with the gears and a continuously variable transmission (CVT) toroidal mechanism, connecting crankshaft and turbocharger shaft, as well as a motor-generator unit connected to the turbocharger shaft of electrically assisted turbocharging. In the latter case, the crankshaft power output had to be depurated of the power from/to the turbocharger shaft during post-processing. Through the motor/generator unit (MGU-H), in this case, the extra power is delivered to, or extracted from, the traction battery, which is also used by the kinetic motor generator unit (MGU-K) delivering power to the wheels, or recovering the braking energy. Figure 1 presents a sketch of the 2014 GT Power model.

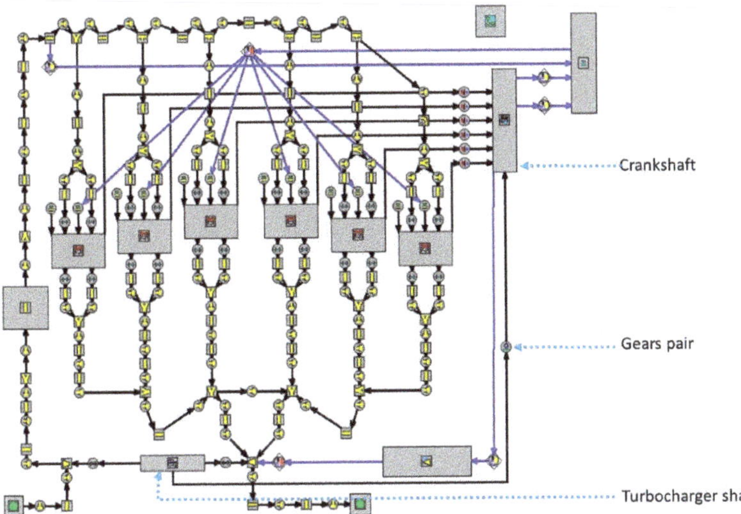


**Figure 1.** Sketch of the engine model featuring two direct injectors per cylinder, and a super-turbocharger. The superturbocharger is modeled by connecting the crankshaft and turbocharger shaft with a junction prescribing speed ratio and mechanical efficiency.

Worthy to note, this model adopts two injectors per cylinder. The turbocharger shaft is connected to the crankshaft through a simple gears pair. The speed ratio across this gears pair and the efficiency is prescribed at every speed and load. This V6 TDI CAE model allows simulation of the operation of super-turbocharged engines, for compression ignition (CI) operation, with diesel or dual fuel diesel-alternative fuels. The intake and exhaust systems are represented through pipes and junctions where the 1D conservation equations of mass, momentum, energy, and species are solved. Complex elements, such as engine cylinders, are represented by special junctions. Single or multiple zones 0D models are used to describe the in-cylinder space that receives intake flows and distribute outflows through the intake and exhaust valves, represented as orifices, and fuel injectors, by solving conservation equations for mass and energy. Ambient junctions are used at the start of the intake system and at the end of the exhaust system. Turbines and compressors are also modeled through special junctions. The model returns the indicated mean effective pressure IMEP. By computing the friction mean effective pressure (FMEP) through a correlation, the brake mean effective pressure is finally computed.

Regarding the super turbocharger, the mechanical efficiency of the connection to the crankshaft is taken as shown in Figures 2 and 3. This efficiency only impacts on the extra energy to or from the turbocharger. The speed of the turbocharger is usually kept at the equilibrium value, i.e., the speed where the turbine power is equal to the compressor power. Apart from selected engine high speeds and loads, and low speeds high loads points, the result is not affected by the value of the transmission efficiency. It is assumed here to use a full toroidal CVT, of the spread of speed ratios 7.930, same as the Torotrack V-Charge described later, that however differs for the external overdrive, and for the fact it connects a compressor shaft rather than a turbocharger shaft to the crankshaft. The CVT works with speed ratios from 0.355:1 (underdrive) up to 2.816:1 (overdrive, 2.816 = 1/0.355). The spread of speed ratios is thus $2.816/0.355 = 2.816^2 = 1/0.355^2 = 7.930$. The external overdrive is here 21.275. The maximum value of the ratio between the speed of the turbocharger and the speed of the engine is, thus, 60, while the minimum ratio is 7.566. The speed of the turbocharger can, thus, be adjusted in between these two values at any speed and load of the engine. Opposite to the Torotrack V-Charge that transmits the full power required to the compressor, this transmission only transmits to or from the turbocharger the difference between compressor and turbine powers, which is usually much less.

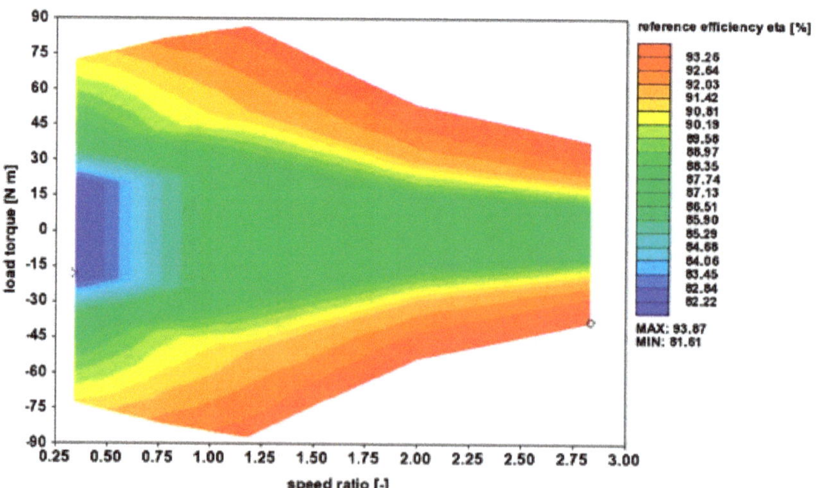


**Figure 2.** Reference efficiency map $\eta(\tau, T_1)$ of the continuously variable transmission (CVT) based transmission. Image reproduced from [30]. Creative Commons Attribution Non-Commercial 4.0 International (CC BY-NC 4.0) License.

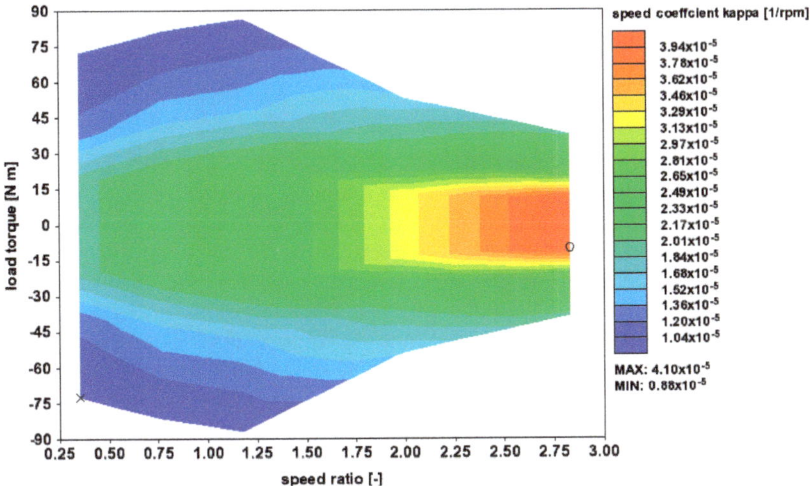


**Figure 3.** Speed coefficient map $(\tau, T_1)$ of the CVT based transmission. Image reproduced from [30]. Creative Commons Attribution Non-Commercial 4.0 International (CC BY-NC 4.0) License.

The design of the super-turbocharger has been recently discussed in [30], and the reader is referred to this work for better details. The proposed toroidal CVT based transmission is a variant of the Torotrack variable-speed supercharging technology [55]. This is a centrifugal compressor linked to the crankshaft through a CVT based mechanism. Here, the toroidal CVT based transmission is used to connect the crankshaft to the turbocharger shaft. The Torotrack V-Charge comprises a traction drive variator the same as the one adopted here from 0.355:1 (underdrive) up to 2.820:1 (overdrive). It is, however, preceded by a 3:1 overdrive from the engine and followed by a 12.67:1 overdrive through traction drive epicyclic, for a total overdrive of 38.010 vs. the 21.575, here considered. The maximum value of the ratio between the speed of the compressor and the speed of the engine is, thus, 107 (vs. 60), while the minimum ratio is 13.50 (vs. 7.566). The Torotrack V-Charge is used for a gasoline engine, having a larger range of engine speeds, and a compressor that also works up to much higher speeds than the turbocharger here used.

With a full-toroidal design, the spread of the CVT speed ratios is limited to around eight. With a half-toroidal design, the spread of the CVT speed ratios is limited to around four [54]. In the case of a diesel engine, the spread of speed is reduced in both the engine (4500 rpm max engine speed) and the turbocharger (150,000 rpm maximum turbocharger speed), and a single full-toroidal CVT may be used. Here, the maximum speed of the engine is 4500 rpm and the maximum speed of the turbocharger is 150,000 rpm. A clutch can be added to prevent the engine crankshaft to run the turbocharger at excessive speeds when reducing the loads. This is not a problem with racing engines, and this is not a problem with a diesel engine. If a larger spread of speed ratio is needed, the option of two half toroidal CVT in series can be considered [30].

Design and analysis of half-toroidal and full-toroidal single-roller CVT are covered in [30], and [56–60].These designs are based on multiple single-roller. Multiple double-roller is considered in [56], and [60–62].Thus, [60] computes the efficiency as a function of speed ratio, load torque, and input speed. By moving from this approach and using similarity, [30] defines for the single-roller, full-toroidal CVT the efficiency map $\eta(\tau, T_1)$ of Figure 2, and the speed correction factor map $k(\tau, T_1)$ of Figure 3; [30] also assumes that the same $\eta$ apply for transmission of power to, and from, the super-turbocharger, obviously with entry conditions on one side or the other of the single-roller full-toroidal CVT. A good design with enhanced clamping control may deliver an area of operating points well above 90%, albeit with efficiencies still dropping below 90% when seal and parasitic losses will dramatically increase at very low torques. The design certainly requires specific research and development with prototyping and testing. The design, and consequently the efficiency map, is very application-specific, not generic. From Figures 2 and 3, it is then

$$\eta(\tau, T_1, N_1) = k(\tau, T_1){\cdot}(N_1 - N_{ref}) + \eta_{ref}(\tau, T_1) \tag{3}$$

with N speed, $T$ torque, and $\tau$ the speed ratio.

The transmission for the super-turbocharger is made up of overdrive, the CVT and another overdrive. The transmission is about the same of the Torotrack V-Charge [55], with differences the different speed ratios for the first and last overdrives, and the fact that the Torotrack V-Charge connects a compressor shaft to the crankshaft, and the super-turbocharger in this paper connects a turbocharger shaft to the crankshaft. As written before, thanks to a different external overdrive, the Torotrack V-Charge works between 13.50 and 107-speed ratios to run the compressor of a gasoline engine. The transmission used here works between 7.566 and 60-speed ratios to run the turbocharger of a diesel engine. The reason for the different range of speed ratios is the different engine—diesel vs. gasoline—and different maps of the turbocharger—the reduced maximum speed here. Better details of the Torotrack V-Charge, including drawings, are proposed in [55] and references there cited.

As shown in Figure 1, this transmission is modeled as a simple connection where a speed ratio and an efficiency of transmission is prescribed. The speed ratio is free to vary between the maximum and minimum, which are 60 and 7.566. It is prescribed close to the maximum value for the close to full load operation, and close to the minimum value approaching idle. Over the most part of the load range, it is equal to the value that provides a balance between turbine power and compressor power. The efficiency of transmission is obtained from Equation (3) and Figures 2 and 3.

Preliminary combustion results are obtained by coupling an empirical diesel Wiebe function to the above-defined injection profile. The diesel Wiebe function prescribes the ignition delay from the start of the first fuel-injected, i.e., the delay in crank-angle degrees between the earlier start of injection of one of the two injectors, and the start of combustion. The diesel Wiebe function then prescribes the premixed fraction, i.e., the fraction of fuel that mixes before the start of combustion and burns in the "premix" portion of the Wiebe function, and the premixed duration, i.e., the duration in crank-angle degrees of the premixed burn. Then, the model also prescribes main and tail fractions, duration, and exponents.

The Wiebe combustion model uses functions similar to the correlations for pre-mixed and diffusion burn regimes as described in [63]. An additional third function has been added to represent the tail, slow late burning at the end of the diffusion burning. The normalized mass fraction burned, starting at 0 and progressing up to the value of the combustion efficiency $\eta_c$, that is typically 1, is given as a function of the crank angle $\theta$ as follows:

$$w = \eta_c{\cdot}\beta_p{\cdot}\left\{1 - e^{[\varphi_p{\cdot}(\theta-SOI-\tau)^{E_p+1}]}\right\} + \eta_c{\cdot}\beta_m{\cdot}\left\{1 - e^{[\varphi_m{\cdot}(\theta-SOI-\tau)^{E_m+1}]}\right\} + \eta_c{\cdot}\beta_t{\cdot}\left\{1 - e^{[\varphi_t{\cdot}(\theta-SOI-\tau)^{E_t+1}]}\right\} \tag{4}$$

where SOI is the Start of Injection, $\tau$ the Ignition Delay, $E_P$ the Premix Exponent, $E_M$ the Main Exponent, $E_T$ the Tail Exponent, $\beta_P$ the Premix Fraction, and $\beta_T$ the Tail Fraction. The Main Fraction $\beta_M = 1 - \beta_P - \beta_T$. The Wiebe Premix, Main and Tail Constants $\varphi_P$, $\varphi_M$ and $\varphi_T$ are expressed as a function of the Premix Duration $\delta_P$ and the Premix Exponent, the Main Duration $\delta_M$ and the Main Exponent, and the Tail Duration $\delta_T$ and the Tail Exponent. The first term represents the premixed burning, the second term the diffusion burning, and the third term the tail burning. The Ignition delay is also computed based on [63]. These Wiebe functions only approximate the typical shape of a direct injection compression ignition, single main injection burn rate engine, which was the design prior to the Fiat Unijet common rail era.

Since at any instant the specified cumulative combustion cannot exceed the specified injected fuel fraction (the combustion rate is limited by the amount of fuel available) the injection events and the combustion events are interrelated. Figure 4 presents a typical normalized burn rate result. The start of combustion follows by the ignition delay the start of the first injection. The premixed combustion is progressing at about a linear rate and it is identified by the sharp increment of the burn rate. Then, there are the main combustion and the tail phases. With reference to old fashion diesel direct injection, the premixed phase is not a very large peak followed by a much less intense burn rate during the main combustion rate. Injection shaping permits a much better heat release rate that maximizes performance and minimizes emissions within the due constraints for maximum pressure and maximum pressure gradient. The major issue of this approach, which is based on the empirical evidence for the diesel-only engine, is the inability to cope with a large amount of methane injected prior to the diesel injection ignition at the actual rate, as an "equivalent" much shorter injection is needed.

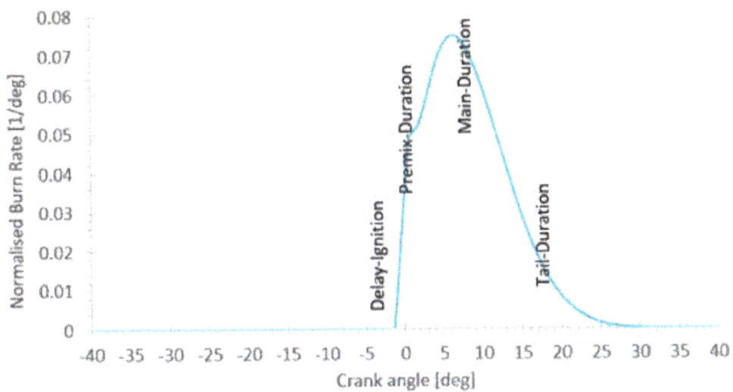


**Figure 4.** Sample non-dimensional burning rate. The relative weight of the "premixed" and "diffusion" combustion phases may be changed significantly in the engine, from basically a "premixed" only combustion of the LNG, which resembles jet ignition or controlled HCCI operation, to a "diffusion"-only combustion of the LNG.

The sample burning rate profile of Figure 4, is obtained by using the diesel Wiebe function. The premixed, main (diffusion) and tail combustion phases are clearly evidenced. This model is applicable to multiple injections of diesel, such as those of Figure 5. However, the model cannot be applied directly to multiple injections of diesel and methane as those depicted in Figure 6. The model refers, in the various sections of the combustion event, premixed, main and post, to the total mass of fuel that may be unavailable at every given time, from every injector. The ignition delay is computed from the first injection. The complex situation of multiple injections of diesel and methane, Figure 6, is therefore impossible to be directly represented, and they may only be represented through "equivalent" injection events.

The heat release rate of Figure 4 is theoretically suitable for a single injection of the diesel, and not for a complex injection of the diesel, such as the one of Figure 5, or a complex injection of the diesel and the LNG, such as the one of Figure 6. Modern diesel engines designed following the Fiat Unijet common rail revolution use injection shaping to shape the heat release rate curve targeting performance and emission parameters within given constraints. The "premixed" injection of the methane may start at any time. With flow rates high enough, the methane will be concentrated at the center of the chamber where the diesel injection ignition will occur. This will then start the combustion of the diesel and the methane in the chamber. This combustion pattern is similar to jet ignition. Then, the "diffusion" injection of the methane will sustain the combustion rate. High flow rates are needed especially during this phase. Westport, [36] to [42,44], has traditionally used a main injection of the LNG following a pilot injection of the diesel, but more recently has also explored the opportunity to inject part of the LNG before, and part of the LNG after the diesel, [39,41,42,44]. This latter activity is both experimental and computational. Westport has also explored the opportunity to have a larger injection pressure for the LNG [44], but only 600 bar. The remarkable difference vs. the Westport studies is the opportunity, thanks to the use of two much higher pressure, smaller and faster injectors, one for the diesel and one for the LNG, to enhance the injection and thus heat release rate shaping capabilities. It is a logical assumption that, having better injectors, the shaping of the heat release rate can be improved. It has been proven with the diesel only, and Westport has partially proven with the diesel and LNG.

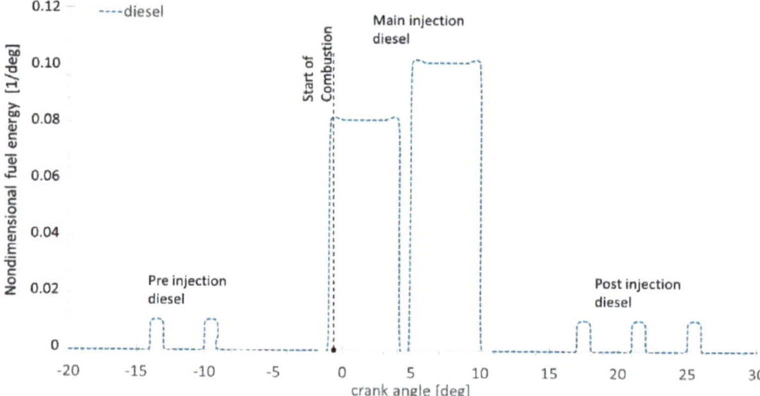


**Figure 5.** Sample injection strategy diesel-only. The post-injection of the diesel is mostly used for particle trap regeneration. The diesel pre-injection typically amounts to less than 5% of the total fuel energy injected, with reduced values reducing with the load and the speed.

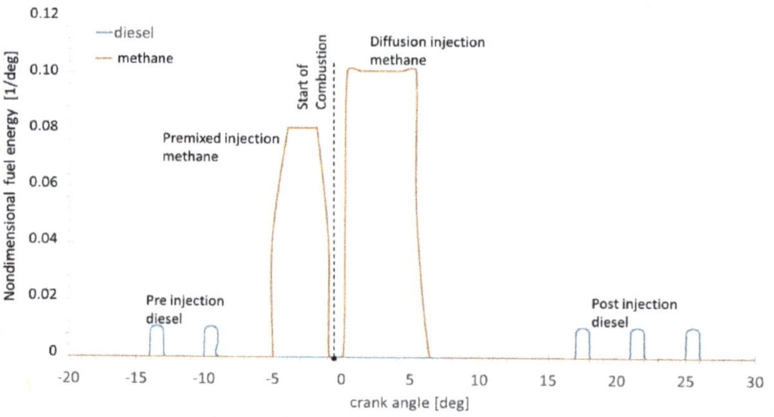


**Figure 6.** Sample injection strategy diesel-methane. The time of the start of combustion is dictated by the ignition delay of the first diesel injection. The post-injection of the diesel is used for particle trap regeneration. The diesel pre-injection typically amounts to less than 5% of the total fuel energy injected, with reduced values reducing with the load and the speed.

Figure 4 is the normalized burning rate. This is the mass of fuel that burns over 1 degree of the crank angle divided by the total mass of fuel injected in a cycle. The integral in d$\theta$ gives a result of 1 if all the fuel burns. An average normalized burning rate of 0.04 1/degrees crank angle. Over a combustion duration of 25 degrees, crank angle returns a total normalized fuel burned of 1. Similarly in Figures 5 and 6 is the normalized energy of fuel injected. This is the fuel energy inflow over 1 degree of the crank angle divided by the total energy of the fuel injected in a cycle.

Post injection was originally developed (see, for example, the Fiat Multijet) for particulate trap regeneration, i.e., it was never used in steady-state map points, but only used sometimes during transients. Other uses have been proposed, for example, reduction of engine-out soot emissions. Post injections are not relevant for performance simulations.

The opportunity of having two separate injections for the diesel and the methane translates in the opportunity to have nearly isochoric combustion about the top dead center of the pilot/pre diesel and the premixed methane-air mixture. This is something like homogeneous charge compression ignition (HCCI) in terms of burning rate, but with the advantages of (1) being controlled, and (2) occurring at the center of the combustion chamber; thus, having reduced heat losses, since a cushion of air surrounds the region where combustion occurs. This rapid premixed combustion phase can be optimized for a trade-off between fuel conversion efficiency and constraints, such as the maximum pressure build-up gradient dp/dθ (θ) and the maximum peak pressure and its location, which also depends on the "diffusion" combustion phase. The following combustion phase is then characterized by a rate controlled by the fuel flow rate, as the methane injected in an environment characterized by extremely high temperatures and pressures with ongoing combustion, vaporizes quickly, and burns as soon as oxygen is found. As the engine is now undertaking the expansion stroke, the pressure and temperature may peak at full load operation, about 14–16 degrees crank angle after the top dead center, then it decreases. Since pressure and temperature reduce during the expansion, and also the availability of oxygen reduces because of the combustion already occurred, the combustion rate is then characterized by a tail following a declining rate main phase.

The injector is modeled through an effective area, opening and closing times with multi-event capability, and a pressure profile. The mass of fuel injected is then adjusted to match the desired brake mean effective pressure (BMEP) output. A trial and error procedure is needed to produce realistic injection and combustion profiles. The fluid is considered a compressible liquid with the option to change the phase to a gas. Polynomial approximations as a function of pressure and temperature are used for all the parameters describing the liquid and gas fluids. While data for the gas methane are available in the template library, the data for the liquid methane need to be defined. This is done by using the data of [45].

For what concerns the detailed model of the injection system, not included in the present work, the approach to follow is to use compressible liquid fuel and vapor fuel models. This permits the dimensioning of all the major components of the injectors with computer-aided engineering (CAE) software tools, similar to those used for engine performance simulations shown in the following sections.

The liquid compressible fuel object describes a compressible liquid that may also undergo a change to the vapor phase for cavitation or boiling. It specifies a vapor fluid object describing the properties of the liquid after it vaporizes and the heat of vaporization. Polynomial functions are describing density and enthalpy as a function of pressure and temperature. The saturation liquid-vapor pressure of the liquid is determined as a function of temperature by integrating the Clausius–Clapeyron equation. Finally, transport properties are also given as a function of pressure and temperature. Properties are approximated as a function of $p$ and T, for example

$$\rho(p, T) = \sum_{i=1}^{n} A_i \cdot p^{\alpha_i} \cdot T^{\beta_i} \tag{5}$$

where A, $\alpha$, and $\beta$ are empirical constants. The gas phase of the liquid is described based on the molecular weight, composition, lower heating value, critical pressure, and temperature and reference entropy, plus polynomial approximation of enthalpy and transport properties function of pressure and temperature. While the gas phase of methane is available in the template properties, the liquid compressible properties require attention. At very high pressures, there are no data in the literature. The injection and combustion models need a better definition for the specific application, based on an experimental campaign. There is a need for novel specific models, as well as for the experimental determination of the parameters embedded in these models. The injector nozzle parameters also impact on the combustion result, through their effect on the injection pressure and velocity. The DI Wiebe combustion model is affected by the values of the injection pressure and velocity. The nozzle discharge coefficient required to achieve the specified injected mass is calculated using the diameter of an individual nozzle hole and the number of holes. A trial-and-error procedure is needed to compute realistic profiles for the injected fuel, and then the combustion. Methane is injected at cryogenic temperatures and high pressure, with high flow rates and high atomization. The methane expands absorbing heat and it quickly mixes with the air. The methane that is injected after the combustion has started vaporizes and mixes even quickly.

## 3. Results

Reference [4] presents the results for the baseline engine and the dual fuel diesel LNG engine with a lower pressure direct injector. The substitutional energy by LNG is proposed there. Here we present the results only for the higher pressure injector (1600 bar), permitting an increased flow rate and thus shorter injection timings. Unfortunately, the present injection and combustion model only works to provide an empirically based estimation of the fuel conversion efficiency. It does not permit to compute the detailed injection, combustion, and pollutant formation processes that may benefit even more from high-pressure fast actuation injection. Better models and the support of dynamometer experiments are needed to fully progress the concept.

Figures 7 and 8 present the fuel conversion efficiency and $\lambda$ maps for the diesel-LNG engine operation, while Figure 9 presents the ratio of the turbocharger speed to the engine speed. Figures 10 and 11 then propose the maximum pressure and the crank angle of maximum pressure. Figure 12 finally presents the substitutional methane energy. It is a design constraint to have peak pressures below 300 bar.

The opportunity to use much higher injection pressures translate into the opportunity to have a much lighter and smaller injector, faster to be actuated, that also delivers a larger flow rate with better atomization. It is unfortunately not possible to fully explore the heat release shaping capabilities of such injectors, because the combustion model, Figure 4 and Equation (4), in theory, is only valid for one simple single injection of the diesel. From the measured pressure curves of one latest diesel, adopting complex injection strategies, there is the opportunity to compute the Wiebe function parameters and an equivalent simple diesel injection to provide about the same heat release rate. The only aspect that can be simulated with the present model for LNG is a faster or slower injection of a given amount of fuel. From Table 3, moving from 300 bar to 1600 bar, $\dot{m} \cdot$LHV increases from 4614 to 15,232 kJ/s if the cross-sectional area remains unchanged. This means that the LNG could be injected during a single main injection phase of length 32% less, while also reducing the injector holes' diameter of 33%, or the cross-sectional areas of the injector of 56%.

A comparison between operation with 300 and 1600 bar injectors is provided. Results obtained by using a 300 bar injector have been recently published in [4]. Peak fuel conversion efficiency there is less than 48%. Peak fuel conversion efficiency here is 50%. Peak fuel conversion efficiency there is obtained with $\lambda = 1.67$. Peak fuel conversion efficiency here is obtained with $\lambda = 1.74$. The BMEP can be further increased simply working richer with the LNG, which is much easier to vaporize, mix with air and burn than the diesel. However, this impact peak pressure. What is done in the model is to simulate a quicker injection and combustion, with the higher injection pressure. The injection and combustion events are differently phased, to avoid excessive peak pressures, while permitting better fuel conversion efficiency, i.e., less fuel for the same BMEP output. No attempt has been made to increase the maximum BMEP for any given speed. Higher injection pressure permits faster actuation, larger flow rates, better atomization, and thus much quicker combustion. The enhanced heat release shaping opportunities are not explored due to the limitations of the combustion model. By only considering a faster main injection of the LNG and quicker combustion, the fuel conversion efficiency increases about two percentage points, and the fuel-to-air ratio similarly reduces for the same BMEP output. The computational model is not optimized, due to the limitations of the injection and combustion model that differs considerably far from the real-world operation. Improvements in real-world engine experiments through appropriate shaping of the heat release rate curve may be larger.

The turbocharger is accelerated to higher speeds at low engine speed and high loads, up to the maximum permitted ratio of 60. The most part of the operating points, far from top loads, is obtained by taking the turbocharger speed as the speed of equilibrium between turbine and compressor powers. Top loads are obtained by using different speeds, with higher compression power at low speeds, and higher turbine powers at high speeds.

Regarding fuel conversion efficiency and λ, these results are constrained by the use of heat release rate profiles, such as the one of Figure 4, which limits the possibilities. The maps of Figures 7 and 8 are computed by assuming the dual-fuel diesel-methane engines have roughly the same combustion behavior of diesel-only engines, that it is not the case. These results are therefore an underestimation of the actual potentials of the proposed solution. It is worth to note the top fuel conversion efficiency, about 48%, is marginally better than the diesel, and also similar to diesel λ values.

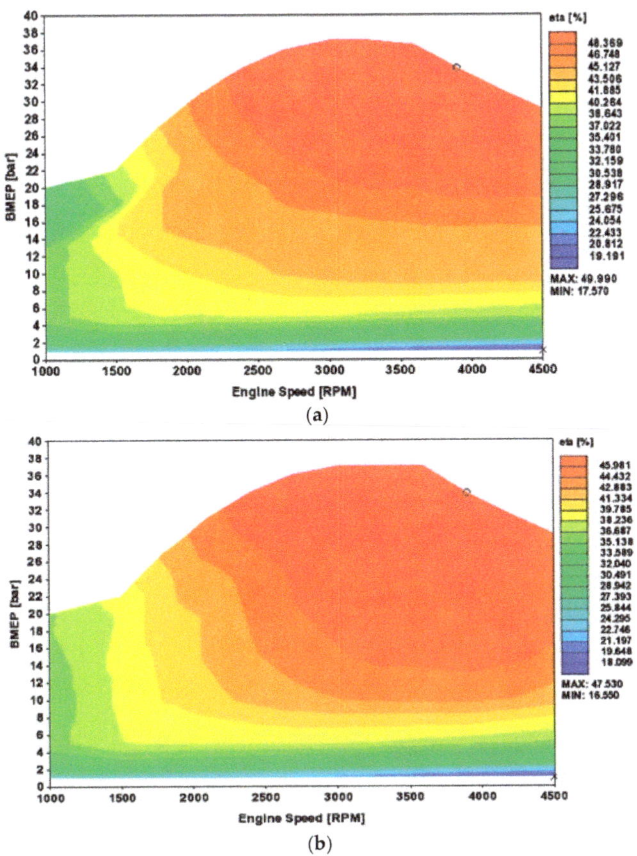

**Figure 7.** Diesel-methane engine steady-state fuel conversion efficiency map; (**a**) 1600 bar injector; (**b**) 300 bar injector. Image (**b**) from [4]. Creative Commons Attribution (CC BY) license 4.0.

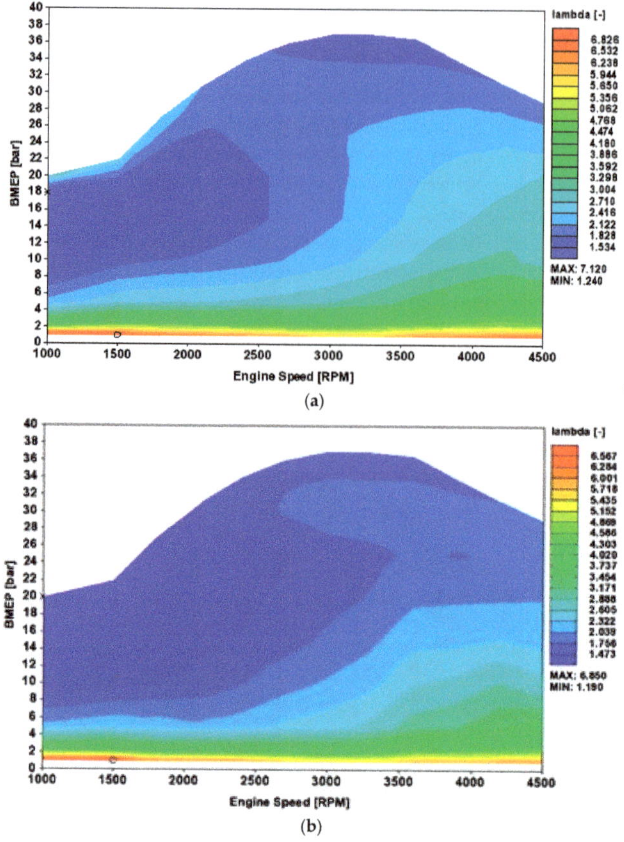

**Figure 8.** Diesel-methane engine steady-state λ map; (**a**) 1600 bar injector; (**b**) 300 bar injector. Image (**b**) from [4]. Creative Commons Attribution (CC BY) license 4.0.

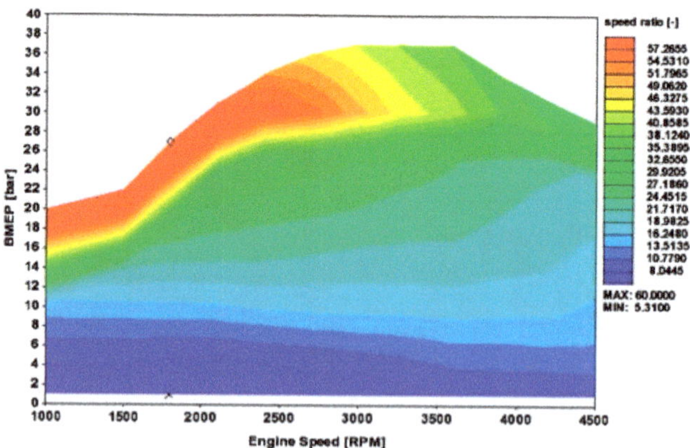

**Figure 9.** Diesel-methane engine steady-state speed ratio (turbocharger speed to engine speed).

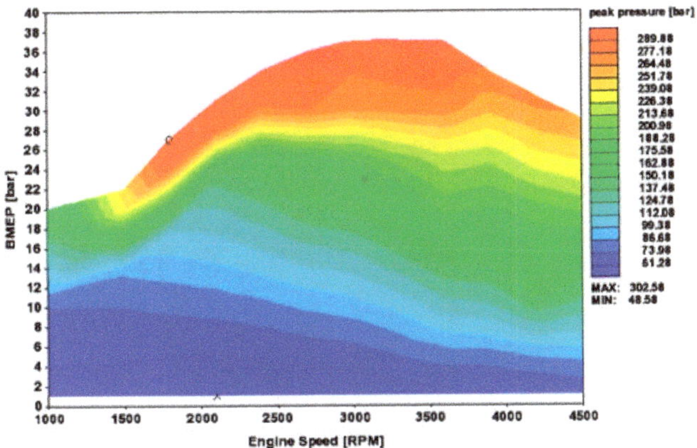

**Figure 10.** Diesel-methane engine steady-state peak pressure.

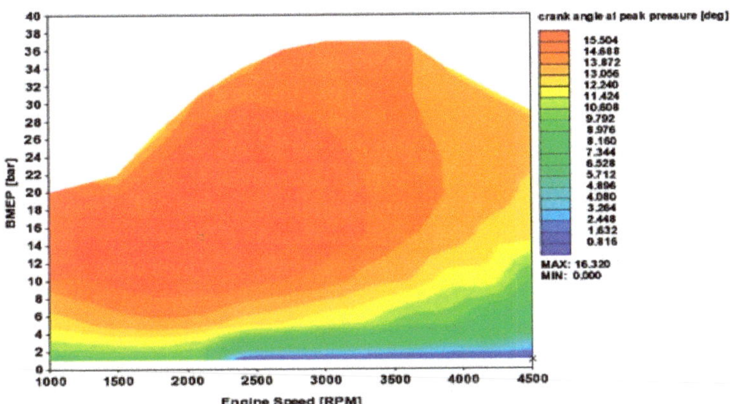

**Figure 11.** Diesel-methane engine steady-state crank angle of peak pressure.

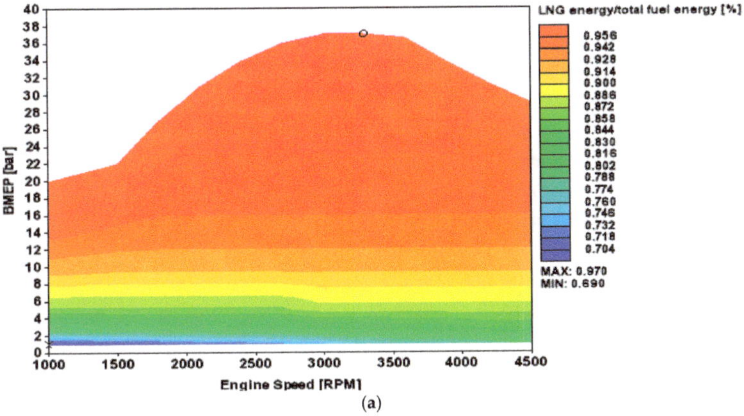

(a)

**Figure 12.** *Cont.*

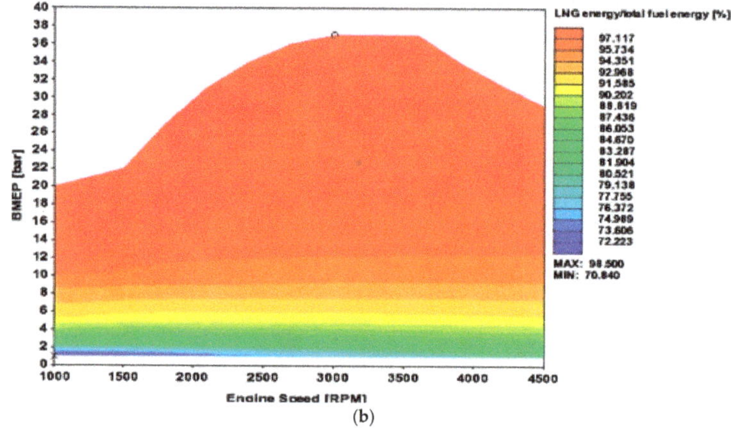



**Figure 12.** Diesel-methane engine steady-state substitutional energy of methane; (**a**) 1600 bar injector; (**b**) 300 bar injector. Image (**b**) from [4]. Creative Commons Attribution (CC BY) license 4.0.

It must be added that while the baseline diesel engine model was validated, the diesel-methane model is not. Both dp/dθ and peak pressure are limited, the same as the diesel. The flow rate in the diesel was also limited in the area of high speeds and top loads by the racing rules, and this is also done in the dual-fuel engine. The full load BMEP of the diesel is set as the full load BMEP of the diesel-methane engine. As in the same BMEP points, the methane-diesel engine works with similar and possibly slightly larger λ and η, and the methane-diesel may work λ than the diesel-only, thanks to the lower injection temperature and the faster vaporization and mixing of the methane, there is certainly the opportunity to also deliver better power densities with methane.

In the simulations, it is assumed LNG has an LHV of 50 MJ/kg fuel, and an A/F stoichiometric of 17.2. This translates into an energy content per unit mass of the mixture of 2.78 MJ/kg mixture. In the case of diesel, LHV is 42.6 MJ/kg, A/F stoichiometric is 14.5, and the energy content per unit mass of mixture is a very close, slightly smaller 2.75 (only 1% difference). For the same λ, thus, there is 1% more energy with the LNG, and a (slightly) larger λ can be adopted for the same MJ/kg. λ may be larger (for the same BMEP point) because η also slightly increases. LNG is injected at 113 K rather than 300 K. However, immediately after injected in an environment characterized by hot partially combusted gases, it vaporizes almost instantaneously, and it quickly mixes with air and burns. Methane is much easier to burn than a heavy hydrocarbon such as diesel fuel.

With reference to a traditional turbocharged engine, the super-turbocharger permits much larger BMEP values in the low-speeds range, an increment from about 10 to about 20 bar BMEP at 1000 rpm, while working with about the same λ. At high speeds and loads, the benefits are up to about one percentage point of better fuel conversion efficiency, working with about the same λ.

It must be mentioned that the opportunity to improve the steady-state output and efficiency with the super-turbocharger depends on the specific engine load and speed. At high engine loads and speeds, the energy, otherwise waste-gated, may often be converted to extra power at the turbine to increase, albeit to a small extent, both engine efficiency and output. At low engine speeds, the compressor can be run at higher speeds to increase the engine output thanks to the increased boost. However, this requires power from the crankshaft.

Concerning the after treatment, the diesel after treatment is obviously maintained. One advantage of the two injectors per cylinder is that the engine may work diesel-only, or diesel-methane. The post injections of the diesel for the regeneration of the particulate filter are maintained. The opportunity to run the engine diesel only, or diesel-LNG, depending on the needs, is a significant advantage during transition periods, where the recharging infrastructure for the LNG cannot be widespread.

To have a particulate filter is relevant to the opportunity to clean up the air of urbanized areas where the ambient concentration of particulates largely exceeds the value at the tailpipe of a diesel or a diesel-methane vehicle.

A validated diesel engine model, developed by using very well-known XAE software tools, extensively used within engine departments for many years, is modified to only accept an additional injector for a second fuel, LNG. The experience by Westport proves dual fuel diesel-LNG engines work very close to diesel-only engines. The proposed results, detailing the benefits of higher injection pressures, fast actuation and enhanced injection shaping capabilities of the LNG and the diesel, are extremely logical, and thus trustworthy.

As higher injection pressures, fast actuation and enhanced injection shaping capabilities have permitted to dramatically improve the diesel engine, there is no reason why similar improvements in the injection of the LNG and the diesel should not be beneficial.

By replacing the Westport dual fuel injector, of limited injection pressure, low rates of actuation, and limited injection shaping capabilities, with two dedicated injectors, one for the diesel, the other for the LNG, both featuring high injection pressures, fast actuation and enhanced injection shaping capabilities, there are certainly significant complications, as the solution is more appropriate for novel engines rather than the conversion of existing engines, but there is really no logical downfall under the aspect of performance.

## 4. Conclusions

Thanks to the super turbocharging, and the use of two dedicated injectors per cylinder, a latest 2850 bar piezo diesel direct injector, and the studied 1600 bar liquid methane (LNG) direct injector, the proposed engine has an excellent steady-state map, with maximum fuel conversion efficiency approaching 50%, and high fuel conversion efficiencies in excess of 40% over the most part of the load range at any speed. Above 2000 rpm, the full load brake mean effective pressure (BMEP) values are in excess of 30 bar. Maximum values are close to 38 bar, and they are only limited by peak pressure and gradient of pressure rise constraints. The engine also has outstanding low-speed torque, and improved high-speed efficiency, thanks to the super-turbocharger. Not shown here, the super-turbocharged engine also has first-rate transient behaviors, in decelerations (energy recovery) as well as accelerations (no turbo-lag). The design of the injector must progress by using experiments and simulations. These latter require a compressible liquid treatment for the LNG, with a further refined description vs. the one here proposed. Regarding the complexity of the two injectors, per cylinder, this is more than compensated by the advantages, and it is definitively not an issue when designing novel engines (positive ignition jet ignition engines accommodate one injector and one jet ignition device per cylinder, compression ignition diesel engines have accommodated one injector and one glow plug per cylinder until very recently). Injection pressures of 2850 bar were already proposed for the diesel almost 10 years ago. The maximum injection pressure proposed so far for the dual fuel diesel-LNG injector is 600 bar. The dual-fuel diesel-LNG injector is large, heavy, and difficult to be actuated. The proposed innovation is the use of two extremely high-pressure injectors, one for the diesel, and one for the LNG, 2850 bar the first, 1600 bar the second, to achieve much faster actuation, much larger flow rates, and much better atomization of the injected fuel for enhanced shaping of the heat release rate, translating in better fuel conversion efficiency.

**Funding:** This research received no external funding.

**Conflicts of Interest:** The authors declare no conflict of interest.

## References

1. IEA. Data and Statistics. 2019. Available online: www.iea.org/data-and-statistics (accessed on 1 January 2020).

2.	Boretti, A. Energy storage requirements to address wind energy variability. *Energy Storage* **2019**, *1*, e77. [CrossRef]

3.	Boretti, A. Production of hydrogen for export from wind and solar energy, natural gas, and coal in Australia. *Int. J. Hydrog. Energy* **2019**. [CrossRef]

4.	Boretti, A. Advances in Diesel-LNG internal combustion engines. *Appl. Sci.* **2020**, *10*, 1296. [CrossRef]

5.	Smajla, I.; Karasalihović Sedlar, D.; Drljača, B.; Jukić, L. Fuel Switch to LNG in Heavy Truck Traffic. *Energies* **2019**, *12*, 515. [CrossRef]

6.	Zhang, C.; Zhou, A.; Shen, Y.; Li, Y.; Shi, Q. Effects of combustion duration characteristic on the brake thermal efficiency and NOx emission of a turbocharged diesel engine fueled with diesel-LNG dual-fuel. *Appl. Therm. Eng.* **2017**, *127*, 312–318. [CrossRef]

7.	Stefana, E.; Marciano, F.; Alberti, M. Qualitative risk assessment of a Dual Fuel (LNG-Diesel) system for heavy-duty trucks. *J. Loss Prev. Process Ind.* **2016**, *39*, 39–58. [CrossRef]

8.	Wang, S.X.; Huang, X.H.; Jiang, S.Y. The experimental study on diesel-LNG dual fuel marine diesel engine. *Ship Sci. Technol.* **2011**, *33*, 79–81.

9.	Le, L.; Phillips, J. LNG: An emerging transport fuel. *Energy News* **2015**, *33*, 15.

10.	Meng, C.; Si, J.P.; Liang, G.X.; Niu, J.H. The Technical Modification and Performance Analysis of Diesel/LNG Dual Fuel Engines. *Adv. Mater. Res.* **2013**, *724*, 1383–1388. [CrossRef]

11.	Unseki, T. Environmentally superior LNG-Fueled vessels. *Mitsubishi Heavy Ind. Tech. Rev.* **2013**, *50*, 37–43.

12.	Bengtsson, S.; Andersson, K.; Fridell, E. A comparative life cycle assessment of marine fuels: Liquefied natural gas and three other fossil fuels. *Proc. Inst. Mech. Eng. Part M J. Eng. Marit. Environ.* **2011**, *225*, 97–110. [CrossRef]

13.	Osorio-Tejada, J.; Llera, E.; Scarpellini, S. LNG: An alternative fuel for road freight transport in Europe. *WIT Trans. Built Environ.* **2015**, *168*, 235–246.

14.	Zheng, J.; Wang, J.; Zhao, Z.; Wang, D.; Huang, Z. Effect of equivalence ratio on combustion and emissions of a dual-fuel natural gas engine ignited with diesel. *Appl. Therm. Eng.* **2019**, *146*, 738–751. [CrossRef]

15.	Constable, G.A.; Gibson, C.J.; Gram, A. *Use of LNG in Heavy-Duty Vehicles*; SAE Technical Paper; No. 891670; SAE International: Warrendale, PA, USA, 1989.

16.	Arteconi, A.; Brandoni, C.; Evangelista, D.; Polonara, F. Life-cycle greenhouse gas analysis of LNG as a heavy vehicle fuel in Europe. *Appl. Energy* **2010**, *87*, 2005–2013. [CrossRef]

17.	Wan, C.; Yan, X.; Zhang, D.; Shi, J.; Fu, S.; Ng, A.K. Emerging LNG-fueled ships in the Chinese shipping industry: A hybrid analysis on its prospects. *Wmu J. Marit. Aff.* **2015**, *14*, 43–59. [CrossRef]

18.	Jacobs, T. Displacing Diesel: The Rising Use of Natural Gas by Onshore Operators. *J. Pet. Technol.* **2013**, *65*, 52–60. [CrossRef]

19.	Sheng-chao, R.; Yi-huai, H. Technical Features of LNG Dual-Fuel Diesel Engine on Inland Ships. *Ship Stand. Eng.* **2014**, *4*, 3.

20.	Yousefi, A.; Guo, H.; Birouk, M. Effect of diesel injection timing on the combustion of natural gas/diesel dual-fuel engine at low-high load and low-high speed conditions. *Fuel* **2019**, *235*, 838–846. [CrossRef]

21.	Yinsheng, P.; Yimin, Z.; Dongbo, C.; Xuling, W.; Guowei, H.; Hongchuan, H. Investigation on Improvement of Exhaust Emission from Diesel Engines with EGR. *Chin. Intern. Combust. Engine Eng.* **2000**, *4*, 6–10.

22.	Wang, S.X.; Zhang, J.; Jiang, S.Y. Experimental Study on the Load Characteristics of Diesel-LNG Hybrid Diesel Engines. *Adv. Mater. Res.* **2012**, *356*, 1375–1378. [CrossRef]

23.	Boretti, A. Advantages of the direct injection of both diesel and hydrogen in dual fuel H2ICE. *Int. J. Hydrog. Energy* **2011**, *36*, 9312–9317. [CrossRef]

24.	Boretti, A. Advances in hydrogen compression ignition internal combustion engines. *Int. J. Hydrog. Energy* **2011**, *36*, 12601–12606. [CrossRef]

25.	Boretti, A. Diesel-like and HCCI-like operation of a truck engine converted to hydrogen. *Int. J. Hydrog. Energy* **2011**, *36*, 15382–15391. [CrossRef]

26.	Boretti, A. Latest concepts for combustion and waste heat recovery systems being considered for hydrogen engines. *Int. J. Hydrog. Energy* **2013**, *38*, 3802–3807. [CrossRef]

27.	Boretti, A. Numerical study of the substitutional diesel fuel energy in a dual fuel diesel-LPG engine with two direct injectors per cylinder. *Fuel Process. Technol.* **2017**, *161*, 41–51. [CrossRef]

28.	Boretti, A. *Prototype Powertrain in Motorsport Endurance Racing*; SAE, PT-185; SAE International: Warrendale, PA, USA, 2018; p. 178. ISBN 978-0-7680-8451-1.

29. Boretti, A. *Advances in Turbocharged Racing Engines*; SAE PT-199; SAE International: Warrendale, PA, USA, 2019; p. 236. ISBN 978-0-7680-0014-6.

30. Boretti, A. Half/Full Toroidal, Single/Double Roller, CVT Based Transmission for a Super-Turbo-Charger. *Proc. Eng. Technol. Innov.* **2019**, *11*, 1.

31. Boretti, A. Super turbocharging the direct injection diesel engine. *Nonlinear Eng.* **2018**, *7*, 17–27. [CrossRef]

32. Boretti, A. Energy flow of a 2018 FIA F1 racing car and proposed changes to the powertrain rules. *Nonlinear Eng.* **2019**, *9*, 28–34. [CrossRef]

33. Green Car Congress. High-Pressure Direct-Injection Hydrogen Engine Achieves Efficiency of 42%; On Par with Turbodiesels. 2009. Available online: www.greencarcongress.com/2009/03/high-pressure-d.html (accessed on 1 January 2020).

34. Autoblog. BMW Cranks up the Efficiency of Hydrogen Internal Combustion Engines. 2009. Available online: www.autoblog.com/2009/03/12/bmw-cranks-up-the-efficiency-of-hydrogen-internal-combustion-eng/ (accessed on 1 January 2020).

35. HyICE. Optimization of the Hydrogen Internal Combustion Engine. 2007. Available online: trimis.ec.europa.eu/sites/default/files/project/documents/20090918_161614_66668_HyICE%20-%20Summary.pdf (accessed on 1 January 2020).

36. Westport.com. 1st Gen. Westport HPDI Technology. 2019. Available online: www.westport.com/old-pages/combustion/hpdi/integration (accessed on 1 January 2020).

37. Westport.com. Westport™ HPDI 2.0. 2019. Available online: www.westport.com/is/core-technologies/hpdi-2 (accessed on 1 January 2020).

38. Mumford, D.; Goudie, D.; Saunders, J. *Potential and Challenges of HPDI*; SAE Technical Paper; No. 2017-01-1928; SAE International: Warrendale, PA, USA, 2017.

39. Florea, R.; Neely, G.; Abidin, Z.; Miwa, J. *Efficiency and Emissions Characteristics of Partially Premixed Dual-Fuel Combustion by Co-Direct Injection of NG and Diesel Fuel (DI2)*; SAE Technical Paper; No. 2016-01-0779; SAE International: Warrendale, PA, USA, 2016.

40. Cong, S.; McTaggart-Cowan, G.; Garner, C. *Effects of Fuel Injection Parameters on Low Temperature Diesel Combustion Stability*; SAE Technical Paper 2010-01-0611; SAE International: Warrendale, PA, USA, 2010.

41. Faghani, E.; Kheirkhah, P.; Mabson, C.W.; McTaggart-Cowan, G.; Kirchen, P.; Rogak, S. *Effect of Injection Strategies on Emissions from a Pilot-Ignited Direct-Injection Natural-Gas Engine—Part II: Slightly Premixed Combustion*; SAE Technical Paper; No. 2017-01-0763; SAE International: Warrendale, PA, USA, 2017.

42. Faghani, E.; Kheirkhah, P.; Mabson, C.W.; McTaggart-Cowan, G.; Kirchen, P.; Rogak, S. *Effect of Injection Strategies on Emissions from a Pilot-Ignited Direct-Injection Natural-Gas Engine—Part I: Late Post Injection*; SAE Technical Paper; No. 2017-01-0774; SAE International: Warrendale, PA, USA, 2017.

43. Pritchard, P. *Fox and McDonald's Introduction to Fluid Mechanics*, 8th ed.; John Wiley & Sons: Hoboken, NJ, USA, 2011.

44. McTaggart-Cowan, G.; Mann, K.; Huang, J.; Singh, A.; Patychuk, B.; Zheng, Z.X.; Munshi, S. Direct injection of natural gas at up to 600 bar in a pilot-ignited heavy-duty engine. *SAE Int. J. Engines* **2015**, *8*, 981–996. [CrossRef]

45. Lemmon, E.W.; McLinden MOand Friend, D.G. Thermophysical Properties of Fluid Systems in NIST Chemistry WebBook. In *NIST Standard Reference Database Number 69*; Linstrom, P.J., Mallard, W.G., Eds.; National Institute of Standards and Technology: Gaithersburg, MD, USA, 2009.

46. Roy, B.N. *Fundamentals of Classical and Statistical Thermodynamics*; John Wiley & Sons: Hoboken, NJ, USA, 2002.

47. Naber, J.D.; Siebers, D.L. Effects of gas density and vaporization on penetration and dispersion of diesel sprays. *SAE Trans.* **1996**, *105*, 82–111.

48. Siebers, D.L. Scaling liquid-phase fuel penetration in diesel sprays based on mixing-limited vaporization. *SAE Trans.* **1999**, *108*, 703–728.

49. Soteriou, C.; Andrews, R.; Smith, M. Further studies of cavitation and atomization in diesel injection. *SAE Trans.* **1999**, *108*, 902–919.

50. Arcoumanis, C.; Flora, H.; Gavaises, M.; Badami, M. Cavitation in real-size multi-hole diesel injector nozzles. *SAE Trans.* **2000**, *109*, 1485–1500.

51. Ejim, C.E.; Fleck, B.A.; Amirfazli, A. Analytical study for atomization of biodiesels and their blends in a typical injector: Surface tension and viscosity effects. *Fuel* **2007**, *86*, 1534–1544. [CrossRef]

52. Suh, H.K.; Chang, S.L. Effect of cavitation in nozzle orifice on the diesel fuel atomization characteristics. *Int. J. Heat Fluid Flow* **2008**, *29*, 1001–1009. [CrossRef]
53. Alternative Fuels Data Center. Fuel Properties Comparison. Available online: https://afdc.energy.gov/fuels/properties (accessed on 1 January 2020).
54. Boretti, A.; Ordys, A.; Al-Zubaidy, S. Dynamic analysis of an LMP1-H racing car by coupling telemetry and lap time simulations. In Proceedings of the International Conference on Applied Mechanics, and Industrial Systems (ICAMIS-Oman-2016), Muscat, Oman, 8–10 December 2016.
55. Hu, B.; Akehurst, S.; Lewis, A.G.; Lu, P.; Millwood, D.; Copeland, C.; Burtt, D. Experimental analysis of the V-Charge variable drive supercharger system on a 1.0 L GTDI engine. *Proc. Inst. Mech. Eng. Part D J. Automob. Eng.* **2018**, *232*, 449–465. [CrossRef]
56. Verbelen, F.; Derammelaere, S.; Sergeant, P.; Stockman, K. A comparison of the full and half toroidal continuously variable transmissions in terms of dynamics of ratio variation and efficiency. *Mech. Mach. Theory* **2018**, *121*, 299–316. [CrossRef]
57. Zhang, Y.; Zhang, X.; Tobler, W. A systematic model for the analysis of contact, side slip and traction of toroidal drives. *J. Mech. Des.* **2000**, *122*, 523–528. [CrossRef]
58. Zou, Z.; Zhang, Y.; Zhang, X.; Tobler, W. Modeling and simulation of traction drive dynamics and control. *J. Mech. Des.* **2001**, *123*, 556–561. [CrossRef]
59. Carbone, G.; Mangialardi, L.; Mantriota, G. A comparison of the performances of full and half toroidal traction drives. *Mech. Mach. Theory* **2004**, *39*, 921–942. [CrossRef]
60. Verbelen, F.; Derammelaere, S.; Sergeant, P.; Stockman, K. Visualizing the efficiency of a continuously variable transmission. In Proceedings of the Energy Efficiency in Motor Driven Systems EEMODS 2017, Rome, Italy, 6–8 September 2017.
61. Carbone, G.; Bottiglione, F.; De Novellis, L.; Mangialardi, L.; Mantriota, G. The Double Roller Full Toroidal Variator: A Promising Solution for KERS Technology. In Proceedings of the FISITA 2012 World Automotive Congress, Beijing, China, 27–30 November 2012; Springer: Berlin/Heidelberg, Germany, 2013; pp. 241–250.
62. De Novellis, L.; Carbone, G.; Mangialardi, L. Traction and efficiency performance of the double roller full-toroidal variator: A comparison with half-and full-toroidal drives. *J. Mech. Des.* **2012**, *134*, 071005. [CrossRef]
63. Watson, N.; Pilley, A.D.; Marzouk, M. *A Combustion Correlation for Diesel Engine Simulation*; SAE Technical Paper; No. 800029; SAE International: Warrendale, PA, USA, 1980.

 *processes*

*Article*

# Effect of Pre-Combustion Chamber Nozzle Parameters on the Performance of a Marine 2-Stroke Dual Fuel Engine

**Hao Guo, Song Zhou *, Majed Shreka and Yongming Feng**

College of Energy and Power Engineering, Harbin Engineering University, Harbin 150001, China;
guohao618@hrbeu.edu.cn (H.G.); majed.shreka@outlook.com (M.S.); fengyongming@hrbeu.edu.cn (Y.F.)
* Correspondence: songzhou@hrbeu.edu.cn; Tel.: +86-138-4506-3167

Received: 4 November 2019; Accepted: 17 November 2019; Published: 21 November 2019

**Abstract:** In recent years and with the increasing rigor of the International Maritime Organization (IMO) emission regulations, the shipping industry has focused more on environment-friendly and efficient power. Low-pressure dual-fuel (LP-DF) engine technology with high efficiency and good emissions has become a promising solution in the development of marine engines. This engine often uses pre-combustion chamber (PCC) to ignite natural gas due to its higher ignition energy. In this paper, a parametric study of the LP-DF engine was proceeded to investigate the design scheme of the PCC. The effect of PCC parameters on engine performance and emissions were studied from two aspects: PCC nozzle diameter and PCC nozzle angle. The results showed that the PCC nozzle diameter affected the propagation of the flame in the combustion chamber. Moreover, suitable PCC nozzle diameters helped to improve flame propagation stability and engine performance and reduce emissions. Furthermore, the angle of the PCC nozzle had a great influence on flame propagation direction, which affected the flame propagation speed and thus the occurrence of knocking. Finally, optimizing the angle of the PCC nozzle was beneficial to the organization of the in-cylinder combustion.

**Keywords:** Computational Fluid Dynamics; two-stroke; dual-fuel engine; simulation; pre-combustion chamber

---

## 1. Introduction

Since January 1, 2016, the Tier III emission standard have been implemented by the International Maritime Organization (IMO) [1,2]. With the fuel Sulphur global limit of 0.5% entering into force on January 1, 2020, the liquefied natural gas (LNG) has gradually become a promising alternative fuel for vessels sailing inside and outside the Emission Control Areas (ECAs) [3,4]. Natural gas, as a clean energy source, has the advantages of higher heating value and lower price—making the marine 2-stroke low-speed gas engines which use it as a fuel significantly economical [5,6]. In addition, the natural gas is free of sulfur, which means that there is almost no formation of Sulfur Oxides (SOx) [7]. However, when natural gas is used as the only fuel of the engine, the power of the gas engine will decrease compared to the same size diesel engine [8]. Using diesel-ignited natural gas is considered as an effective way to solve the power reduction problem [9]. Under such measures, the problem of power reduction when using natural gas can be better solved, and the latest emission regulations can be fulfilled [10].

The marine low-pressure dual-fuel engine uses the low-pressure injection technology in which the natural gas is injected into the cylinder at a low-pressure after the scavenging port is closed. After that, a small amount of diesel fuel is sprayed in the pre-chamber when the piston reaches the top dead center (TDC), which is used to ignite the gas/air mixture. In addition, this engine uses the Otto cycle principles to reduce the peak in-cylinder combustion pressure and temperature—resulting in low Nitrogen Oxides (NOx) emissions [11]. The main advantages of this engine include high efficiency at high load, high mean effective pressure, and low NOx emissions [12].

At present, the marine low-pressure dual-fuel engine is represented by WinGD's RT-Flex50DF engine. Compared with traditional low-speed 2-stroke diesel engines using heavy fuel oil (HFO) or Light Fuel Oil (LFO), the LP-DF engine reduced the particulate matter (PM) to almost 98% and the SOx emissions by nearly 99% [13]. Besides, this engine decreased the NOx emissions by about 90%, which means that the IMO Tier III emission standard can be fulfilled without the use of after-treatment devices. The power and the thermal efficiency of the dual-fuel engine are close to those of conventional low-speed marine engines, but its emissions performance has been significantly improved. Furthermore, the marine DF engine can easily switch between the gas mode and the diesel mode to achieve smooth operation under full working conditions. Therefore, the DF engine technology is gradually gaining attention from various shipping companies due to its huge advantages and potential [14].

Papagiannakis et al. used experimental methods to study the effect of air/fuel ratio on the thermal efficiency and the emissions of a DF engine. They reported that the engine efficiency was lower than that of the diesel engine in the DF mode after decreasing the air/fuel ratio ($\lambda$). Besides, the engine efficiency improved at medium and low load with the increase of the diesel injection quantity [15].

The pre-ignition caused by lubricating oil has become more apparent with the increase in the engine mean effective pressure, which considers one of the critical issues nowadays affecting most of the premixed combustion engines. Hirose et al. observed the pre-ignition of a 2-stroke low-speed premixed gas engine through experimental methods. He determined the effect of lubricating oil on pre-ignition through visualization techniques and high-speed cameras. He found that the self-ignition temperature of lubricating oil was different from the ignition temperature of pilot fuel. Only by reducing the temperature to avoid pre-ignition, the proper premixed gas equivalent ratio played an important role in the stability of 2-stroke premixed combustion [16].

The pilot fuel injection parameters are important for the DF engines. Alla et al. carried out experimental work on a single-cylinder machine to study the effect of fuel injection timing and pilot fuel injection quantity on the performance of DF engines through experiments on a single cylinder machine. They reported that that by increasing the injection time of the pilot fuel significantly improved the efficiency and decreased the emissions at low loads. However, increasing the fuel injection amount under high load caused knocking [17].

Duan et al. studied the performance, the knock characteristics, and the combustion of a high compression ratio and lean-burn heavy-duty spark ignition (SI) engine fueled with n-butane and liquefied methane gas blend. Results indicated that the heat release rate, the in-cylinder pressure, and the cumulative heat release amount increased with the increased n-butane energy share. Once the n-butane energy ratio exceeded 5% at 1400 r/min and full-load, light knock occurred at this operating condition [18].

In recent years, computational flow dynamic (CFD) technology has developed rapidly. Numerical simulation methods can effectively reduce costs and shorten the research and development cycle of new engine [19,20]. Yousefi et al. simulated the combustion and emission characteristics of a premixed natural gas DF engine by coupling the CFD software with chemical reaction kinetics. They found that the use of PCC structure reduced the unburned methane emissions by an average of 46% compared to dual fuel engines without PCC [21]. Cernik et al. proposed a quasi-dimensional combustion model for a large-bore two-stroke dual-fuel marine engine. The diffusion combustion of the pilot fuel and the propagation of the premixed gas flame front were described in detail [22].

Moreover, the pilot fuel affects the power and emissions of the 2-stroke DF engine, which prompted some researchers to study its impacts such as Amin et al. who simulated the in-cylinder combustion process of a 2-stroke DF engine through the coupling of 3D CFD and chemical kinetics. They reported that increasing the amount of pilot fuel increased the ignition delay and the peak in-cylinder pressure and when the amount of pilot fuel is high, the emissions of NOx and Carbon Monoxide (CO) increase [23].

The PCC design features such as the volume ratio, the nozzle length/diameter ratio, and the pilot diesel spray direction influence the DF engine combustion and emissions. Liu et al. used the traditional CFD tool STAR-CD to study the LP-DF engine and the relevant numerical model was validated by experimental data. The results showed that increasing the PCC volume and shortening the nozzle length was beneficial to the combustion process. Besides, NO emission was mainly formed from the combustion in the PCC while shortening the nozzle length caused the No emissions to increase [24].

Maghbouli et al. established a model of the dual-fuel engine by integrating CHEMKIN chemical solver with KIVA-3V. He studied the combustion process of the DF engine under knocking conditions and introduced a new knock intensity factor. He reported that the exhaust gas recirculation technique could effectively reduce knocking [25]. Jha et al. used CFD software to simulate the effect of in-cylinder swirl on the combustion of the DF engine. The study showed that increasing the in-cylinder swirl ratio from 0 to 1.5 increased the in-cylinder pressure and the heat release rate [26]. Furthermore, optimizing the eddy current is a feasible strategy to improve the combustion efficiency and reduce the engine's Hydrocarbon (HC) and CO emissions at low load [27].

In a large 2-stroke marine engine, it is difficult to organize a strong swirl in the cylinder due to the large bore [28]. Therefore, this study aims at improving the airflow movement and the combustion efficiency in a marine DF engine by optimizing the PCC parameters. GT-SUITE and CONVERGE were selected as the simulation software to establish an effective one-dimensional (1D) and three-dimensional (3D) simulation model of a marine LP-DF engine (Section 2). The effect of PCC parameters on the engine performance and emissions were studied from two aspects: PCC nozzle diameter and PCC nozzle angle (Section 3). It is hoped that the conclusions and contents can provide reference information for subsequent work.

## 2. Model Description

### 2.1. LP-DF Engine Basic Parameters

In this study, the RT-Flex50DF 2-stroke engine (WinGD) was investigated. It is a camshaft-less low-speed 2-stroke engine consisting of five cylinders connected in an in-line arrangement, one air cooler unit, one turbocharger unit, and two auxiliary blowers. The PCC nozzle diameter of this engine is normally 16 mm with an angle of 65°. Natural gas with an injection pressure lower than 1.6 MPa is admitted to the cylinders right before the air inlet valve. Besides, the gas admission valves are electronically actuated and controlled by the engine control system to give exactly the correct amount of gas to each cylinder-thereby, the combustion in each cylinder can be fully and individually controlled.

Since the premixed lean-burn combustion of the Otto cycle is realized in the cylinder, the RT-flex50DF engine is fully compliant with the IMO Tier III NOx emissions limits without requiring the use of any after-treatment systems. Simultaneously, three technologies are used to reduce the HC emissions including the pre-chamber technology for best ignition and combustion stability, the valve timing optimization to avoid the escape of natural gas, and the combustion chamber shape adjustment to avoid flameout. The basic parameters of the RT-Flex50DF engine are shown in Table 1 [29].

**Table 1.** WinGD RT-Flex50DF engine dimensions.

| Parameter | Value |
| --- | --- |
| Bore | 500 mm |
| Stroke | 2050 mm |
| Cylinder Number | 5–8 |
| Speed | 124 r/min |
| Power | 8640 kW |
| Compression Ratio | 12 |
| Brake Specific Pilot Fuel Consumption (BSPC) (DF Mode) * | 1.8 g/kWh |
| Brake Specific Gas Consumption (BSGC) (DF Mode) * | 142.7 g/kWh |

* All other reference conditions refer to ISO standard (ISO 3046-1). The following tolerances for BSPC and BSGC are taken into account: +5% for 100–85% engine power.

The GT-Suite software, which is a renowned 1D simulation program for engine analysis and modelling, was employed for the LP-DF engine simulation. As shown in Figure 1, the developed 1D model of the LP-DF engine includes blocks for the scavenging receiver, the cylinders, the scavenging ports, the intake and exhaust valves, the intake and exhaust ports, the waste gate, the turbocharger, the crank train, the combustion chamber, and the natural gas nozzles. The 1D GT-Suite model was built to simulate the steady-state conditions of the 5RT-Flex50DF, which helped to obtain more accurate boundary conditions and initial conditions for the 3D simulation. This GT model uses the user defined combustion heat release rate to simulate the DF engine combustion. The heat release rate was determined from the experimental data under 75% engine load, which can accurately predict the performance of the LP-DF engine. Moreover, the required boundary conditions and the initial conditions for the 3D CFD calculation were obtained based on the verified model.

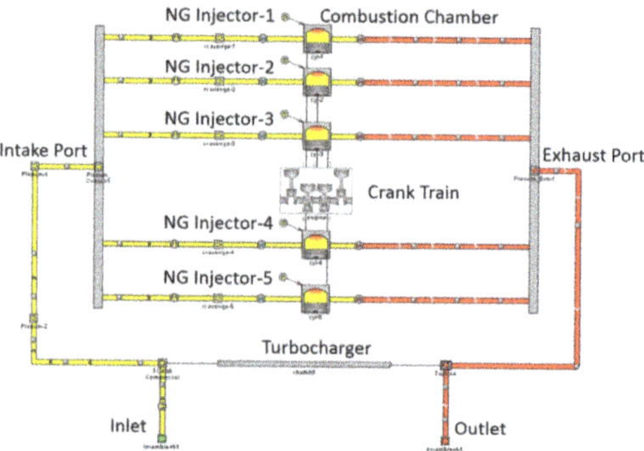


**Figure 1.** The 1D model of the 5RT-Flex50DF engine.

As shown in Figure 2, the LP-DF engine optimization analysis requires 1D and 3D simulation coupling. First, the computer-aided design (CAD) software Catia was used to design different PCC model schemes and the 3D model was imported into the CFD simulation software CONVERGE. The adaptive mesh refinement of the CFD domain was applied on the model with focusing on the PCC mesh, the natural gas nozzles, the pilot fuel injection nozzles, and the flame front surface. After that, the initial conditions and the boundary conditions in the 3D simulation were given by the GT-power simulation results. Finally, the efficient and stable working range of the LP-DF engine was studied based on the 3D and 1D simulation results.

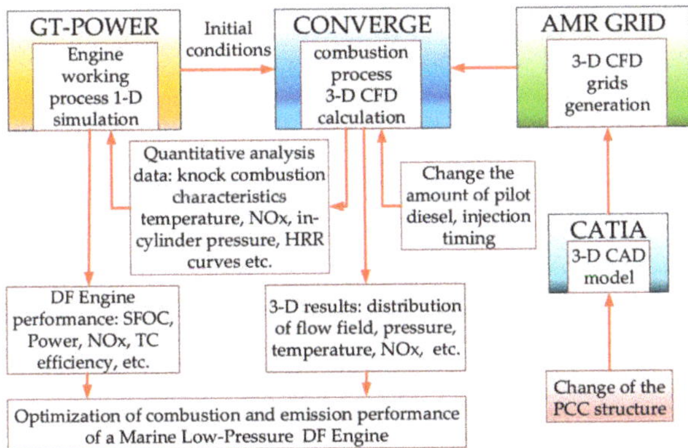

**Figure 2.** Calculation simulation flow chart.

## 2.2. CFD Model Verification

As shown in Figure 3, the CAD software Catia is used for 3D modeling. The PCC was arranged on both sides of the main combustion chamber (MCC) in the LP-DF engine model while the pilot fuel injection nozzles were installed on the top of the PCC. When the pilot fuel was injected into the PCC, a high-speed jet flame was formed by rapid spontaneous combustion under high temperature and high pressure at TDC. After that, the flame passed through the PCC nozzle to ignite the lean gas mixture in the MCC. The two small ellipsoidal structures on the top of the MCC are pre-combustion chambers. Furthermore, two natural gas admission valves (GAV) were symmetrically distributed in the lower middle of the DF engine cylinder. Furthermore, the Natural Gas composition used during the CFD simulations mainly included 95.1% methane, 2.53% ethane and other gases while the natural gas fuel lower calorific value was 47.64 MJ/kg.

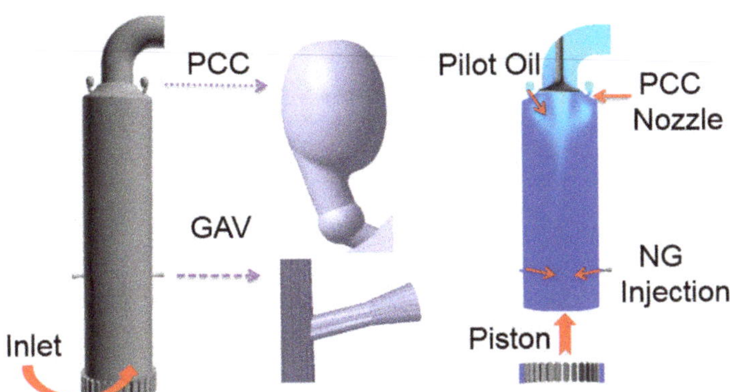


**Figure 3.** 3D analysis domain of a low-pressure dual-fuel (LP-DF) engine.

After the mesh adaptive refinement processing, the maximum number of 3D calculation domains was about 630,000. The KH-RT model was used as the breakup model in the spray model while the NTC model was selected as the pilot oil droplet collision model. Besides, the standard K-$\varepsilon$ model was chosen as the turbulence model selected for the CFD. Moreover, the chemical reaction kinetic model of SAGE was used as the combustion model and the extended Zeldovich NOx emission model was used as the emission model [30].

As shown in Figure 4, the calculation results are compared with the measured data and the simulation error is less than 3.6%, which meets the accuracy requirements of the CFD numerical calculation. From Figure 4, the NOx emissions calculation results under different working conditions are in good agreement with the measured data. Moreover, the accuracy of the simulation model and the related parameters are verified, which can be used for the following performance and combustion simulation work.

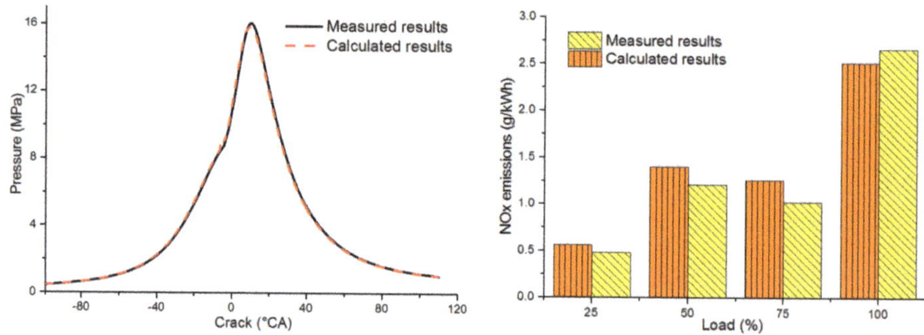


**Figure 4.** Marine DF engine model verification.

## 3. Results and Discussion

### 3.1. Performance Characteristics of A Marine LP-DF Engine

The marine 2-stroke LP-DF engine can change its working mode freely, as it can switch from diesel mode using HFO or LFO as fuel to dual-fuel mode using natural gas as fuel. These two modes have their working characteristics. The limitation of knocking is not considered during the operation of the engine in the diesel mode. During the dual-fuel mode, the engine uses Otto cycle combustion and the high-speed flame jet from the PCC easily ignites the pre-mixture, which limits the problems of knocking and misfire.

In WinGD RT-Flex50DF lean-burn engine, the excess air ratio can be very high (typically 2.2). As shown in Figure 5, the stable working window of the dual-fuel engine is very narrow [31]. Once the mixture is too lean, the engine is prone to misfire whilst the mixture is prone to self-ignition when it is too rich [32]. Therefore, the in-cylinder excess air ratio must be precisely controlled by adjusting some parameters such as the valve timing and the fuel injection timing to ensure the efficient and stable operation of the LP-DF engine. Since the same specific heat quantity released by combustion is used to heat a large mass of air, the peak temperature and consequently the NOx emissions are lower.

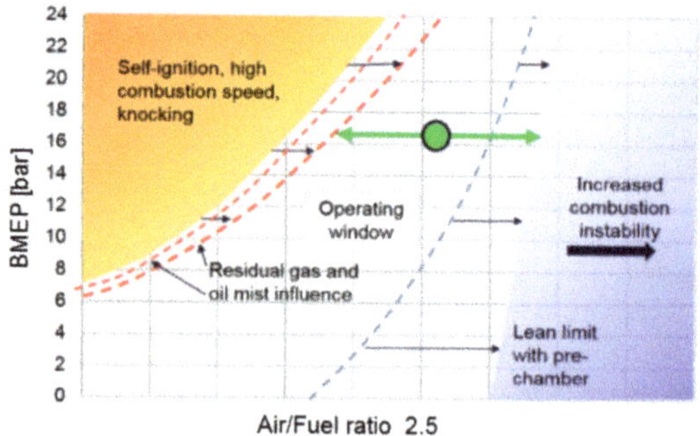


**Figure 5.** Lean-burn DF engine Otto combustion limits.

Figure 6 depicts that the compression pressure of the LP-DF engine in dual-fuel mode is lower than that of the diesel mode, which reduces the compression ratio in order to avoid knocking. In addition, the opening timing of the exhaust valve in dual-fuel mode is adjusted. From Figure 6, the maximum combustion pressure in diesel mode is about 4.8 MPa lower than that in the dual-fuel mode because of the IMO TierII NOx emission limit requirement, which delays the diesel injection timing.

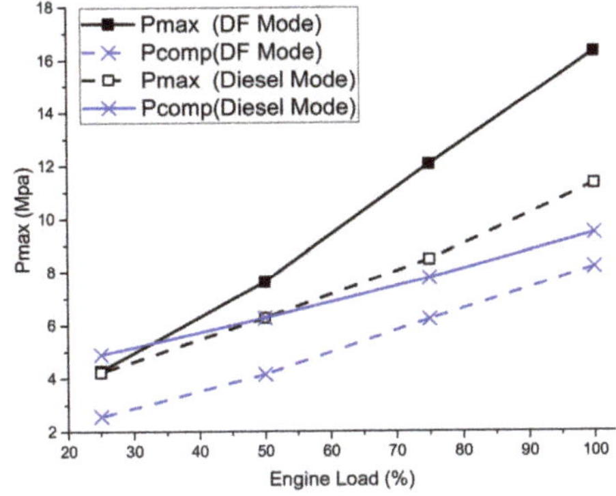


**Figure 6.** Comparison of dual fuel mode and diesel mode.

As shown in Figure 7, the average in-cylinder combustion temperature in the dual-fuel mode is higher than that in the diesel mode and the peak combustion temperature is about 283 K higher. Besides, the peak combustion temperature occurs earlier in the dual-fuel mode. Moreover, the combustion duration of the dual-fuel mode is longer due to the lower flame propagation speed of natural gas. Based on the above analysis, the primary task of the diesel mode design is to reach the IMO NOx emission limit and for this reason, the power performance under high load has been reduced. Due to the limitation of knocking, the primary task of the DF mode in the LP-DF engine is to ensure the stable operation of the engine.

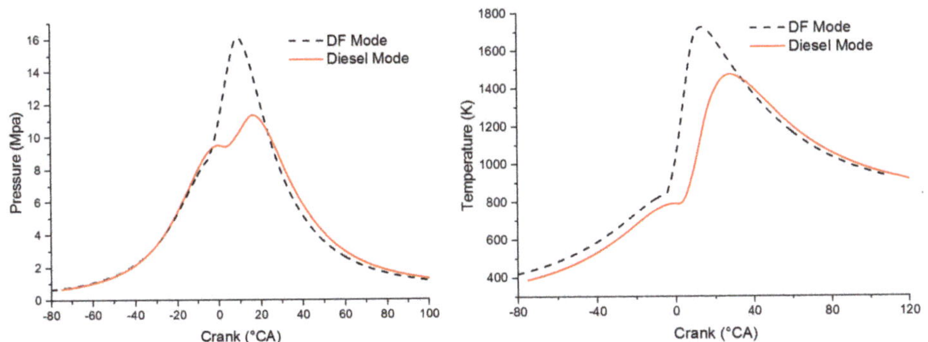


**Figure 7.** Comparison of pressure and temperature in different working modes.

### 3.2. The Effect of PCC Nozzle Diameter

To study the influence of the diameter of the PCC nozzle on the performance and emissions of the dual-fuel engine, three schemes with a different PCC nozzle diameter of 10 mm, 16 mm, and 24 mm, respectively were set for comparison. As shown in Figure 8, the boundary condition parameters of the LP-DF engine are kept the same while the diameter of the PCC is changed to set the simulation cases.

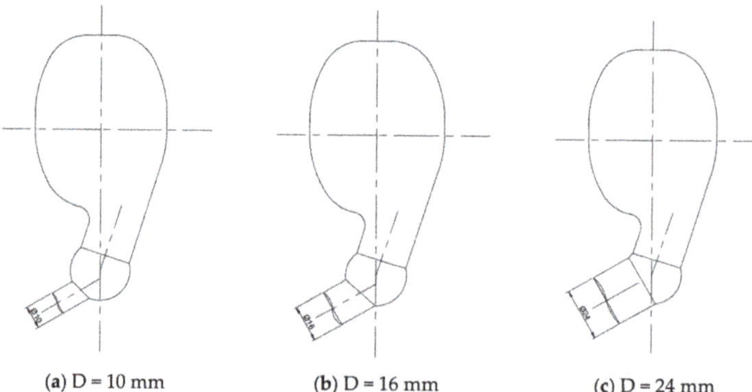

(a) D = 10 mm      (b) D = 16 mm      (c) D = 24 mm

**Figure 8.** Different schemes of pre-combustion chamber (PCC) nozzle diameter.

### 3.2.1. Influence on Combustion

The diameter of the PCC nozzle not only affects the fuel mixing and the combustion in the pre-chamber but also the flame propagation in the main combustion chamber. Figure 9 shows the temperature distribution in the combustion chamber at different nozzle diameters. By studying the temperature distribution of these three groups of combustion processes, it can be seen that the time at which the flame propagates throughout the combustion chamber varies with different nozzle diameters. When the nozzle diameter is D = 10 mm, the flame propagates throughout the entire combustion chamber at 9°CA after Top Dead Center (aTDC) while for D = 24 mm, the time at which the flame propagates throughout the combustion chamber is 11°CA aTDC. In addition, the flame front spreads to the bottom of the combustion chamber earlier when the flame jet speed is higher. Conversely, the flame front does not propagate to the bottom of the combustion chamber for the PCC nozzle with a small diameter when the flame jet speed is slow, which indicates that the flame is more easily affected by the in-cylinder airflow. Moreover, the flame propagates throughout the entire combustion chamber at 6°CA aTDC for the PCC nozzle diameter of D = 16 mm. Since the starting time of combustion are the same in the three cases, the flame propagation speed in the cylinder is the fastest at D = 16 mm.

In Figure 9, the jet flame gradually becomes stronger as the diameter of the PCC nozzle increases at −4°CA before Top Dead Center (bTDC). Comparing the flame distributions of the three diameters, the flame distribution is asymmetrical under the action of turbulence when the nozzle diameter is small at −3°CA bTDC. Besides, the flame jet is too thin to withstand the effects of in-cylinder gas turbulence. Under the same pilot fuel injection conditions, the flame basically propagates to the bottom of the combustion chamber when the nozzle diameter is 16 mm at −3°CA bTDC but does not reach the bottom of the combustion chamber for the other two diameters.

Figure 10 presents the flame velocity vector distribution under different PCC nozzle diameters at −3°CA bTDC. From Figure 10, the jet flame speed is large at D = 16 mm and relatively small at D = 20 mm. At D = 10 mm, the flame is greatly affected by the turbulence, which means that the flame intensity is the weakest in the three simulation cases.

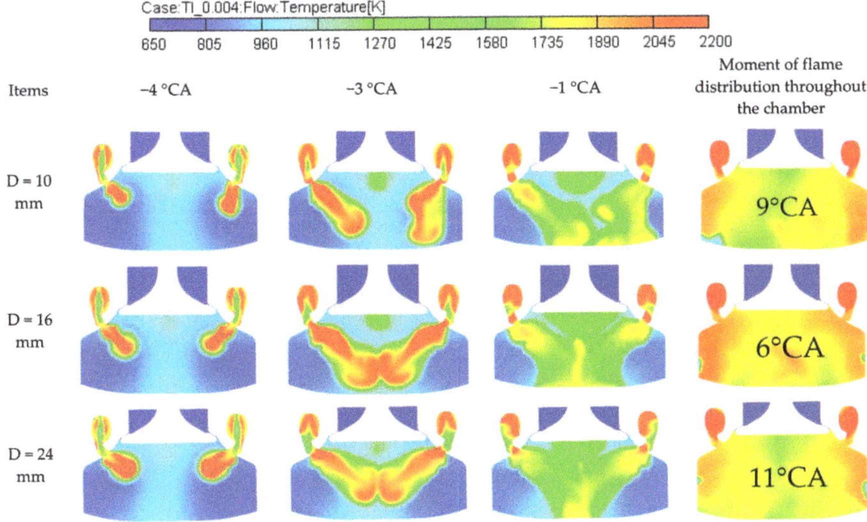



**Figure 9.** Temperature distribution in the combustion chamber.

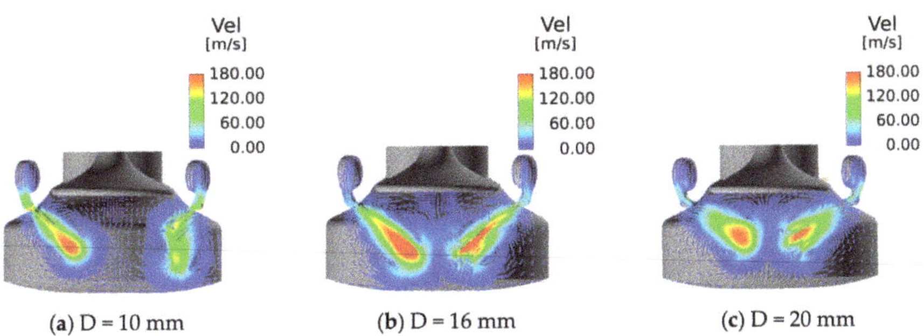



**Figure 10.** Flame velocity vector distribution.

### 3.2.2. Influence on Performance

Figure 11 shows the calculated in-cylinder pressure under different PCC nozzle diameters at 100% load. It can be seen that the peak in-cylinder pressure gradually decreases as the diameter of the nozzle increases while the crank-angle location of the in-cylinder pressure is substantially unaffected by the PCC nozzle diameter. This is because when the diameter of the nozzle is too large, the jet flame velocity from the pre-combustion chamber decreases, which means that the flame needs more time to distribute inside the combustion chamber—making the combustion duration longer. Therefore, peak pressure decreases with the increase of the PCC nozzle diameter.

The Rate of Heat Release (ROHR) curves are also depicted in Figure 11 under different nozzle diameters. From Figure 11, the heat release rate is essentially the same at D = 10 mm and D = 16 mm. Besides, the peak heat release rate is relatively lower, and the combustion duration is relatively long at D = 24 mm—indicating that the combustion is relatively poorer inside the cylinder.

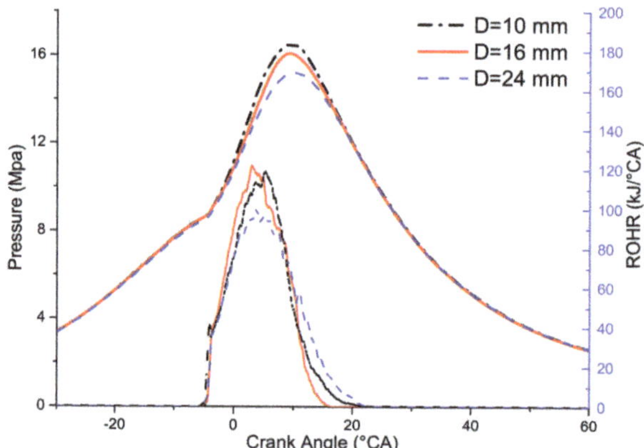

**Figure 11.** Pressure and heat release rate at different PCC nozzle diameters.

Figure 12 illustrates the mean temperature in the cylinder under different PCC nozzle diameters. From Figure 12, the peak in-cylinder temperature is the highest at D = 16 mm and there are no many differences between the in-cylinder temperatures values for D = 10 mm and D = 16 mm. Moreover, the peak in-cylinder temperature is the lowest for the nozzle with the large diameter D = 24 mm and the peak temperature appears later than the other two cases (D = 10 mm and D = 16 mm). When the fuel injection conditions are kept unchanged, the pre-chamber nozzle has an important influence on the flow of the in-cylinder mixture and the combustion in the MCC. This is because after the same quality of the pilot fuel is ignited, the jet flame enters the MCC through the pre-chamber nozzle and the parameters such as the diameter and the angle of the pre-chamber directly affect the angle, the position, and the speed of the flame.

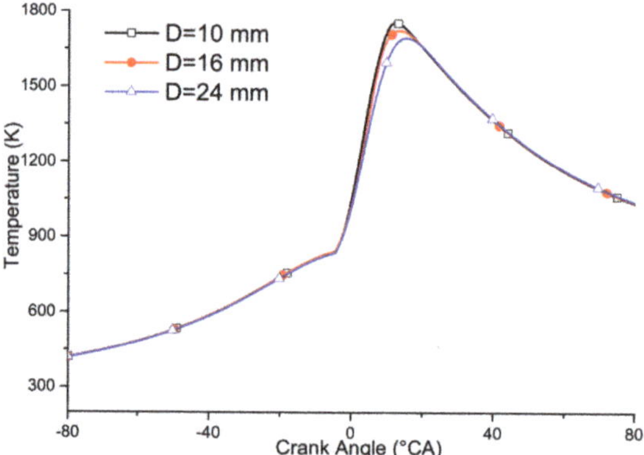

**Figure 12.** Mean in-cylinder temperature at different PCC nozzle diameters.

Figure 13 depicts the effects of different PCC nozzle diameters on the amount of NOx emissions. From Figure 13, NOx emissions are the lowest at D = 16 mm because the flame propagation speed is the fastest and the combustion duration is the shortest for this diameter as can be seen from the temperature distribution in Figure 9. Conversely, the NOx emissions are the highest at D = 24 mm compared to the other cases because of the longer combustion duration and the slower flame spread, which increases the formation of NOx emissions. The high-temperature environment has a great influence on NOx emissions.

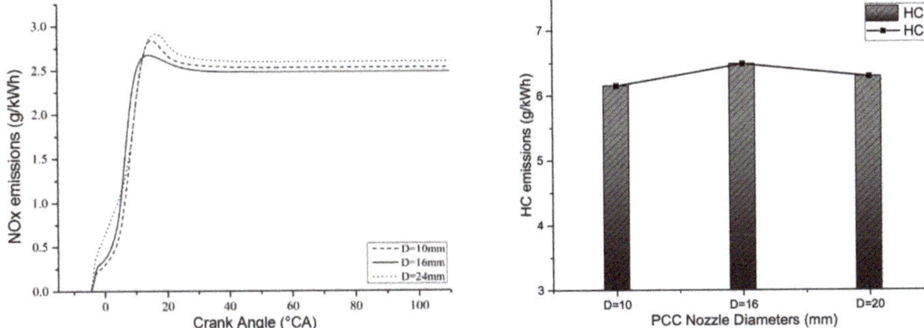

**Figure 13.** NOx and HC emissions under different PCC nozzle diameters.

The HC emissions of the LP-DF engine under different PCC nozzle diameters are shown in Figure 13. It can be seen from Figure 13 that the HC emissions at D = 16 mm are slightly higher than those for D = 10 mm and D = 24 mm. This is because the flame propagation speed is relatively slow, the combustion duration is longer, and the combustion is more complete when the PCC nozzle diameter is D = 10 mm and D = 24 mm, thereby the HC emissions are relatively less.

The diameter of the pre-chamber nozzle affects the velocity of the flame jet and the propagation of the flame in the combustion chamber. The changes in the in-cylinder pressure and the heat release rate are the same when the nozzle diameters are D = 10 mm and D = 16 mm, which indicates that the diameter of the PCC nozzle influences the engine performance and emission characteristics when it changes within a certain range. However, the large nozzle diameter affects the heat release duration of the mixture and the propagation speed of the flame in the combustion chamber, which is not conducive to the rapid combustion of the mixture and affects the LP-DF engine emissions characteristics. Appropriate PCC nozzle diameters can improve engine performance and reduce emissions.

### 3.3. The Effect of PCC Nozzle Angle

In order to study the effects of the PCC nozzle angle on the performance and emissions of the LP-DF engine, the engine boundary condition parameters were assumed to remain unchanged. As shown in Figure 14, four geometric schemes are designed for the PCC nozzle angles of 60°, 65°, 70°, and 75° (the angle between the PCC nozzle and the vertical direction). The current PCC nozzle angle of the LP-DF engine is 65°.

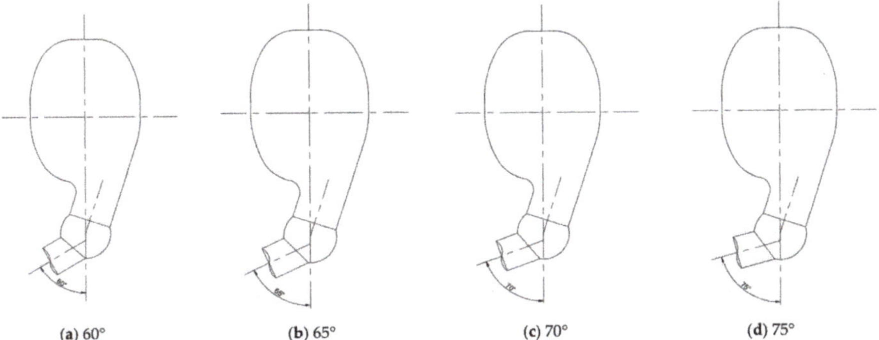

**(a)** 60°   **(b)** 65°   **(c)** 70°   **(d)** 75°

**Figure 14.** Different schemes of PCC nozzle angles.

### 3.3.1. Influence on Combustion

The flame propagation process in the MCC under different PCC nozzle angles was simulated. Because of the same injection conditions of pilot oil, the combustion in the PCC under different nozzle angles was the same, only the propagation of the flame after entering the MCC was studied. Figure 15 shows the temperature distribution in the combustion chamber of four PCC nozzle angles. By comparing the four cases of the temperature distribution, it can be seen that the time at which the flame propagates throughout the combustion chamber is different. When the angle between the PCC nozzle and the vertical direction is 65°, the flame propagates throughout the combustion chamber at 6°CA aTDC and the flame propagation speed is the fastest. At 60° PCC nozzle angle, the flame propagates throughout the chamber at 8°CA aTDC and when the PCC nozzle angle increases to 70° and 75°, the flame propagates throughout the combustion chamber at 10°CA aTDC.

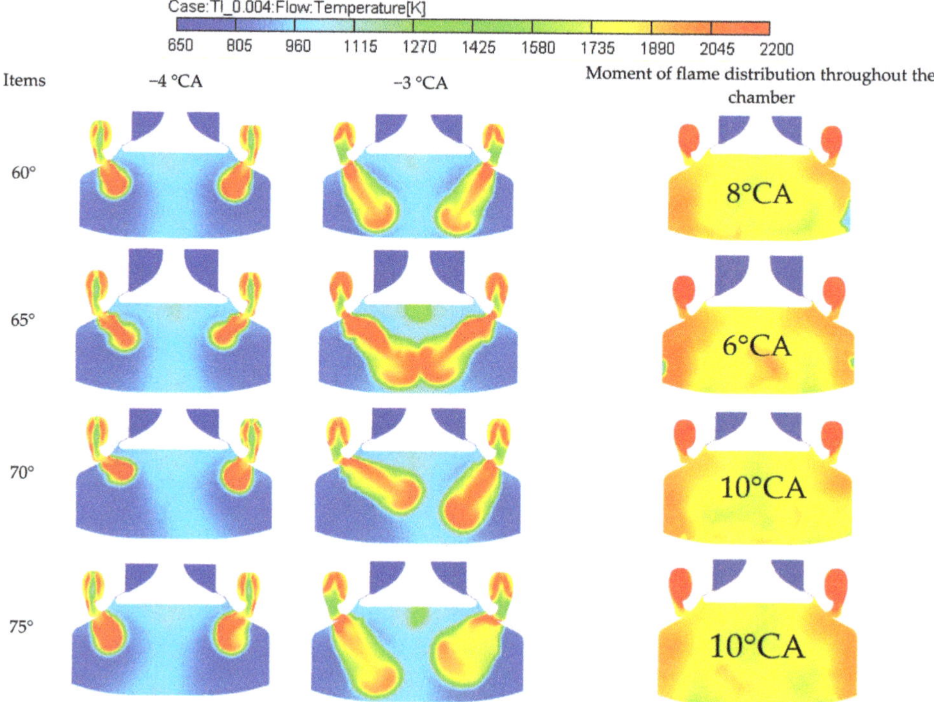


**Figure 15.** Flame distribution in the combustion chamber.

From Figure 15, the direction of flame propagation is consistent with the nozzle angle when the nozzle angle is 60° and 65° and the jet flame is symmetrically distributed in the cylinder. Besides, the flame propagation direction does not propagate along with the direction of the PCC nozzle at 70° and 75° nozzle angle and the flame is distributed asymmetrically in the cylinder. This indicates indicated that when the angle between the PCC nozzle and the vertical direction is too large, the flame passing through the nozzle is affected by the turbulence of the mixture and cannot maintain propagation in the same direction as the PCC nozzle. Moreover, the flame spreads throughout the whole combustion chamber at 60° slightly later than when the nozzle angle is 65°, which means that the small nozzle angle is not conducive to the flame propagation in the combustion chamber and affects the burning speed.

Figure 16 shows the flame velocity vector distribution under different PCC nozzle angles at 3°CA bTDC. It can be seen that the velocity distribution is uniform and the flame is relatively less affected by the in-cylinder turbulence when the nozzle angle is 60° and 65°. In addition, the flame propagation velocity is large at 65° PCC nozzle angle comparing to that of other PCC nozzle angles.

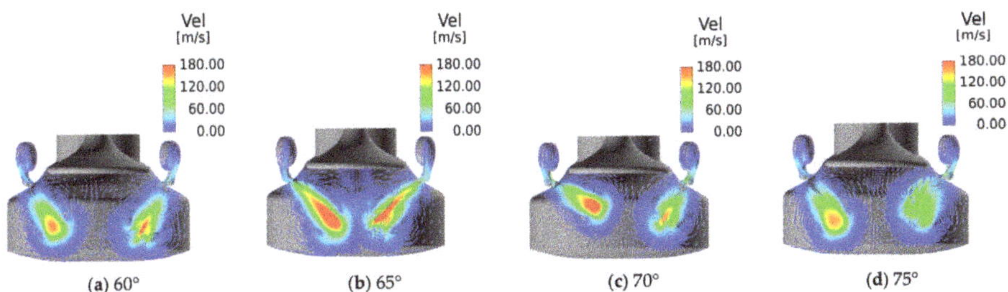



**Figure 16.** Flame velocity vector distribution.

### 3.3.2. Influence on Performance

Figure 17 shows the in-cylinder pressure for different pre-chamber nozzle angles. From Figure 17, the maximum in-cylinder pressure increases first and then decrease as the nozzle angle increases. The highest combustion pressure in the cylinder occurs at 65° nozzle angle and the peak pressure appears slightly earlier than the peak pressure at other angles. The change in heat release rate at different nozzle angles is shown in Figure 17. It can be seen from Figure 17 that the peak of the heat release rate obtained at 65° PCC nozzle angle is significantly higher than the heat release rate of the other three angles. Besides, the slope of the heat release rate curve is large and the heat release time of the mixture gas is relatively short at 65°, which means that the mixture burns more quickly at this nozzle's angle.

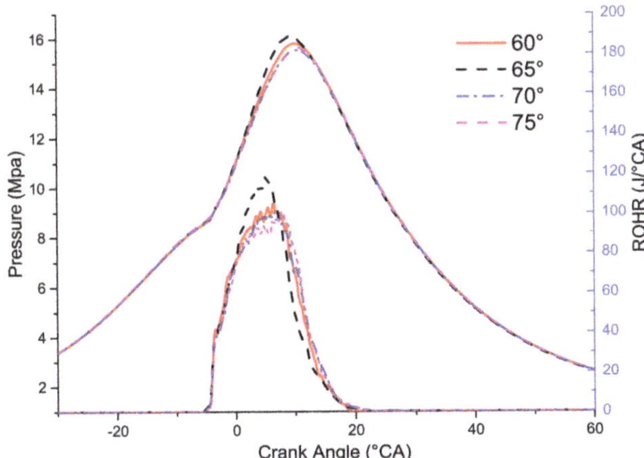



**Figure 17.** In-cylinder pressure and heat release rate at different nozzle angles.

As can be seen from Figure 18, the maximum in-cylinder temperature first increases and then decreases as the angle of the PCC nozzle increases. At 65° nozzle angle, the average temperature in the cylinder is the highest and the temperature peak occurs slightly earlier than other conditions. Consequently, the angle of the pre-chamber nozzle determines the direction of flame propagation when the flame enters the main combustion chamber, and the difference in the direction of propagation affects the combustion rate of the mixture.

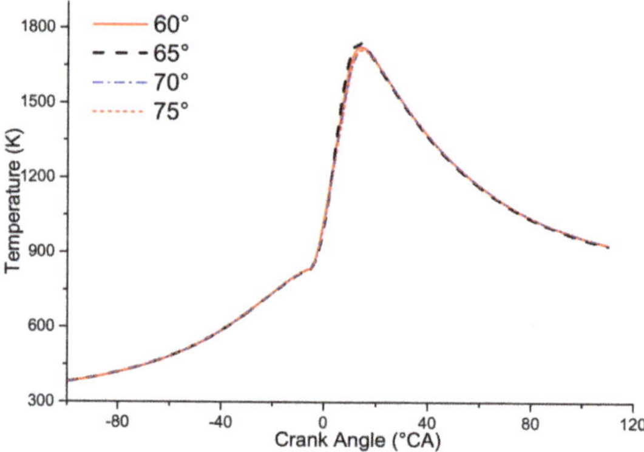

**Figure 18.** Mean in-cylinder temperature at different PCC nozzle angles.

Figure 19 shows the variation of the amount of NOx emissions under different PCC nozzle angles. From Figure 19, the NOx emissions under the four PCC nozzle angles do not change too much. The amount of NOx emissions generated at a 60° nozzle angle is slightly lower than that of the other conditions. Although the combustion condition at 65° PCC nozzle angle is better than the other three cases, the in-cylinder temperature is slightly higher. Therefore, the amount of NOx emissions is higher. Nevertheless, the NOx emissions are not significantly affected by the angle of the pre-chamber nozzle. Furthermore, the HC emissions generated at the nozzle angles of 60°, 70°, and 75° are relatively lower than that formed at 65°. This is can be explained by the fact that the flame propagation speed is relatively slow and the combustion is more complete and lasts longer at these three angles, thereby the HC emissions are relatively small.

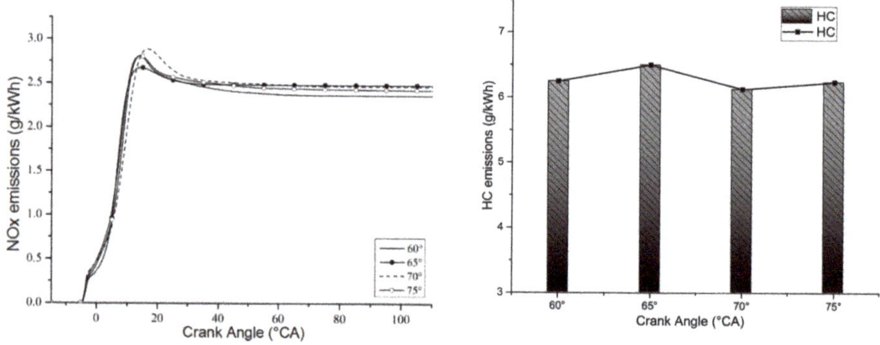

**Figure 19.** NOx and HC emissions at different PCC nozzle angles.

Through the above analysis, the direction of jet flame propagation was affected by the angle of the PCC nozzle while different flame propagation directions affected the flame diffusion and combustion speed. Besides, the flame propagation was easily influenced by the turbulence of the in-cylinder mixture when the angle of the PCC nozzle was too large and the PCC nozzle could not maintain the flame propagation—resulting in the instability of flame propagation. Furthermore, the velocity of the jet flame was affected by the small nozzle angle, which was not conducive to the rapid combustion of the mixture.

## 4. Conclusions

In this paper, the effect of different PCC chamber parameters on the performance and emissions of a marine LP-DF engine was investigated using the 3D CFD simulation: PCC nozzle diameter and PCC nozzle angle. The main conclusions are listed as follows:

(1)   The pre-combustion chamber effectively organized the airflow in the cylinder and increased the flame propagation speed to achieve efficient lean combustion. The diameter of the PCC nozzle affected the propagation of the flame in the combustion chamber. Suitable PCC nozzle diameters helped to improve the flame propagation stability and engine performance and reduce emissions.

(2)   The angle of the PCC nozzle affected the direction of the flame propagation, which affected the flame propagation speed and thus the occurrence of knocking. Optimizing the angle of the PCC nozzle was beneficial to the organization of the in-cylinder combustion and helped to increase the flame propagation speed.

(3)   The 3D CFD simulation could be a powerful tool for LP-DF engine design. Future studies could include other aspects such as the PCC volume ratio, the PCC nozzle length, and the variable exhaust valve timing.

**Author Contributions:** H.G. and S.Z. contributed to the case study and the original manuscript. M.S. translated the original manuscript. Y.F. checked the results of the whole manuscript.

**Funding:** This research was supported by the China financial support of Marine Low-Speed Engine Project-Phase I and the National Key R&D Program of China (grant no. 2016YFC0205202).

**Conflicts of Interest:** The authors declare no conflict of interest.

## Abbreviations

| | |
|---|---|
| IMO | International Maritime Organization |
| ECAs | Emission Control Areas |
| NOx | Nitrogen Oxides |
| SOx | Sulfur Oxides |
| CO | Carbon Monoxide |
| PM | Particulate Matter |
| HFO | Heavy Fuel Oil |
| LFO | Light Fuel Oil |
| LP-DF | Low-Pressure Dual-Fuel |
| HC | Hydrocarbon |
| ISO | International Organization for Standardization |
| CFD | Computational Fluid Dynamics |
| 1D | One-Dimensional |
| 3D | Three-Dimensional |
| BSPC | Brake Specific Pilot Fuel Consumption |
| BSGC | Brake Specific Gas Consumption |
| CAD | Computer Aided Design |
| PCC | Pre-Combustion Chamber |
| MCC | Main Combustion Chamber |
| GAV | Gas Admission Valve |
| λ | Air-Fuel Ratio |
| ROHR | Rate of Heat Release |
| TDC | Top Dead Center |
| °CA | Crank Angle Degree |
| bTDC | Before Top Dead Center |
| aTDC | After Top Dead Center |

## References


1. Christen, C.; Brand, D. IMO Tier III: Gas and Dual Fuel Engines as a Clean and Efficient Solution. In Proceedings of the CIMAC Congress, Shanghai, China, 13–17 May 2013; p. 187.
2. Mehdi, G.; Zhou, S.; Zhu, Y.; Shah, A.H.; Chand, K. Numerical Investigation of SCR Mixer Design Optimization for Improved Performance. *Processes* **2019**, *7*, 168. [CrossRef]
3. Kronholm, M. Demanding icebreaking LNG-powered icebreaking features the industry's most advanced technology. In Proceedings of the CIMAC Congress, Helsinki, Finland, 6–10 June 2016; p. 226.
4. Ahmed, S.A.; Zhou, S.; Zhu, Y.; Feng, Y.; Malik, A.; Ahmad, N. Influence of Injection Timing on Performance and Exhaust Emission of CI Engine Fuelled with Butanol-Diesel Using a 1D GT-Power Model. *Processes* **2019**, *6*, 299. [CrossRef]
5. Kezirian, M.T.; Phoenix, S.L. Natural Gas Hydrate as a Storage Mechanism for Safe, Sustainable and Economical Production from Offshore Petroleum Reserves. *Energies* **2017**, *10*, 828. [CrossRef]
6. Youfeng, L.; Liangjun, X. Development of low-fuel consumption and low-emission locomotive engine. In Proceedings of the CIMAC Congress, Vancouver, BC, Canada, 10–14 June 2019; p. 51.
7. Xiang, L.; Song, E.; Ding, Y. A Two-Zone Combustion Model for Knocking Prediction of Marine Natural Gas SI Engines. *Energies* **2018**, *11*, 561. [CrossRef]
8. Tozzi, L.; Emmanuella, S.; Greg, B. Novel Pre-Combustion Chamber Technology for Large Bore Natural Gas Engines. In Proceedings of the CIMAC Congress, Helsinki, Finland, 6–10 June 2016; p. 259.
9. Mohr, H.; Baufeld, T. Improvement of Dual Fuel Engine Technology for Current and Future Applications. In Proceedings of the CIMAC Congress, Shanghai, China, 13–17 May 2013; p. 412.
10. Delneri, D.; Sirch, G. Enhanced Flexibility in Gas Engine Operation for Marine and Power Generation Demanding Applications. In Proceedings of the CIMAC Congress, Vancouver, BC, Canada, 10–14 June 2019; p. 93.
11. Ilari, K.; Ulf, A.; Grant, G. Product Technology Development for Increased Customer Benefits. In Proceedings of the CIMAC Congress, Helsinki, Finland, 6–10 June 2016; p. 285.
12. Yasueda, S.; Kuboyama, T.; Matsumura, M.; Moriyoshi, Y.; Doyen, V.; Martin, J.B. The examination on the main contributing factors of lube oil pre-ignition. In Proceedings of the CIMAC Congress, Helsinki, Finland, 6–10 June 2016; p. 147.
13. Zheng, X.; Yang, J. An Investigation into the Gas-Mode Start Strategies for a Marine Medium-Speed Micro-Pilot-Ignition Dual-Fuel Engine. In Proceedings of the CIMAC Congress, Vancouver, BC, Canada, 10–14 June 2019; p. 71.
14. Alla, G.H.A.; Soliman, H.A.; Badr, O.A. Effect of injection timing on the performance of diesel engine. *Energy Convers. Manag.* **2002**, *43*, 269–277. [CrossRef]
15. Papagiannakis, R.G.; Rakopoulos, C.D.; Hountalas, D.T.; Rakopoulos, D.C. Emission characteristics of high speed, dual fuel, compression ignition engine operating in a wide range of natural gas/diesel fuel proportions. *Fuel* **2010**, *89*, 1397–1406. [CrossRef]
16. Hirose, T.; Masuda, Y. Technical Challenge for the 2-Stroke Premixed Combustion Gas Engine. Pre-ignition Behavior and Overcoming Technique. In Proceedings of the CIMAC Congress, Shanghai, China, 13–17 May 2013; p. 185.
17. Alla, G.A.; Soliman, O.A. Effect of pilot fuel quantity on the performance of a dual fuel engine. *Energy Convers. Manag.* **2000**, *41*, 559–572. [CrossRef]
18. Duan, X.; Liu, J.; Yao, J. Performance, combustion and knock assessment of a high compression ratio and lean-burn heavy-duty spark-ignition engine fuelled with n-butane and liquefied methane gas blend. *Energy* **2018**, *158*, 256–268. [CrossRef]
19. Guo, H.; Song, Z. Study on the Influence of Prechamber Structure on the Knock of a Marine Low-Speed Dual-fuel Engine. In Proceedings of the CIMAC Congress, Vancouver, BC, Canada, 10–14 June 2019; p. 50.
20. Hokimoto, S.; Kuboyama, T. Combustion analysis in a natural gas engine with pre-chamber by three-dimensional numerical simulation. *Trans. Jpn. Soc. Mech. Eng.* **2015**, *47*, 28–37.
21. Yousefi, A.; Birouk, M. Investigation of natural gas energy fraction and injection timing on the performance and emissions of a dual-fuel engine with pre-combustion chamber under low engine load. *Appl. Energy* **2017**, *189*, 492–505. [CrossRef]

22. Cernik, F.; Macek, J.; Dahnz, C.; Hensel, S. *Dual Fuel Combustion Model for a Large Low-Speed 2-Stroke Engine*; SAE Technical Paper: Prague, Czech Republic, 2016.

23. Maghbouli, A.; Saray, R.K.; Shafee, S. Numerical study of combustion and emission characteristics of dual-fuel engines using 3D-CFD models coupled with chemical kinetics. *Fuel* **2013**, *106*, 98–105. [CrossRef]

24. Liu, T.; Gui, Y. Pre-Combustion Chamber Design Scheme Analysis for a Typical Marine Low Speed Duel Fuel Engine. In Proceedings of the CIMAC Congress, Vancouver, BC, Canada, 10–14 June 2019; p. 100.

25. Maghbouli, A.; Shafee, S.K.; Saray, R.; Yang, W. A Multi-Dimensional CFD-Chemical Kinetics Approach in Detection and Reduction of Knocking Combustion in Diesel-Natural Gas Dual-Fuel Engines Using Local Heat Release Analysis. *SAE Int. J. Engines* **2013**, *6*, 777–787. [CrossRef]

26. Jha, P.R.; Srinivasan, K.K.; Krishnan, S.R. Influence of Swirl Ratio on Diesel-Methane Dual Fuel Combustion: A CFD Investigation. In Proceedings of the ASME 2017 Internal Combustion Engine Division Fall Technical Conference, Seattle, WA, USA, 15–18 October 2017.

27. Nylund, I.; Ott, M. Development of a Dual Fuel Technology for Slow-speed Engines. In Proceedings of the CIMAC Congress, Shanghai, China, 13–17 May 2013; p. 284.

28. Christoforos, M.; Gerasimos, T. Numerical investigation of a premixed combustion large marine two-stroke dual fuel engine for optimising engine settings via parametric runs. *Energy Convers. Manag.* **2018**, *160*, 48–59.

29. Takahiro, K.; Takahide, A. Study on Mixture Formation Process in Two Stroke Low Speed Premixed Gas Fueled Engine. In Proceedings of the CIMAC Congress, Helsinki, Finland, 6–10 June 2016; p. 207.

30. Zhou, S.; Gao, R.F.; Feng, Y.M. Evaluation of Miller cycle and fuel injection direction strategies for low NOx emission in marine two-stroke engine. *Int. J. Hydrog. Energy* **2017**, *42*, 20351–20360. [CrossRef]

31. Ott, M.; Nylund, I. The 2-stroke Low-Pressure Dual-Fuel Technology: From Concept to Reality. In Proceedings of the CIMAC Congress, Helsinki, Finland, 6–10 June 2016; p. 233.

32. Flot, P.; MESLATI, A.; Digneton, M. Improving Efficiency and Emissions of Otto Gas Engines, by Continuously Monitoring Fuel Gas Quality. In Proceedings of the CIMAC Congress, Vancouver, BC, Canada, 10–14 June 2019; p. 26.

MDPI

St. Alban-Anlage 66

4052 Basel

Switzerland

Tel. +41 61 683 77 34

Fax +41 61 302 89 18

www.mdpi.com

*Processes* Editorial Office

E-mail: processes@mdpi.com

www.mdpi.com/journal/processes